Elementary Differential Geometry

Revised Second Edition

Elementary Differential Geometry

Revised Second Edition

Barrett O'Neill

Department of Mathematics
University of California, Los Angeles

ELSEVIER

AMSTERDAM • BOSTON • HEIDELBERG
LONDON • NEW YORK OXFORD • PARIS
SAN DIEGO • SAN FRANCISCO
SINGAPORE • SYDNEY • TOKYO

An imprint of Elsevier

Acquisitions Editor: Pamela Chester
Project Manager: Heather Furrow
Editorial Assistant: Kelly Weaver
Marketing Manager: Linda Beattie
Cover Design: Eric DeCicco
Composition: SNP Best-set Typesetter Ltd., Hong Kong
Cover Printer: Phoenix Color Corp
Interior Printer: Maple Press

Academic Press is an imprint of Elsevier
30 Corporate Drive, Suite 400, Burlington, MA 01803, USA
525 B Street, Suite 1900, San Diego, California 92101-4495, USA
84 Theobald's Road, London WC1X 8RR, UK

This book is printed on acid-free paper. ∞

Library of Congress Cataloging-in-Publication Data
O'Neill, Barrett.
 Elementary differential geometry / Barrett O'Neill.—Rev. 2nd ed.
 p. cm.
 Includes bibliographical references and index.
 ISBN-13: *978-0-12-088735-4 (acid-free paper)
 ISBN-10: 0-12-088735-5 (acid-free paper)
 1. Geometry, Differential. I. Title.
QA641.O5 2006
516.3′6—dc22   2005057176

British Library Cataloguing-in-Publication Data
A catalogue record for this book is available from the British Library.

ISBN 13: 978-0-12-088735-4
ISBN 10: 0-12-088735-5

For information on all Elsevier Academic Press Publications
visit our Web site at www.books.elsevier.com

Printed in the United States of America
06 07 08 09   10 9 8 7 6 5 4 3 2 1

Contents

3. Euclidean Geometry

4. Calculus on a Surface

5. Shape Operators

6. Geometry of Surfaces in R³

7. Riemannian Geometry

8. Global Structure of Surfaces

Preface to the Revised Second Edition

This book is an elementary account of the geometry of curves and surfaces. It is written for students who have completed standard courses in calculus and linear algebra, and its aim is to introduce some of the main ideas of differential geometry.

The language of the book is established in Chapter 1 by a review of the core content of differential calculus, emphasizing linearity. Chapter 2 describes the method of *moving frames*, which is introduced, as in elementary calculus, to study curves in space. (This method turns out to apply with equal efficiency to surfaces.) Chapter 3 investigates the rigid motions of space, in terms of which congruence of curves and surfaces is defined in the same way as congruence of triangles in the plane.

Chapter 4 requires special comment. One weakness of classical differential geometry is its lack of any adequate definition of *surface*. In this chapter we decide just what a surface is, and show that every surface has a differential and integral calculus of its own, strictly analogous to the familiar calculus of the plane. This exposition provides an introduction to the notion of *differentiable manifold*, which is the foundation for those branches of mathematics and its applications that are based on the calculus.

The next two chapters are devoted to the geometry of surfaces in 3-space. Chapter 5 measures the *shape* of a surface and derives basic geometric invariants, notably Gaussian curvature. Intuitive and computational aspects are stressed to give geometrical meaning to the theory in Chapter 6.

In the final two chapters, although our methods are unchanged, there is a radical shift of viewpoint. Roughly speaking, we study the geometry of a surface *as seen by its inhabitants*, with no assumption that the surface can be found in ordinary three-dimensional space. Chapter 7 is dominated by *curvature* and culminates in the Gauss-Bonnet theorem and its geometric and topological consequences. In particular, we use the Gauss-Bonnet theorem to

ix

prove the Poincaré-Hopf theorem, which relates the singularities of a vector field on M to the topology of M.

Chapter 8 studies the local and global properties of geodesics. Full development of the global properties requires the notion of covering surface. With it, we can give a comprehensive survey of the surfaces of constant Gaussian curvature and prove the theorems of Bonnet and Hadamard on, respectively, positive and nonnegative curvature.

No branch of mathematics makes a more direct appeal to the intuition than geometry. I have sought to emphasize this by a large number of illustrations that form an integral part of the text.

Each chapter of the book is divided into sections, and in each section a single sequence of numbers designates collectively the theorems, lemmas, examples, and so on. Each section ends with a set of exercises; these range from routine checks of comprehension to moderately challenging problems.

In this revision, the structure of the text, including the numbering of its contents, remains the same, but there are many changes around this framework. The most significant are, first, correction of all known errors; second, a better way of referencing exercises (the most common reference); third, general improvement of the exercises. These improvements include deletion of a few unreasonably difficult exercises, simplification of others, and fuller answers to odd-numbered ones.

In teaching from earlier versions of this book, I have usually covered the background material in Chapter 1 rather rapidly and not devoted any classroom time to Chapter 3. A short course in the geometry of curves and surfaces in 3-space might consist of Chapter 2 (omit Sec. 8), Chapter 4 (omit Sec. 8), Chapter 5, Chapter 6 (covering Secs. 6–9 lightly), and a leap to Section 6 of Chapter 7: the Gauss-Bonnet theorem. This is essentially the content of a traditional undergraduate course in differential geometry, with clarification of the notions of surface and mapping.

Such a course, however, neglects the shift of viewpoint mentioned earlier, in which the geometric concept of surface evolved from a *shape* in 3-space to an independent entity—a two-dimensional *Riemannian manifold*.

This development is important from a practical viewpoint since it makes surface theory applicable throughout the range of scientific applications where 2-parameter objects appear that meet the requisite conditions—for example, in the four-dimensional manifolds of general relativity.

Such a surface is logically simpler than a surface in 3-space since it is constructed (at the start of Chapter 7) by discarding effects of Euclidean space. However, readers can neglect this transition and—as suggested for the Gauss-Bonnet theorem—proceed directly to most of the topics considered in the final two chapters, for example, properties of geodesics (length-minimization and completeness), singularities of vector fields, and the theorems of Bonnet and Hadamard.

For readers with access to a computer containing either the *Mathematica* or *Maple* computation system, I have included some forty computer exercises. These offer an opportunity to amplify the text in various ways.

Previous computer experience is not required. The Appendix contains a summary of the syntaxes of the most recent versions of *Mathematica* and *Maple*, together with a list of explicit computer commands covering the basic geometry of curves and surfaces. Further commands appear in the answers to exercises.

It is important to go, step by step, through the hand calculation of the Gaussian curvature of a parametrized surface, but once this is understood, repetition becomes tedious. A surface in $\mathbf{R}^3$ given only by a formula is seldom easy to sketch. But using computer commands, a picture of a surface can be drawn and its curvature computed, often in no more than a few seconds. Analogous remarks hold for space curves.

Among other applications appearing in the exercises, the most valuable, since unreachable for humans, is the numerical solution of differential equations—and the plotting of these solutions.

This book would not have been possible without generous contributions by Allen B. Altman and Joseph E. Borzellino.

Barrett O'Neill

Elementary Differential Geometry

Revised Second Edition

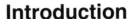

Introduction

This book presupposes a reasonable knowledge of elementary calculus and linear algebra. It is a working knowledge of the fundamentals that is actually required. The reader will, for example, frequently be called upon to *use* the chain rule for differentiation, but its proof need not concern us.

Calculus deals mostly with real-valued functions of one or more variables, linear algebra with functions (linear transformations) from one vector space to another. We shall need functions of these and other types, so we give here general definitions that cover all types.

A *set* S is a collection of objects that are called the *elements* of S. A set A is a *subset* of S provided each element of A is also an element of S.

A *function f* from a set D to a set R is a rule that assigns to each element x of D a unique element $f(x)$ of R. The element $f(x)$ is called the *value* of f at x. The set D is called the *domain* of f; the set R is sometimes called the *range* of f. If we wish to emphasize the domain and range of a function f, the notation $f: D \to R$ is used. Note that the function is denoted by a single letter, say f, while $f(x)$ is merely a value of f.

Many different terms are used for functions—mappings, transformations, correspondences, operators, and so on. A function can be described in various ways, the simplest case being an explicit formula such as

$$f(x) = 3x^2 + 1,$$

which we may also write as $x \to 3x^2 + 1$.

If both f_1 and f_2 are functions from D to R, then $f_1 = f_2$ means that $f_1(x) = f_2(x)$ for all x in D. This is not a definition, but a logical consequence of the definition of *function*.

Let $f: D \to R$ and $g: E \to S$ be functions. In general, the *image* of f is the subset of R consisting of all elements of the form $f(x)$; it is usually denoted by $f(D)$. If this image happens to be a subset of the domain E of g, it is possible to combine these two functions to get the *composite function* $g(f)$: $D \to S$. By definition, $g(f)$ is the function whose value at each element x of D is the element $g(f(x))$ of S.

If f: $D \to R$ is a function and A is a subset of D, then the *restriction* of f to A is the function $f|A$: $A \to R$ defined by the same rule as f, but applied only to elements of A. This seems a rather minor change, but the function $f|A$ may have properties quite different from f itself.

Here are two vital properties that a function may possess. A function f: $D \to R$ is *one-to-one* provided that if x and y are any elements of D such that $x \neq y$, then $f(x) \neq f(y)$. A function f: $D \to R$ is *onto* (or *carries D onto R*) provided that for every element y of R there is at least one element x of D such that $f(x) = y$. In short, the image of f is the entire set R. For example, consider the following functions, each of which has the real numbers as both domain and range:

(1) The function $x \to x^3$ is both one-to-one and onto.
(2) The exponential function $x \to e^x$ is one-to-one, but not onto.
(3) The function $x \to x^3 + x^2$ is onto, but not one-to-one.
(4) The sine function $x \to \sin x$ is neither one-to-one nor onto.

If a function f: $D \to R$ is both one-to-one and onto, then for each element y of R there is one and only one element x such that $f(x) = y$. By defining $f^{-1}(y) = x$ for all x and y so related, we obtain a function f^{-1}: $R \to D$ called the *inverse* of f. Note that the function f^{-1} is also one-to-one and onto, and that *its* inverse function is the original function f.

Here is a short list of the main notations used throughout the book, in order of their appearance in Chapter 1:

p, q	points	(Section 1.1)
f, g	real-valued functions	(Section 1.1)
v, w	tangent vectors	(Section 1.2)
V, W	vector fields	(Section 1.2)
α, β	curves	(Section 1.4)
ϕ, ψ	differential forms	(Section 1.5)
F, G	mappings	(Section 1.7)

In Chapter 1 we define these concepts for Euclidean 3-space. (Extension to arbitrary dimensions is virtually automatic.) In Chapter 4 we show how these concepts can be adapted to a surface.

A few references are given to the brief bibliography at the end of the book; these are indicated by initials in square brackets.

Chapter 1
Calculus on Euclidean Space

As mentioned in the Preface, the purpose of this initial chapter is to establish the mathematical language used throughout the book. Much of what we do is simply a review of that part of elementary calculus dealing with differentiation of functions of three variables and with curves in space. Our definitions have been formulated so that they will apply smoothly to the later study of surfaces.

1.1 Euclidean Space

Three-dimensional space is often used in mathematics without being formally defined. Looking at the corner of a room, one can picture the familiar process by which rectangular coordinate axes are introduced and three numbers are measured to describe the position of each point. A precise definition that realizes this intuitive picture may be obtained by this device: instead of saying that three numbers *describe the position* of a point, we define them to *be* a point.

1.1 Definition *Euclidean 3-space* $\mathbf{R}^3$ is the set of all ordered triples of real numbers. Such a triple $\mathbf{p} = (p_1, p_2, p_3)$ is called a *point* of $\mathbf{R}^3$.

In linear algebra, it is shown that $\mathbf{R}^3$ is, in a natural way, a vector space over the real numbers. In fact, if $\mathbf{p} = (p_1, p_2, p_3)$ and $\mathbf{q} = (q_1, q_2, q_3)$ are points of $\mathbf{R}^3$, their *sum* is the point

$$\mathbf{p} + \mathbf{q} = (p_1 + q_1, p_2 + q_2, p_3 + q_3).$$

3

The *scalar multiple* of a point $\mathbf{p} = (p_1, p_2, p_3)$ by a number a is the point

$$a\mathbf{p} = (ap_1, ap_2, ap_3).$$

It is easy to check that these two operations satisfy the axioms for a vector space. The point $\mathbf{0} = (0, 0, 0)$ is called the *origin* of $\mathbf{R}^3$.

Differential calculus deals with another aspect of $\mathbf{R}^3$ starting with the notion of differentiable real-valued functions on $\mathbf{R}^3$. We recall some fundamentals.

1.2 Definition Let x, y, and z be the real-valued functions on $\mathbf{R}^3$ such that for each point $\mathbf{p} = (p_1, p_2, p_3)$

$$x(\mathbf{p}) = p_1, \quad y(\mathbf{p}) = p_2, \quad z(\mathbf{p}) = p_3.$$

These functions x, y, z are called the *natural coordinate functions* of $\mathbf{R}^3$. We shall also use index notation for these functions, writing

$$x_1 = x, \quad x_2 = y, \quad x_3 = z.$$

Thus the value of the function x_i on a point $\mathbf{p}$ is the number p_i, and so we have the identity $\mathbf{p} = (p_1, p_2, p_3) = (x_1(\mathbf{p}), x_2(\mathbf{p}), x_3(\mathbf{p}))$ for each point $\mathbf{p}$ of $\mathbf{R}^3$. Elementary calculus does not always make a sharp distinction between the *numbers* p_1, p_2, p_3 and the *functions* x_1, x_2, x_3. Indeed the analogous distinction on the real line may seem pedantic, but for higher-dimensional spaces such as $\mathbf{R}^3$, its absence leads to serious ambiguities. (Essentially the same distinction is being made when we denote a function on $\mathbf{R}^3$ by a single letter f, reserving $f(\mathbf{p})$ for its value at the point $\mathbf{p}$.)

We assume that the reader is familiar with partial differentiation and its basic properties, in particular the chain rule for differentiation of a composite function. We shall work mostly with first-order partial derivatives $\partial f/\partial x$, $\partial f/\partial y$, $\partial f/\partial z$ and second-order partial derivatives $\partial^2 f/\partial x^2$, $\partial^2 f/\partial x \partial y$, ... In a few situations, third- and even fourth-order derivatives may occur, but to avoid worrying about exactly how many derivatives we can take in any given context, we establish the following definition.

1.3 Definition A real-valued function f on $\mathbf{R}^3$ is *differentiable* (or *infinitely differentiable*, or *smooth*, or *of class C^∞*) provided all partial derivatives of f, of all orders, exist and are continuous.

Differentiable real-valued functions f and g may be added and multiplied in a familiar way to yield functions that are again differentiable and real-valued. We simply add and multiply their values at each point—the formulas read

$$(f + g)(\mathbf{p}) = f(\mathbf{p}) + g(\mathbf{p}), \quad (fg)(\mathbf{p}) = f(\mathbf{p})g(\mathbf{p}).$$

The phrase "differentiable real-valued function" is unpleasantly long. Hence we make the convention that *unless the context indicates otherwise*, "function" shall mean "real-valued function," and (unless the issue is explicitly raised) the functions we deal with will be assumed to be differentiable. We do not intend to overwork this convention; for the sake of emphasis the words "differentiable" and "real-valued" will still appear fairly frequently.

Differentiation is always a *local* operation: To compute the value of the function $\partial f/\partial x$ at a point $\mathbf{p}$ of $\mathbf{R}^3$, it is sufficient to know the values of f at all points $\mathbf{q}$ of $\mathbf{R}^3$ that are sufficiently near $\mathbf{p}$. Thus, Definition 1.3 is unduly restrictive; the domain of f need not be the whole of $\mathbf{R}^3$, but need only be an *open set* of $\mathbf{R}^3$. By an *open set* $\mathcal{O}$ of $\mathbf{R}^3$ we mean a subset of $\mathbf{R}^3$ such that if a point $\mathbf{p}$ is in $\mathcal{O}$, then so is every other point of $\mathbf{R}^3$ that is sufficiently near $\mathbf{p}$. (A more precise definition is given in Chapter 2.) For example, the set of all points $\mathbf{p} = (p_1, p_2, p_3)$ in $\mathbf{R}^3$ such that $p_1 > 0$ is an open set, and the function $yz \log x$ defined on this set is certainly differentiable, even though its domain is not the whole of $\mathbf{R}^3$. Generally speaking, the results in this chapter remain valid if $\mathbf{R}^3$ is replaced by an arbitrary open set $\mathcal{O}$ of $\mathbf{R}^3$.

We are dealing with *three-dimensional* Euclidean space only because this is the dimension we use most often in later work. It would be just as easy to work with *Euclidean n-space* $\mathbf{R}^n$, for which the points are n-tuples $\mathbf{p} = (p_1, \ldots, p_n)$ and which has n natural coordinate functions $x_1, \ldots, x_n$. All the results in this chapter are valid for Euclidean spaces of arbitrary dimensions, although we shall rarely take advantage of this except in the case of the *Euclidean plane* $\mathbf{R}^2$. In particular, the results are valid for the *real line* $\mathbf{R}^1 = \mathbf{R}$. Many of the concepts introduced are designed to deal with higher dimensions, however, and are thus apt to be overelaborate when reduced to dimension 1.

Exercises

1. Let $f = x^2 y$ and $g = y \sin z$ be functions on $\mathbf{R}^3$. Express the following functions in terms of x, y, z:

(a) fg^2.

(b) $\dfrac{\partial f}{\partial x} g + \dfrac{\partial g}{\partial y} f$.

(c) $\dfrac{\partial^2 (fg)}{\partial y \partial z}$.

(d) $\dfrac{\partial}{\partial y} (\sin f)$.

2. Find the value of the function $f = x^2y - y^2z$ at each point:
(a) (1, 1, 1).   (b) (3, −1, ½).
(c) $(a, 1, 1 - a)$.   (d) (t, t^2, t^3).

3. Express $\partial f/\partial x$ in terms of x, y, and z if
(a) $f = x \sin(xy) + y \cos(xz)$.
(b) $f = \sin g$, $g = e^h$, $h = x^2 + y^2 + z^2$.

4. If g_1, g_2, g_3, and h are real-valued functions on $\mathbf{R}^3$, then

$$f = h(g_1, g_2, g_3)$$

is the function such that

$$f(\mathbf{p}) = h(g_1(\mathbf{p}), g_2(\mathbf{p}), g_3(\mathbf{p})) \quad \text{for all } \mathbf{p}.\dagger$$

Express $\partial f/\partial x$ in terms of x, y, and z, if $h = x^2 - yz$ and
(a) $f = h(x + y, y^2, x + z)$.   (b) $f = h(e^z, e^{x+y}, e^x)$.
(c) $f = h(x, -x, x)$.

1.2 Tangent Vectors

Intuitively, a vector in $\mathbf{R}^3$ is an oriented line segment, or "arrow." Vectors are used widely in physics and engineering to describe forces, velocities, angular momenta, and many other concepts. To obtain a definition that is both practical and precise, we shall describe an "arrow" in $\mathbf{R}^3$ by giving its starting point $\mathbf{p}$ and the change, or vector $\mathbf{v}$, necessary to reach its end point $\mathbf{p} + \mathbf{v}$. Strictly speaking, $\mathbf{v}$ is just a point of $\mathbf{R}^3$.

2.1 Definition‡   A *tangent vector* $\mathbf{v}_p$ to $\mathbf{R}^3$ consists of two points of $\mathbf{R}^3$: its *vector part* $\mathbf{v}$ and its *point of application* $\mathbf{p}$.

We shall always picture $\mathbf{v}_p$ as *the arrow from the point* $\mathbf{p}$ *to the point* $\mathbf{p} + \mathbf{v}$. For example, if $\mathbf{p} = (1, 1, 3)$ and $\mathbf{v} = (2, 3, 2)$, then $\mathbf{v}_p$ runs from (1, 1, 3) to (3, 4, 5) as in Fig. 1.1.

We emphasize that tangent vectors are equal, $\mathbf{v}_p = \mathbf{w}_q$, if and only if they have the same vector part, $\mathbf{v} = \mathbf{w}$, and the same point of application, $\mathbf{p} = \mathbf{q}$.

---

† A consequence is the identity $f = f(x, y, z)$.
‡ The term "tangent" in this definition will acquire a more direct geometric meaning in Chapter 4.

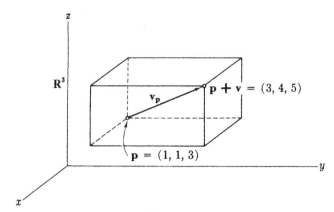

FIG. 1.1

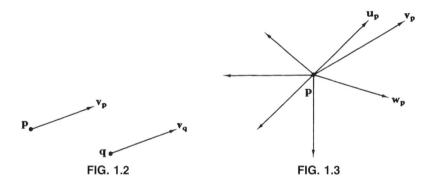

FIG. 1.2   **FIG. 1.3**

Tangent vectors $\mathbf{v}_p$ and $\mathbf{v}_q$ with the same vector part, but different points of application, are said to be *parallel* (Fig. 1.2). It is essential to recognize that $\mathbf{v}_p$ and $\mathbf{v}_q$ are different tangent vectors if $\mathbf{p} \neq \mathbf{q}$. In physics the concept of moment of a force shows this clearly enough: The same force $\mathbf{v}$ applied at different points $\mathbf{p}$ and $\mathbf{q}$ of a rigid body can produce quite different rotational effects.

2.2 Definition Let $\mathbf{p}$ be a point of $\mathbf{R}^3$. The set $T_p(\mathbf{R}^3)$ consisting of all tangent vectors that have $\mathbf{p}$ as point of application is called the *tangent space* of $\mathbf{R}^3$ at $\mathbf{p}$ (Fig. 1.3).

We emphasize that $\mathbf{R}^3$ has a different tangent space at each and every one of its points.

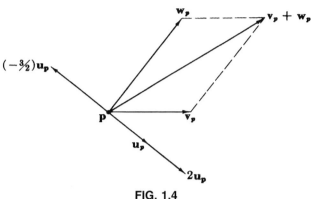

FIG. 1.4

Since all the tangent vectors in a given tangent space have the same point of application, we can borrow the vector addition and scalar multiplication of $\mathbf{R}^3$ to turn $T_p(\mathbf{R}^3)$ into a vector space. Explicitly, we define $\mathbf{v}_p + \mathbf{w}_p$ to be $(\mathbf{v} + \mathbf{w})_p$. and if c is a number we define $c(\mathbf{v}_p)$ to be $(c\mathbf{v})_p$. This is just the usual "parallelogram law" for addition of vectors, and scalar multiplication by c merely stretches a tangent vector by the factor c—reversing its direction if $c < 0$ (Fig. 1.4).

These operations on the tangent space $T_p(\mathbf{R}^3)$ make it a vector space iso-morphic to $\mathbf{R}^3$ itself. Indeed, it follows immediately from the definitions above that for a fixed point $\mathbf{p}$, the function $\mathbf{v} \to \mathbf{v}_p$ is a linear isomorphism from $\mathbf{R}^3$ to $T_p(\mathbf{R}^3)$—that is, a linear transformation that is one-to-one and onto.

A standard concept in physics and engineering is that of a force field. The gravitational force field of the earth, for example, assigns to each point of space a force (vector) directed at the center of the earth.

2.3 Definition   A *vector field* V on $\mathbf{R}^3$ is a function that assigns to each point $\mathbf{p}$ of $\mathbf{R}^3$ a tangent vector $V(\mathbf{p})$ to $\mathbf{R}^3$ at $\mathbf{p}$.

Roughly speaking, a vector field is just a big collection of arrows, one at each point of $\mathbf{R}^3$.

There is a natural algebra of vector fields. To describe it, we first reexam-ine the familiar notion of addition of real-valued functions f and g. It is pos-sible to add f and g because it is possible to add their values at each point. The same is true of vector fields V and W. At each point $\mathbf{p}$, the values $V(\mathbf{p})$ and $W(\mathbf{p})$ are in the same vector space—the tangent space $T_p(\mathbf{R}^3)$—hence we can add $V(\mathbf{p})$ and $W(\mathbf{p})$. Consequently, we can add V and W by adding their values at each point. The formula for this addition is thus the same as for addition of functions,

$$(V + W)(\mathbf{p}) = V(\mathbf{p}) + W(\mathbf{p}).$$

This scheme occurs over and over again. We shall call it the *pointwise principle:* If a certain operation can be performed on the values of two functions at each point, then that operation can be extended to the functions themselves; simply apply it to their values at each point.

For example, we invoke the pointwise principle to extend the operation of *scalar multiplication* (on the tangent spaces of $\mathbf{R}^3$). If f is a real-valued function on $\mathbf{R}^3$ and V is a vector field on $\mathbf{R}^3$, then fV is defined to be the vector field on $\mathbf{R}^3$ such that

$$(fV)(\mathbf{p}) = f(\mathbf{p})V(\mathbf{p}) \quad \text{for all } \mathbf{p}.$$

Our aim now is to determine in a concrete way just what vector fields look like. For this purpose we introduce three special vector fields that will serve as a "basis" for all vector fields.

2.4 Definition   Let U_1, U_2, and U_3 be the vector fields on $\mathbf{R}^3$ such that

$$U_1(\mathbf{p}) = (1, 0, 0)_p$$
$$U_2(\mathbf{p}) = (0, 1, 0)_p$$
$$U_3(\mathbf{p}) = (0, 0, 1)_p$$

for each point $\mathbf{p}$ of $\mathbf{R}^3$ (Fig. 1.5). We call U_1, U_2, U_3—collectively—the *natural frame field* on $\mathbf{R}^3$.

Thus, U_i ($i = 1, 2, 3$) is the unit vector field in the positive x_i direction.

2.5 Lemma   If V is a vector field on $\mathbf{R}^3$, there are three uniquely determined real-valued functions, v_1, v_2, v_3 on $\mathbf{R}^3$ such that

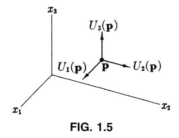

FIG. 1.5

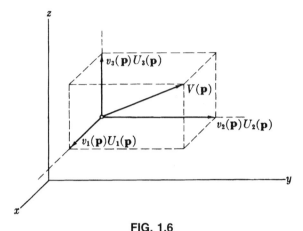

FIG. 1.6

$$V = v_1 U_1 + v_2 U_2 + v_3 U_3.$$

The functions v_1, v_2, v_3 are called the *Euclidean coordinate functions of V.*

Proof. By definition, the vector field V assigns to each point $\mathbf{p}$ a tangent vector $V(\mathbf{p})$ at $\mathbf{p}$. Thus, the vector part of $V(\mathbf{p})$ depends on $\mathbf{p}$, so we write it $(v_1(\mathbf{p}), v_2(\mathbf{p}), v_3(\mathbf{p}))$. (This defines v_1, v_2, and v_3 as real-valued *functions* on $\mathbf{R}^3$.) Hence

$$V(\mathbf{p}) = \left(v_1(\mathbf{p}), v_2(\mathbf{p}), v_3(\mathbf{p})\right)_p$$
$$= v_1(\mathbf{p})(1, 0, 0)_p + v_2(\mathbf{p})(0, 1, 0)_p + v_3(\mathbf{p})(0, 0, 1)_p$$
$$= v_1(\mathbf{p})U_1(\mathbf{p}) + v_2(\mathbf{p})U_2(\mathbf{p}) + v_3(\mathbf{p})U_3(\mathbf{p})$$

for each point $\mathbf{p}$ (Fig. 1.6). By our (pointwise principle) definitions, this means that the vector fields V and $\sum v_i U_i$ have the same (tangent vector) value at each point. Hence $V = \sum v_i U_i$.   ◆

This last sentence uses two of our standard conventions: $\sum v_i U_i$ means sum over $i = 1, 2, 3$; the symbol (◆) indicates the end of a proof.

The tangent-vector identity $(a_1, a_2, a_3)_p = \sum a_i U_i(\mathbf{p})$ appearing in this proof will be used very often.

Computations involving vector fields may always be expressed in terms of their Euclidean coordinate functions. For example, addition and multiplication by a function, are expressed in terms of coordinates by

$$\sum v_i U_i + \sum w_i U_i = \sum (v_i + w_i) U_i,$$
$$f\left(\sum v_i U_i\right) = \sum (f v_i) U_i.$$

Since this is differential calculus, we shall naturally require that the various objects we deal with be differentiable. A vector field V is *differentiable* provided its Euclidean coordinate functions are differentiable (in the sense of Definition 1.3). From now on, we shall understand "vector field" to mean "differentiable vector field."

Exercises

1. Let $\mathbf{v} = (-2, 1, -1)$ and $\mathbf{w} = (0, 1, 3)$.
   (a) At an arbitrary point $\mathbf{p}$, express the tangent vector $3\mathbf{v}_p - 2\mathbf{w}_p$ as a linear combination of $U_1(\mathbf{p})$, $U_2(\mathbf{p})$, $U_3(\mathbf{p})$.
   (b) For $\mathbf{p} = (1, 1, 0)$, make an accurate sketch showing the four tangent vectors $\mathbf{v}_p$, $\mathbf{w}_p$, $-2\mathbf{v}_p$, and $\mathbf{v}_p + \mathbf{w}_p$.

2. Let $V = xU_1 + yU_2$ and $W = 2x^2U_2 - U_3$. Compute the vector field $W - xV$, and find its value at the point $\mathbf{p} = (-1, 0, 2)$.

3. In each case, express the given vector field V in the standard form $\sum v_i U_i$.
   (a) $2z^2 U_1 = 7V + xyU_3$.
   (b) $V(\mathbf{p}) = (p_1, p_3 - p_1, 0)_p$ for all $\mathbf{p}$.
   (c) $V = 2(xU_1 + yU_2) - x(U_1 - y^2U_3)$.
   (d) At each point $\mathbf{p}$, $V(\mathbf{p})$ is the vector from the point (p_1, p_2, p_3) to the point $(1 + p_1, p_2 p_3, p_2)$.
   (e) At each point $\mathbf{p}$, $V(\mathbf{p})$ is the vector from $\mathbf{p}$ to the origin.

4. If $V = y^2 U_1 - x^2 U_3$ and $W = x^2 U_1 - zU_2$, find functions f and g such that the vector field $fV + gW$ can be expressed in terms of U_2 and U_3 only.

5. Let $V_1 = U_1 - xU_3$, $V_2 = U_2$, and $V_3 = xU_1 + U_3$.
   (a) Prove that the vectors $V_1(\mathbf{p})$, $V_2(\mathbf{p})$, $V_3(\mathbf{p})$ are linearly independent at each point of $\mathbf{R}^3$.
   (b) Express the vector field $xU_1 + yU_2 + zU_3$ as a linear combination of V_1, V_2, V_3.

1.3 Directional Derivatives

Associated with each tangent vector $\mathbf{v}_p$ to $\mathbf{R}^3$ is the straight line $t \to \mathbf{p} + t\mathbf{v}$ (see Example 4.2). If f is a differentiable function on $\mathbf{R}^3$, then $t \to f(\mathbf{p} + t\mathbf{v})$ is an ordinary differentiable function on the real line. Evidently the derivative of this function at $t = 0$ tells the initial rate of change of f as $\mathbf{p}$ moves in the $\mathbf{v}$ direction

3.1 Definition   Let f be a differentiable real-valued function on $\mathbf{R}^3$, and let $\mathbf{v}_p$ be a tangent vector to $\mathbf{R}^3$. Then the number

$$\mathbf{v}_p[f] = \frac{d}{dt}(f(\mathbf{p} + t\mathbf{v}))|_{t=0}$$

is called the *derivative of f with respect* to $\mathbf{v}_p$.

This definition appears in elementary calculus with the additional restriction that $\mathbf{v}_p$ be a unit vector. Even though we do not impose this restriction, we shall nevertheless refer to $\mathbf{v}_p[f]$ as a *directional derivative*.

For example, we compute $\mathbf{v}_p[f]$ for the function $f = x^2yz$, with $\mathbf{p} = (1, 1, 0)$ and $\mathbf{v} = (1, 0, -3)$. Then

$$\mathbf{p} + t\mathbf{v} = (1, 1, 0) + t(1, 0, -3) = (1 + t, 1, -3t)$$

describes the line through $\mathbf{p}$ in the $\mathbf{v}$ direction. Evaluating f along this line, we get

$$f(\mathbf{p} + t\mathbf{v}) = (1 + t)^2 \cdot 1 \cdot (-3t) = -3t - 6t^2 - 3t^3.$$

Now,

$$\frac{d}{dt}(f(\mathbf{p} + t\mathbf{v})) = -3 - 12t - 9t^2;$$

hence at $t = 0$, we find $\mathbf{v}_p[f] = -3$. Thus, in particular, the function f is initially decreasing as $\mathbf{p}$ moves in the $\mathbf{v}$ direction.

The following lemma shows how to compute $\mathbf{v}_p[f]$ in general, in terms of the partial derivatives of f at the point $\mathbf{p}$.

3.2 Lemma   If $\mathbf{v}_p = (v_1, v_2, v_3)_p$ is a tangent vector to $\mathbf{R}^3$, then

$$\mathbf{v}_p[f] = \sum v_i \frac{\partial f}{\partial x_i}(\mathbf{p}).$$

Proof.   Let $\mathbf{p} = (p_1, p_2, p_3)$; then

$$\mathbf{p} + t\mathbf{v} = (p_1 + tv_1, p_2 + tv_2, p_3 + tv_3).$$

We use the chain rule to compute the derivative at $t = 0$ of the function

$$f(\mathbf{p} + t\mathbf{v}) = f(p_1 + tv_1, p_2 + tv_2, p_3 + tv_3).$$

Since

$$\frac{d}{dt}(p_i + tv_i) = v_i,$$

we obtain

$$\mathbf{v}_p[f] = \frac{d}{dt}(f(\mathbf{p} + t\mathbf{v}))|_{t=0} = \sum \frac{\partial f}{\partial x_i}(\mathbf{p})v_i. \qquad \blacklozenge$$

Using this lemma, we recompute $\mathbf{v}_p[f]$ for the example above. Since $f = x^2yz$, we have

$$\frac{\partial f}{\partial x} = 2xyz, \quad \frac{\partial f}{\partial y} = x^2z, \quad \frac{\partial f}{\partial z} = x^2y.$$

Thus, at the point $\mathbf{p} = (1, 1, 0)$,

$$\frac{\partial f}{\partial x}(\mathbf{p}) = 0, \quad \frac{\partial f}{\partial y}(\mathbf{p}) = 0, \quad \text{and} \quad \frac{\partial f}{\partial z}(\mathbf{p}) = 1.$$

Then by the lemma,

$$\mathbf{v}_p[f] = 0 + 0 + (-3)1 = -3,$$

as before.

The main properties of this notion of derivative are as follows.

3.3 Theorem Let f and g be functions on $\mathbf{R}^3$, $\mathbf{v}_p$ and $\mathbf{w}_p$ tangent vectors, a and b numbers. Then

(1) $(a\mathbf{v}_p + b\mathbf{w}_p)[f] = a\mathbf{v}_p[f] + b\mathbf{w}_p[f]$.
(2) $\mathbf{v}_p[af + bg] = a\mathbf{v}_p[f] + b\mathbf{v}_p[g]$.
(3) $\mathbf{v}_p[fg] = \mathbf{v}_p[f]\cdot g(\mathbf{p}) + f(\mathbf{p})\cdot\mathbf{v}_p[g]$.

Proof. All three properties may be deduced easily from the preceding lemma. For example, we prove (3). By the lemma, if $\mathbf{v} = (v_1, v_2, v_3)$, then

$$\mathbf{v}_p[fg] = \sum v_i \frac{\partial(fg)}{\partial x_i}(\mathbf{p}).$$

But

$$\frac{\partial(fg)}{\partial x_i} = \frac{\partial f}{\partial x_i}\cdot g + f\cdot\frac{\partial g}{\partial x_i}.$$

Hence

$$\mathbf{v}_p[fg] = \sum v_i\left(\frac{\partial f}{\partial x_i}(\mathbf{p})\cdot g(\mathbf{p}) + f(\mathbf{p})\cdot\frac{\partial g}{\partial x_i}(\mathbf{p})\right)$$

$$= \left(\sum v_i \frac{\partial f}{\partial x_i}(\mathbf{p})\right)g(\mathbf{p}) + f(\mathbf{p})\left(\sum v_i \frac{\partial g}{\partial x_i}(\mathbf{p})\right)$$

$$= \mathbf{v}_p[f]\cdot g(\mathbf{p}) + f(\mathbf{p})\cdot\mathbf{v}_p[g]. \qquad \blacklozenge$$

The first two properties in the preceding theorem may be summarized by saying that $\mathbf{v}_p[f]$ is *linear* in $\mathbf{v}_p$ and in f. The third property, as its proof makes clear, is essentially just the usual Leibniz rule for differentiation of a product. *No matter what form differentiation may take, it will always have suitable* linear *and* Leibnizian *properties.*

We now use the pointwise principle to define the *operation of a vector field V on a function* f. The result is the real-valued function $V[f]$ whose value at each point $\mathbf{p}$ is the number $V(\mathbf{p})[f]$, that is, the derivative of f with respect to the tangent vector $V(\mathbf{p})$ at $\mathbf{p}$. This process should be no surprise, since for a function f on the real line, one begins by defining the derivative of f *at a point*—then the derivative *function* df/dx is the function whose value at each point is the derivative at that point. Evidently, the definition of $V[f]$ is strictly analogous. In particular, if U_1, U_2, U_3 is the natural frame field on $\mathbf{R}^3$, then $U_i[f] = \partial f/\partial x_i$. This is an immediate consequence of Lemma 3.2. For example, $U_1(\mathbf{p}) = (1, 0, 0)_p$; hence

$$U_1(\mathbf{p})[f] = \frac{d}{dt}\left(f(p_1 + t, p_2, p_3)\right)\big|_{t=0},$$

which is precisely the definition of $(\partial f/\partial x_1)(\mathbf{p})$. This is true for all points $\mathbf{p} = (p_1, p_2, p_3)$; hence $U_1[f] = \partial f/\partial x_1$.

We shall use this notion of directional derivative more in the case of vector fields than for individual tangent vectors.

3.4 Corollary If V and W are vector fields on $\mathbf{R}^3$ and f, g, h are real-valued functions, then

(1) $(fV + gW)[h] = fV[h] + gW[h]$.
(2) $V[af + bg] = aV[f] + bV[g]$, for all real numbers a and b.
(3) $V[fg] = V[f]\cdot g + f\cdot V[g]$.

Proof. The pointwise principle guarantees that to derive these properties from Theorem 3.3 we need only be careful about the placement of parentheses. For example, we prove the third formula. By definition, the value of the function $V[fg]$ at $\mathbf{p}$ is $V(\mathbf{p})[fg]$. But by Theorem 3.3 this is

$$V(\mathbf{p})[f]\cdot g(\mathbf{p}) + f(\mathbf{p})\cdot V(\mathbf{p})[g] = V[f](\mathbf{p})\cdot g(\mathbf{p}) + f(\mathbf{p})\cdot V[g](\mathbf{p})$$

$$= (V[f]\cdot g + f\cdot V[g])(\mathbf{p}). \qquad \blacklozenge$$

If the use of parentheses here seems extravagant, we remind the reader that a meticulous proof of Leibniz's formula

$$\frac{d}{dx}(fg) = \frac{df}{dx} \cdot g + f \cdot \frac{dg}{dx}$$

must involve the same shifting of parentheses.

Note that the linearity of $V[f]$ in V and f is for *functions* as "scalars" in the first formula in Corollary 3.4 but only for *numbers* as "scalars" in the second. This stems from the fact that fV signifies merely multiplication, but $V[f]$ is differentiation.

The identity $U_i[f] = \partial f/\partial x_i$ makes it a simple matter to carry out explicit computations. For example, if $V = xU_1 - y^2U_3$ and $f = x^2y + z^3$, then

$$V[f] = xU_1[x^2y] + xU_1[z^3] - y^2U_3[x^2y] - y^2U_3[z^3]$$
$$= x(2xy) + 0 - 0 - y^2(3z^2) = 2x^2y - 3y^2z^2.$$

3.5 Remark Since the subscript notation $\mathbf{v}_p$ for a tangent vector is somewhat cumbersome, from now on we shall frequently omit the point of application $\mathbf{p}$ from the notation. This can cause no confusion, since $\mathbf{v}$ and $\mathbf{w}$ will always denote tangent vectors, and $\mathbf{p}$ and $\mathbf{q}$ points of $\mathbf{R}^3$. In many situations (for example, Definition 3.1) the point of application is crucial, and will be indicated by using either the old notation $\mathbf{v}_p$ or the phrase "a tangent vector $\mathbf{v}$ to $\mathbf{R}^3$ at $\mathbf{p}$."

Exercises

1. Let $\mathbf{v}_p$ be the tangent vector to $\mathbf{R}^3$ with $\mathbf{v} = (2, -1, 3)$ and $\mathbf{p} = (2, 0, -1)$. Working directly from the definition, compute the directional derivative $\mathbf{v}_p[f]$, where
    (a) $f = y^2z$.   (b) $f = x^7$.
    (c) $f = e^x \cos y$.

2. Compute the derivatives in Exercise 1 using Lemma 3.2.

3. Let $V = y^2U_1 - xU_3$, and let $f = xy$, $g = z^3$. Compute the functions
    (a) $V[f]$.   (b) $V[g]$.
    (c) $V[fg]$.   (d) $fV[g] - gV[f]$.
    (e) $V[f^2 + g^2]$.   (f) $V[V[f]]$.

4. Prove the identity $V = \sum V[x_i]U_i$, where x_1, x_2, x_3 are the natural coordinate functions. (*Hint:* Evaluate $V = \sum v_iU_i$ on x_j.)

5. If $V[f] = W[f]$ for every function f on $\mathbf{R}^3$, prove that $V = W$.

1.4 Curves in $\mathbf{R}^3$

Let I be an open interval in the real line R. We shall interpret this liberally
to include not only the usual finite open interval $a < t < b$ (a, b real numbers),
but also the infinite types $a < t$ (a half-line to $+\infty$), $t < b$ (a half-line to $-\infty$),
and also the whole real line.

One can picture a curve in $\mathbf{R}^3$ as a trip taken by a moving point α. At each
"time" t in some open interval, α is located at the point

$$\alpha(t) = (\alpha_1(t), \alpha_2(t), \alpha_3(t))$$

in $\mathbf{R}^3$. In rigorous terms then, α is a function from I to $\mathbf{R}^3$, and the real-valued
functions α_1, α_2, α_3 are its *Euclidean coordinate functions*. Thus we write
$\alpha = (\alpha_1, \alpha, \alpha_3)$, meaning, of course, that

$$\alpha(t) = (\alpha_1(t), \alpha_2(t), \alpha_3(t)) \quad \text{for all } t \text{ in } I.$$

We define the function α to be *differentiable* provided its (real-valued) coordinate functions are differentiable in the usual sense.

4.1 Definition A *curve* in $\mathbf{R}^3$ is a differentiable function $\alpha: I \to \mathbf{R}^3$ from
an open interval I into $\mathbf{R}^3$.

We shall give several examples of curves, which will be used in Chapter 2
to experiment with results on the geometry of curves.

4.2 Example (1) *Straight line.* A line is the simplest type of curve in
Euclidean space; its coordinate functions are linear (in the sense $t \to at + b$,
not in the homogeneous sense $t \to at$). Explicitly, the curve $\alpha: \mathbf{R} \to \mathbf{R}^3$ such
that

$$\alpha(t) = \mathbf{p} + t\mathbf{q} = (p_1 + tq_1, p_2 + tq_2, p_3 + tq_3) \quad (\mathbf{q} \neq 0)$$

is the *straight line* through the point $\mathbf{p} = \alpha(0)$ in the $\mathbf{q}$ direction.

(2) *Helix.* (Fig. 1.7). The curve $t \to (a\cos t, a\sin t, 0)$ travels around a circle
of radius $a > 0$ in the xy plane of $\mathbf{R}^3$. If we allow this curve to rise (or fall)
at a constant rate, we obtain a *helix* $\alpha: \mathbf{R} \to \mathbf{R}^3$, given by the formula

$$\alpha(t) = (a\cos t, a\sin t, bt)$$

where $a > 0$, $b \neq 0$.

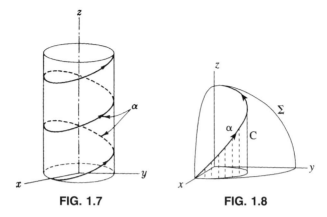

FIG. 1.7   **FIG. 1.8**

(3) The curve

$$\alpha(t) = \left(1 + \cos t, \sin t, 2 \sin \frac{t}{2}\right) \quad \text{for all } t$$

has a noteworthy property: Let C be the cylinder in $\mathbf{R}^3$ over the circle in the xy plane with center at $(1, 0, 0)$ and radius 1. Then α perpetually travels the route sliced from C by the sphere $\sum$ with radius 2 and center at the origin. A segment of this route is shown in Fig. 1.8.

(4) The curve $\alpha\colon \mathbf{R} \to \mathbf{R}^3$ such that

$$\alpha(t) = \left(e^t, e^{-t}, \sqrt{2}\,t\right)$$

shares with the helix in (2) the property of rising constantly. However, it lies over the hyperbola $xy = 1$ in the xy plane instead of over a circle.

(5) The 3-*curve* $\alpha\colon \mathbf{R} \to \mathbf{R}^3$ is defined by

$$\alpha(t) = \left(3t - t^3, 3t^2, 3t + t^3\right).$$

If the coordinate functions of a curve are simple enough, its shape in $\mathbf{R}^3$ can be found, at least approximately, by plotting a few points. We could get a reasonable picture of curve α for $0 \leq t \leq 1$ by computing $\alpha(t)$ for $t = 0, \frac{1}{10}, \frac{1}{2}, \frac{9}{10}, 1$.

If we visualize a curve α in $\mathbf{R}^3$ as a moving point, then at every time t there is a tangent vector at the point $\alpha(t)$ that gives the instantaneous velocity of α at that time. ◆

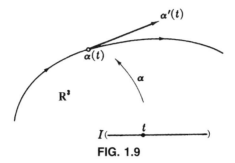

FIG. 1.9

4.3 Definition   Let $\alpha: I \to \mathbf{R}^3$ be a curve in $\mathbf{R}^3$ with $\alpha = (\alpha_1, \alpha_2, \alpha_3)$. For each number t in I, the *velocity vector of α at t* is the tangent vector

$$\alpha'(t) = \left(\frac{d\alpha_1}{dt}(t), \frac{d\alpha_2}{dt}(t), \frac{d\alpha_3}{dt}(t) \right)_{\alpha(t)}$$

at the point $\alpha(t)$ in $\mathbf{R}^3$ (Fig. 1.9).

This definition can be interpreted geometrically as follows. The derivative at t of a real-valued function f on $\mathbf{R}$ is given by

$$\frac{df}{dt}(t) = \lim_{\Delta t \to 0} \frac{f(t + \Delta t) - f(t)}{\Delta t}.$$

This formula still makes sense if f is replaced by a curve $\alpha = (\alpha_1, \alpha_2, \alpha_3)$. In fact,

$$\frac{1}{\Delta t}(\alpha(t + \Delta t) - \alpha(t)) = \left(\frac{\alpha_1(t + \Delta t) - \alpha_1(t)}{\Delta t}, \right.$$
$$\left. \frac{\alpha_2(t + \Delta t) - \alpha_2(t)}{\Delta t}, \frac{\alpha_3(t + \Delta t) - \alpha_3(t)}{\Delta t} \right).$$

This is the vector from $\alpha(t)$ to $\alpha(t + \Delta t)$, scalar multiplied by $1/\Delta t$ (Fig. 1.10).

Now, as Δt gets smaller, $\alpha(t + \Delta t)$ approaches $\alpha(t)$, and in the limit as $\Delta t \to 0$, we get a vector *tangent* to the curve α at the point $\alpha(t)$, namely,

$$\left(\frac{d\alpha_1}{dt}(t), \frac{d\alpha_2}{dt}(t), \frac{d\alpha_3}{dt}(t) \right).$$

As the figure suggests, the point of application of this vector must be the point $\alpha(t)$. Thus the standard limit operation for derivatives gives rise to our definition of the velocity of a curve.

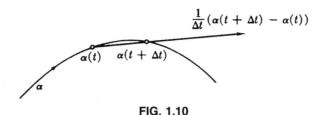

FIG. 1.10

An application of the identity

$$(v_1, v_2, v_3)_p = \sum v_i U_i(\mathbf{p})$$

to the velocity vector $\alpha'(t)$ at t yields the alternative formula

$$\alpha'(t) = \sum \frac{d\alpha_i}{dt}(t)U_i(\alpha(t)).$$

For example, the velocity of the straight line $\alpha(t) = \mathbf{p} + t\mathbf{q}$ is

$$\alpha'(t) = (q_1, q_2, q_3)_{\alpha(t)} = \mathbf{q}_{\alpha(t)}.$$

The fact that α is straight is reflected in the fact that all its velocity vectors are parallel; only the point of application changes as t changes.

For the helix

$$\alpha(t) = (a \cos t, a \sin t, bt),$$

the velocity is

$$\alpha'(t) = (-a \sin t, a \cos t, b)_{\alpha(t)}.$$

The fact that the helix rises constantly is shown by the constancy of the z coordinate of $\alpha'(t)$.

Given any curve, it is easy to construct new curves that follow the same route.

4.4 Definition   Let $\alpha: I \to \mathbf{R}^3$ be a curve. If $h: J \to I$ is a differentiable function on an open interval J, then the composite function

$$\beta = \alpha(h): J \to \mathbf{R}^3$$

is a curve called a *reparametrization* of α by h.

For each $s \in J$, the new curve β is at the point $\beta(s) = \alpha(h(s))$ reached by α at $h(s)$ in I (Fig. 1.11). Thus β represents a different trip over at least part of the route of α.

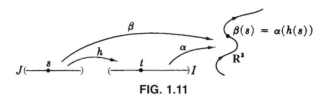

FIG. 1.11

To compute the coordinates of β, simply substitute $t = h(s)$ into the coordinates $\alpha_1(t)$, $\alpha_2(t)$, $\alpha_3(t)$ of α. For example, suppose

$$\alpha(t) = \left(\sqrt{t}, t\sqrt{t}, 1 - t\right) \text{ on } I: 0 < t < 4.$$

If $h(s) = s^2$ on $J: 0 < s < 2$, then the reparametrized curve is

$$\beta(s) = \alpha(h(s)) = \alpha(s^2) = (s, s^3, 1 - s^2).$$

The following lemma relates the velocities of a curve and of a reparametrization.

4.5 Lemma   If β is the reparametrization of α by h, then

$$\beta'(s) = (dh/ds)(s)\alpha'(h(s)).$$

Proof.   If $\alpha = (\alpha_1, \alpha_2, \alpha_3)$, then

$$\beta(s) = \alpha(h(s)) = (\alpha_1(h(s)), \alpha_2(h(s)), \alpha_3(h(s))).$$

Using the "prime" notation for derivatives, the chain rule for a composition of real-valued functions f and g reads $(g(f))' = g'(f) \cdot f'$. Thus, in the case at hand,

$$\alpha_i(h)'(s) = \alpha_i'(h(s)) \cdot h'(s).$$

By the definition of velocity, this yields

$$\beta'(s) = \alpha(h)'(s)$$
$$= (\alpha_1'(h(s)) \cdot h'(s), \alpha_2'(h(s)) \cdot h'(s), \alpha_3'(h(s)) \cdot h'(s))$$
$$= h'(s)\alpha'(h(s)). \qquad \blacklozenge$$

According to this lemma, to obtain the velocity of a reparametrization of α by h, first reparametrize α' by h, then scalar multiply by the derivative of h.

Since velocities are tangent vectors, we can take the derivative of a function with respect to a velocity.

4.6 Lemma   Let α be a curve in $\mathbf{R}^3$ and let f be a differentiable function on $\mathbf{R}^3$. Then

$$\alpha'(t)[f] = \frac{d(f(\alpha))}{dt}(t).$$

Proof.   Since

$$\alpha' = \left(\frac{d\alpha_1}{dt}, \frac{d\alpha_2}{dt}, \frac{d\alpha_3}{dt}\right)_\alpha,$$

we conclude from Lemma 3.2 that

$$\alpha'(t)[f] = \sum \frac{\partial f}{\partial x_i}(\alpha(t))\frac{d\alpha_i}{dt}(t).$$

But the composite function $f(\alpha)$ may be written $f(\alpha_1, \alpha_2, \alpha_3)$, and the chain rule then gives exactly the same result for the derivative of $f(\alpha)$. ◆

By definition, $\alpha'(t)[f]$ is the rate of change of f along the line through $\alpha(t)$ in the $\alpha'(t)$ direction. (If $\alpha'(t) \neq 0$, this is the tangent line to α at $\alpha(t)$; see Exercise 9.) The lemma shows that this rate of change is the same as that of f along the curve α itself.

Since a curve $\alpha\colon I \to \mathbf{R}^3$ is a function, it makes sense to say that α is one-to-one; that is, $\alpha(t) = \alpha(t_1)$ only if $t = t_1$. Another special property of curves is periodicity: A curve $\alpha\colon \mathbf{R} \to \mathbf{R}^3$ is *periodic* if there is a number $p > 0$ such that $\alpha(t + p) = \alpha(t)$ for all t—and the smallest such number p is then called the *period* of α.

From the viewpoint of calculus, the most important condition on a curve α is that it be *regular*, that is, have all velocity vectors different from zero. Such a curve can have no corners or cusps.

The following remarks about curves (offered without proof) describe another familiar way to formulate the concept of "curve." If f is a differentiable real-valued function on $\mathbf{R}^2$, let

$$C\colon f = a$$

be the set of all points $\mathbf{p}$ in $\mathbf{R}^2$ such that $f(\mathbf{p}) = a$. Now, if the partial derivatives $\partial f/\partial x$ and $\partial f/\partial y$ are never simultaneously zero at any point of C, then C consists of one or more separate "components," which we shall call *Curves*.† For example, $C\colon x^2 + y^2 = r^2$ is the circle of radius r centered at the

---

† The capital C distinguishes this notion from a (parametrized) curve $\alpha\colon I \to \mathbf{R}^2$.

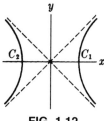

FIG. 1.12

origin of $\mathbf{R}^2$, and the hyperbola C: $x^2 - y^2 = r^2$ splits into two Curves ("branches") C_1 and C_2 as shown in Fig. 1.12.

Every Curve C is the route of many regular curves, called *parametrizations* of C. For example, the curve

$$\alpha(t) = (r \cos t, r \sin t)$$

is a well-known periodic parametrization of the circle given above, and for $r > 0$ the one-to-one curve

$$\beta(t) = (r \cosh t, r \sinh t)$$

parametrizes the branch $x > 0$ of the hyperbola.

Exercises

1.   Compute the velocity vector of the curve in Example 4.2(3) for arbitrary t and for $t = 0$, $t = \pi/2$, $t = \pi$, visualizing those on Fig. 1.8.

2.   Find the unique curve such that $\alpha(0) = (1, 0, 5)$ and $\alpha'(t) = (t^2, t, e^t)$.

3.   Find the coordinate functions of the curve $\beta = \alpha(h)$, where α is the curve in Example 4.2(3) and $h(s) = \cos^{-1}(s)$ on J: $0 < s < 1$.

4.   Reparametrize the curve α in Example 4.2(4) using $h(s) = \log s$ on J: $s > 0$. Check the equation in Lemma 4.5 in this case by calculating each side separately.

5.   Find the equation of the straight line through the points $(1, -3, -1)$ and $(6, 2, 1)$. Does this line meet the line through the points $(-1, 1, 0)$ and $(-5, -1, -1)$?

6.   Deduce from Lemma 4.6 that in the definition of directional derivative (Def. 3.1) the straight line $t \to \mathbf{p} + t\mathbf{v}$ can be replaced by any curve α with *initial velocity* $\mathbf{v}_p$, that is, such that $\alpha(0) = \mathbf{p}$ and $\alpha'(0) = \mathbf{v}_p$.

7. (*Continuation.*)

(a) Show that the curves with coordinate functions

$$(t, 1 + t^2, t), \quad (\sin t, \cos t, t), \quad (\sinh t, \cosh t, t)$$

all have the same initial velocity v_p.

(b) If $f = x^2 - y^2 + z^2$, compute $v_p[f]$ by calculating $d(f(\alpha))/dt$ at $t = 0$, using each of three curves in (a).

8. Sketch the following Curves in $\mathbf{R}^2$, and find parametrizations for each.

(a) $C: 4x^2 + y^2 = 1$,   (b) $C: 3x + 4y = 1$,

(c) $C: y = e^x$.

9. For a fixed t, the *tangent line* to a regular curve α at the point $\alpha(t)$ is the straight line $u \to \alpha(t) + u\alpha'(t)$, where we delete the point of application of $\alpha'(t)$. Find the tangent line to the helix $\alpha(t) = (2\cos t, 2\sin t, t)$ at the points $\alpha(0)$ and $\alpha(\pi/4)$.

1.5   1-Forms

If f is a real-valued function on $\mathbf{R}^3$, then in elementary calculus the differential of f is usually defined as

$$df = \frac{\partial f}{\partial x} dx + \frac{\partial f}{\partial y} dy + \frac{\partial f}{\partial z} dz.$$

It is not always made clear exactly what this formal expression means. In this section we give a rigorous treatment using the notion of 1-form, and forms tend to appear at crucial moments in later work.

5.1 Definition   A *1-form* ϕ on $\mathbf{R}^3$ is a real-valued function on the set of all tangent vectors to $\mathbf{R}^3$ such that ϕ is linear at each point, that is,

$$\phi(a\mathbf{v} + b\mathbf{w}) = a\phi(\mathbf{v}) + b\phi(\mathbf{w})$$

for any numbers a, b and tangent vectors $\mathbf{v}$, $\mathbf{w}$ at the same point of $\mathbf{R}^3$.

We emphasize that for every tangent vector $\mathbf{v}$, a 1-form ϕ defines a real number $\phi(\mathbf{v})$; and for each point $\mathbf{p}$ in $\mathbf{R}^3$, the resulting function $\phi_p\colon T_p(\mathbf{R}^3) \to \mathbf{R}$ is linear. Thus at each point $\mathbf{p}$, ϕ_p is an element of the *dual space* of $T_p(\mathbf{R}^3)$. In this sense the notion of 1-form is dual to that of vector field.

The sum of 1-forms ϕ and ψ is defined in the usual pointwise fashion:

$$(\phi + \psi)(\mathbf{v}) = \phi(\mathbf{v}) + \psi(\mathbf{v}) \quad \text{for all tangent vectors } \mathbf{v}.$$

Similarly, if f is a real-valued function on $\mathbf{R}^3$ and ϕ is a 1-form, then $f\phi$ is the 1-form such that

$$(f\phi)(\mathbf{v}_p) = f(\mathbf{p})\phi(\mathbf{v}_p)$$

for all tangent vectors $\mathbf{v}_p$.

There is also a natural way to *evaluate a 1-form ϕ on a vector field V* to obtain a real-valued function $\phi(V)$: At each point $\mathbf{p}$ the value of $\phi(V)$ is the number $\phi(V(\mathbf{p}))$. Thus a 1-form may also be viewed as a machine that converts vector fields into real-valued functions. If $\phi(V)$ is differentiable whenever V is, we say that ϕ is *differentiable*. As with vector fields, we shall always assume that the 1-forms we deal with are differentiable.

A routine check of definitions shows that $\phi(V)$ is linear in both ϕ and V; that is,

$$\phi(fV + gW) = f\phi(V) + g\phi(W)$$

and

$$(f\phi + g\psi)(V) = f\phi(V) + g\psi(V),$$

where f and g are functions.

Using the notion of directional derivative, we now define a most important way to convert functions into 1-forms.

5.2 Definition   If f is a differentiable real-valued function on $\mathbf{R}^3$, the *differential df* of f is the 1-form such that

$$df(\mathbf{v}_p) = \mathbf{v}_p[f] \quad \text{for all tangent vectors } \mathbf{v}_p.$$

In fact, df is a 1-form, since by definition it is a real-valued function on tangent vectors, and by (1) of Theorem 3.3 it is linear at each point $\mathbf{p}$. Clearly, df knows all rates of change of f in all directions on $\mathbf{R}^3$, so it is not surprising that differentials are fundamental to the calculus on $\mathbf{R}^3$.

Our task now is to show that these rather abstract definitions lead to familiar results when expressed in terms of coordinates.

5.3 Example   1-Forms on $\mathbf{R}^3$. (1) The differentials dx_1, dx_2, dx_3 of the natural coordinate functions. Using Lemma 3.2 we find

$$dx_i(\mathbf{v}_p) = \mathbf{v}_p[x_i] = \sum_j v_j \frac{\partial x_i}{\partial x_j}(\mathbf{p}) = \sum_j v_j \delta_{ij} = v_i,$$

where δ_{ij} is the Kronecker delta (0 if $i \neq j$, 1 if $i = j$). Thus *the value of dx_i on an arbitrary tangent vector $\mathbf{v}_p$ is the ith coordinate v_i of its vector part*—and does not depend on the point of application $\mathbf{p}$.

(2) The 1-form $\psi = f_1 dx_1 + f_2 dx_2 + f_3 dx_3$. Since dx_i is a 1-form, our definitions show that ψ is also a 1-form for any functions f_1, f_2, f_3. The value of ψ on an arbitrary tangent vector $\mathbf{v}_p$ is

$$\psi(\mathbf{v}_p) = \left(\sum f_i dx_i\right)(\mathbf{v}_p) = \sum f_i(\mathbf{p}) dx_i(\mathbf{v}_p) = \sum f_i(\mathbf{p}) v_i,$$

The first of these examples shows that the 1-forms dx_1, dx_2, dx_3 are the analogues for tangent vectors of the natural coordinate functions x_1, x_2, x_3 for points. Alternatively, we can view dx_1, dx_2, dx_3 as the "duals" of the natural unit vector fields U_1, U_2, U_3. In fact, it follows immediately from (1) above that the function $dx_i(U_j)$ has the constant value δ_{ij}.

We now show that every 1-form can be written in the concrete manner given in (2) above.

5.4 Lemma   If ϕ is a 1-form on $\mathbf{R}^3$, then $\phi = \sum f_i dx_i$, where $f_i = \phi(U_i)$. These functions f_1, f_2, f_3 are called the *Euclidean coordinate functions* of ϕ.

Proof.   By definition, a 1-form is a function on tangent vectors; thus ϕ and $\sum f_i dx_i$ are equal if and only if they have the same value on every tangent vector $\mathbf{v}_p = \sum v_i U_i(\mathbf{p})$. In (2) of Example 5.3 we saw that

$$\left(\sum f_i dx_i\right)(\mathbf{v}_p) = \sum f_i(\mathbf{p}) v_i.$$

On the other hand,

$$\phi(\mathbf{v}_p) = \phi\left(\sum v_i U_i(\mathbf{p})\right) = \sum v_i \phi(U_i(\mathbf{p})) = \sum v_i f_i(\mathbf{p})$$

since $f_i = \phi(U_i)$. Thus ϕ and $\sum f_i dx_i$ do have the same value on every tangent vector.   ◆

This lemma shows that a 1-form on $\mathbf{R}^3$ is nothing more than an expression $f\,dx + g\,dy + h\,dz$, and such expressions are now rigorously defined as functions on tangent vectors. Let us now show that the definition of differential of a function (Definition 5.2) agrees with the informal definition given at the start of this section.

5.5 Corollary   If f is a differentiable function on $\mathbf{R}^3$, then

$$df = \sum \frac{\partial f}{\partial x_i} dx_i.$$

Proof. The value of $\sum(\partial f/\partial x_i)dx_i$ on an arbitrary tangent vector $\mathbf{v}_p$ is $\sum(\partial f/\partial x_i)(\mathbf{p})v_i$. By Lemma 3.2, $df(\mathbf{v}_p) = \mathbf{v}_p[f]$ is the same. Thus the 1-forms df and $\sum(\partial f/\partial x_i)\,dx_i$ are equal.   ◆

Using either this result or the definition of d, it is immediate that

$$d(f + g) = df + dg.$$

Finally, we determine the effect of d on *products* of functions and on *compositions* of functions.

5.6 Lemma Let fg be the product of differentiable functions f and g on $\mathbf{R}^3$. Then

$$d(fg) = g\,df + f\,dg.$$

Proof. Using Corollary 5.5, we obtain

$$d(fg) = \sum \frac{\partial(fg)}{\partial x_i} dx_i = \sum \left(\frac{\partial f}{\partial x_i} g + f \frac{\partial g}{\partial x_i} \right) dx_i$$

$$= g\left(\sum \frac{\partial f}{\partial x_i} dx_i \right) + f\left(\sum \frac{\partial g}{\partial x_i} dx_i \right) = g\,df + f\,dg.   ◆$$

5.7 Lemma Let $f: \mathbf{R}^3 \to \mathbf{R}$ and $h: \mathbf{R} \to \mathbf{R}$ be differentiable functions, so the composite function $h(f): \mathbf{R}^3 \to \mathbf{R}$ is also differentiable. Then

$$d(h(f)) = h'(f)\,df.$$

Proof. (The prime here is just the ordinary derivative, so $h'(f)$ is again a composite function, from $\mathbf{R}^3$ to $\mathbf{R}$.) The usual chain rule for a composite function such as $h(f)$ reads

$$\frac{\partial(h(f))}{\partial x_i} = h'(f)\frac{\partial f}{\partial x_i}.$$

Hence

$$d(h(f)) = \sum \frac{\partial(h(f))}{\partial x_i} dx_i = \sum h'(f)\frac{\partial f}{\partial x_i} dx_i = h'(f)\,df.   ◆$$

To compute df for a given function f it is almost always simpler to use these properties of d rather than substitute in the formula of Corollary 5.5. Then from df we immediately get the partial derivatives of f, and, in fact, *all its directional derivatives*. For example, suppose

$$f = (x^2 - 1)y + (y^2 + 2)z.$$

Then by Lemmas 5.6 and 5.7,

$$df = (2x\,dx)y + (x^2 - 1)dy + (2y\,dy)z + (y^2 + 2)dz$$

$$= \underbrace{2xy\,dx}_{\partial f/\partial x} + \underbrace{(x^2 + 2yz - 1)dy}_{\partial f/\partial y} + \underbrace{(y^2 + 2)dz}_{\partial f/\partial z}.$$

Now use the rules above to evaluate this expression on a tangent vector $\mathbf{v}_p$. The result is

$$\mathbf{v}_p[f] = df(\mathbf{v}_p) = 2p_1 p_2 v_1 + (p_1^2 + 2p_2 p_3 - 1)v_2 + (p_2^2 + 1)v_3.$$

Exercises

1. Let $\mathbf{v} = (1, 2, -3)$ and $\mathbf{p} = (0, -2, 1)$. Evaluate the following 1-forms on the tangent vector $\mathbf{v}_p$.
(a) $y^2\,dx$.   (b) $z\,dy - y\,dz$.
(c) $(z^2 - 1)dx - dy + x^2\,dz$.

2. If $\phi = \sum f_i dx_i$ and $V = \sum v_i U_i$, show that the 1-form ϕ evaluated on the vector field V is the function $\phi(V) = \sum f_i v_i$.

3. Evaluate the 1-form $\phi = x^2\,dx - y^2\,dz$ on the vector fields
$V = xU_1 + yU_2 + zU_3$,
$W = xy\,(U_1 - U_3) + yz\,(U_1 - U_2)$, and $(1/x)V + (1/y)W$.

4. Express the following differentials in terms of df:
(a) $d(f^5)$.   (b) $d(\sqrt{f})$, where $f > 0$.
(c) $d(\log(1 + f^2))$.

5. Express the differentials of the following functions in the standard form $\sum f_i\,dx_i$.
(a) $(x^2 + y^2 + z^2)^{1/2}$.   (b) $\tan^{-1}(y/x)$.

6. In each case compute the differential of f and find the directional derivative $\mathbf{v}_p[f]$, for $\mathbf{v}_p$ as in Exercise 1.
(a) $f = xy^2 - yz^2$.   (b) $f = xe^{yz}$.
(c) $f = \sin(xy)\cos(xz)$.

7.   Which of the following are 1-forms? In each case ϕ is the function on tangent vectors such that the value of ϕ on $(v_1, v_2, v_3)_p$ is

(a) $v_1 - v_3$.
(b) $p_1 - p_3$.
(c) $v_1 p_3 + v_2 p_1$.
(d) $v_p[x^2 + y^2]$.
(e) 0.
(f) $(p_1)^2$.

In case ϕ is a 1-form, express it as $\sum f_i \, dx_i$.

8.   Prove Lemma 5.6 directly from the definition of d.

9.   A 1-form ϕ is *zero* at a point $\mathbf{p}$ provided $\phi(v_p) = 0$ for all tangent vectors at $\mathbf{p}$. A point at which its differential df is zero is called a *critical point* of the function f. Prove that $\mathbf{p}$ is a critical point of f if and only if

$$\frac{\partial f}{\partial x}(\mathbf{p}) = \frac{\partial f}{\partial y}(\mathbf{p}) = \frac{\partial f}{\partial z}(\mathbf{p}) = 0.$$

Find all critical points of $f = (1 - x^2)y + (1 - y^2)z$.

(*Hint:* Find the partial derivatives of f by computing df.)

10.   (*Continuation.*) Prove that the local maxima and local minima of f are critical points of f. (f has a *local maximum* at $\mathbf{p}$ if $f(\mathbf{q}) \leq f(\mathbf{p})$ for all $\mathbf{q}$ near $\mathbf{p}$.)

11.   It is sometimes asserted that df is the linear approximation of Δf.
(a) Explain the sense in which $(df)(v_p)$ is a linear approximation of $f(\mathbf{p} + \mathbf{v}) - f(\mathbf{p})$.
(b) Compute exact and approximate values of $f(0.9, 1.6, 1.2) - f(1, 1.5, 1)$, where $f = x^2 y/z$.

1.6   Differential Forms

The 1-forms on $\mathbf{R}^3$ are part of a larger system called the *differential forms* on $\mathbf{R}^3$. We shall not give as rigorous an account of differential forms as we did of 1-forms since our use of the full system on $\mathbf{R}^3$ is limited. However, the *properties* established here are valid whenever differential forms are used.

Roughly speaking, a *differential form* on $\mathbf{R}^3$ is an expression obtained by adding and multiplying real-valued functions and the differentials dx_1, dx_2, dx_3 of the natural coordinate functions of $\mathbf{R}^3$. These two operations obey the usual associative and distributive laws; however, the multiplication is not commutative. Instead, it obeys the

alternation rule: $dx_i \, dx_j = -dx_j \, dx_i$ $(1 \leq i, j \leq 3)$.

This rule appears—although rather inconspicuously—in elementary calculus (see Exercise 9).

A consequence of the alternation rule is the fact that "repeats are zero," that is, $dx_i\, dx_i = 0$, since if $i = j$ the alternation rule reads

$$dx_i\, dx_i = -dx_i\, dx_i.$$

If each summand of a differential form contains p dx_i's ($p = 0, 1, 2, 3$), the form is called a *p-form*, and is said to have *degree p*. Thus, shifting to dx, dy, dz, we find

A 0-form is just a differentiable function f.

A 1-form is an expression $f\, dx + g\, dy + h\, dz$, just as in the preceding section.

A 2-form is an expression $f\, dx\, dy + g\, dx\, dz + h\, dy\, dz$.

A 3-form is an expression $f\, dx\, dy\, dz$.

We already know how to add 1-forms: simply add corresponding coefficient functions. Thus, in index notation,

$$\sum f_i dx_i + \sum g_i dx_i = \sum (f_i + g_i) dx_i.$$

The corresponding rule holds for 2-forms or 3-forms.

On three-dimensional Euclidean space, all p-forms with $p > 3$ are zero. This is a consequence of the alternation rule, for a product of more than three dx_i's must contain some dx_i twice, but repeats are zero, as noted above. For example, $dx\, dy\, dx\, dz = -dx\, dx\, dy\, dz = 0$, since $dx\, dx = 0$. As a reminder that the alternation rule is to be used, we denote this multiplication of forms by a *wedge* $\wedge$. (However, we do not bother with the wedge when only products of dx, dy, dz are involved.)

6.1 **Example**   Computation of wedge products.

(1) Let

$$\phi = x\, dx - y\, dy \quad \text{and} \quad \psi = z\, dx + x\, dz.$$

Then

$$\phi \wedge \psi = (x\, dx - y\, dy) \wedge (z\, dx + x\, dz)$$

$$= xz\, dx\, dx + x^2\, dx\, dz - yz\, dy\, dx - yx\, dy\, dz.$$

But $dx\, dx = 0$ and $dy\, dx = -dx\, dy$. Thus

$$\phi \wedge \psi = yz\, dx\, dy + x^2 dx\, dz - xy\, dy\, dz.$$

In general, the product of two 1-forms is a 2-form.

(2) Let ϕ and ψ be the 1-forms given above and let $\theta = z\,dy$. Then

$$\theta \wedge \phi \wedge \psi = yz^2\,dy\,dx\,dy + x^2z\,dy\,dx\,dz - xyz\,dy\,dy\,dz.$$

Since $dy\,dx\,dy$ and $dy\,dy\,dz$ each contain repeats, both are zero. Thus

$$\theta \wedge \phi \wedge \psi = -x^2z\,dx\,dy\,dz.$$

(3) Let ϕ be as above, and let η be the 2-form $y\,dx\,dz + x\,dy\,dz$. Omitting forms containing repeats, we find

$$\phi \wedge \eta = x^2\,dx\,dy\,dz - y^2\,dy\,dx\,dz = (x^2 + y^2)\,dx\,dy\,dz.$$

It should be clear from these examples that the wedge product of a p-form and a q-form is a $(p + q)$-form. Thus such a product is automatically zero whenever $p + q > 3$.

6.2 Lemma   If ϕ and ψ are 1-forms, then

$$\phi \wedge \psi = -\psi \wedge \phi.$$

Proof.   Write

$$\phi = \sum f_i\,dx_i, \quad \psi = \sum g_i\,dx_i.$$

Then by the alternation rule,

$$\phi \wedge \psi = \sum f_i g_j\,dx_i\,dx_j = -\sum g_j f_i\,dx_j\,dx_i = -\psi \wedge \phi. \qquad \blacklozenge$$

In the language of differential forms, the operator d of Definition 5.2 converts a 0-form f into a 1-form df. It is easy to generalize to an operator (also denoted by d) that converts a p-form η into a $(p + 1)$-form $d\eta$: One simply applies d (of Definition 5.2) to the coefficient functions of η. For example, here is the case $p = 1$.

6.3 Definition   If $\phi = \sum f_i\,dx_i$ is a 1-form on $\mathbf{R}^3$, the *exterior derivative* of ϕ is the 2-form $d\phi = \sum df_i \wedge dx_i$.

If we expand the preceding definition using Corollary 5.5, we obtain the following interesting formula for the exterior derivative of

$$\phi = f_1\,dx_1 + f_2\,dx_2 + f_3\,dx_3:$$

$$d\phi = \left(\frac{\partial f_2}{\partial x_1} - \frac{\partial f_1}{\partial x_2}\right)dx_1\,dx_2 + \left(\frac{\partial f_3}{\partial x_1} - \frac{\partial f_1}{\partial x_3}\right)dx_1\,dx_3 + \left(\frac{\partial f_3}{\partial x_2} - \frac{\partial f_2}{\partial x_3}\right)dx_2\,dx_3.$$

There is no need to memorize this formula; it is more reliable simply to apply the definition in each case. For example, suppose

$$\phi = xy \, dx + x^2 \, dz.$$

Then

$$d\phi = d(xy) \wedge dx + d(x^2) \wedge dz$$
$$= (y \, dx + x \, dy) \wedge dx + (2x \, dx) \wedge dz$$
$$= -x \, dx \, dy + 2x \, dx \, dz.$$

It is easy to check that the general exterior derivative enjoys the same linearity property as the particular case in Definition 5.2; that is,

$$d(a\phi + b\psi) = a \, d\phi + b \, d\psi,$$

where ϕ and ψ are arbitrary forms and a and b are numbers.

The exterior derivative and the wedge product work together nicely:

6.4 Theorem   Let f and g be functions, ϕ and ψ 1-forms. Then

(1) $d(fg) = df \, g + f \, dg.$
(2) $d(f\phi) = df \wedge \phi + f \, d\phi.$
(3) $d(\phi \wedge \psi) = d\phi \wedge \psi - \phi \wedge d\psi.$†

Proof.   The first formula is just Lemma 5.6. We include it to show the family resemblance of all three formulas. The proof of (2) is a simpler version of that of (3), so we outline a proof of the latter—watching to see where the minus sign comes from.

It suffices to prove the formula when $\phi = f \, du$, $\psi = g \, dv$, where u and v are any of the coordinate functions x_1, x_2, x_3. In fact, every 1-form is a sum of such terms, so the general case will follow by the linearity of d and the algebra of wedge products.

For example, let us try the typical case $\phi = f \, dx$, $\psi = g \, dy$. Since repeats kill, there is no use writing down terms that are bound to be eliminated. Hence

$$d(\phi \wedge \psi) = d(fg \, dx \, dy) = \frac{\partial(fg)}{\partial z} dz \, dx \, dy = \left(f \frac{\partial g}{\partial z} + g \frac{\partial f}{\partial z} \right) dx \, dy \, dz. \quad (*)$$

---

† As usual, multiplication takes precedence over addition or subtraction, so this expression should be read as $(d\phi \wedge \psi) - (\phi \wedge d\psi)$.

Now,

$$d\phi \wedge \psi = d(f\ dx) \wedge g\ dy = \frac{\partial f}{\partial z} dz \wedge dx \wedge g\ dy = g\frac{\partial f}{\partial z} dx\ dy\ dz.$$

But

$$\phi \wedge d\psi = f\ dx \wedge d(g\ dy) = f\ dx \wedge \frac{\partial g}{\partial z} dz \wedge dy = -f\frac{\partial g}{\partial z} dx\ dy\ dz,$$

since $dx\ dz\ dy = -dx\ dy\ dz$. Thus we must *subtract* this last equation from its predecessor to get (*).   ◆

One way to remember the minus sign in equation (3) of the theorem is to treat d as if it were a 1-form. To reach ψ, d must change places with ϕ, hence the minus sign is consistent with the alternation rule in Lemma 6.2.

Differential forms, and the associated notions of wedge product and exterior derivative, provide the means of expressing quite complicated relations among the partial derivatives in a highly efficient way. The wedge product saves much useless labor by discarding, right at the start, terms that will eventually disappear. But the exterior derivative d is the key. Exercise 8 shows, for example, how it replaces all three of the differentiation operations of classical vector analysis.

Exercises

1. Let $\phi = yz\ dx + dz$, $\psi = \sin z\ dx + \cos z\ dy$, $\xi = dy + z\ dz$. Find the standard expressions (in terms of $dxdy$, . . .) for
  (a) $\phi \wedge \psi$, $\psi \wedge \xi$, $\xi \wedge \phi$.   (b) $d\phi$, $d\psi$, $d\xi$.

2. Let $\phi = dx/y$ and $\psi = z\ dy$. Check the Leibnizian formula (3) of Theorem 6.4 in this case by computing each term separately.

3. For any function f show that $d(df) = 0$. Deduce that $d(f\ dg) = df \wedge dg$.

4. Simplify the following forms:
  (a) $d(f\ dg + g\ df)$.   (b) $d((f - g)\ (df + dg))$.
  (c) $d(f\ dg \wedge g\ df)$.   (d) $d(gf\ df) + d(f\ dg)$.

5. For any three 1-forms $\phi_i = \sum_j f_{ij}dx_j$ $(1 \leq i \leq 3)$, prove

$$\phi_1 \wedge \phi_2 \wedge \phi_3 = \begin{vmatrix} f_{11} & f_{12} & f_{13} \\ f_{21} & f_{22} & f_{23} \\ f_{31} & f_{32} & f_{33} \end{vmatrix} dx_1\ dx_2\ dx_3.$$

6. If r, ϑ, z are the cylindrical coordinate functions on $\mathbf{R}^3$, then $x = r\cos\vartheta$, $y = r\sin\vartheta$, $z = z$. Compute the *volume element dx dy dz* of $\mathbf{R}^3$ in cylindrical coordinates. (That is, express $dx\,dy\,dz$ in terms of the functions r, ϑ, z, and their differentials.)

7. For a 2-form

$$\eta = f\,dx\,dy + g\,dx\,dz + h\,dy\,dz,$$

the *exterior derivative dη* is defined to be the 3-form obtained by replacing f, g, and h by their differentials. Prove that for any 1-form ϕ, $d(d\phi) = 0$.

Exercises 3 and 7 show that $d^2 = 0$, that is, for any form ξ, $d(d\xi) = 0$. (If ξ is a 2-form, then $d(d\xi) = 0$, since its degree exceeds 3.)

8. Classical *vector analysis* avoids the use of differential forms on $\mathbf{R}^3$ by converting 1-forms and 2-forms into vector fields by means of the following one-to-one correspondences:

$$\overset{(1)}{\sum f_i\,dx_i} \leftrightarrow \overset{(2)}{\sum f_i U_i} \leftrightarrow f_3\,dx_1\,dx_2 - f_2\,dx_1\,dx_3 + f_1\,dx_2\,dx_3.$$

Vector analysis uses three basic operations based on partial differentiation:

Gradient of a function f:

$$\operatorname{grad} f = \sum \frac{\partial f}{\partial x_i} U_i.$$

Curl of a vector field $V = \sum f_i U_i$:

$$\operatorname{curl} V = \left(\frac{\partial f_3}{\partial x_2} - \frac{\partial f_2}{\partial x_3}\right) U_1 + \left(\frac{\partial f_1}{\partial x_3} - \frac{\partial f_3}{\partial x_1}\right) U_2 + \left(\frac{\partial f_2}{\partial x_1} - \frac{\partial f_1}{\partial x_2}\right) U_3.$$

Divergence of a vector field $V = \sum f_i U_i$:

$$\operatorname{div} V = \sum \frac{\partial f_i}{\partial x_i}.$$

Prove that all three operations may be expressed by exterior derivatives as follows:

(a) $\overset{(1)}{df} \leftrightarrow \operatorname{grad} f$.

(b) If $\overset{(1)}{\phi} \leftrightarrow V$, then $\overset{(2)}{d\phi} \leftrightarrow \operatorname{curl} V$.

(c) If $\overset{(1)}{\eta} \leftrightarrow V$, then $d\eta = (\operatorname{div} V)\,dx\,dy\,dz$.

9.   Let f and g be real-valued functions on $\mathbf{R}^2$. Prove that

$$df \wedge dg = \begin{vmatrix} \dfrac{\partial f}{\partial x} & \dfrac{\partial f}{\partial y} \\ \dfrac{\partial g}{\partial x} & \dfrac{\partial g}{\partial y} \end{vmatrix} dx\ dy.$$

This formula appears in elementary calculus; show that it implies the alternation rule.

1.7 Mappings

In this section we discuss functions from $\mathbf{R}^n$ to $\mathbf{R}^m$. If $n = 3$ and $m = 1$, then such a function is just a real-valued function on $\mathbf{R}^3$. If $n = 1$ and $m = 3$, it is a curve in $\mathbf{R}^3$. Although our results will necessarily be stated for arbitrary m and n, we are primarily interested in only three other cases:

$$\mathbf{R}^2 \to \mathbf{R}^2, \quad \mathbf{R}^2 \to \mathbf{R}^3, \quad \mathbf{R}^3 \to \mathbf{R}^3.$$

The fundamental observation about a function $F \colon \mathbf{R}^n \to \mathbf{R}^m$ is that it can be completely described by m real-valued functions on $\mathbf{R}^n$. (We saw this already in Section 4 for $n = 1$, $m = 3$.)

7.1 Definition   Given a function $F \colon \mathbf{R}^n \to \mathbf{R}^m$, let $f_1,\ f_2, \ldots, f_m$, denote the real-valued functions on $\mathbf{R}^n$ such that

$$F(\mathbf{p}) = (f_1(\mathbf{p}),\, f_2(\mathbf{p}), \ldots, f_m(\mathbf{p}))$$

for all points $\mathbf{p}$ in $\mathbf{R}^n$. These functions are called the *Euclidean coordinate functions* of F, and we write $F = (f_1, f_2, \ldots, f_m)$.

The function F is *differentiable* provided its coordinate functions are differentiable in the usual sense. A differentiable function $F \colon \mathbf{R}^n \to \mathbf{R}^m$ is called a *mapping* from $\mathbf{R}^n$ to $\mathbf{R}^m$.

Note that the coordinate functions of F are the composite functions $f_i = x_i(F)$, where $x_1, \ldots, x_m$ are the coordinate functions of $\mathbf{R}^m$.

Mappings may be described in many different ways. For example, suppose $F \colon \mathbf{R}^3 \to \mathbf{R}^3$ is the mapping $F = (x^2, yz, xy)$. Thus

$$F(\mathbf{p}) = \left(x(\mathbf{p})^2,\, y(\mathbf{p})z(\mathbf{p}),\, x(\mathbf{p})y(\mathbf{p})\right) \quad \text{for all } \mathbf{p}.$$

Now, $\mathbf{p} = (p_1, p_2, p_3)$, and by definition of the coordinate functions,

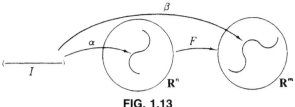

FIG. 1.13

$$x(\mathbf{p}) = p_1, \; y(\mathbf{p}) = p_2, \; z(\mathbf{p}) = p_3.$$

Hence we obtain the following *pointwise* formula for F:

$$F(p_1, p_2, p_3) = (p_1^2, p_2 p_3, p_1 p_2) \quad \text{for all } p_1, p_2, p_3.$$

Thus, for example,

$$F(1, -2, 0) = (1, 0, -2) \quad \text{and} \quad F(-3, 1, 3) = (9, 3, -3).$$

In principle, one could deduce the theory of curves from the general theory of mappings. But curves are reasonably simple, while a mapping, even in the case $\mathbf{R}^2 \to \mathbf{R}^2$, can be quite complicated. Hence we reverse this process and use curves, at every stage, to gain an understanding of mappings.

7.2 Definition   If $\alpha: I \to \mathbf{R}^n$ is a curve in $\mathbf{R}^n$ and $F: \mathbf{R}^n \to \mathbf{R}^m$ is a mapping, then the composite function $\beta = F(\alpha): I \to \mathbf{R}^m$ is a curve in $\mathbf{R}^m$ called the *image of α under F* (Fig. 1.13).

7.3 Example   Mappings. (1) Consider the mapping $F: \mathbf{R}^3 \to \mathbf{R}^3$ such that

$$F = (x - y, x + y, 2z).$$

In pointwise terms then,

$$F(p_1, p_2, p_3) = (p_1 - p_2, p_1 + p_2, 2p_3) \quad \text{for all } p_1, p_2, p_3.$$

Only when a mapping is quite simple can one hope to get a good idea of its behavior by merely computing its values at some finite number of points. But this function *is* quite simple —it is a *linear* transformation from $\mathbf{R}^3$ to $\mathbf{R}^3$.

Thus by a well-known theorem of linear algebra, F is completely determined by its values at three (linearly independent) points, say the *unit points*

$$\mathbf{u}_1 = (1, 0, 0), \quad \mathbf{u}_2 = (0, 1, 0), \quad \mathbf{u}_3 = (0, 0, 1).$$

(2) The mapping $F: \mathbf{R}^2 \to \mathbf{R}^2$ such that $F(u, v) = (u^2 - v^2, 2uv)$. (Here u and v are the coordinate functions of $\mathbf{R}^2$.) To analyze this mapping, we examine its effect on the curve $\alpha(t) = (r\cos t, r\sin t)$, where $0 \leq t \leq 2\pi$. This curve takes one counterclockwise trip around the circle of radius r with center at the origin. The image curve is

$$\begin{aligned}
\beta(t) &= F(\alpha(t)) \\
&= F(r\cos t, r\sin t) \\
&= (r^2\cos^2 t - r^2\sin^2 t, 2r^2\cos t\sin t)
\end{aligned}$$

with $0 \leq t \leq 2\pi$. Using the trigonometric identities

$$\cos 2t = \cos^2 t - \sin^2 t, \quad \sin 2t = 2\sin t\cos t,$$

we find for $\beta = F(\alpha)$ the formula

$$\beta(t) = (r^2\cos 2t, r^2\sin 2t),$$

with $0 \leq t \leq 2\pi$. This curve takes *two* counterclockwise trips around the circle of radius r^2 centered at the origin (Fig. 1.14).

Thus the effect of F is to wrap the plane $\mathbf{R}^2$ smoothly around itself twice—leaving the origin fixed, since $F(0, 0) = (0, 0)$. In this process, each circle of radius r is wrapped twice around the circle of radius r^2.

Generally speaking, differential calculus deals with approximation of smooth objects by linear objects. The best-known case is the approximation of a differentiable real-valued function f near x by the linear function $\Delta x \to f'(x)\,\Delta x$, which gives the tangent line at x to the graph of f. Our goal now is to define an analogous linear approximation for a mapping $F: \mathbf{R}^n \to \mathbf{R}^m$ near a point $\mathbf{p}$ of $\mathbf{R}^n$.

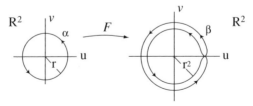

FIG. 1.14

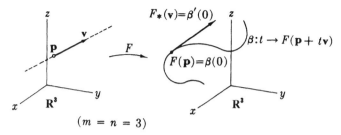

$(m = n = 3)$

FIG. 1.15

Since $\mathbf{R}^n$ is filled by the radial lines $\alpha(t) = \mathbf{p} + t\mathbf{v}$ starting at $\mathbf{p}$, $\mathbf{R}^m$ is filled by their image curves $\beta(t) = F(\mathbf{p} + t\mathbf{v})$ starting at $F(\mathbf{p})$ (Fig. 1.15). So we approximate F near $\mathbf{p}$ by the map F_* that sends each initial velocity $\alpha'(0) = \mathbf{v}_p$ to the initial velocity $\beta'(0)$.

7.4 Definition   Let $F: \mathbf{R}^n \to \mathbf{R}^m$ be a mapping. If $\mathbf{v}$ is a tangent vector to $\mathbf{R}^n$ at $\mathbf{p}$, let $F_*(\mathbf{v})$ be the initial velocity of the curve $t \to F(\mathbf{p} + t\mathbf{v})$. The resulting function F_* sends tangent vectors to $\mathbf{R}^n$ to tangent vectors to $\mathbf{R}^m$, and is called the *tangent map* of F.

The tangent map can be described explicitly as follows.

7.5 Proposition   Let $F = (f_1, f_2, \ldots, f_m)$ be a mapping from $\mathbf{R}^n$ to $\mathbf{R}^m$. If $\mathbf{v}$ is a tangent vector to $\mathbf{R}^n$ at $\mathbf{p}$, then

$$F_*(\mathbf{v}) = (\mathbf{v}[f_1], \ldots, \mathbf{v}[f_m]) \quad \text{at } F(\mathbf{p}).$$

Proof.   For definiteness, take $m = 3$. Then

$$\beta(t) = F(\mathbf{p} + t\mathbf{v}) = (f_1(\mathbf{p} + t\mathbf{v}), f_2(\mathbf{p} + t\mathbf{v}), f_3(\mathbf{p} + t\mathbf{v})).$$

By definition, $F_*(\mathbf{v}) = \beta'(0)$. To get $\beta'(0)$, we take the derivatives, at $t = 0$, of the coordinate functions of β (Definition 4.3). But

$$\frac{d}{dt}(f_i(\mathbf{p} + t\mathbf{v}))\big|_{t=0} = \mathbf{v}[f_i].$$

Thus

$$F_*(\mathbf{v}) = (\mathbf{v}[f_1], \; \mathbf{v}[f_2], \; \mathbf{v}[f_3])\big|_{\beta(0)},$$

and $\beta(0) = F(\mathbf{p})$.   ◆

Fix a point $\mathbf{p}$ of $\mathbf{R}^n$. The definition of tangent map shows that F_* sends tangent vectors at $\mathbf{p}$ to tangent vectors at $F(\mathbf{p})$. Thus for each $\mathbf{p}$ in $\mathbf{R}^n$, the function F_* gives rise to a function

$$F_{*p} \colon T_p(\mathbf{R}^n) \to T_{F(p)}(\mathbf{R}^m)$$

called the tangent map of F at $\mathbf{p}$. (Compare the analogous situation in elementary calculus where a function $f \colon \mathbf{R} \to \mathbf{R}$ has a derivative function $f' \colon \mathbf{R} \to \mathbf{R}$ that at each point t of $\mathbf{R}$ gives the derivative of f at t.)

7.6 Corollary   If $F \colon \mathbf{R}^n \to \mathbf{R}^m$ is a mapping, then at each point $\mathbf{p}$ of $\mathbf{R}^n$ the tangent map $F_{*p} \colon T_p(\mathbf{R}^n) \to T_{F(p)}(\mathbf{R}^m)$ is a linear transformation.

Proof.   We must show that for tangent vectors $\mathbf{v}$ and $\mathbf{w}$ at $\mathbf{p}$ and numbers a, b,

$$F_*(a\mathbf{v} + b\mathbf{w}) = aF_*(\mathbf{v}) + bF_*(\mathbf{w}).$$

This follows immediately from the preceding proposition by using the linearity in assertion (1) of Theorem 3.3.   ◆

In fact, *the tangent map F_{*p} at $\mathbf{p}$ is the linear transformation that best approximates F near $\mathbf{p}$.* This idea is fully developed in advanced calculus, where it is used to prove Theorem 7.10.

Another consequence of the proposition is that *mappings preserve velocities of curves.* Explicitly:

7.7 Corollary   Let $F \colon \mathbf{R}^n \to \mathbf{R}^m$ be a mapping. If $\beta = F(\alpha)$ is the image of a curve α in $\mathbf{R}^n$, then $\beta' = F_*(\alpha')$.

Proof.   Again, set $m = 3$. If $F = (f_1, f_2, f_3)$, then

$$\beta = F(\alpha) = (f_1(\alpha), f_2(\alpha), f_3(\alpha)).$$

Hence Theorem 7.5 gives

$$F_*(\alpha') = (\alpha'[f_1], \alpha'[f_2], \alpha'[f_3]).$$

But by Lemma 4.6,

$$\alpha'[f_i] = \frac{df_i(\alpha)}{dt}.$$

Hence

$$F_*(\alpha'(t)) = \left(\frac{df_1(\alpha)}{dt}(t), \frac{df_2(\alpha)}{dt}(t), \frac{df_3(\alpha)}{dt}(t)\right)_{\beta(t)} = \beta'(t). \qquad \blacklozenge$$

Let $\{U_j\}$ ($1 \le j \le n$) and $\{\overline{U}_i\}$ ($1 \le i \le m$) be the natural frame fields of $\mathbf{R}^n$ and $\mathbf{R}^m$, respectively (Def. 2.4). Then:

7.8 Corollary If $F = (f_1, \ldots, f_m)$ is a mapping from $\mathbf{R}^n$ to $\mathbf{R}^m$, then

$$F_*(U_j(\mathbf{p})) = \sum_{i=1}^{m} \frac{\partial f_i}{\partial x_j}(\mathbf{p})\overline{U}_i(F(\mathbf{p})) \quad (1 \le j \le n).$$

Proof. This follows directly from Proposition 7.5, since $U_j[f_i] = \dfrac{\partial f_i}{\partial x_j}$.

$\blacklozenge$

The matrix appearing in the preceding formula,

$$\left(\frac{\partial f_i}{\partial x_j}(\mathbf{p})\right)_{1 \le i \le m, 1 \le j \le n}$$

is called the *Jacobian matrix* of F at $\mathbf{p}$. (When $m = n = 1$; it reduces to a single number: the derivative of F at $\mathbf{p}$.)

Just as the derivative of a function is used to gain information about the function, the tangent map F_* can be used in the study of a mapping F.

7.9 Definition A mapping $F: \mathbf{R}^n \to \mathbf{R}^m$ is *regular* provided that at every point $\mathbf{p}$ of $\mathbf{R}^n$ the tangent map F_{*p} is one-to-one.

Since tangent maps are linear transformations, standard results of linear algebra show that the following conditions are equivalent:

(1) F_{*p} is one-to-one.
(2) $F_*(\mathbf{v}_p) = 0$ implies $\mathbf{v}_p = 0$.
(3) The Jacobian matrix of F at $\mathbf{p}$ has rank n, the dimension of the domain $\mathbf{R}^n$ of F.

The following noteworthy property of linear transformations $T: V \to W$ will be useful in dealing with tangent maps. If the vector spaces V and W have the same dimension, then T is one-to-one if and only if it is onto, so either property is equivalent to T being a linear isomorphism.

A mapping that has a (differentiable) inverse mapping is called a *diffeomorphism*. The results of this section all remain valid when Euclidean spaces $\mathbf{R}^n$ are replaced by open sets of Euclidean spaces, so we can speak of a diffeomorphism from one open set to another.

We state without proof one of the fundamental results of advanced calculus.

7.10 Theorem   Let $F: \mathbf{R}^n \to \mathbf{R}^n$ be a mapping between Euclidean spaces of the same dimension. If F_{*p} is one-to-one at a point $\mathbf{p}$, there is an open set $\mathscr{U}$ containing $\mathbf{p}$ such that F restricted to $\mathscr{U}$ is a diffeomorphism of $\mathscr{U}$ onto an open set $\mathscr{V}$.

This is called the *inverse function theorem* since it asserts that the restricted mapping $\mathscr{U} \to \mathscr{V}$ has a differentiable inverse mapping $\mathscr{V} \to \mathscr{U}$. Exercise 6 gives a suggestion of its importance.

Exercises

In the first four exercises F denotes the mapping $F(u, v) = (u^2 - v^2, 2uv)$ in Example 7.3.

1.   Find all points $\mathbf{p}$ such that
  (a) $F(\mathbf{p}) = (0, 0)$.   (b) $F(\mathbf{p}) = (8, 6)$.
  (c) $F(\mathbf{p}) = \mathbf{p}$.

2.   (a) Sketch the horizontal line $v = 1$ and its image under F (a parabola).
  (b) Do the same for the vertical $u = 1$.
  (c) Describe the image of the unit square $0 \leq u, v \leq 1$ under F.

3.   Let $\mathbf{v} = (v_1, v_2)$ be a tangent vector to $\mathbf{R}^2$ at $\mathbf{p} = (p_1, p_2)$. Apply Definition 7.4 directly to express $F_*(\mathbf{v})$ in terms of the coordinates of $\mathbf{v}$ and $\mathbf{p}$.

4.   Find a formula for the Jacobian matrix of F at all points, and deduce that F_{*p} is a linear isomorphism at every point of $\mathbf{R}^2$ except the origin.

5.   If $F: \mathbf{R}^n \to \mathbf{R}^m$ is a linear transformation, prove that $F_*(\mathbf{v}_p) = F(\mathbf{v})_{F(p)}$.

6.   (a) Give an example to demonstrate that a one-to-one and onto mapping need not be a diffeomorphism. (*Hint:* Take $m = n = 1$.)

(b) Prove that if a one-to-one and onto mapping $F: \mathbf{R}^n \to \mathbf{R}^n$ is regular, then it is a diffeomorphism.

7. Prove that a mapping $F: \mathbf{R}^n \to \mathbf{R}^m$ preserves directional derivatives in this sense: If $\mathbf{v}_p$ is a tangent vector to $\mathbf{R}^n$ and g is a differentiable function on $\mathbf{R}^m$, then $F_*(\mathbf{v}_p)[g] = \mathbf{v}_p[g(F)]$.

8. In the definition of tangent map (Def. 7.4), the straight line $t \to \mathbf{p} + t\mathbf{v}$ can be replaced by any curve α with initial velocity $\mathbf{v}_p$.

9. Let $F: \mathbf{R}^n \to \mathbf{R}^m$ and $G: \mathbf{R}^m \to \mathbf{R}^p$ be mappings. Prove:
(a) Their composition $GF: \mathbf{R}^n \to \mathbf{R}^p$ is a (differentiable) mapping. (Take $m = p = 2$ for simplicity.)
(b) $(GF)_* = G_*F_*$. (*Hint:* Use the preceding exercise.)
This concise formula is the general *chain rule*. Unless dimensions are small, it becomes formidable when expressed in terms of Jacobian matrices.
(c) If F is a diffeomorphism, then so is its inverse mapping F^{-1}.

10. Show (in two ways) that the map $F: \mathbf{R}^2 \to \mathbf{R}^2$ such that $F(u, v) = (ve^u, 2u)$ is a diffeomorphism:
(a) Prove that it is one-to-one, onto, and regular;
(b) Find a formula for its inverse $F^{-1}: \mathbf{R}^2 \to \mathbf{R}^2$ and observe that F^{-1} is differentiable. Verify the formula by checking that both $F\,F^{-1}$ and $F^{-1}\,F$ are identity maps.

1.8 Summary

Starting from the familiar notion of real-valued functions and using linear algebra at every stage, we have constructed a variety of mathematical objects. The basic notion of tangent vector led to vector fields, which dualized to 1-forms—which in turn led to arbitrary differential forms. The notions of curve and differentiable function were generalized to that of a mapping $F: \mathbf{R}^n \to \mathbf{R}^m$.

Then, starting from the usual notion of the derivative of a real-valued function, we proceeded to construct appropriate differentiation operations for these objects: the directional derivative of a function, the exterior derivative of a form, the velocity of a curve, the tangent map of a mapping. These operations all reduced to (ordinary or partial) derivatives of real-valued coordinate functions, but it is noteworthy that in most cases the *definitions* of these operations did not involve coordinates. (This could be achieved in all cases.) Generally speaking, these differentiation operations all exhibited in one form or another the characteristic linear and Leibnizian properties of ordinary differentiation.

Most of these concepts are probably already familiar to the reader, at least in special cases. But we now have careful definitions and a catalogue of basic properties that will enable us to begin the exploration of differential geometry.

Chapter 2
Frame Fields

Roughly speaking, geometry begins with the measurement of distances and angles. We shall see that the geometry of Euclidean space can be derived from the *dot product*, the natural inner product on Euclidean space.

Much of this chapter is devoted to the geometry of curves in $\mathbf{R}^3$. We emphasize this topic not only because of its intrinsic importance, but also because the basic method used to investigate curves has proved effective throughout differential geometry. A curve in $\mathbf{R}^3$ is studied by assigning at each point a certain *frame*—that is, set of three orthogonal unit vectors. The rate of change of these vectors along the curve is then expressed in terms of the vectors themselves by the celebrated *Frenet formulas* (Theorem 3.2). In a real sense, the theory of curves in $\mathbf{R}^3$ is merely a corollary of these fundamental formulas.

Later on we shall use this "method of moving frames" to study a *surface* in $\mathbf{R}^3$. The general idea is to think of a surface as a kind of two-dimensional curve and follow the Frenet approach as closely as possible. To carry out this scheme we shall need the generalization (Theorem 7.2) of the Frenet formulas devised by E. Cartan. It was Cartan who, in the early 1900s, first realized the full power of this method not only in differential geometry but also in a variety of related fields.

2.1 Dot Product

We begin by reviewing some basic facts about the natural inner product on the vector space $\mathbf{R}^3$.

1.1 Definition   The dot product of points $\mathbf{p} = (p_1, p_2, p_3)$ and $\mathbf{q} = (q_1, q_2, q_3)$ in $\mathbf{R}^3$ is the number

$$\mathbf{p} \bullet \mathbf{q} = p_1 q_1 + p_2 q_2 + p_3 q_3.$$

The dot product is an inner product since it has the following three properties:

(1) Bilinearity:

$$(a\mathbf{p} + b\mathbf{q}) \bullet \mathbf{r} = a\mathbf{p} \bullet \mathbf{r} + b\mathbf{q} \bullet \mathbf{r},$$

$$\mathbf{r} \bullet (a\mathbf{p} + b\mathbf{q}) = a\mathbf{r} \bullet \mathbf{p} + b\mathbf{r} \bullet \mathbf{q}.$$

(2) Symmetry: $\mathbf{p} \bullet \mathbf{q} = \mathbf{q} \bullet \mathbf{p}$.
(3) Positive definiteness: $\mathbf{p} \bullet \mathbf{p} \geq 0$, and $\mathbf{p} \bullet \mathbf{p} = 0$ if and only if $\mathbf{p} = 0$.
(Here $\mathbf{p}$, $\mathbf{q}$, and $\mathbf{r}$ are arbitrary points of $\mathbf{R}^3$, and a and b are numbers.)

The *norm* of a point $\mathbf{p} = (p_1, p_2, p_3)$ is the number

$$\| \mathbf{p} \| = (\mathbf{p} \bullet \mathbf{p})^{1/2} = (p_1^2 + p_2^2 + p_3^2)^{1/2}.$$

The norm is thus a real-valued function on $\mathbf{R}^3$; it has the fundamental properties $\| \mathbf{p} + \mathbf{q} \| \leq \| \mathbf{p} \| + \| \mathbf{q} \|$ and $\| a\mathbf{p} \| = | a | \| \mathbf{p} \|$, where $| a |$ is the absolute value of the number a.

In terms of the norm we get a compact version of the usual distance formula in $\mathbf{R}^3$.

1.2 Definition   If $\mathbf{p}$ and $\mathbf{q}$ are points of $\mathbf{R}^3$, the *Euclidean distance* from $\mathbf{p}$ to $\mathbf{q}$ is the number

$$d(\mathbf{p}, \mathbf{q}) = \| \mathbf{p} - \mathbf{q} \|.$$

In fact, since

$$\mathbf{p} - \mathbf{q} = (p_1 - q_1, p_2 - q_2, p_3 - q_3),$$

expansion of the norm gives the well-known formula (Fig. 2.1)

$$d(\mathbf{p}, \mathbf{q}) = \left((p_1 - q_1)^2 + (p_2 - q_2)^2 + (p_3 - q_3)^2\right)^{1/2}.$$

Euclidean distance may be used to give a more precise definition of open sets (Chapter 1, Section 1). First, if $\mathbf{p}$ is a point of $\mathbf{R}^3$ and $\varepsilon > 0$ is a number, the ε *neighborhood* $\mathcal{N}_\varepsilon$ of $\mathbf{p}$ in $\mathbf{R}^3$ is the set of all points $\mathbf{q}$ of $\mathbf{R}^3$ such that $d(\mathbf{p}, \mathbf{q}) < \varepsilon$. Then a subset $\mathcal{O}$ of $\mathbf{R}^3$ is *open* provided that each point of $\mathcal{O}$ has an ε neighborhood that is entirely contained in $\mathcal{O}$. In short, all points near enough to a point of an open set are also in the set. This definition is valid with $\mathbf{R}^3$ replaced by $\mathbf{R}^n$—or indeed any set furnished with a reasonable distance function.

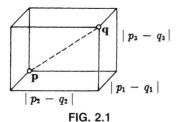

FIG. 2.1

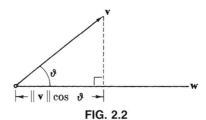

FIG. 2.2

We saw in Chapter 1 that for each point **p** of $\mathbf{R}^3$ there is a *canonical isomorphism* $\mathbf{v} \to \mathbf{v}_p$ from $\mathbf{R}^3$ onto the tangent space $T_p(\mathbf{R}^3)$ at **p**. These isomorphisms lie at the heart of Euclidean geometry—using them, the dot product on $\mathbf{R}^3$ itself may be transferred to each of its tangent spaces.

1.3 Definition   The *dot product* of tangent vectors $\mathbf{v}_p$ and $\mathbf{w}_p$ at the same point of $\mathbf{R}^3$ is the number $\mathbf{v}_p \bullet \mathbf{w}_p = \mathbf{v} \bullet \mathbf{w}$.

For example, $(1, 0, -1)_p \bullet (3, -3, 7)_p = 1(3) + 0(-3) + (-1)7 = -4$. Evidently this definition provides a dot product on each tangent space $T_p(\mathbf{R}^3)$ with the same properties as the original dot product on $\mathbf{R}^3$. In particular, each tangent vector $\mathbf{v}_p$ to $\mathbf{R}^3$ has *norm* (or *length*) $\| \mathbf{v}_p \| = \| \mathbf{v} \|$.

A fundamental result of linear algebra is the Schwarz inequality $| \mathbf{v} \bullet \mathbf{w} | \leq \| \mathbf{v} \| \| \mathbf{w} \|$. This permits us to define the cosine of the angle ϑ between $\mathbf{v}$ and $\mathbf{w}$ by the equation (Fig. 2.2).

$$\mathbf{v} \bullet \mathbf{w} = \|\mathbf{v}\| \|\mathbf{w}\| \cos \vartheta.$$

Thus the dot product of two vectors is the product of their lengths times the cosine of the angle between them. (The angle ϑ is not uniquely determined unless further restrictions are imposed, say $0 \leq \vartheta \leq \pi$.)

In particular, if $\vartheta = \pi/2$, then $\mathbf{v} \bullet \mathbf{w} = 0$. Thus we shall define two vectors to be *orthogonal* provided their dot product is zero. A vector of length 1 is called a *unit vector*.

1.4 Definition   A set e_1, e_2, e_3 of three mutually orthogonal unit vectors tangent to $\mathbf{R}^3$ at **p** is called a *frame* at the point **p**.

Thus e_1, e_2, e_3 is a frame if and only if

$$e_1 \bullet e_1 = e_2 \bullet e_2 = e_3 \bullet e_3 = 1,$$

$$e_1 \bullet e_2 = e_1 \bullet e_3 = e_2 \bullet e_3 = 0.$$

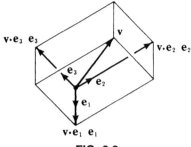

FIG. 2.3

By the symmetry of the dot product, the second row of equations is, of course, the same as

$$\mathbf{e}_2 \cdot \mathbf{e}_1 = \mathbf{e}_3 \cdot \mathbf{e}_1 = \mathbf{e}_3 \cdot \mathbf{e}_2 = 0.$$

Using index notation, all nine equations may be concisely expressed as $\mathbf{e}_i \cdot \mathbf{e}_j = \delta_{ij}$ for $1 \leq i\,j \leq 3$, where δ_{ij} is the Kronecker delta (0 if $i \neq j$, 1 if $i = j$). For example, at each point $\mathbf{p}$ of $\mathbf{R}^3$, the vectors $U_1(\mathbf{p})$, $U_2(\mathbf{p})$, $U_3(\mathbf{p})$ of Definition 2.4 in Chapter 1 constitute a frame at $\mathbf{p}$.

1.5 Theorem   Let $\mathbf{e}_1$, $\mathbf{e}_2$, $\mathbf{e}_3$ be a frame at a point $\mathbf{p}$ of $\mathbf{R}^3$. If $\mathbf{v}$ is any tangent vector to $\mathbf{R}^3$ at $\mathbf{p}$, then (Fig. 2.3)

$$\mathbf{v} = (\mathbf{v} \cdot \mathbf{e}_1)\mathbf{e}_1 + (\mathbf{v} \cdot \mathbf{e}_2)\mathbf{e}_2 + (\mathbf{v} \cdot \mathbf{e}_3)\mathbf{e}_3.$$

Proof.   First we show that the vectors $\mathbf{e}_1$, $\mathbf{e}_2$, $\mathbf{e}_3$ are linearly independent. Suppose $\sum a_i \mathbf{e}_i = 0$. Then

$$0 = \left(\sum a_i \mathbf{e}_i\right) \cdot \mathbf{e}_j = \sum a_i \mathbf{e}_i \cdot \mathbf{e}_j = \sum a_i \delta_{ij} = a_j,$$

where all sums are over $i = 1, 2, 3$. Thus

$$a_1 = a_2 = a_3 = 0,$$

as required. Now, the tangent space $T_p(\mathbf{R}^3)$ has dimension 3, since it is linearly isomorphic to $\mathbf{R}^3$. Thus by a well-known theorem of linear algebra, the three independent vectors $\mathbf{e}_1$, $\mathbf{e}_2$, $\mathbf{e}_3$ form a basis for $T_p(\mathbf{R}^3)$. Hence for each vector $\mathbf{v}$ there are three (unique) numbers c_1, c_2, c_3 such that

$$\mathbf{v} = \sum c_i \mathbf{e}_i.$$

But

$$\mathbf{v} \cdot \mathbf{e}_j = \left(\sum c_i \mathbf{e}_i\right) \cdot \mathbf{e}_j = \sum c_i \delta_{ij} = c_j,$$

and thus

$$\mathbf{v} = \sum (\mathbf{v} \cdot \mathbf{e}_j)\mathbf{e}_j. \qquad \blacklozenge$$

This result (valid in any inner-product space) is one of the great labor-saving devices in mathematics. For to find the coordinates of a vector $\mathbf{v}$ with respect to an *arbitrary* basis, one must in general solve a set of nonhomogeneous linear equations, a task that even in dimension 3 is not always entirely trivial. But the theorem shows that to find the coordinates of $\mathbf{v}$ with respect to a frame (that is, an *orthonormal* basis) it suffices merely to compute the three dot products $\mathbf{v} \cdot \mathbf{e}_1$, $\mathbf{v} \cdot \mathbf{e}_2$, $\mathbf{v} \cdot \mathbf{e}_3$. We call this process *orthonormal expansion* of $\mathbf{v}$ in terms of the frame $\mathbf{e}_1$, $\mathbf{e}_2$, $\mathbf{e}_3$. In the special case of the natural frame $U_1(\mathbf{p})$, $U_2(\mathbf{p})$, $U_3(\mathbf{p})$, the identity

$$\mathbf{v} = (v_1, v_2, v_3) = \sum v_i U_i(\mathbf{p})$$

is an orthonormal expansion, and the dot product is defined in terms of these *Euclidean coordinates* by $\mathbf{v} \cdot \mathbf{w} = \sum v_i w_i$. If we use instead an arbitrary frame $\mathbf{e}_1$, $\mathbf{e}_2$, $\mathbf{e}_3$, then each vector $\mathbf{v}$ has new coordinates $a_i = \mathbf{v} \cdot \mathbf{e}_i$ relative to this frame, but *the dot product is still given by the same simple formula*

$$\mathbf{v} \cdot \mathbf{w} = \sum a_i b_i$$

since

$$\mathbf{v} \cdot \mathbf{w} = \left(\sum a_i \mathbf{e}_i\right) \cdot \left(\sum b_i \mathbf{e}_i\right) = \sum_{i,j} a_i b_j \, \mathbf{e}_i \cdot \mathbf{e}_j$$

$$= \sum_{i,j} a_i b_j \delta_{ij} = \sum a_i b_i.$$

When applied to more complicated geometric situations, the advantage of using frames becomes enormous, and this is why they appear so frequently throughout this book.

The notion of frame is very close to that of orthogonal matrix.

1.6 Definition Let $\mathbf{e}_1$, $\mathbf{e}_2$, $\mathbf{e}_3$ be a frame at a point $\mathbf{p}$ of $\mathbf{R}^3$. The 3×3 matrix A whose rows are the Euclidean coordinates of these three vectors is called the *attitude matrix* of the frame.

Explicitly, if

$$\mathbf{e}_1 = (a_{11}, a_{12}, a_{13})_p,$$

$$\mathbf{e}_2 = (a_{21}, a_{22}, a_{23})_p,$$

$$\mathbf{e}_3 = (a_{31}, a_{32}, a_{33})_p,$$

then

$$A = \begin{pmatrix} a_{11} & a_{12} & a_{13} \\ a_{21} & a_{22} & a_{23} \\ a_{31} & a_{32} & a_{33} \end{pmatrix}.$$

Thus A does describe the "attitude" of the frame in $\mathbf{R}^3$, although not its point of application.

Evidently the rows of A are orthonormal, since

$$\sum_k a_{ik}a_{jk} = \mathbf{e}_i \cdot \mathbf{e}_j = \delta_{ij} \quad \text{for } 1 \leq i, j \leq 3.$$

By definition, this means that A is an *orthogonal* matrix.

In terms of matrix multiplication, these equations may be written $A\,{}^tA = I$, where I is the 3×3 identity matrix and ${}^tA$ is the *transpose* of A:

$$^tA = \begin{pmatrix} a_{11} & a_{21} & a_{31} \\ a_{12} & a_{22} & a_{32} \\ a_{13} & a_{23} & a_{33} \end{pmatrix}.$$

It follows by a standard theorem of linear algebra that ${}^tAA = I$, so that ${}^tA = A^{-1}$, the *inverse* of A.

There is another product on $\mathbf{R}^3$, closely related to the wedge product of 1-forms and second in importance only to the dot product. We shall transfer it immediately to each tangent space of $\mathbf{R}^3$.

1.7 Definition   If $\mathbf{v}$ and $\mathbf{w}$ are tangent vectors to $\mathbf{R}^3$ at the same point $\mathbf{p}$, then the *cross product* of $\mathbf{v}$ and $\mathbf{w}$ is the tangent vector

$$\mathbf{v} \times \mathbf{w} = \begin{vmatrix} U_1(\mathbf{p}) & U_2(\mathbf{p}) & U_3(\mathbf{p}) \\ v_1 & v_2 & v_3 \\ w_1 & w_2 & w_3 \end{vmatrix}.$$

This formal determinant is to be expanded along its first row. For example, if $\mathbf{v} = (1, 0, -1)_p$ and $\mathbf{w} = (2, 2, -7)_p$, then

$$\mathbf{v} \times \mathbf{w} = \begin{vmatrix} U_1(\mathbf{p}) & U_2(\mathbf{p}) & U_3(\mathbf{p}) \\ 1 & 0 & -1 \\ 2 & 2 & -7 \end{vmatrix}$$

$$= 2U_1(\mathbf{p}) + 5U_2(\mathbf{p}) + 2U_3(\mathbf{p}) = (2, 5, 2)_p.$$

Familiar properties of determinants show that the cross product $\mathbf{v} \times \mathbf{w}$ is *linear* in $\mathbf{v}$ and in $\mathbf{w}$, and satisfies the *alternation rule*

$$\mathbf{v} \times \mathbf{w} = -\mathbf{w} \times \mathbf{v}.$$

Hence, in particular, $\mathbf{v} \times \mathbf{v} = 0$. The geometric usefulness of the cross product is based mostly on this fact:

1.8 Lemma The cross product $\mathbf{v} \times \mathbf{w}$ is orthogonal to both $\mathbf{v}$ and $\mathbf{w}$, and has length such that

$$\| \mathbf{v} \times \mathbf{w} \|^2 = (\mathbf{v} \cdot \mathbf{v})(\mathbf{w} \cdot \mathbf{w}) - (\mathbf{v} \cdot \mathbf{w})^2.$$

Proof. Let $\mathbf{v} \times \mathbf{w} = \sum c_i U_i(\mathbf{p})$. Then the dot product $\mathbf{v} \cdot (\mathbf{v} \times \mathbf{w})$ is just $\sum v_i c_i$. But by the definition of cross product, the Euclidean coordinates c_1, c_2, c_3 of $\mathbf{v} \times \mathbf{w}$ are such that

$$\mathbf{v} \cdot (\mathbf{v} \times \mathbf{w}) = \begin{vmatrix} v_1 & v_2 & v_3 \\ v_1 & v_2 & v_3 \\ w_1 & w_2 & w_3 \end{vmatrix}.$$

This determinant is zero, since two of its rows are the same; thus $\mathbf{v} \times \mathbf{w}$ is orthogonal to $\mathbf{v}$, and similarly, to $\mathbf{w}$.

Rather than use tricks to prove the length formula, we give a brute-force computation. Now,

$$
\begin{aligned}
(\mathbf{v} \cdot \mathbf{v})(\mathbf{w} \cdot \mathbf{w}) - (\mathbf{v} \cdot \mathbf{w})^2 &= \left(\sum v_i^2\right)\left(\sum w_j^2\right) - \left(\sum v_i w_i\right)^2 \\
&= \sum_{i,j} v_i^2 w_j^2 - \left\{ \sum v_i^2 w_i^2 + 2\sum_{i<j} v_i w_i v_j w_j \right\} \\
&= \sum_{i \neq j} v_i^2 w_j^2 - 2\sum_{i<j} v_i w_i v_j w_j.
\end{aligned}
$$

On the other hand,

$$
\begin{aligned}
\| \mathbf{v} \times \mathbf{w} \|^2 &= (\mathbf{v} \times \mathbf{w}) \cdot (\mathbf{v} \times \mathbf{w}) = \sum c_i^2 \\
&= (v_2 w_3 - v_3 w_2)^2 + (v_3 w_1 - v_1 w_3)^2 + (v_1 w_2 - v_2 w_1)^2,
\end{aligned}
$$

and expanding these squares gives the same result as above. ◆

A more intuitive description of the length of a cross product is

$$\| \mathbf{v} \times \mathbf{w} \| = \| \mathbf{v} \| \| \mathbf{w} \| \sin \vartheta,$$

where $0 \leq \vartheta \leq \pi$ is the smaller of the two angles from $\mathbf{v}$ to $\mathbf{w}$. The direction of $\mathbf{v} \times \mathbf{w}$ on the line orthogonal to $\mathbf{v}$ and $\mathbf{w}$ is given, for practical purposes, by this "right-hand rule": If the fingers of the right hand point in the

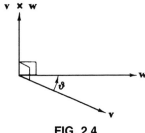

FIG. 2.4

direction of the shortest rotation of **v** to **w**, then the thumb points in the direction of **v** × **w** (Fig. 2.4).

Combining the dot and cross product, we get the *triple scalar product*, which assigns to any three vectors **u**, **v**, **w** the number **u** • **v** × **w** (Exercise 4). Parentheses are unnecessary: **u** • (**v** × **w**) is the only possible meaning.

Exercises

1. Let **v** = (1, 2, −1) and **w** = (−1, 0, 3) be tangent vectors at a point of $\mathbf{R}^3$. Compute:

(a) **v** • **w**.   (b) **v** × **w**.
(c) **v**/‖ **v** ‖, **w**/‖ **w** ‖.   (d) ‖ **v** × **w** ‖.
(e) the cosine of the angle between **v** and **w**.

2. Prove that Euclidean distance has the properties
(a) $d(\mathbf{p}, \mathbf{q}) \geq 0$; $d(\mathbf{p}, \mathbf{q}) = 0$ if and only if $\mathbf{p} = \mathbf{q}$,
(b) $d(\mathbf{p}, \mathbf{q}) = d(\mathbf{q}, \mathbf{p})$,
(c) $d(\mathbf{p}, \mathbf{q}) + d(\mathbf{q}, \mathbf{r}) \geq d(\mathbf{p}, \mathbf{r})$, for any points **p**, **q**, **r** in $\mathbf{R}^3$.

3. Prove that the tangent vectors

$$\mathbf{e}_1 = \frac{(1, 2, 1)}{\sqrt{6}}, \quad \mathbf{e}_2 = \frac{(-2, 0, 2)}{\sqrt{8}}, \quad \mathbf{e}_3 = \frac{(1, -1, 1)}{\sqrt{3}}$$

constitute a frame. Express **v** = (6, 1, −1) as a linear combination of these vectors. (Check the result by direct computation.)

4. Let **u** = (u_1, u_2, u_3), **v** = (v_1, v_2, v_3), **w** = (w_1, w_2, w_3). Prove that

(a) $\mathbf{u} \bullet \mathbf{v} \times \mathbf{w} = \begin{vmatrix} u_1 & u_2 & u_3 \\ v_1 & v_2 & v_3 \\ w_1 & w_2 & w_3 \end{vmatrix}.$

(b) $\mathbf{u} \cdot \mathbf{v} \times \mathbf{w} \neq 0$ if and only if $\mathbf{u}$, $\mathbf{v}$, and $\mathbf{w}$ are linearly independent.
(c) If any two vectors in $\mathbf{u} \cdot \mathbf{v} \times \mathbf{w}$ are reversed, the product changes sign.
(d) $\mathbf{u} \cdot \mathbf{v} \times \mathbf{w} = \mathbf{u} \times \mathbf{v} \cdot \mathbf{w}$.

5. Prove that $\mathbf{v} \times \mathbf{w} \neq 0$ if and only if $\mathbf{v}$ and $\mathbf{w}$ are linearly independent, and show that $\| \mathbf{v} \times \mathbf{w} \|$ is the area of the parallelogram with sides $\mathbf{v}$ and $\mathbf{w}$.

6. If e_1, e_2, e_3 is a frame, show that

$$\mathbf{e}_1 \cdot \mathbf{e}_2 \times \mathbf{e}_3 = \pm 1.$$

Deduce that any 3×3 orthogonal matrix has determinant ±1.

7. If $\mathbf{u}$ is a unit vector, then the *component* of $\mathbf{v}$ in the $\mathbf{u}$ direction is

$$(\mathbf{v} \cdot \mathbf{u})\mathbf{u} = \| \mathbf{v} \| \cos \vartheta \, \mathbf{u}.$$

Show that $\mathbf{v}$ has a unique expression $\mathbf{v} = \mathbf{v}_1 + \mathbf{v}_2$, where $\mathbf{v}_1 \cdot \mathbf{v}_2 = 0$ and $\mathbf{v}_1$ is the component of $\mathbf{v}$ in the $\mathbf{u}$ direction.

8. Prove: The volume of the parallelepiped with sides $\mathbf{u}$, $\mathbf{v}$, $\mathbf{w}$ is $\pm \mathbf{u} \cdot \mathbf{v} \times \mathbf{w}$ (Fig. 2.5). (*Hint:* Use the indicated unit vector $\mathbf{e} = \mathbf{v} \times \mathbf{w}/\| \mathbf{v} \times \mathbf{w} \|$.)

9. Prove, using ε-neighborhoods, that each of the following subsets of $\mathbf{R}^3$ is open:
(a) All points $\mathbf{p}$ such that $\| \mathbf{p} \| < 1$.
(b) All $\mathbf{p}$ such that $p_3 > 0$. (*Hint:* $| p_i - q_i | \leq d(\mathbf{p}, \mathbf{q})$.)

10. In each case, let S be the set of all points $\mathbf{p}$ that satisfy the given condition. Describe S, and decide whether it is *open.*
(a) $p_1^2 + p_2^2 + p_3^2 = 1$.   (b) $p_3 \neq 0$.
(c) $p_1 = p_2 \neq p_3$.   (d) $p_1^2 + p_2^2 < 9$.

11. If f is a differentiable function on $\mathbf{R}^3$, show that the gradient

$$\nabla f = \sum \frac{\partial f}{\partial x_i} U_i$$

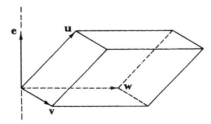

FIG. 2.5

(Ex. 8 of Sec. 1.6) has the following properties:

(a) $\mathbf{v}[f] = (df)(\mathbf{v}) = \mathbf{v} \bullet (\nabla f)(\mathbf{p})$ for any tangent vector at $\mathbf{p}$.

(b) The norm $\| (\nabla f)(\mathbf{p}) \| = \left[\sum (\partial f / \partial x_i)^2 (\mathbf{p}) \right]^{1/2}$ of (∇f) $(\mathbf{p})$ is the maximum of the directional derivatives $\mathbf{u}[f]$ for all *unit* vectors at $\mathbf{p}$. Furthermore, if $(\nabla f)(\mathbf{p}) \neq 0$, the unit vector for which the maximum occurs is

$$(\nabla f)(\mathbf{p})/\| (\nabla f)(\mathbf{p}) \|.$$

The notations grad f, curl V, and div V (in the exercise referred to) are often replaced by ∇f, $\nabla \times V$, and $\nabla \bullet V$, respectively.

12. *Angle functions.* Let f and g be differentiable real-valued functions on an interval I. Suppose that $f^2 + g^2 = 1$ and that ϑ_0 is a number such that $f(0) = \cos \vartheta_0$, $g(0) = \sin \vartheta_0$. If ϑ is the function such that

$$\vartheta(t) = \vartheta_0 + \int_0^t (fg' - gf')\,du,$$

prove that

$$f = \cos \vartheta, \quad g = \sin \vartheta.$$

Hint: We want $(f - \cos \vartheta)^2 + (g - \sin \vartheta)^2 = 0$, so show that its derivative is zero.

The point of this exercise is that ϑ is a differentiable function, unambiguously defined on the whole interval I.

2.2 Curves

We begin the geometric study of curves by reviewing some familiar definitions. Let $\alpha\colon I \to \mathbf{R}^3$ be a curve. In Chapter 1, Section 4, we defined the velocity vector $\alpha'(t)$ of α at t. Now we define the *speed* of α at t to be the length $v(t) = \| \alpha'(t) \|$ of the velocity vector. Thus speed is a real-valued function on the interval I. In terms of Euclidean coordinates $\alpha = (\alpha_1, \alpha_2, \alpha_3)$, we have

$$\alpha'(t) = \left(\frac{d\alpha_1}{dt}(t), \frac{d\alpha_2}{dt}(t), \frac{d\alpha_3}{dt}(t) \right).$$

Hence the speed function v of α is given by the usual formula

$$v = \| \alpha' \| = \left(\left(\frac{d\alpha_1}{dt} \right)^2 + \left(\frac{d\alpha_2}{dt} \right)^2 + \left(\frac{d\alpha_3}{dt} \right)^2 \right)^{1/2}.$$

In physics, the distance traveled by a moving point is determined by integrating its speed with respect to time. Thus we define the *arc length* of α from $t = a$ to $t = b$ to be the number

$$\int_a^b \| \alpha'(t) \| \, dt.$$

Substituting the formula for $\| \alpha' \|$ given above, we get the usual formula for arc length. This length involves only the restriction of α (defined on some open interval) to the *closed* interval $[a, b]$: $a \leq t \leq b$. Such a restriction $\sigma: [a, b] \to \mathbf{R}^3$ is called a *curve segment*, and its length is denoted by $L(\sigma)$. Note that the velocity of σ is well defined at the endpoints a and b of $[a, b]$.

Sometimes one is interested only in the route followed by a curve and not in the particular speed at which it traverses its route. One way to ignore the speed of a curve α is to reparametrize to a curve β that has *unit speed* $\| \beta' \| = 1$. Then β represents a "standard trip" along the route of α.

2.1 Theorem If α is a regular curve in $\mathbf{R}^3$, then there exists a reparametrization β of α such that β has unit speed.

Proof. Fix a number a in the domain I of $\alpha: I \to \mathbf{R}^3$, and consider the *arc length function*

$$s(t) = \int_a^b \| \alpha'(u) \| \, du.$$

(The resulting reparametrization is said to be *based at* $t = a$.) Thus the derivative ds/dt of the function $s = s(t)$ is the speed function $v = \| \alpha' \|$ of α. Since α is regular, by definition α' is never zero; hence $ds/dt > 0$. By a standard theorem of calculus, the function s has an inverse function $t = t(s)$, whose derivative dt/ds at $s = s(t)$ is the reciprocal of ds/dt at $t = t(s)$. In particular, $dt/ds > 0$.

Now let β be the reparametrization $\beta(s) = \alpha(t(s))$ of α. We assert that β has unit speed. In fact, by Lemma 4.5 of Chapter 1,

$$\beta'(s) = \frac{dt}{ds}(s)\alpha'(t(s)).$$

Hence, by the preceding remarks, the speed of β is

$$\| \beta'(s) \| = \frac{dt}{ds}(s) \| \alpha'(t(s)) \| = \frac{dt}{ds}(s)\frac{ds}{dt}(t(s)) = 1. \qquad \blacklozenge$$

We shall use the notation of this proof frequently in later work. The unit-speed curve β is sometimes said to have *arc-length parametrization*, since the arc length of β from $s = a$ to $s = b$ ($a < b$) is just $b - a$.

For example, consider the helix α in Example 4.2 of Chapter 1. Since $\alpha(t) = (a \cos t, a \sin t, bt)$, the velocity α' is given by the formula

$$\alpha'(t) = (-a \sin t, a \cos t, b).$$

Hence

$$\| \alpha'(t) \|^2 = \alpha'(t) \cdot \alpha'(t) = a^2 \sin^2 t + a^2 \cos^2 t + b^2 = a^2 + b^2.$$

Thus α has *constant* speed $c = \| \alpha' \| = (a^2 + b^2)^{1/2}$. If we measure arc length from $t = 0$, then

$$s(t) = \int_0^t c \, du = ct.$$

Hence, $t(s) = s/c$. Substituting in the formula for α, we get the unit-speed reparametrization

$$\beta(s) = \alpha\left(\frac{s}{c}\right) = \left(a \cos \frac{s}{c}, a \sin \frac{s}{c}, \frac{bs}{c} \right).$$

It is easy to check directly that $\| \beta'(s) \| = 1$ for all s.

A reparametrization $\alpha(h)$ of a curve α is *orientation-preserving* if $h' \geq 0$ and *orientation-reversing* if $h' \leq 0$. In the latter case, $\alpha(h)$ still follows the route of α but in the opposite direction. By definition, a unit-speed reparametrization is always orientation-preserving since $ds/dt > 0$ for a regular curve.

In the *theory* of curves we will frequently reparametrize regular curves to obtain unit speed; however, it is rarely possible to do this in practice. The problem is basic calculus: Even when the coordinate functions of the curve are rather simple, the speed function cannot usually be integrated explicitly— at least in terms of familiar functions.

The general notion of vector field (Definition 2.3 of Chapter 1) can be adapted to curves as follows.

2.2 Definition   A *vector field* Y *on curve* α: $I \to \mathbf{R}^3$ is a function that assigns to each number t in I a tangent vector $Y(t)$ to $\mathbf{R}^3$ at the point $\alpha(t)$.

We have already met such vector fields: For any curve α, its velocity α' evidently satisfies this definition. Note that unlike α', arbitrary vector fields on α need not be tangent to α, but may point in any direction (Fig. 2.6).

The properties of vector fields on curves are analogous to those of vector fields on $\mathbf{R}^3$. For example, if Y is a vector field on α: $I \to \mathbf{R}^3$, then for each t in I we can write

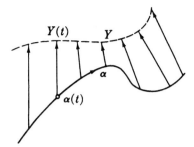

FIG. 2.6

$$Y(t) = (y_1(t),\, y_2(t),\, y_3(t))_{\alpha(t)} = \sum y_i(t)U_i(\alpha(t)).$$

We have thus defined real-valued functions y_1, y_2, y_3 on I, called the *Euclidean coordinate functions* of Y. These will always be assumed to be differentiable. Note that the composite function $t \to U_i(\alpha(t))$ is a vector field on α. Where it seems safe to do so, we shall often write merely U_i instead of $U_i(\alpha(t))$.

The operations of addition, scalar multiplication, dot product, and cross product of vector fields (on the same curve) are all defined in the usual pointwise fashion. Thus if

$$Y(t) = t^2 U_1 - t U_3, \quad Z(t) = (1 - t^2)U_2 + t U_3,$$

and $f(t) = (t + 1)/t$, we obtain the vector fields

$$(Y + Z)(t) = t^2 U_1 + (1 - t^2)U_2,$$

$$(fY)(t) = t(t + 1)U_1 - (t + 1)U_3,$$

$$(Y \times Z)(t) = \begin{vmatrix} U_1 & U_2 & U_3 \\ t^2 & 0 & -t \\ 0 & 1-t^2 & t \end{vmatrix}$$

$$= t(1 - t^2)U_1 - t^3 U_2 + t^2(1 - t^2)U_3$$

and the real-valued function

$$(Y \cdot Z)(t) = -t^2.$$

To differentiate a vector field on α one simply differentiates its Euclidean coordinate functions, thus obtaining a new vector field on α. Explicitly, if $Y = \sum y_i U_i$, then $Y' = \sum \dfrac{dy_i}{dt} U_i$. Thus, for Y as above, we get

$$Y' = 2t U_1 - U_3, \quad Y'' = 2U_1, \quad \text{and} \quad Y''' = 0.$$

In particular, the derivative α'' of the velocity α' of α is called the *acceleration* of α. Thus if $\alpha = (\alpha_1, \alpha_2, \alpha_3)$, the acceleration α'' is the vector field

$$\alpha'' = \left(\frac{d^2\alpha_1}{dt^2}, \frac{d^2\alpha_2}{dt^2}, \frac{d^2\alpha_3}{dt^2} \right)_\alpha$$

on α. By contrast with velocity, acceleration is generally not tangent to the curve.

As we mentioned earlier, in whatever form it appears, differentiation always has suitable linearity and Leibnizian properties. In the case of vector fields on a curve, it is easy to prove the linearity property

$$(aY + bZ)' = aY' + bZ'$$

(a and b numbers) and the Leibnizian properties

$$(fY)' = \frac{df}{dt}Y + fY' \quad \text{and} \quad (Y \cdot Z)' = Y' \cdot Z + Y \cdot Z'.$$

If the function $Y \cdot Z$ is constant, the last formula shows that

$$Y' \cdot Z + Y \cdot Z' = 0.$$

This observation will be used frequently in later work. In particular, if Y has constant length $\| Y \|$, then Y and Y' are orthogonal at each point, since $\| Y \|^2 = Y \cdot Y$ constant implies $2Y \cdot Y' = 0$.

Recall that tangent vectors are parallel if they have the same vector parts. We say that a vector field Y on a curve is *parallel* provided all its (tangent vector) values are parallel. In this case, if the common vector part is (c_1, c_2, c_3), then

$$Y(t) = (c_1, c_2, c_3)_{\alpha(t)} = \sum c_i U_i \quad \text{for all } t.$$

Thus parallelism for a vector field is equivalent to the constancy of its Euclidean coordinate functions.

Vanishing of derivatives is always important in calculus; here are three simple cases.

2.3 Lemma   (1) A curve α is constant if and only if its velocity is zero, $\alpha' = 0$.

(2) A nonconstant curve α is a straight line if and only if its acceleration is zero, $\alpha'' = 0$.

(3) A vector field Y on a curve is parallel if and only if its derivative is zero, $Y' = 0$.

Proof.   In each case it suffices to look at the Euclidean coordinate functions. For example, we shall prove (2). If $\alpha = (\alpha_1, \alpha_2, \alpha_3)$, then

$$\alpha'' = \left(\frac{d^2\alpha_1}{dt^2}, \frac{d^2\alpha_2}{dt^2}, \frac{d^2\alpha_3}{dt^2} \right).$$

Thus $\alpha'' = 0$ if and only if each $d^2\alpha_i/dt^2 = 0$. By elementary calculus, this is equivalent to the existence of constants p_i and q_i such that

$$\alpha_i(t) = p_i + tq_i, \quad \text{for } i = 1, 2, 3.$$

Thus $\alpha(t) = \mathbf{p} + t\mathbf{q}$, and α is a straight line as defined in Example 4.2 of Chapter 1. (Note that nonconstancy implies $\mathbf{q} \neq 0$.)   ◆

Exercises

1. For the curve $\alpha(t) = (2t, t^2, t^3/3)$,
(a) find the velocity, speed, and acceleration for arbitrary t, and at $t = 1$;
(b) find the arc length function $s = s(t)$ (based at $t = 0$), and determine the arc length of α from $t = -1$ to $t = +1$.

2. Show that a curve has constant speed if and only if its acceleration is everywhere orthogonal to its velocity.

3. Show that the curve $\alpha(t) = (\cosh t, \sinh t, t)$ has arc length function $s(t) = \sqrt{2} \sinh t$, and find a unit-speed reparametrization of α.

4. Consider the curve $\alpha(t) = (2t, t^2, \log t)$ on $I: t > 0$. Show that this curve passes through the points $\mathbf{p} = (2, 1, 0)$ and $\mathbf{q} = (4, 4, \log 2)$, and find its arc length between these points.

5. Suppose that β_1 and β_2 are unit-speed reparametrizations of the same curve α. Show that there is a number s_0 such that $\beta_2(s) = \beta_1(s + s_0)$ for all s. What is the geometric significance of s_0?

6. Let Y be a vector field on the helix $\alpha(t) = (\cos t, \sin t, t)$. In each of the following cases, express Y in the form $\sum y_i U_i$:
(a) $Y(t)$ is the vector from $\alpha(t)$ to the origin of $\mathbf{R}^3$.
(b) $Y(t) = \alpha'(t) - \alpha''(t)$.
(c) $Y(t)$ has unit length and is orthogonal to both $\alpha'(t)$ and $\alpha''(t)$.
(d) $Y(t)$ is the vector from $\alpha(t)$ to $\alpha(t + \pi)$.

7. A reparametrization $\alpha(h): [c, d] \to \mathbf{R}^3$ of a curve segment $\alpha: [a, b] \to \mathbf{R}^3$ is *monotone* provided either

(i) $h' \geq 0$, $h(c) = a$, $h(d) = b$ or   (ii) $h' \leq 0$, $h(c) = b$, $h(d) = a$.

Prove that monotone reparametrization does not change arc length.

8. Let Y be a vector field on a curve α. If $\alpha(h)$ is a reparametrization of α, show that the reparametrization $Y(h)$ is a vector field on $\alpha(h)$, and prove the chain rule $Y(h)' = h' Y'(h)$.

9. (*Numerical integration.*) The curve segments

$$\alpha(t) = (\sin t, \ t^2 \cos t, \ \sin 2t), \quad \beta(t) = (t^2 \sin t, \ t^2, \ t^2(1 + \cos t)),$$

defined on $0 \le t \le \pi$, run from the origin 0 to $(0, \pi^2, 0)$. Which is shorter? (See Integration in the Appendix.)

10. Let α, $\beta: I \to \mathbf{R}^3$ be curves such that $\alpha'(t)$ and $\beta'(t)$ are parallel (same Euclidean coordinates) at each t. Prove that α and β are *parallel* in the sense that there is a point $\mathbf{p}$ in $\mathbf{R}^3$ such that $\beta(t) = \alpha(t) + \mathbf{p}$ for all t.

11. Prove that a straight line is the shortest distance between two points in $\mathbf{R}^3$. Use the following scheme; let $\alpha: [a, b] \to \mathbf{R}^3$ be an arbitrary curve segment from $\mathbf{p} = \alpha(a)$ to $\mathbf{q} = \alpha(b)$. Let $\mathbf{u} = (\mathbf{q} - \mathbf{p})/\| \mathbf{q} - \mathbf{p} \|$.
  (a) If σ is a straight line segment from $\mathbf{p}$ to $\mathbf{q}$, say

$$\sigma(t) = (1 - t)\mathbf{p} + t\mathbf{q} \quad (0 \le t \le 1),$$

show that $L(\sigma) = d(\mathbf{p}, \mathbf{q})$.
  (b) From $\| \alpha' \| \ge \alpha' \cdot \mathbf{u}$, deduce $L(\alpha) \ge d(\mathbf{p}, \mathbf{q})$, where $L(\alpha)$ is the length of α and d is Euclidean distance.
  (c) Furthermore, show that if $L(\alpha) = d(\mathbf{p}, \mathbf{q})$, then (but for parametrization) α is a straight line segment. (*Hint:* write $\alpha' = (\alpha' \cdot \mathbf{u})\mathbf{u} + Y$, where $Y \cdot \mathbf{u} = 0$.)

2.3 The Frenet Formulas

We now derive mathematical measurements of the turning and twisting of a curve in $\mathbf{R}^3$. Throughout this section we deal only with *unit-speed* curves; in the next we extend the results to arbitrary regular curves.

  Let $\beta: I \to \mathbf{R}^3$ be a unit-speed curve, so $\| \beta'(s) \| = 1$ for each s in I. Then $T = \beta'$ is called the *unit tangent* vector field on β. Since T has constant length 1, its derivative $T' = \beta''$ measures the way the curve is turning in $\mathbf{R}^3$. We call T' the *curvature* vector field of β. Differentiation of $T \cdot T = 1$ gives $2T' \cdot T = 0$, so T' is always orthogonal to T, that is, *normal* to β.

  The length of the curvature vector field T' gives a numerical measurement of the turning of β. The real-valued function κ such that $\kappa(s) = \| T'(s) \|$ for

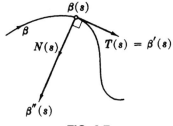

FIG. 2.7

all s in I is called the *curvature* function of β. Thus $\kappa \geqq 0$, and the larger κ is, the sharper the turning of β.

To carry this analysis further, we impose the restriction that κ is never zero so $\kappa > 0$. The unit-vector field $N = T'/\kappa$ on β then tells the *direction* in which β is turning at each point. N is called the *principal normal* vector field of β (Fig. 2.7). The vector field $B = T \times N$ on β is called the *binormal* vector field of β.

3.1 Lemma   Let β be a unit-speed curve in $\mathbf{R}^3$ with $\kappa > 0$. Then the three vector fields T, N, and B on β are unit vector fields that are mutually orthogonal at each point. We call T, N, B the *Frenet frame field* on β.

Proof.   By definition $\| T \| = 1$. Since $\kappa = \| T' \| > 0$,

$$\| N \| = (1/\kappa) \| T' \| = 1.$$

We saw above that T and N are orthogonal—that is, $T \bullet N = 0$. Then by applying Lemma 1.8 at each point, we conclude that $\| B \| = 1$, and B is orthogonal to both T and N.   ◆

In summary, we have $T = \beta'$, $N = T'/\kappa$, and $B = T \times N$, satisfying $T \bullet T = N \bullet N = B \bullet B = 1$, with all other dot products zero.

The key to the successful study of the geometry of a curve β is to use its Frenet frame field T, N, B whenever possible, instead of the natural frame field U_1, U_2, U_3. The Frenet frame field of β is full of information about β, whereas the natural frame field contains none at all.

The first and most important use of this idea is to express the *derivatives T', N', B' in terms of T, N, B.* Since $T = \beta'$, we have $T' = \beta'' = \kappa N$. Next consider B'. We claim that B' is, at each point, a scalar multiple of N. To prove this, it suffices by orthonormal expansion to show that $B' \bullet B = 0$ and $B' \bullet T = 0$. The former holds since B is a unit vector. To prove the latter, differentiate $B \bullet T = 0$, obtaining $B' \bullet T + B \bullet T' = 0$; then

$$B' \bullet T = -B \bullet T' = -B \bullet \kappa N = 0.$$

Thus we can now define the *torsion* function τ of the curve β to be the real-valued function on the interval I such that $B' = -\tau N$. (The minus sign is traditional.) By contrast with curvature, there is no restriction on the values of τ—it may be positive, negative, or zero at various points of I. We shall presently show that τ does measure the torsion, or twisting, of the curve β.

3.2 Theorem   (Frenet formulas). If $\beta: I \to \mathbf{R}^3$ is a unit-speed curve with curvature $\kappa > 0$ and torsion τ, then

$$
\begin{aligned}
T' &= & \kappa N, \\
N' &= -\kappa T & & + \tau B, \\
B' &= & -\tau N.
\end{aligned}
$$

Proof.   As we saw above, the first and third formulas are essentially just the definitions of curvature and torsion. To prove the second, we use orthonormal expansion to express N' in terms of T, N, B:

$$
N' = N' \cdot T \, T + N' \cdot N \, N + N' \cdot B \, B.
$$

These coefficients are easily found. Differentiating $N \cdot T = 0$, we get $N' \cdot T + N \cdot T' = 0$; hence

$$
N' \cdot T = -N \cdot T' = -N \cdot \kappa N = -\kappa.
$$

As usual, $N' \cdot N = 0$, since N is a unit vector field. Finally,

$$
N' \cdot B = -N \cdot B' = -N \cdot (-\tau N) = \tau. \qquad \blacklozenge
$$

3.3 Example   We compute the Frenet frame T, N, B and the curvature and torsion functions of the unit-speed helix

$$
\beta(s) = \left(a \cos \frac{s}{c}, a \sin \frac{s}{c}, \frac{bs}{c} \right),
$$

where $c = (a^2 + b^2)^{1/2}$ and $a > 0$. Now

$$
T(s) = \beta'(s) = \left(-\frac{a}{c} \sin \frac{s}{c}, \frac{a}{c} \cos \frac{s}{c}, \frac{b}{c} \right).
$$

Hence

$$
T'(s) = \left(-\frac{a}{c^2} \cos \frac{s}{c}, -\frac{a}{c^2} \sin \frac{s}{c}, 0 \right).
$$

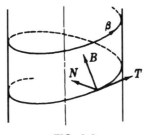

FIG. 2.8

Thus

$$\kappa(s) = \|T'(s)\| = \frac{a}{c^2} = \frac{a}{a^2 + b^2} > 0.$$

Since $T' = \kappa N$, we get

$$N(s) = \left(-\cos\frac{s}{c}, -\sin\frac{s}{c}, 0\right).$$

Note that regardless of what values a and b have, N always points straight in toward the axis of the cylinder on which β lies (Fig. 2.8).

Applying the definition of cross product to $B = T \times N$ gives

$$B(s) = \left(\frac{b}{c}\sin\frac{s}{c}, -\frac{b}{c}\cos\frac{s}{c}, \frac{a}{c}\right).$$

It remains to compute torsion. Now,

$$B'(s) = \left(\frac{b}{c^2}\cos\frac{s}{c}, \frac{b}{c^2}\sin\frac{s}{c}, 0\right),$$

and by definition, $B' = -\tau N$. Comparing the formulas for B' and N, we conclude that

$$\tau(s) = \frac{b}{c^2} = \frac{b}{a^2 + b^2}.$$

So the torsion of the helix is also constant.

Note that when the parameter b is zero, the helix reduces to a circle of radius a. The curvature of this circle is $\kappa = 1/a$ (so the smaller the radius, the larger the curvature), and the torsion is identically zero.

This example is a very special one—in general (as the examples in the exercises show) neither the curvature nor the torsion functions of a curve need be constant.

3.4 Remark We have emphasized all along the distinction between a tangent vector and a point of $\mathbf{R}^3$. However, Euclidean space has, as we have seen, the remarkable property that given a point $\mathbf{p}$, there is a natural one-to-one correspondence between points (v_1, v_2, v_3) and tangent vectors $(v_1, v_2, v_3)_{\mathbf{p}}$ at $\mathbf{p}$. Thus one can transform points into tangent vectors (and vice versa) by means of this canonical isomorphism. In the next two sections particularly, it will often be convenient to switch quietly from one to the other without change of notation. Since *corresponding objects have the same Euclidean coordinates*, this switching can have no effect on scalar multiplication, addition, dot products, differentiation, or any other operation defined in terms of Euclidean coordinates.

Thus a vector field $Y = (y_1, y_2, y_3)_{\beta}$ on a curve β becomes itself a curve (y_1, y_2, y_3) in $\mathbf{R}^3$. In particular, if Y is parallel, its Euclidean coordinate functions are constant, so Y is identified with a single point of $\mathbf{R}^3$.

A *plane* in $\mathbf{R}^3$ can be described as the union of all the perpendiculars to a given line at a given point. In vector language then, the *plane through* $\mathbf{p}$ *orthogonal to* $\mathbf{q} \neq 0$ consists of all points $\mathbf{r}$ in $\mathbf{R}^3$ such that $(\mathbf{r} - \mathbf{p}) \bullet \mathbf{q} = 0$. By the remark above, we may picture $\mathbf{q}$ as a tangent vector at $\mathbf{p}$ as shown in Fig. 2.9.

We can now give an informative approximation of a given curve near an arbitrary point on the curve. The goal is to show how curvature and torsion influence the shape of the curve. To derive this approximation we use a Taylor approximation of the curve—and express this in terms of the Frenet frame at the selected point.

For simplicity, we shall consider the unit-speed curve $\beta = (\beta_1, \beta_2, \beta_3)$ near the point $\beta(0)$. For s small, each coordinate $\beta_i(s)$ is closely approximated by the initial terms of its Taylor series:

$$\beta_i(s) \sim \beta_i(0) + \frac{d\beta_i}{ds}(0)s + \frac{d^2\beta_i}{ds^2}(0)\frac{s^2}{2} + \frac{d^3\beta_i}{ds^3}(0)\frac{s^3}{6}.$$

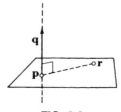

FIG. 2.9

Hence

$$\beta(s) \sim \beta(0) + s\beta'(0) + \frac{s^2}{2}\beta''(0) + \frac{s^3}{6}\beta'''(0).$$

But $\beta'(0) = T_0$, and $\beta''(0) = \kappa_0 N_0$, where the subscript indicates evaluation at $s = 0$, and we assume $\kappa_0 \neq 0$. Now

$$\beta''' = (\kappa N)' = \frac{d\kappa}{ds}N + \kappa N'.$$

Thus by the Frenet formula for N', we get

$$\beta'''(0) = -\kappa_0^2 T_0 + \frac{d\kappa}{ds}(0)N_0 + \kappa_0\tau_0 B_0.$$

Finally, substitute these derivatives into the approximation of $\beta(s)$ given above, and keep only the dominant term in each component (that is, the one containing the smallest power of s). The result is

$$\beta(s) \sim \beta(0) + sT_0 + \kappa_0\frac{s^2}{2}N_0 + \kappa_0\tau_0\frac{s^3}{6}B_0.$$

Denoting the right side by $\hat{\beta}(s)$, we obtain a curve $\hat{\beta}$ called the *Frenet approximation* of β near $s = 0$. We emphasize that β has a different Frenet approximation near each of its points; if 0 is replaced by an arbitrary number s_0, then s is replaced by $s - s_0$, as usual in Taylor expansions.

Let us now examine the Frenet approximation given above. The first term in the expression for $\hat{\beta}$ is just the point $\beta(0)$. The first two terms give the *tangent line* $s \to \beta(0) + sT_0$ of β at $\beta(0)$—the best linear approximation of β near $\beta(0)$. The first three terms give the parabola

$$s \to \beta(0) + sT_0 + \kappa_0(s^2/2)N_0,$$

which is the best quadratic approximation of β near $\beta(0)$. Note that this parabola lies in the plane through $\beta(0)$ orthogonal to B_0, the *osculating plane* of β at $\beta(0)$. This parabola has the same shape as the parabola $y = \kappa_0 x^2/2$ in the xy plane, and is completely determined by the curvature κ_0 of β at $s = 0$.

Finally, the torsion τ_0, which appears in the last and smallest term of $\hat{\beta}$, controls the motion of β orthogonal to its osculating plane at $\beta(0)$, as shown in Fig. 2.10.

On the basis of this discussion, it is a reasonable guess that *if a unit-speed curve has curvature identically zero, then it is a straight line.* In fact, this follows immediately from (2) of Lemma 2.3, since $\kappa = \|T'\| = \|\beta''\|$, so that $\kappa = 0$ if and only if $\beta'' = 0$. Thus curvature does measure deviation from straightness.

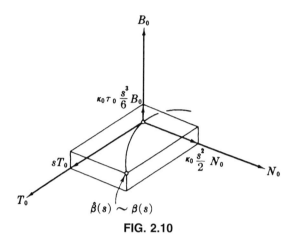

FIG. 2.10

A *plane curve* in $\mathbf{R}^3$ is a curve that lies in a single plane of $\mathbf{R}^3$. Evidently a plane curve does not twist in as interesting a way as even the simple helix in Example 3.3. The discussion above shows that for s small the curve β tends to stay in its osculating plane at $\beta(0)$; it is $\tau_0 \neq 0$ that causes β to twist out of the osculating plane. Thus if the torsion of β is identically zero, we may well suspect that β never leaves this plane.

3.5 Corollary Let β be a unit-speed curve in $\mathbf{R}^3$ with $\kappa > 0$. Then β is a plane curve if and only if $\tau = 0$.

Proof. Suppose β is a plane curve. Then by the remarks above, there exist points $\mathbf{p}$ and $\mathbf{q}$ such that $(\beta(s) - \mathbf{p}) \cdot \mathbf{q} = 0$ for all s. Differentiation yields

$$\beta'(s) \cdot \mathbf{q} = \beta''(s) \cdot \mathbf{q} = 0 \quad \text{for all } s.$$

Thus $\mathbf{q}$ is always orthogonal to $T = \beta'$ and $N = \beta''/\kappa$. But B is also orthogonal to T and N, so, since B has unit length, $B = \pm\mathbf{q}/\|\mathbf{q}\|$. Thus $B' = 0$, and by definition $\tau = 0$ (Fig. 2.11).

Conversely, suppose $\tau = 0$. Thus $B' = 0$; that is, B is parallel and may thus be identified (by Remark 3.4) with a *point* of $\mathbf{R}^3$. We assert that β lies in the plane through $\beta(0)$ orthogonal to B. To prove this, consider the real-valued function

$$f(s) = (\beta(s) - \beta(0)) \cdot B \quad \text{for all } s.$$

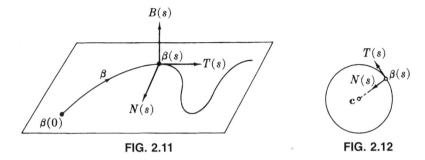

FIG. 2.11   FIG. 2.12

Then

$$\frac{df}{ds} = \beta' \cdot B = T \cdot B = 0.$$

But obviously, $f(0) = 0$, so f is identically zero. Thus

$$(\beta(s) - \beta(0)) \cdot B \quad \text{for all } s,$$

which shows that β lies entirely in this plane orthogonal to the (parallel) binormal of β.   ◆

We saw at the end of Example 3.3 that a circle of radius a has curvature $1/a$ and torsion zero. Furthermore, the formula given there for the principal normal shows that for a circle, N always points toward its center. This suggests how to prove the following converse.

3.6 Lemma If β is a unit-speed curve with constant curvature $\kappa > 0$ and torsion zero, then β is part of a circle of radius $1/\kappa$.

Proof. Since $\tau = 0$, β is a plane curve. What we must now show is that every point of β is at distance $1/\kappa$ from some fixed point—which will thus be the center of the circle. Consider the curve $\gamma = \beta + (1/\kappa)N$. Using the hypothesis on β, and (as usual) a Frenet formula, we find

$$\gamma' = \beta' + \frac{1}{\kappa} N' = T + \frac{1}{\kappa}(-\kappa T) = 0.$$

Hence the curve γ is constant; that is, $\beta(s) + (1/\kappa)N(s)$ has the same value, say $\mathbf{c}$, for all s (see Fig. 2.12). But the distance from $\mathbf{c}$ to $\beta(s)$ is

$$d(\mathbf{c}, \beta(s)) = \|\mathbf{c} - \beta(s)\| = \left\|\frac{1}{\kappa} N(s)\right\| = \frac{1}{\kappa}.$$ ◆

In principle, every geometric problem about curves can be solved by means of the Frenet formulas. In simple cases it may be just enough to record the data of the problem in convenient form, differentiate, and use the Frenet formulas. For example, suppose β is a unit-speed curve that lies entirely in the sphere $\sum$ of radius a centered at the origin of $\mathbf{R}^3$. To stay in the sphere, β must curve; in fact it is a reasonable guess that the minimum possible curvature occurs when β is on a great circle of $\sum$. Such a circle has radius a, so we conjecture that *a spherical curve β has curvature $\kappa \geq 1/a$, where a is the radius of its sphere.*

To prove this, observe that since every point of $\sum$ has distance a from the origin, we have $\beta \cdot \beta = a^2$. Differentiation yields $2\beta' \cdot \beta = 0$, that is, $\beta \cdot T = 0$. Another differentiation gives $\beta' \cdot T + \beta \cdot T' = 0$, and by using a Frenet formula we get $T \cdot T + \kappa\beta \cdot N = 0$; hence

$$\kappa\beta \cdot N = -1.$$

By the Schwarz inequality,

$$|\beta \cdot N| \leq \| \beta \| \| N \| = a,$$

and since $\kappa \geq 0$ we obtain the required result:

$$\kappa = |\kappa| = \frac{1}{|\beta \cdot N|} \geq \frac{1}{a}.$$

Continuation of this procedure leads to a necessary and sufficient condition (expressed in terms of curvature and torsion) for a curve to be *spherical*, that is, lie on some sphere in $\mathbf{R}^3$ (Exercise 10).

Exercises

1. Compute the *Frenet apparatus* κ, τ, T, N, B of the unit-speed curve $\beta(s) = (\tfrac{4}{5}\cos s, 1 - \sin s, -\tfrac{3}{5}\cos s)$. Show that this curve is a circle; find its center and radius.

2. Consider the curve

$$\beta(s) = \left(\frac{(1+s)^{3/2}}{3}, \frac{(1-s)^{3/2}}{3}, \frac{s}{\sqrt{2}} \right)$$

defined on I: $-1 < s < 1$. Show that β has unit speed, and compute its Frenet apparatus.

3. For the helix in Example 3.3, check the Frenet formulas by direct substitution of the computed values of κ, τ, T, N, B.

4. Prove that

$$T = N \times B = -B \times N,$$

$$N = B \times T = -T \times B,$$

$$B = T \times N = -N \times T.$$

(A formal proof uses properties of the cross product established in the Exercises of Section 1—but one can recall these formulas by using the right-hand rule given at the end of that section.)

5. If A is the vector field $\tau T + \kappa B$ on a unit-speed curve β, show that the Frenet formulas become

$$T' = A \times T,$$

$$N' = A \times N,$$

$$B' = A \times B.$$

6. A unit-speed parametrization of a circle may be written

$$\gamma(s) = \mathbf{c} + r \cos \frac{s}{r} \, \mathbf{e}_1 + r \sin \frac{s}{r} \, \mathbf{e}_2,$$

where $\mathbf{e}_i \bullet \mathbf{e}_j = \delta_{ij}$.

If β is a unit-speed curve with $\kappa(0) > 0$, prove that there is one and only one circle γ that approximates β near $\beta(0)$ in the sense that

$$\gamma(0) = \beta(0), \quad \gamma'(0) = \beta'(0), \quad \text{and} \quad \gamma''(0) = \beta''(0).$$

Show that γ lies in the osculating plane of β at $\beta(0)$ and find its center $\mathbf{c}$ and radius r (see Fig. 2.13). The circle γ is called the *osculating circle* and $\mathbf{c}$ the *center of curvature* of β at $\beta(0)$. (The same results hold when 0 is replaced by any number s.)

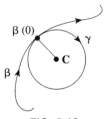

FIG. 2.13

7. If α and a reparametrization $\bar{\alpha} = \alpha(h)$ are both unit-speed curves, show that
(a) $h(s) = \pm s + s_0$ for some number s_0;
(b) $\bar{T} = \pm T(h)$,
$\bar{N} = N(h), \quad \bar{\kappa} = \kappa(h), \quad \bar{\tau} = \tau(h),$
$\bar{B} = \pm B(h),$

where the sign ($\pm$) is the same as that in (a), and we assume $\kappa > 0$. Thus even in the orientation-reversing case, the principal normals N and $\bar{N}$ still point in the same direction.

8. *Curves in the plane.* For a unit-speed curve $\beta(s) = (x(s), y(s))$ in $\mathbf{R}^2$, the *unit tangent* is $T = \beta' = (x', y')$ as usual, but the *unit normal* N is defined by rotating T through $+90°$, so $N = (-y', x')$. Thus T' *and* N are collinear, and the *plane curvature* $\tilde{\kappa}$ of β is defined by the Frenet equation $T' = \tilde{\kappa}N$.
(a) Prove that $\tilde{\kappa} = T' \bullet N$ and $N' = -\tilde{\kappa}T$.
(b) The *slope angle* $\varphi(s)$ of β is the differentiable function such that

$$T = (\cos \varphi, \sin \varphi) = \cos \varphi \, U_x + \sin \varphi \, U_y.$$

(The existence of φ derives from Ex. 12 of Sec. 1.) Show that $\tilde{\kappa} = \varphi'$.
(c) Find the curvature $\tilde{\kappa}$ of the following plane curves.

(i) $(r\cos \dfrac{t}{r}, r\sin \dfrac{t}{r})$, counterclockwise circle.

(ii) $(r\cos(-\dfrac{t}{r}), r\sin(-\dfrac{t}{r}))$, clockwise circle.

(d) Show that if $\tilde{\kappa}$ does not change sign, then $|\tilde{\kappa}|$ is the usual $\mathbf{R}^3$ curvature κ. (For such comparisons we can always regard $\mathbf{R}^2$ as, say, the xy plane in $\mathbf{R}^3$.)

9. Let $\tilde{\beta}$ be the Frenet approximation of a unit-speed curve β with $\tau \neq 0$ near $s = 0$.
If, say, the B_0 component of $\tilde{\beta}$ is removed, the resulting curve is the *orthogonal projection* of $\tilde{\beta}$ in the T_0N_0 plane. It is the view of $\beta \approx \tilde{\beta}$ that one gets by looking toward $\beta(0) = \tilde{\beta}(0)$ directly along the vector B_0.
Sketch the general shape of the orthogonal projections of $\tilde{\beta}$ near $s = 0$ in each of the planes T_0N_0 (*osculating plane*), T_0B_0 (*rectifying plane*), and N_0B_0 (*normal plane*). These views of $\beta \approx \tilde{\beta}$ can be confirmed experimentally using a bent piece of wire. For computer views, see Exercise 15 of Section 4.

10. *Spherical curves.* Let α be a unit-speed curve with $\kappa > 0, \tau \neq 0$.
(a) If α lies on a sphere of center $\mathbf{c}$ and radius r, show that

$$\alpha - \mathbf{c} = -\rho N - \rho'\sigma B,$$

where $\rho = 1/\kappa$ and $\sigma = 1/\tau$. Thus $r^2 = \rho^2 + (\rho'\sigma)^2$.

(b) Conversely, if $\rho^2 + (\rho'\sigma)^2$ has constant value r^2 and $\rho' \neq 0$, show that α lies on a sphere of radius r.

(*Hint:* For (b), show that the "center curve" $\gamma = \alpha + \rho N + \rho'\sigma B$—suggested by (a)—is constant.)

11. Let β, $\bar{\beta}$: $I \to \mathbf{R}^3$ be unit-speed curves with nonvanishing curvature and torsion. If $T = \bar{T}$, then β and $\bar{\beta}$ are parallel (Ex. 10 of Sec. 2). If $B = \bar{B}$, prove that $\bar{\beta}$ is parallel to either β or the curve $s \to -\beta(s)$.

2.4 Arbitrary-Speed Curves

It is a simple matter to adapt the results of the previous section to the study of a regular curve α: $I \to \mathbf{R}^3$ that does not necessarily have unit speed. We merely transfer to α the Frenet apparatus of a unit-speed reparametrization $\bar{\alpha}$ of α. Explicitly, if s is an arc length function for α as in Theorem 2.1, then

$$\alpha(t) = \bar{\alpha}(s(t)) \quad \text{for all } t,$$

or, in functional notation, $\alpha = \bar{\alpha}(s)$, as suggested by Fig. 2.14. Now if $\bar{\kappa} > 0$, $\bar{\tau}$, $\bar{T}$, $\bar{N}$, and $\bar{B}$ are defined for $\bar{\alpha}$ as in Section 3, we define for α the

curvature function: $\kappa = \bar{\kappa}(s)$,
torsion function: $\tau = \bar{\tau}(s)$,
unit tangent vector field: $T = \bar{T}(s)$,
principal normal vector field: $N = \bar{N}(s)$,
binormal vector field: $B = \bar{B}(s)$.

In general κ and $\bar{\kappa}$ are different functions, defined on different intervals. But *they give exactly the same description of the turning of the common route of α and $\bar{\alpha}$, since at any point $\alpha(t) = \bar{\alpha}(s(t))$ the numbers $\kappa(t)$ and $\bar{\kappa}(s(t))$ are*

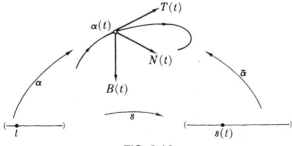

FIG. 2.14

by definition the same. Similarly with the rest of the Frenet apparatus; since only a change of parametrization is involved, its fundamental geometric meaning is the same as before. In particular, T, N, B is again a frame field on α linked to the shape of α as indicated in the discussion of Frenet approximations.

For purely theoretical work, this simple transference is often all that is needed. Data about α converts into data about the unit-speed reparametrization $\bar{\alpha}$; results about $\bar{\alpha}$ convert to results about α. For example, if α is a regular curve with $\tau = 0$, then by the definition above $\bar{\alpha}$ has $\bar{\tau} = 0$; by Corollary 3.5, $\bar{\alpha}$ is a plane curve, so obviously α is too.

However, for explicit numerical computations—and occasionally for the theory as well—this transference is impractical, since it is rarely possible to find explicit formulas for $\bar{\alpha}$. (For example, try to find a unit-speed parametrization for the curve $\alpha(t) = (t, t^2, t^3)$.)

The Frenet formulas are valid only for unit-speed curves; they tell the rate of change of the frame field T, N, B *with respect to arc length*. However, the speed v of the curve is the proper correction factor in the general case.

4.1 Lemma   If α is a regular curve in $\mathbf{R}^3$ with $\kappa > 0$, then

$$
\begin{aligned}
T' &= & \kappa v N, & \\
N' &= -\kappa v T & & + \ \tau v B, \\
B' &= & -\tau v N.
\end{aligned}
$$

Proof.   Let $\bar{\alpha}$ be a unit-speed reparametrization of α. Then by definition, $T = \bar{T}(s)$, where s is an arc length function for α. The chain rule as applied to differentiation of vector fields (Exercise 7 of Section 2) gives

$$ T' = \bar{T}'(s)\frac{ds}{dt}. $$

By the usual Frenet equations, $\bar{T}' = \bar{\kappa}\bar{N}$. Substituting the function s in this equation yields

$$ \bar{T}'(s) = \bar{\kappa}(s)\bar{N}(s) = \kappa N $$

by the definition of κ and N in the arbitrary-speed case. Since ds/dt is the speed function v of α, these two equations combine to yield $T' = \kappa v N$. The formulas for N' and B' are derived in the same way.   ◆

There is a commonly used notation for the calculus that completely ignores change of parametrization. For example, the same letter would designate both a curve α and its unit-speed parametrization $\bar{\alpha}$, and similarly with the

Frenet apparatus of these two curves. Differences in derivatives are handled by writing, say, dT/dt for T', but dT/ds for either $\overline{T}'$ or its reparametrization $\overline{T}'(s)$. If these conventions were used, the proof above would combine the chain rule $dT/dt = (dT/ds)(ds/dt)$ and the Frenet formula $dT/ds = \kappa N$ to give $dT/dt = \kappa v N$.

Only for a *constant-speed* curve is acceleration always orthogonal to velocity, since $\beta' \cdot \beta'$ constant is equivalent to $(\beta' \cdot \beta')' = 2\beta' \cdot \beta'' = 0$. In the general case, we analyze velocity and acceleration by expressing them in terms of the Frenet frame field.

4.2 Lemma   If α is a regular curve with speed function v, then the velocity and acceleration of α are given by (Fig. 2.15.)

$$\alpha' = vT, \quad \alpha'' = \frac{dv}{dt}T + \kappa v^2 N.$$

Proof.   Since $\alpha = \overline{\alpha}(s)$, where s is the arc length function of α, we find, using Lemma 4.5 of Chapter 1, that

$$\alpha' = \overline{\alpha}'(s)\frac{ds}{dt} = v\overline{T}(s) = vT.$$

Then a second differentiation yields

$$\alpha'' = \frac{dv}{dt}T + vT' = \frac{dv}{dt}T + \kappa v^2 N,$$

where we use Lemma 4.1.   ◆

The formula $\alpha' = vT$ is to be expected since α' and T are each tangent to the curve and T has a unit length, while $\|\alpha'\| = v$. The formula for acceleration is more interesting. By definition, α'' is the rate of change of the

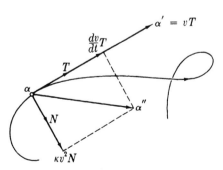

FIG. 2.15

velocity α', and in general both the length and the direction of α' are changing. The *tangential component* $(dv/dt)T$ of α'' measures the rate of change of the length of α' (that is, of the speed of α). The *normal component* $\kappa v^2 N$ measures the rate of change of the direction of α'. Newton's laws of motion show that these components may be experienced as forces. For example, in a car that is speeding up or slowing down on a straight road, the only force one feels is due to $(dv/dt)T$. If one takes an unbanked curve at speed v, the resulting sideways force is due to $\kappa v^2 N$. Here κ measures how sharply the *road* turns; the effect of speed is given by v^2, so 60 miles per hour is four times as unsettling as 30.

We now find effectively computable expressions for the Frenet apparatus.

4.3 Theorem   Let α be a regular curve in $\mathbf{R}^3$. Then

$$T = \alpha'/\|\alpha'\|,$$

$$N = B \times T, \qquad\qquad \kappa = \|\alpha' \times \alpha''\|/\|\alpha'\|^3,$$
$$B = \alpha' \times \alpha''/\|\alpha' \times \alpha''\|, \quad \tau = (\alpha' \times \alpha'') \cdot \alpha'''/\|\alpha' \times \alpha''\|^2.$$

Proof.   Since $v = \|\alpha'\| > 0$, the formula $T = \alpha'/\|\alpha'\|$ is equivalent to $\alpha' = vT$. From the preceding lemma we get

$$\alpha' \times \alpha'' = (vT) \times \left(\frac{dv}{dt} T + \kappa v^2 N \right)$$

$$= v \frac{dv}{dt} T \times T + \kappa v^3 T \times N = \kappa v^3 B,$$

since $T \times T = 0$. Taking norms we find

$$\|\alpha' \times \alpha''\| = \|\kappa v^3 B\| = \kappa v^3$$

because $\|B\| = 1$, $\kappa \geq 0$, and $v > 0$. Indeed, *this equation shows that for regular curves*, $\|\alpha' \times \alpha''\| > 0$ *is equivalent to the usual condition* $\kappa > 0$. (Thus for $\kappa > 0$, α' and α'' are linearly independent and determine the osculating plane at each point, as do T and N.) Then

$$B = \frac{\alpha' \times \alpha''}{\kappa v^3} = \frac{\alpha' \times \alpha''}{\|\alpha' \times \alpha''\|}.$$

Since $N = B \times T$ is true for any Frenet frame field (Exercise 4 of Section 3), only the formula for torsion remains to be proved.

To find the dot product $(\alpha' \times \alpha'') \cdot \alpha'''$ we express everything in terms of T, N, B. We already know that $\alpha' \times \alpha'' = \kappa v^3 B$. Thus, since $0 = T \cdot B = N \cdot B$, *we need only find the B component of* α'''. But

$$\alpha''' = \left(\frac{dv}{dt} T + \kappa v^2 N \right)' = \kappa v^2 N' + \cdots$$

$$= \kappa v^3 \tau B + \cdots,$$

where we use Lemma 4.1. Consequently, $(\alpha' \times \alpha'') \bullet \alpha''' = \kappa^2 v^6 \tau$, and since $\| \alpha' \times \alpha'' \| = \kappa v^3$, we have the required formula for τ. ◆

The triple scalar product in this formula for τ could (by Exercise 4 of Section 1) also be written $\alpha' \bullet \alpha'' \times \alpha'''$. But we need $\alpha' \times \alpha''$ anyway, so it is more efficient to find $(\alpha' \times \alpha'') \bullet \alpha'''$.

4.4 Example We compute the Frenet apparatus of the 3-*curve*

$$\alpha(t) = (3t - t^3, 3t^2, 3t + t^3).$$

The derivatives are

$$\alpha'(t) = 3(1 - t^2, 2t, 1 + t^2),$$

$$\alpha''(t) = 6(-t, 1, t),$$

$$\alpha'''(t) = 6(-1, 0, 1).$$

Now,

$$\alpha'(t) \bullet \alpha'(t) = 18(1 + 2t^2 + t^4),$$

so

$$v(t) = \| \alpha'(t) \| = \sqrt{18} (1 + t^2).$$

Applying the definition of cross product yields

$$\alpha'(t) \times \alpha''(t) = 18 \begin{vmatrix} U_1 & U_2 & U_3 \\ 1 - t^2 & 2t & 1 + t^2 \\ -t & 1 & t \end{vmatrix} = 18(-1 + t^2, -2t, 1 + t^2).$$

Dotting this vector with itself, we get

$$(18)^2 \left[(-1 + t^2)^2 + 4t^2 + (1 + t^2)^2 \right] = 2(18)^2 (1 + t^2)^2.$$

Hence

$$\| \alpha'(t) \times \alpha''(t) \| = 18\sqrt{2} (1 + t^2).$$

The expressions above for $\alpha' \times \alpha''$ and α''' yield

$$(\alpha' \times \alpha'') \bullet \alpha''' = 6 \bullet 18 \bullet 2.$$

It remains only to substitute this data into the formulas in Theorem 4.3, with N being computed by another cross product. The final results are

$$T = \frac{(1 - t^2,\, 2t,\, 1 + t^2)}{\sqrt{2}(1 + t^2)},$$

$$N = \frac{(-2t,\, 1 - t^2,\, 0)}{1 + t^2},$$

$$B = \frac{(-1 + t^2,\, -2t,\, 1 + t^2)}{\sqrt{2}(1 + t^2)},$$

$$\kappa = \tau = \frac{1}{3(1 + t^2)^2}.$$

Alternatively, we could use the identity in Lemma 1.8 to compute $\| \alpha' \times \alpha'' \|$ and express

$$(\alpha' \times \alpha'') \cdot \alpha''' = \alpha' \cdot (\alpha'' \times \alpha''')$$

as a determinant by Exercise 4 of Section 1.

To summarize, we now have the Frenet apparatus for an arbitrary regular curve α, namely, its curvature, torsion, and Frenet frame field. This apparatus satisfies the extended Frenet formulas with speed factor v and can be computed by Theorem 4.3. If $v = 1$, that is, if α is a unit-speed curve, the results of Section 3 are recovered.

Let us consider some applications of the Frenet formulas. There are a number of natural ways in which a given curve β gives rise to a new curve $\tilde{\beta}$ whose geometric properties illuminate some aspect of the behavior of β.

For example, the *spherical image* of a unit-speed curve β is the curve $\sigma \approx T$ with the same Euclidean coordinates as $T = \beta'$. Geometrically, σ is gotten by moving each $T(s)$ to the origin of $\mathbf{R}^3$, as suggested in Fig. 2.16. Thus σ lies on the unit sphere Σ, and the *motion* of σ represents the *turning* of β.

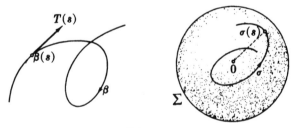

FIG. 2.16

For instance, if β is the helix in Example 3.3, the formula there for T shows that

$$\sigma(s) = \left(-\frac{a}{c}\sin\frac{s}{c}, \frac{a}{c}\cos\frac{s}{c}, \frac{b}{c} \right).$$

So the spherical image of a helix lies on the circle cut from Σ by the plane $z = b/c$.

Although the original curve β has unit speed, we cannot expect that σ does also. In fact, $\sigma = T$ implies $\sigma' = T' = \kappa N$, so the *speed* of σ equals the *curvature* κ of β. Thus to compute the curvature of σ, we must use the extended Frenet formulas in Theorem 4.3. From

$$\sigma'' = (\kappa N)' = \frac{d\kappa}{ds}N + \kappa N' = -\kappa^2 T + \frac{d\kappa}{ds}N + \kappa\tau B,$$

we get

$$\sigma' \times \sigma'' = -\kappa^3 N \times T + \kappa^2\tau N \times B = \kappa^2(\kappa B + \tau T).$$

By Theorem 4.3 the curvature of the spherical image σ is

$$\kappa_\sigma = \frac{\|\sigma' \times \sigma''\|}{v^3} = \frac{\sqrt{\kappa^2 + \tau^2}}{\kappa} = \left(1 + \left(\frac{\tau}{\kappa}\right)^2\right)^{1/2} \geq 1$$

and thus depends only on the ratio of torsion to curvature for the original curve β.

Here is a closely related application in which this ratio τ/κ turns out to be decisive.

4.5 Definition A regular curve α in $\mathbf{R}^3$ is a *cylindrical helix* provided the unit tangent T of α has constant angle ϑ with some fixed unit vector $\mathbf{u}$; that is, $T(t) \cdot \mathbf{u} = \cos\vartheta$ for all t.

This condition is not altered by reparametrization, so for theoretical purposes we need only deal with a cylindrical helix β that has unit speed. So suppose β is a unit-speed curve with $T \cdot \mathbf{u} = \cos\vartheta$. If we pick a reference point, say $\beta(0)$, on β, then the real-valued function

$$h(s) = (\beta(s) - \beta(0)) \cdot \mathbf{u}$$

tells how far $\beta(s)$ has "risen" in the $\mathbf{u}$ direction since leaving $\beta(0)$ (Fig. 2.17). But

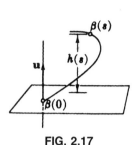

FIG. 2.17

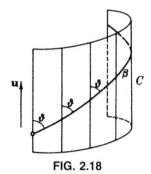

FIG. 2.18

$$\frac{dh}{ds} = \beta' \bullet \mathbf{u} = T \bullet \mathbf{u} = \cos \vartheta,$$

so β is rising at a constant rate *relative to arc length*, and $h(s) = s \cos \vartheta$. If we shift to an arbitrary parametrization, this formula becomes

$$h(t) = s(t) \cos \vartheta,$$

where s is the arc length function.

By drawing a line through each point of β in the $\mathbf{u}$ direction, we construct a cylinder C on which β moves in such a way as to cut each such line at constant angle ϑ, as in Fig. 2.18. In the special case when this cylinder is circular, β is evidently a helix of the type defined in Example 3.3.

It turns out to be quite easy to identify cylindrical helices.

4.6 Theorem A regular curve α with $\kappa > 0$ is a cylindrical helix if and only if the ratio τ/κ is constant.

Proof. It suffices to consider the case where α has unit speed. If α is a cylindrical helix with $T \bullet \mathbf{u} = \cos \vartheta$, then

$$0 = (T \bullet \mathbf{u})' = T' \bullet \mathbf{u} = \kappa N \bullet \mathbf{u}.$$

Since $\kappa > 0$, we conclude that $N \bullet \mathbf{u} = 0$. Thus for each s, $\mathbf{u}$ lies in the plane determined by $T(s)$ and $B(s)$. Orthonormal expansion yields

$$\mathbf{u} = \cos \vartheta \, T + \sin \vartheta \, B.$$

As usual we differentiate and apply Frenet formulas to obtain

$$0 = (\kappa \cos \vartheta - \tau \sin \vartheta)N.$$

Hence $\tau \sin \vartheta = \kappa \cos \vartheta$, so that τ/κ has constant value $\cot \vartheta$.

Conversely, suppose that τ/κ is constant. Choose an angle ϑ such that $\cot \vartheta = \tau/\kappa$. If

$$U = \cos \vartheta \, T + \sin \vartheta \, B,$$

we find

$$U' = (\kappa \cos \vartheta - \tau \sin \vartheta)N = 0.$$

This parallel vector field U then determines (as in Remark 3.4) a unit vector **u** such that $T \cdot \mathbf{u} = \cos \vartheta$, so α is a cylindrical helix.   ◆

In Exercise 9 this information about cylindrical helices is used to show that *circular* helices are characterized by constancy of curvature and torsion (see also Corollary 5.5 of Chapter 3).

Simple hypotheses on a regular curve in $\mathbf{R}^3$ thus have the following effects ($\Leftrightarrow$ means "if and only if"):

$\kappa = 0$	$\Leftrightarrow$	straight line,
$\tau = 0$	$\Leftrightarrow$	plane curve,
κ const > 0 and $\tau = 0$	$\Leftrightarrow$	circle,
κ const > 0 and τ const > 0	$\Leftrightarrow$	circular helix,
τ/κ const $\neq 0$	$\Leftrightarrow$	cylindrical helix.

Exercises

Computer commands that produce the Frenet apparatus, κ, τ, T, N, B, of a curve are given in the Appendix. Their use is optional in the following exercises.

1.   For the curve $\alpha(t) = (2t, t^2, t^3/3)$,
   (a) Compute the Frenet apparatus.
   (b) Sketch the curve for $-4 \leq t \leq 4$, showing T, N, B at $t = 2$.
   (c) Find the limiting values of T, N, and B as $t \to -\infty$ and $t \to \infty$.

2.   Express the curvature and torsion of the curve $\alpha(t) = (\cosh t, \sinh t, t)$ in terms of arc length s measured from $t = 0$.

3. The curve $\alpha(t) = (t\cos t, t\sin t, t)$ lies on a double cone and passes through the vertex at $t = 0$.
  (a) Find the Frenet apparatus of α at $t = 0$.
  (b) Sketch the curve for $-2\pi \le t \le 2\pi$, showing T, N, B at $t = 0$.

4. Show that the curvature of a regular curve in $\mathbf{R}^3$ is given by

$$\kappa^2 v^4 = \| \alpha'' \|^2 - (dv/dt)^2.$$

5. If α is a curve with constant speed $c > 0$, show that

$$T = \alpha'/c, \quad N = \alpha''/\| \alpha'' \|, \quad B = \alpha' \times \alpha''/(c\| \alpha'' \|),$$

$$\kappa = \frac{\| \alpha'' \|}{c^2}, \quad \tau = \frac{\alpha' \times \alpha'' \cdot \alpha'''}{c^2 \| \alpha'' \|^2},$$

where for N, B, τ, we assume α'' never zero, that is, $\kappa > 0$.

6. (a) If α is a cylindrical helix, prove that its unit vector $\mathbf{u}$ (Thm. 4.5) is

$$\mathbf{u} = \frac{\tau}{\sqrt{\kappa^2 + \tau^2}} T + \frac{\kappa}{\sqrt{\kappa^2 + \tau^2}} B,$$

and the coefficients here are $\cos \vartheta$ and $\sin \vartheta$ (for ϑ as in Def. 4.5).
  (b) Check (a) for the cylindrical helix in Example 4.2 of Chapter 1.

7. Let $\alpha: I \to \mathbf{R}^3$ be a cylindrical helix with unit vector $\mathbf{u}$. For $t_0 \in I$, the curve

$$\gamma(t) = \alpha(t) - ((\alpha(t) - \alpha(t_0)) \cdot \mathbf{u}) \mathbf{u}$$

is called a *cross-sectional curve* of the cylinder on which α lies. Prove:
  (a) γ lies in the plane through $\alpha(t_0)$ orthogonal to $\mathbf{u}$.
  (b) The curvature of γ is $\kappa/\sin^2 \vartheta$, where κ is the curvature of α.

8. Verify that the following curves are cylindrical helices and, for each, find the unit vector $\mathbf{u}$, angle ϑ, and cross-sectional curve σ.
  (a) The curve in Exercise 1.   (b) The curve in Example 4.4.
  (c) The curve in Exercise 2.

9. If α is a curve with $\kappa > 0$ and τ both constant, show that α is a circular helix.

10. (a) Prove that a curve is a cylindrical helix if and only if its spherical image is part of a circle.
  (b) Sketch the spherical image of the cylindrical helix in Exercise 1. Is it a complete circle? Find its center.

11. If α is a curve with $\kappa > 0$, its *central curve* $\alpha^* = \alpha + (1/\kappa)N$ consists of all centers of curvature of α (Ex. 6 of Sec. 3). For nonzero numbers a and b, let β_{ab} be the helix in Example 3.3.

(a) Show that the central curve of β_{ab} is the helix $\beta_{\hat{a}b}$, where $\hat{a} = -b^2/a$.

(b) Deduce that the central curve of $\beta_{\hat{a}b}$ is the original helix β_{ab}.

(c) (*Computer graphics.*) Plot three complete turns of the mutually central helices $\beta_{2,1}$ and $\beta_{-1/2,1}$ in the same figure.

12. If $\alpha(t) = (x(t), y(t))$ is a regular curve in $\mathbf{R}^2$, show that its plane curvature (Ex. 8 of Sec. 3) is given by

$$\tilde{\kappa} = \frac{\alpha'' \cdot J(\alpha')}{v^3} = \frac{x'y'' - x''y'}{(x'^2 + y'^2)^{3/2}},$$

where J is the rotation operator $J(a, b) = (-b, a)$.

13. (*Continuation.*) For a plane curve α with $\tilde{\kappa} \neq 0$, the central curve $\alpha^* = \alpha + (1/\tilde{\kappa})N$ is called the *evolute* of α. Thus α^* gives a direct pointwise description of the turning of α.

(a) Show that

$$\alpha^* = \alpha + \frac{\alpha' \cdot \alpha'}{\alpha'' \cdot J(\alpha')} J(\alpha').$$

(b) Find a formula for the line segment λ_t from $\alpha(t)$ to $\alpha^*(t)$. This segment is the radius (line) of the approximating circle to α near $\alpha(t)$ (Ex. 6 of Sec. 3)

(c) Prove that λ_t is normal to α at $\alpha(t)$ and tangent to α^* at $\alpha^*(t)$. (*Hint:* It can be assumed that α has unit speed.)

14. (*Continuation, Computer graphics.*) In each case, plot the given plane curve and its evolute on the same figure, showing some of the construction lines λ_t.

(a) The ellipse $a(t) = (2\cos t, \sin t)$.

(b) The cycloid $\alpha(t) = (t + \sin t, 1 + \cos t)$ for $-2\pi \leq t \leq 2\pi$. (Here the evolute bears an unexpected relation to the original curve.)

15. (*Computer continuation of Ex. 9 of Sec. 3.*)

(a) Write the commands that, given a regular curve α with $\kappa(0) > 0$, plot— on a small interval $-\varepsilon \leq t \leq \varepsilon$—the orthogonal projection of α into the osculating, rectifying, and normal planes at $\alpha(0)$. Show the projections as curves in $\mathbf{R}^2$.

(b) Test (a) on the curves (3), (4), (5) in Example 4.2 of Chapter 1 and those in Example 4.3 of Chapter 3. Compare results.

The following exercise shows that the condition $\kappa > 0$ cannot be avoided in a detailed study of the geometry of curves in $\mathbf{R}^3$ for even if κ is zero at only a single point, the geometric character of the curve can change radically at that point. (This difficulty does not arise for curves in the plane.)

16. It is shown in advanced calculus that the function

$$f(t) = \begin{cases} 0 & \text{if } t \le 0, \\ e^{-1/t^2} & \text{if } t > 0. \end{cases}$$

is infinitely differentiable (has continuous derivatives of all orders). Thus

$$\alpha(t) = (t,\ f(t),\ f(-t))$$

is a well-defined differentiable curve.
   (a) Sketch α on an interval $-a \le t \le a$.
   (b) Show that the curvature of α is zero only at $t = 0$.
   (c) What are the osculating planes of α for $t < 0$ and $t > 0$?

In the following exercise, a global geometric invariant of curves is gotten by integrating a local invariant.

17. The *total curvature* of a unit-speed curve $\alpha: I \to \mathbf{R}^3$ is $\int_I \kappa(s)ds$. If α is merely regular, the formula becomes $\int_I \kappa(t)v(t)dt$. Find the total curvature of the following curves:
   (a) The curve in Example 4.4.
   (b) The helix in Example 3.3.
   (c) The curve in Exercise 2.
   (d) The ellipse $\alpha(t) = (a\cos t, b\sin t)$ on $0 \le t \le 2\pi$.

18. One definition of convexity for a smoothly closed plane curve is that its curvature κ is positive (hence its plane curvature $\tilde{\kappa}$ is either always positive or always negative). Prove that a convex closed plane curve has total curvature 2π. (*Hint:* Consider its spherical image.)
   A theorem of Fenchel asserts that every regular closed curve α in $\mathbf{R}^3$ has total curvature $\ge 2\pi$. Surprisingly, this has an easy proof in terms of surface theory (see Sec. 8 of Ch. 6).

19. (*Computer.*)
   (a) Plot the curve

$$\tau(t) = (4 \cos 2t + 2 \cos t,\ 4 \sin 2t - 2 \sin t,\ \sin 3t) \quad \text{on } 0 \le t \le 2\pi.$$

Even looking at this curve from different viewpoints may not make its crossing pattern clear, but Exercise 21 of Section 5.4 will show that τ is a *trefoil knot*.

(Intuitively, a simple closed curve in $\mathbf{R}^3$ is a *knot* provided it cannot be continuously deformed—always remaining simply closed—until it becomes a circle.) The Fary-Milnor theorem asserts that every knot has total curvature strictly greater than 4π. Show:

(b) The plane curve obtained from τ by removing the z-component $\sin 3t$ has total curvature exactly 4π. (This curve is not simply closed, and hence is not a knot.)

(c) τ can be deformed to a knot that has (numerically estimated) total curvature less than 4.01π.

20. (*Computer.*)

(a) Write a command that, given an arbitrary regular curve, returns the test function in Exercise 10 of Section 3 whose constancy implies that the curve lies on a sphere. (Plotting this function provides a good test for constancy and does not require simplifying it.) (*Hint:* To allow for arbitrary parametrization, replace derivatives $f'(s)$ by $f'(t)v(t)$, where $v(t) = ds/dt$.)

(b) In each case, decide whether the curve lies on a sphere, and if so, find its radius and center:

(i) $\alpha(t) = (2\sin t, \sin 2t, 2\sin^2 t)$;

(ii) $\beta(t) = (\cos^2 t, \sin 2t, 2\sin t)$;

(iii) $\gamma(t) = (\cos t, 1 + \sin t, 2\sin \dfrac{t}{2})$.

21. Prove that the cubic curve $\gamma(t) = (at, bt^2, ct^3)$, $abc \neq 0$, is a cylindrical helix if and only if $3ac = \pm 2b^2$. (Computer optional.)

2.5 Covariant Derivatives

In Chapter 1 the definition of a new object (curve, differential form, mapping, . . .) was usually followed by an appropriate notion of *derivative* of that object. To see how to define the derivative of a vector field on a Euclidean space, we mimic the definition of the derivative $\mathbf{v}[f]$ of a function f relative to a tangent vector $\mathbf{v}$ at a point $\mathbf{p}$ (Definition 3.1 of Chapter 1). In fact, replacing f by a vector field W on $\mathbf{R}^3$ gives a vector field $t \rightarrow W(\mathbf{p} + t\mathbf{v})$ on the curve $t \rightarrow \mathbf{p} + t\mathbf{v}$. The derivative of such a vector field was defined in Section 2. Then the derivative of W with respect to $\mathbf{v}$ will be the derivative of $t \rightarrow W(\mathbf{p} + t\mathbf{v})$ at $t = 0$.

5.1 Definition Let W be a vector field on $\mathbf{R}^3$, and let $\mathbf{v}$ be a tangent vector field to $\mathbf{R}^3$ at the point $\mathbf{p}$. Then the *covariant derivative* of W with respect to $\mathbf{v}$ is the tangent vector

$$\nabla_{\mathbf{v}} W = W(\mathbf{p} + t\mathbf{v})'(0)$$

at the point $\mathbf{p}$.

Evidently $\nabla_v W$ *measures the initial rate of change of* $W(\mathbf{p})$ *as* $\mathbf{p}$ *moves in the* $\mathbf{v}$ *direction.* (The term "covariant" derives from the generalization of this notion discussed in Chapter 7.)

For example, suppose $W = x^2 U_1 + yz U_3$, and

$$\mathbf{v} = (-1, 0, 2) \quad \text{at} \quad \mathbf{p} = (2, 1, 0).$$

Then

$$\mathbf{p} + t\mathbf{v} = (2 - t, 1, 2t),$$

so

$$W(\mathbf{p} + t\mathbf{v}) = (2 - t)^2 U_1 + 2t U_3,$$

where strictly speaking U_1 and U_3 are also evaluated at $\mathbf{p} + t\mathbf{v}$. Thus,

$$\nabla_v W = W(\mathbf{p} + t\mathbf{v})'(0) = -4U_1(\mathbf{p}) + 2U_3(\mathbf{p}).$$

5.2 Lemma   If $W = \sum w_i U_i$ is a vector field on $\mathbf{R}^3$, and $\mathbf{v}$ is a tangent vector at $\mathbf{p}$, then

$$\nabla_v W = \sum \mathbf{v}[w_i] U_i(\mathbf{p}).$$

Proof.   We have

$$W(\mathbf{p} + t\mathbf{v}) = \sum w_i(\mathbf{p} + t\mathbf{v}) U_i(\mathbf{p} + t\mathbf{v})$$

for the restriction of W to the curve $t \to \mathbf{p} + t\mathbf{v}$. To differentiate such a vector field (at $t = 0$), one simply differentiates its Euclidean coordinates (at $t = 0$). But by the definition of directional derivative (Definition 3.1 of Chapter 1), the derivative of $w_i(\mathbf{p} + t\mathbf{v})$ at $t = 0$ is precisely $\mathbf{v}[w_i]$. Thus

$$\nabla_v W = W(\mathbf{p} + t\mathbf{v})'(0) = \sum \mathbf{v}[w_i] U_i(\mathbf{p}). \qquad \blacklozenge$$

In short, *to apply* ∇_v *to a vector field, apply* $\mathbf{v}$ *to its Euclidean coordinates.* Thus the following linearity and Leibnizian properties of covariant derivative follow easily from the corresponding properties (Theorem 3.3 of Chapter 1) of directional derivatives.

5.3 Theorem   Let $\mathbf{v}$ and $\mathbf{w}$ be tangent vectors to $\mathbf{R}^3$ at $\mathbf{p}$, and let Y and Z be vector fields on $\mathbf{R}^3$. Then for numbers a, b and functions f,

(1) $\nabla_{av+bw} Y = a\nabla_v Y + b\nabla_w Y$.

(2) $\nabla_v(aY + bZ) = a\nabla_v Y + b\nabla_v Z$.

(3) $\nabla_v(fY) = \mathbf{v}[f] Y(\mathbf{p}) + f(\mathbf{p})\nabla_v Y$.

(4) $\mathbf{v}[Y \bullet Z] = \nabla_v Y \bullet Z(\mathbf{p}) + Y(\mathbf{p}) \bullet \nabla_v Z$.

Proof.   For example, let us prove (4). If

$$Y = \sum y_i U_i \quad \text{and} \quad Z = \sum z_i U_i,$$

then

$$Y \cdot Z = \sum y_i z_i.$$

Hence by Theorem 3.3 of Chapter 1,

$$\mathbf{v}[Y \cdot Z] = \mathbf{v}\Big[\sum y_i z_i\Big] = \sum \mathbf{v}[y_i]z_i(\mathbf{p}) + \sum y_i(\mathbf{p})\mathbf{v}[z_i].$$

But by the preceding lemma,

$$\nabla_v Y = \sum \mathbf{v}[y_i]U_i(\mathbf{p}) \quad \text{and} \quad \nabla_v Z = \sum \mathbf{v}[z_i]U_i(\mathbf{p}).$$

Thus the two sums displayed above are precisely $\nabla_v Y \cdot Z(\mathbf{p})$ and $Y(\mathbf{p}) \cdot \nabla_v Z$.
◆

Using the pointwise principle (Chapter 1, Section 2), we can take the covariant derivative of a vector field W with respect to a *vector field V*, rather than a single tangent vector $\mathbf{v}$. The result is the *vector field $\nabla_V W$* whose value at each point $\mathbf{p}$ is $\nabla_{V(p)} W$. Thus $\nabla_V W$ consists of all the covariant derivatives of W with respect to the vectors of V. It follows immediately from the lemma above that if $W = \sum w_i U_i$, then

$$\nabla_V W = \sum V[w_i]U_i.$$

Coordinate computations are easy using the basic identity $U_i[f] = \partial f/\partial x_i$. For example, suppose $V = (y - x)U_1 + xy U_3$ and (as in the example above) $W = x^2 U_1 + yz U_3$. Then

$$V[x^2] = (y - x)U_1[x^2] = 2x(y - x),$$
$$V[yz] = xy U_3[yz] = xy^2.$$

Hence

$$\nabla_V W = 2x(y - x)U_1 + xy^2 U_3.$$

For the covariant derivative $\nabla_V W$ as expressed entirely in terms of vector fields, the properties in the preceding theorem take the following form.

5.4 Corollary   Let V, W, Y, and Z be vector fields on $\mathbf{R}^3$. Then
(1) $\nabla_{fV+gW} Y = f\nabla_V Y + g\nabla_W Y$, for all functions f and g.
(2) $\nabla_V(aY + bZ) = a\nabla_V Y + b\nabla_V Z$, for all numbers a and b.

(3) $\nabla_V(fY) = V[f]Y + f\nabla_V Y$, for all functions f.
(4) $V[Y \bullet Z] = \nabla_V Y \bullet Z + Y \bullet \nabla_V Z$.

We shall omit the proof, which is an exercise in the use of parentheses based on the (pointwise principle) definition $(\nabla_V Y)(\mathbf{p}) = \nabla_{V(p)} Y$.

Note that $\nabla_V Y$ does not behave symmetrically with respect to V and Y. This is to be expected, since it is Y that is being differentiated, while the role of V is merely algebraic. In particular, $\nabla_{fV} Y$ is $f\nabla_V Y$, but $\nabla_V(fY)$ is not $f\nabla_V Y$. There is an extra term arising from the differentiation of f by V.

Exercises

1. Consider the tangent vector $\mathbf{v} = (1, -1, 2)$ at the point $\mathbf{p} = (1, 3, -1)$. Compute $\nabla_v W$ directly from the definition, where
  (a) $W = x^2 U_1 + y U_2$.   (b) $W = x U_1 + x^2 U_2 - z^2 U_3$.

2. Let $V = -y U_1 + x U_3$ and $W = \cos x U_1 + \sin x U_2$. Express the following covariant derivatives in terms of U_1, U_2, U_3:
  (a) $\nabla_V W$.   (b) $\nabla_V V$.
  (c) $\nabla_V(z^2 W)$.   (d) $\nabla_W(V)$.
  (e) $\nabla_V(\nabla_v W)$.   (f) $\nabla_V(xV - zW)$.

3. If W is a vector field with constant length $\|W\|$, prove that for any vector field V, the covariant derivative $\nabla_V W$ is everywhere orthogonal to W.

4. Let X be the special vector field $\sum x_i U_i$, where x_1, x_2, x_3 are the natural coordinate functions of $\mathbf{R}^3$. Prove that $\nabla_V X = V$ for every vector field V.

5. Let W be a vector field defined on a region containing a regular curve α. Then $t \rightarrow W(\alpha(t))$ is a vector field on α called the *restriction* of W to α and denoted by W_α.
  (a) Prove that $\nabla_{\alpha'(t)} W = (W_\alpha)'(t)$.
  (b) Deduce that the straight line in Definition 5.1 may be replaced by *any* curve with initial velocity $\mathbf{v}$. Thus the derivative Y' of a vector field Y on a curve α is (almost) $\nabla_{\alpha'} Y$.

2.6 Frame Fields

When the Frenet formulas were discovered (by Frenet in 1847, and independently by Serret in 1851), the theory of *surfaces* in $\mathbf{R}^3$ was already a richly developed branch of geometry. The success of the Frenet approach to curves led Darboux (around 1880) to adapt this "method of moving frames" to the study of surfaces. Then, as we mentioned earlier, it was Cartan who brought the method to full generality. His essential idea was very simple: To each point of the object under study (a curve, a surface, Euclidean space itself, ...) assign a frame; then using orthonormal expansion express the rate of change of the frame in terms of the frame itself. This, of course, is just what the Frenet formulas do in the case of a curve.

In the next three sections we shall carry out this scheme for the Euclidean space $\mathbf{R}^3$. We shall see that geometry of curves and surfaces in $\mathbf{R}^3$ is not merely an analogue, but actually a *corollary*, of these basic results. Since the main application (to surface theory) comes only in Chapter 6, these sections may be postponed, and read later as a preliminary to that chapter.

By means of the pointwise principle (Chapter 1, Section 2) we can automatically extend operations on individual tangent vectors to operations on vector fields. For example, if V and W are vector fields on $\mathbf{R}^3$, then the *dot product* $V \cdot W$ of V and W is the (differentiable) real-valued function on $\mathbf{R}$ whose value at each point $\mathbf{p}$ is $V(\mathbf{p}) \cdot W(\mathbf{p})$. The *norm* $\|V\|$ of V is the real-valued function on $\mathbf{R}^3$ whose value at $\mathbf{p}$ is $\|V(\mathbf{p})\|$. Thus $\|V\| = (V \cdot V)^{1/2}$. By contrast with $V \cdot W$, the norm function $\|V\|$ need not be differentiable at points for which $V(\mathbf{p}) = 0$, since the square-root function is badly behaved at 0.

In Chapter 1 we called the three vector fields U_1, U_2, U_3 the natural frame field on $\mathbf{R}^3$. Here is a simple but crucial generalization.

6.1 Definition   Vector fields E_1, E_2, E_3 on $\mathbf{R}^3$ constitute a *frame field* on $\mathbf{R}^3$ provided

$$E_i \cdot E_j = \delta_{ij} \quad (1 \le i, j \le 3),$$

where δ_{ij} is the Kronecker delta.

Thus at each point $\mathbf{p}$ the vectors $E_1(\mathbf{p})$, $E_2(\mathbf{p})$, $E_3(\mathbf{p})$ do in fact form a frame (Definition 1.4) since they have unit length and are mutually orthogonal.

In elementary calculus, frame fields are usually derived from coordinate systems, as in the following cases.

6.2 Example   (1) *The cylindrical frame field* (Fig. 2.19). Let r, ϑ, z be the usual cylindrical coordinate functions on $\mathbf{R}^3$. We shall pick a unit vector field in the direction in which each coordinate increases (when the other two are held constant). For r, this is evidently

$$E_1 = \cos \vartheta \, U_1 + \sin \vartheta \, U_2,$$

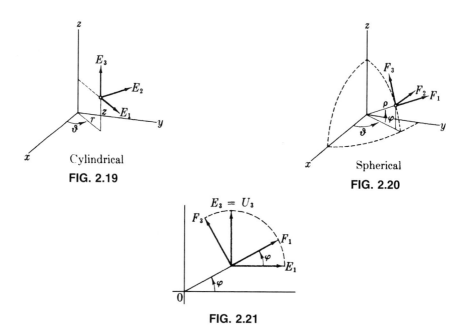

Cylindrical

FIG. 2.19

Spherical

FIG. 2.20

FIG. 2.21

pointing straight out from the z axis. Then

$$E_2 = -\sin \vartheta \, U_1 + \cos \vartheta \, U_2$$

points in the direction of increasing ϑ as in Fig. 2.19. Finally, the direction of increase of z is, of course, straight up, so

$$E_3 = U_3.$$

It is easy to check that $E_i \bullet E_j = \delta_{ij}$, so this is a frame field (defined on all of $\mathbf{R}^3$ except the z axis). We call it the *cylindrical frame field* on $\mathbf{R}^3$.

(2) *The spherical frame field on* $\mathbf{R}^3$ (Fig. 2.20). In a similar way, a frame field F_1, F_2, F_3 can be derived from the spherical coordinate functions ρ, ϑ, φ on $\mathbf{R}^3$. As indicated in the figure, we shall measure φ up from the xy plane rather than (as is usually done) down from the z axis.

Let E_1, E_2, E_3 be the cylindrical frame field. For spherical coordinates, the unit vector field F_2 in the direction of increasing ϑ is the same as above, so $F_2 = E_2$. The unit vector field F_1, in the direction of increasing ρ, points straight out from the origin; hence it can be expressed as

$$F_1 = \cos \varphi \, E_1 + \sin \varphi \, E_3$$

(Fig. 2.21). Similarly, the vector field for increasing φ is

$$F_3 = -\sin \varphi \ E_1 + \cos \varphi \ E_3.$$

Thus the formulas for E_1, E_2, E_3 in (1) yield

$$F_1 = \cos \varphi (\cos \vartheta \ U_1 + \sin \vartheta \ U_2) + \sin \varphi \ U_3,$$

$$F_2 = -\sin \vartheta \ U_1 + \cos \vartheta \ U_2,$$

$$F_3 = -\sin \varphi (\cos \vartheta \ U_1 + \sin \vartheta \ U_2) + \cos \varphi \ U_3.$$

By repeated use of the identity $\sin^2 + \cos^2 = 1$, we check that F_1, F_2, F_3 is a frame field—the *spherical frame field* on $\mathbf{R}^3$. (Its actual domain of definition is $\mathbf{R}^3$ minus the z axis, as in the cylindrical case.)

The following useful result is an immediate consequence of orthonormal expansion.

6.3 Lemma   Let E_1, E_2, E_3 be a frame field on $\mathbf{R}^3$.

(1) If V is a vector field on $\mathbf{R}^3$, then $V = \sum f_i E_i$, where the functions $f_i = V \bullet E_i$ are called the *coordinate functions* of V with respect to E_1, E_2, E_3.

(2) If $V = \sum f_i E_i$ and $W = \sum g_i E_i$, then $V \bullet W = \sum f_i g_i$. In particular, $\| V \| = (\sum f_i^2)^{1/2}$.

Thus a given vector field V has a different set of coordinate functions with respect to each choice of a frame field E_1, E_2, E_3. The *Euclidean* coordinate functions (Lemma 2.5 of Chapter 1), of course, come from the natural frame field U_1, U_2, U_3. In Chapter 1, we used this natural frame field exclusively, but now we shall gradually shift to arbitrary frame fields. The reason is clear: In studying curves and surfaces in $\mathbf{R}^3$, we shall then be able to choose a frame field *specifically adapted to the problem at hand.* Not only does this simplify computations, but it gives a clearer understanding of geometry than if we had insisted on using the same frame field in every situation.

Exercises

1.   If V and W are vector fields on $\mathbf{R}^3$ that are linearly independent at each point, show that

$$E_1 = \frac{V}{\| V \|}, \quad E_2 = \frac{\tilde{W}}{\| \tilde{W} \|}, \quad E_3 = E_1 \times E_2$$

is a frame field, where $\tilde{W} = W - (W \bullet E_1)E_1$.

2. Express each of the following vector fields (i) in terms of the cylindrical frame field (with coefficients in terms of r, ϑ, z) and (ii) in terms of the spherical frame field (with coefficients in terms of ρ, ϑ, φ):

(a) U_1.
(b) $\cos\vartheta\, U_1 + \sin\vartheta\, U_2 + U_3$.
(c) $xU_1 + yU_2 + zU_3$.

3. Find a frame field E_1, E_2, E_3 such that

$$E_1 = \cos x\, U_1 + \sin x \cos z\, U_2 + \sin x \sin z\, U_3.$$

2.7 Connection Forms

Once more we state the essential point: The power of the Frenet formulas stems not from the fact that they tell what the derivatives T', N', B' are, but from the fact that *they express these derivatives in terms of T, N, B*—and thereby define curvature and torsion. We shall now do the same thing with an arbitrary frame field E_1, E_2, E_3 on $\mathbf{R}^3$; namely, *express the covariant derivatives of these vector fields in terms of the vector fields themselves.* We begin with the covariant derivative with respect to an arbitrary tangent vector $\mathbf{v}$ at a point $\mathbf{p}$. Then

$$\nabla_v E_1 = c_{11}E_1(\mathbf{p}) + c_{12}E_2(\mathbf{p}) + c_{13}E_3(\mathbf{p}),$$

$$\nabla_v E_2 = c_{21}E_1(\mathbf{p}) + c_{22}E_2(\mathbf{p}) + c_{23}E_3(\mathbf{p}),$$

$$\nabla_v E_3 = c_{31}E_1(\mathbf{p}) + c_{32}E_2(\mathbf{p}) + c_{33}E_3(\mathbf{p}),$$

where by orthonormal expansion the coefficients of these equations are

$$c_{ij} = \nabla_v E_i \bullet E_j(\mathbf{p}) \quad \text{for } 1 \leqq i, j \leqq 3.$$

These coefficients c_{ij}, depend on the particular tangent vector $\mathbf{v}$, so a better notation for them is

$$\omega_{ij}(\mathbf{v}) = \nabla_v E_i \bullet E_j(\mathbf{p}), \quad (1 \leqq i, j \leqq 3).$$

Thus for each choice of i and j, ω_{ij} is a real-valued function defined on all tangent vectors. But we have met that kind of function before.

7.1 Lemma   Let E_1, E_2, E_3 be a frame field on $\mathbf{R}^3$. For each tangent vector $\mathbf{v}$ to $\mathbf{R}^3$ at the point $\mathbf{p}$, let

$$\omega_{ij}(\mathbf{v}) = \nabla_v E_i \bullet E_j(\mathbf{p}), \quad (1 \leqq i, j \leqq 3).$$

Then each ω_{ij} is a 1-form, and $\omega_{ij} = -\omega_{ji}$. These 1-forms are called the *connection forms* of the frame field E_1, E_2, E_3.

Proof. By definition, ω_{ij} is a real-valued function on tangent vectors, so to verify that ω_{ij} is a 1-form (Def. 5.1 of Ch. 1), it suffices to check the linearity condition. Using Theorem 5.3, we get

$$\omega_{ij}(a\mathbf{v} + b\mathbf{w}) = \nabla_{a\mathbf{v}+b\mathbf{w}}E_i \bullet E_j(\mathbf{p})$$
$$= (a\nabla_\mathbf{v}E_i + b\nabla_\mathbf{w}E_i) \bullet E_j(\mathbf{p})$$
$$= a\nabla_\mathbf{v}E_i \bullet E_j(\mathbf{p}) + b\nabla_\mathbf{w}E_i \bullet E_j(\mathbf{p})$$
$$= a\omega_{ij}(\mathbf{v}) + b\omega_{ij}(\mathbf{w}).$$

To prove that $\omega_{ij} = -\omega_{ji}$ we must show that $\omega_{ij}(\mathbf{v}) = -\omega_{ji}(\mathbf{v})$ for every tangent vector $\mathbf{v}$. By definition of frame field, $E_i \bullet E_j = \delta_{ij}$, and since each Kronecker delta has constant value 0 or 1, the Leibnizian formula (4) of Theorem 5.3 yields

$$0 = \mathbf{v}[E_i \bullet E_j] = \nabla_\mathbf{v}E_i \bullet E_j(\mathbf{p}) + E_i(\mathbf{p}) \bullet \nabla_\mathbf{v}E_j.$$

By the symmetry of the dot product, the two vectors in this last term may be reversed, so we have found that $0 = \omega_{ij}(\mathbf{v}) + \omega_{ji}(\mathbf{v})$. ◆

The geometric significance of the connection forms is no mystery. The definition $\omega_{ij}(\mathbf{v}) = \nabla_\mathbf{v}E_i \bullet E_j(\mathbf{p})$ shows that $\omega_{ij}(\mathbf{v})$ *is the initial rate at which E_i rotates toward E_j as $\mathbf{p}$ moves in the $\mathbf{v}$ direction.* Thus the 1-forms ω_{ij} contain this information for *all* tangent vectors to $\mathbf{R}^3$.

The following basic result is little more than a rephrasing of the definition of connection forms.

7.2 Theorem   Let ω_{ij} $(1 \le i, j \le 3)$ be the connection forms of a frame field E_1, E_2, E_3 on $\mathbf{R}^3$. Then for any vector field V on $\mathbf{R}^3$,

$$\nabla_V E_i = \sum_j \omega_{ij}(V)E_j, \quad (1 \le i \le 3).$$

We call these the *connection equations* of the frame field E_1, E_2, E_3.

Proof. For fixed i, both sides of this equation are vector fields. Thus we must show that at each point $\mathbf{p}$,

$$\nabla_{V(\mathbf{p})}E_i = \sum_j \omega_{ij}(V(\mathbf{p}))E_j(\mathbf{p}).$$

But as we have already seen, the very definition of connection form makes this equation a consequence of orthonormal expansion. ◆

When $i = j$, the skew-symmetry condition $\omega_{ij} = -\omega_{ji}$ becomes $\omega_{ii} = -\omega_{ii}$; thus

$$\omega_{11} = \omega_{22} = \omega_{33} = 0.$$

Hence this condition has the effect of reducing the nine 1-forms ω_{ij} for $1 \leq i, j \leq 3$ to essentially only three, say ω_{12}, ω_{13}, ω_{23}. It is perhaps best to regard the connection forms ω_{ij} as the entries of a skew-symmetric matrix of 1-forms,

$$\omega = \begin{pmatrix} \omega_{11} & \omega_{12} & \omega_{13} \\ \omega_{21} & \omega_{22} & \omega_{23} \\ \omega_{31} & \omega_{32} & \omega_{33} \end{pmatrix} = \begin{pmatrix} 0 & \omega_{12} & \omega_{13} \\ -\omega_{12} & 0 & \omega_{23} \\ -\omega_{13} & -\omega_{23} & 0 \end{pmatrix}.$$

Thus in expanded form, the connection equations (Theorem 7.2) become

$$\begin{aligned} \nabla_V E_1 &= & \omega_{12}(V)E_2 + \omega_{13}(V)E_3, \\ \nabla_V E_2 &= -\omega_{12}(V)E_1 & + \omega_{23}(V)E_3, \\ \nabla_V E_3 &= -\omega_{13}(V)E_1 - \omega_{23}(V)E_2. \end{aligned} \qquad (*)$$

showing an obvious relation to the Frenet formulas

$$\begin{aligned} T' &= & \kappa N, \\ N' &= -\kappa N & + \tau B, \\ B' &= & -\tau N. \end{aligned}$$

The absence from the Frenet formulas of terms corresponding to $\omega_{13}(V)E_3$ and $-\omega_{13}(V)E_1$ is a consequence of the special way the Frenet frame field is fitted to its curve. Having gotten $T(\sim E_1)$, we chose $N(\sim E_2)$ so that the derivative T' would be a scalar multiple of N alone and not involve $B(\sim E_3)$.

Another difference between the Frenet formulas and the equations above stems from the fact that $\mathbf{R}^3$ has three dimensions, while a curve has but one. The coefficients—curvature κ and torsion τ—in the Frenet formulas measure the rate of change of the frame field T, N, B only along its curve, that is, in the direction of T alone. But the coefficients in the connection equations must be able to make this measurement for E_1, E_2, E_3 with respect to *arbitrary* vector fields in $\mathbf{R}^3$. This is why the connection forms are 1-forms and not just functions.

These formal differences aside, a more fundamental distinction stands out. It is because a Frenet frame field is specially fitted to its curve that the Frenet formulas give information about that curve. Since the frame field E_1, E_2, E_3 used above is completely arbitrary, the connection equations give no direct information about $\mathbf{R}^3$, but only information about the "rate of rotation" of that particular frame field. This is not a weakness, but a strength, since as indicated earlier, if we can fit a frame field to a geometric problem arising in $\mathbf{R}^3$, then the connection equations will give direct information about that problem. Thus, these equations play a fundamental role in all the differential geometry of $\mathbf{R}^3$. For example, the Frenet formulas can be deduced from them (Exercise 8).

Given an arbitrary frame field E_1, E_2, E_3 on $\mathbf{R}^3$, it is fairly easy to find an explicit formula for its connection forms. First use orthonormal expansion to express the vector fields E_1, E_2, E_3 in terms of the natural frame field U_1, U_2, U_3 on $\mathbf{R}^3$:

$$E_1 = a_{11}U_1 + a_{12}U_2 + a_{13}U_3,$$

$$E_2 = a_{21}U_1 + a_{22}U_2 + a_{23}U_3,$$

$$E_3 = a_{31}U_1 + a_{32}U_2 + a_{33}U_3.$$

Here each $a_{ij} = E_i \bullet U_j$ is a real-valued function on $\mathbf{R}^3$. The matrix

$$A = \left(a_{ij}\right) = \begin{pmatrix} a_{11} & a_{12} & a_{13} \\ a_{21} & a_{22} & a_{23} \\ a_{31} & a_{32} & a_{33} \end{pmatrix}$$

with these functions as entries is called the *attitude matrix* of the frame field E_1, E_2, E_3. In fact, at each point $\mathbf{p}$, the numerical matrix

$$A(\mathbf{p}) = \left(a_{ij}(\mathbf{p})\right)$$

is exactly the attitude matrix of the frame $E_1(\mathbf{p})$, $E_2(\mathbf{p})$, $E_3(\mathbf{p})$ as in Definition 1.6. Since attitude matrices are orthogonal, the transpose ${}^tA$ of A is equal to its inverse A^{-1}.

Define the differential of $A = (a_{ij})$ to be $dA = (da_{ij})$, so dA is a matrix whose entries are 1-forms. We can now give a simple expression for the connection forms in terms of the attitude matrix.

7.3 Theorem If $A = (a_{ij})$ is the attitude matrix and $\omega = (\omega_{ij})$ the matrix of connection forms of a frame field E_1, E_2, E_3, then

$$\omega = dA \,{}^tA \quad \text{(matrix multiplication)},$$

or equivalently,

$$\omega_{ij} = \sum_k a_{jk}\, da_{ik} \quad \text{for} \quad 1 \le i,\, j \le 3.$$

Since the proof is routine, it may be more informative to illustrate the result by an example. For the cylindrical frame field in Example 6.2, we found the attitude matrix

$$A = \begin{pmatrix} \cos\vartheta & \sin\vartheta & 0 \\ -\sin\vartheta & \cos\vartheta & 0 \\ 0 & 0 & 1 \end{pmatrix}.$$

Thus

$$\omega = dA\ {}^tA = \begin{pmatrix} -\sin\vartheta\, d\vartheta & \cos\vartheta\, d\vartheta & 0 \\ -\cos\vartheta\, d\vartheta & -\sin\vartheta\, d\vartheta & 0 \\ 0 & 0 & 0 \end{pmatrix} \begin{pmatrix} \cos\vartheta & -\sin\vartheta & 0 \\ \sin\vartheta & \cos\vartheta & 0 \\ 0 & 0 & 1 \end{pmatrix}$$

$$= \begin{pmatrix} 0 & d\vartheta & 0 \\ -d\vartheta & 0 & 0 \\ 0 & 0 & 0 \end{pmatrix}.$$

Since $\omega_{12} = d\vartheta$ is the only nonzero connection form (except, of course, $\omega_{21} = -\omega_{12}$), the connection equations (∗) reduce to

$$\nabla_V E_1 = d\vartheta(V)E_2 = V[\vartheta]E_2,$$
$$\nabla_V E_2 = -d\vartheta(V)E_1 = -V[\vartheta]E_1,$$
$$\nabla_V E_3 = 0.$$

These equations have immediate geometrical significance. Because V is arbitrary, the third equation says that the vector field E_3 is parallel. We knew this already since in the cylindrical frame field, E_3 is just U_3.

The first two equations tell us that the covariant derivatives of E_1 and E_2 with respect to a vector field V depend only on the rate of change of the angle ϑ in the V direction.

For example, the definition of ϑ shows that $V[\vartheta] = 0$ whenever V is a vector field that at each point is tangent to a plane through the z axis. Thus for a vector field of this type the connection equations above predict that $\nabla_V E_1 = \nabla_V E_2 = 0$. In fact, it is clear from Fig. 2.19 that E_1 and E_2 do remain parallel on any plane through the z axis.

Exercises

1. For any function f, show that the vector fields

$$E_1 = (\sin f \; U_1 + U_2 - \cos f \; U_3)/\sqrt{2},$$
$$E_2 = (\sin f \; U_1 - U_2 - \cos f \; U_3)/\sqrt{2},$$
$$E_3 = \cos f \; U_1 + \sin f \; U_3$$

form a frame field, and find its connection forms.

2. Find the connection forms of the natural frame field U_1, U_2, U_3.

3. For any function f, show that

$$A = \begin{pmatrix} \cos^2 f & \cos f \sin f & \sin f \\ \sin f \cos f & \sin^2 f & -\cos f \\ -\sin f & \cos f & 0 \end{pmatrix}$$

is the attitude matrix of a frame field, and compute its connection forms.

4. Prove that the connection forms of the spherical frame field are

$$\omega_{12} = \cos \varphi \; d\vartheta, \quad \omega_{13} = d\varphi, \quad \omega_{23} = \sin \varphi \; d\vartheta.$$

5. If E_1, E_2, E_3 is a frame field and $W = \sum f_i E_i$, prove the *covariant derivative formula*:

$$\nabla_V W = \sum_j \left\{ V[f_j] + \sum_i f_i \omega_{ij}(V) \right\} E_j.$$

6. Let E_1, E_2, E_3 be the cylindrical frame field. If V is a vector field such that $V[r] = r$ and $V[\vartheta] = 1$, compute $\nabla_V (r \cos \vartheta E_1 + r \sin \vartheta E_3)$.

7. (*Computer.*) (a) Write a computer command that, given the attitude matrix A of a frame field on $\mathbf{R}^3$, returns the matrix $\omega = dA \; {}^t A$ of its connection forms. (*Hint:* For *Maple*, use the differential operator d from the package *difforms*. For *Mathematica*, use the total differential Dt.) (b) Test part (a) on the cylindrical frame field and on the spherical frame field (Ex. 4).

8. Let β be a unit-speed curve in $\mathbf{R}^3$ with $\kappa > 0$, and suppose that E_1, E_2, E_3 is a frame field on $\mathbf{R}^3$ such that the restriction of these vector fields to β gives the Frenet-frame field T, N, B of β. Prove that

$$\omega_{12}(T) = \kappa, \quad \omega_{13}(T) = 0, \quad \omega_{23}(T) = \tau.$$

Then deduce the Frenet formulas from the connection equations. (*Hint:* Ex. 5 of Sec. 5.)

2.8 The Structural Equations

We have seen that 1-forms—the connection forms—give the simplest description of the rate of rotation of a frame field. Furthermore, the frame field itself can be described in terms of 1-forms.

8.1 Definition   If E_1, E_2, E_3 is a frame field on $\mathbf{R}^3$, then the *dual* 1-*forms* θ_1, θ_2, θ_3 of the frame field are the 1-forms such that

$$\theta_i(\mathbf{v}) = \mathbf{v} \cdot E_i(\mathbf{p})$$

for each tangent vector $\mathbf{v}$ to $\mathbf{R}^3$ at $\mathbf{p}$.

Note that θ_i is linear on the tangent vectors at each point; hence it *is* a 1-form. In particular, $\theta_i(E_j) = \delta_{ij}$, so readers familiar with the notion of dual vector spaces will recognize that at each point, θ_1, θ_2, θ_3 gives the dual basis of E_1, E_2, E_3.

In the case of the natural frame field U_1, U_2, U_3, the dual forms are just dx_1, dx_2, dx_3. In fact, from Example 5.3 of Chapter 1 we get

$$dx_i(\mathbf{v}) = v_i = \mathbf{v} \cdot U_i(\mathbf{p})$$

for each tangent vector $\mathbf{v}$; hence $dx_i = \theta_i$.

Using dual forms, the orthonormal expansion formula in Lemma 6.3 may be written $V = \sum \theta_i(V)E_i$. In the characteristic fashion of duality, this formula becomes the following lemma.

8.2 Lemma   Let θ_1, θ_2, θ_3 be the dual 1-forms of a frame field E_1, E_2, E_3. Then any 1-form ϕ on $\mathbf{R}^3$ has a unique expression

$$\phi = \sum \phi(E_i)\theta_i.$$

Proof.   Two 1-forms are the same if they have the same value on any vector field V. But

$$\left(\sum \phi(E_i)\theta_i\right)(V) = \sum \phi(E_i)\theta_i(V)$$
$$= \phi\left(\sum \theta_i(V)E_i\right) = \phi(V). \qquad \blacklozenge$$

Thus ϕ is expressed in terms of dual forms of E_1, E_2, E_3 by evaluating it on E_1, E_2, E_3. This useful fact is the generalization to arbitrary frame fields of Lemma 5.4 of Chapter 1.

We compared a frame field E_1, E_2, E_3 to the natural frame field by means of its attitude matrix $A = (a_{ij})$, for which

$$E_i = \sum a_{ij}U_j \quad (1 \le i \le 3).$$

The dual formulation is just

$$\theta_i = \sum a_{ij}dx_j$$

with *the same coefficients.* In fact, by the preceding lemma,

$$\theta_i = \sum \theta_i(U_j)dx_j.$$

But

$$\theta_i(U_j) = E_i \cdot U_j = \left(\sum a_{ik}U_k\right) \cdot U_j = \sum a_{ik}\delta_{kj} = a_{ij}.$$

These formulas for E_i and θ_i show plainly that θ_1, θ_2, θ_3 is merely the dual description of the frame field E_1, E_2, E_3.

In calculus, when a new function appears on the scene, it is natural to ask what its derivative is. Similarly with 1-forms—having associated with each frame field its dual forms and connection forms, it is reasonable to ask what their exterior derivatives are. The answer is given by two neat sets of equations discovered by Cartan.

8.3 Theorem   (Cartan structural equations.) Let E_1, E_2, E_3 be a frame field on $\mathbf{R}^3$ with dual forms θ_1, θ_2, θ_3 and connection forms ω_{ij} $(1 \le i, j \le 3)$. The exterior derivatives of these forms satisfy

(1) the *first structural equations:*

$$d\theta_i = \sum_j \omega_{ij} \wedge \theta_j \quad (1 \le i \le 3);$$

(2) the *second structural equations:*

$$d\omega_{ij} = \sum_k \omega_{ik} \wedge \omega_{kj} \quad (1 \le i, j \le 3).$$

Because θ_i is the dual of E_i, the first structural equations may be easily recognized as the dual of the connection equations. Only later experience will show that the second structural equations mean that $\mathbf{R}^3$ is flat—roughly speaking, in the same sense that the plane $\mathbf{R}^2$ is flat.

The most efficient proof of the structural equations requires some preliminary remarks. In the Cartan approach, the fundamental objects are not individual forms, but rather *matrices whose entries are forms*. We have already seen that the simplest description of the connection forms ω_{ij} of a frame field is as a single skew-symmetric matrix ω with entries ω_{ij}. Then, for example, ω is expressed in terms of the attitude matrix A of the frame field by the matrix equation $\omega = dA\,{}^{t}A$. (Here, as always, to apply d to a matrix, apply it to each entry of the matrix.)

Similarly, the dual forms of a frame field can be described by a single $n \times 1$ matrix θ with entries θ_i. If ξ is the $n \times 1$ matrix whose entries are the natural coordinates x_i of $\mathbf{R}^3$, then

$$\theta = \begin{pmatrix} \theta_1 \\ \theta_2 \\ \theta_3 \end{pmatrix} \quad \text{and} \quad d\xi = \begin{pmatrix} dx_1 \\ dx_2 \\ dx_3 \end{pmatrix},$$

so the formula $\theta_i = \sum a_{ij}\, dx_j$ above can be written as

$$\theta = A\, d\xi.$$

For such matrices of forms, matrix multiplication is defined as usual, but of course when *entries* are multiplied it is by the wedge product.

The proof of Theorem 8.3 is now quite simple. Recall that since the attitude matrix A is orthogonal, ${}^{t}A A$ is the identity matrix I, which can be inserted in any matrix formula without effect.

Proof of the First Structural Equation. Since $d^2 = 0$, we evidently have $d(d\xi) = 0$, so

$$d\theta = d(A\, d\xi) = dA \cdot d\xi = dA\, {}^{t}A \cdot A\, d\xi = \omega\theta.$$

Expressed in terms of entries, this is indeed the version in (1) of Theorem 8.3.

Proof of the Second Structural Equation. For functions f and g,

$$d(df\ g) = d(g\, df) = dg \wedge df = -df \wedge dg.$$

Thus, using the transpose rule ${}^{t}(AB) = {}^{t}B\,{}^{t}A$, we get

$$d\omega = d(dA\, {}^{t}A) = -dA \cdot d({}^{t}A) = -dA\, {}^{t}A \cdot A\, {}^{t}(dA) = -\omega\, {}^{t}\omega = \omega\omega,$$

where the last step uses the skew-symmetry of ω. Again, in terms of entries, this is the version in (2) of Theorem 8.3.   ◆

8.4 Example Structural equations for the spherical frame field (Example 6.2). The dual forms and connection forms are

$$\theta_1 = d\rho, \qquad\qquad \omega_{12} = \cos \varphi \, d\vartheta,$$
$$\theta_2 = \rho \cos \varphi \, d\vartheta, \quad \omega_{13} = d\varphi,$$
$$\theta_3 = \rho \, d\varphi, \qquad\qquad \omega_{23} = \sin \varphi \, d\vartheta.$$

Let us check, say, the first structural equation

$$d\theta_3 = \sum \omega_{3j} \wedge \theta_j = \omega_{31} \wedge \theta_1 + \omega_{32} \wedge \theta_2.$$

Using the skew-symmetry $\omega_{ij} = -\omega_{ji}$ and the general properties of forms developed in Chapter 1, we get

$$\omega_{31} \wedge \theta_1 = -d\varphi \wedge d\rho = d\rho \wedge d\varphi,$$
$$\omega_{32} \wedge \theta_2 = (-\sin \varphi \, d\vartheta) \wedge (\rho \cos \varphi \, d\vartheta) = 0$$

(the latter since $d\vartheta \wedge d\vartheta = 0$). The sum of these terms is, correctly,

$$d\theta_3 = d(\rho \, d\varphi) = d\rho \wedge d\varphi.$$

Second structural equations involve only one wedge product. For example, since $\omega_{11} = \omega_{22} = 0$,

$$d\omega_{12} = \sum \omega_{1k} \wedge \omega_{k2} = \omega_{13} \wedge \omega_{32}.$$

In this case,

$$\omega_{13} \wedge \omega_{32} = d\varphi \wedge (-\sin \varphi \, d\vartheta) = -\sin \varphi \, d\varphi \wedge d\vartheta.$$

which is the same as

$$d\omega_{12} = d(\cos \varphi \, d\vartheta) = d(\cos \varphi) \wedge d\vartheta = -\sin \varphi \, d\varphi \wedge d\vartheta.$$

To derive the expressions given above for the dual 1-forms, first compute dx_1, dx_2, dx_3 by differentiating the well-known equations

$$x_1 = \rho \cos \varphi \cos \vartheta,$$
$$x_2 = \rho \cos \varphi \sin \vartheta,$$
$$x_3 = \rho \sin \vartheta.$$

Then substitute in the formula $\theta_i = \sum a_{ij} \, dx_j$, where $A = (a_{ij})$ is the attitude matrix from Example 6.2. This result, somewhat disguised, is derived in elementary calculus by a familiar plausibility argument: If at each point the spherical coordinates ρ, ϑ, φ are altered by increments $d\rho$, $d\vartheta$, $d\varphi$, then the sides of the resulting infinitesimal box (Fig. 2.22) are $d\rho$, $\rho \cos \varphi \, d\vartheta$, $\rho \, d\varphi$. These are exactly the formulas for θ_1, θ_2, θ_3.

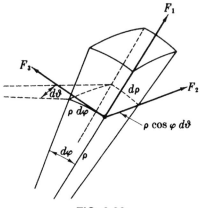

FIG. 2.22

The structural equations provide a powerful method for dealing with geo-metrical problems in $\mathbf{R}^3$: Select a frame field well adapted to the problem at hand; find its dual 1-forms and connection forms; apply the structural equations; interpret the results. We will use this method later to study the geometry of surfaces in $\mathbf{R}^3$.

Exercises

1.   For a 1-form $\phi = \sum f_i \theta_i$, prove

$$d\phi = \sum_j \left\{ df_j + \sum_i f_i \omega_{ij} \right\} \wedge \theta_j.$$

(Compare Ex. 5 of Sec. 7.)

2.   Check all the structural equations of the spherical frame field.

3.   For the cylindrical frame field E_1, E_2, E_3.
(a) Starting from the basic cylindrical equations $x = r \cos \vartheta$, $y = r \sin \vartheta$, $z = z$, show that the dual 1-forms are

$$\theta_1 = dr, \quad \theta_2 = r \, d\vartheta, \quad \theta_3 = dz.$$

(b) Deduce that $E_1[r] = 1$, $E_2[\vartheta] = 1/r$, $E_3[z] = 1$ and that the other six possibilities $E_1[\vartheta]$, ... are all zero.
(c) For a function $f(r, \vartheta, z)$, show that

$$E_1[f] = \frac{\partial f}{\partial r}, \quad E_2[f] = \frac{1}{r}\frac{\partial f}{\partial \vartheta}, \quad E_3[f] = \frac{\partial f}{\partial z}.$$

4. Frame fields on $\mathbf{R}^2$. Given a frame field E_1, E_2 on $\mathbf{R}^2$ there is an angle function ψ such that

$$E_1 = \cos \psi \, U_1 + \sin \psi \, U_2,$$

$$E_2 = -\sin \psi \, U_1 + \cos \psi \, U_2.$$

(a) Express the connection form and dual 1-forms in terms of ψ and the natural coordinates x, y.

(b) What are the structural equations in this case? Check that the results in part (a) satisfy these equations.

(*Hint:* Defining $E_3 = U_3$ gives a frame field on $\mathbf{R}^3$.)

2.9 Summary

We have accomplished the aims set at the beginning of this chapter. The idea of a moving frame has been expressed rigorously as a *frame field*—either on a curve in $\mathbf{R}^3$ or on an open set of $\mathbf{R}^3$ itself. In the case of a curve, we used only the Frenet frame field T, N, B of the curve. Expressing the derivatives of these vector fields in terms of the vector fields themselves, we discovered the *curvature* and *torsion* of the curve. It is already clear that curvature and torsion tell a lot about the geometry of a curve; we shall find in Chapter 3 that they tell everything. In the case of an open set of $\mathbf{R}^3$, we dealt with an arbitrary frame field E_1, E_2, E_3. Cartan's generalization (Theorem 7.2) of the Frenet formulas followed the same pattern of expressing the (covariant) derivatives of these vector fields in terms of the vector fields themselves. Omitting the vector field V from the notation in Theorem 7.2, we have

	Cartan		*Frenet*	
$\nabla E_1 = $	$\omega_{12}E_2 + \omega_{13}E_3,$	$T' = $	$\kappa N,$	
$\nabla E_2 = -\omega_{12}E_1$	$+ \omega_{23}E_3,$	$N' = -\kappa T$	$+ \tau B,$	
$\nabla E_3 = -\omega_{13}E_1 - \omega_{23}E_2,$		$B' = $	$-\tau N.$	

Cartan's equations are not conspicuously more complicated than Frenet's, because the notion of 1-form is available for the coefficients ω_{ij}, the connection forms.

Chapter 3
Euclidean Geometry

We recall some familiar features of plane geometry. First of all, two triangles are *congruent* if there is a rigid motion of the plane that carries one triangle exactly onto the other. Corresponding angles of congruent triangles are equal, corresponding sides have the same length, the areas enclosed are equal, and so on. Indeed, any geometric property of a given triangle is automatically shared by every congruent triangle. Conversely, there are a number of simple ways in which one can decide whether two given triangles are congruent—for example, if for each the same three numbers occur as lengths of sides.

In this chapter we shall investigate the rigid motions (isometries) of Euclidean space, and see how these remarks about triangles can be extended to other geometric objects.

3.1 Isometries of R³

An isometry, or rigid motion, of Euclidean space is a mapping that preserves the Euclidean distance d between points (Definition 1.2, Chapter 2).

1.1 Definition An *isometry* of $\mathbf{R}^3$ is a mapping $F: \mathbf{R}^3 \to \mathbf{R}^3$ such that

$$d(F(\mathbf{p}), F(\mathbf{q})) = d(\mathbf{p}, \mathbf{q})$$

for all points $\mathbf{p}, \mathbf{q}$ in $\mathbf{R}^3$.

1.2 Example (1) *Translations.* Fix a point $\mathbf{a}$ in $\mathbf{R}^3$ and let T be the mapping that adds $\mathbf{a}$ to every point of $\mathbf{R}^3$. Thus $T(\mathbf{p}) = \mathbf{p} + \mathbf{a}$ for all

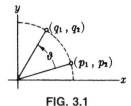

FIG. 3.1

points **p**. *T* is called *translation* by **a**. It is easy to see that T is an isometry, since

$$d(T(\mathbf{p}),\ T(\mathbf{q})) = d(\mathbf{p} + \mathbf{a},\ \mathbf{q} + \mathbf{a})$$

$$= \|(\mathbf{p} + \mathbf{a}) - (\mathbf{q} + \mathbf{a})\|$$

$$= \|\mathbf{p} - \mathbf{q}\| = d(\mathbf{p},\ \mathbf{q}).$$

(2) *Rotation around a coordinate axis.* A rotation of the xy plane through an angle ϑ carries the point (p_1, p_2) to the point (q_1, q_2) with coordinates (Fig. 3.1)

$$q_1 = p_1 \cos \vartheta - p_2 \sin \vartheta,$$

$$q_2 = p_1 \sin \vartheta + p_2 \cos \vartheta.$$

Thus a *rotation C* of three-dimensional Euclidean space $\mathbf{R}^3$ around the z axis, through an angle ϑ, has the formula

$$C(\mathbf{p}) = C(p_1, p_2, p_3) = (p_1 \cos \vartheta - p_2 \sin \vartheta,\ p_1 \sin \vartheta + p_2 \cos \vartheta,\ p_3).$$

Evidently, the mapping C is a linear transformation. A straightforward computation shows that C preserves Euclidean distance, so it is an isometry.

Recall that if F and G are mappings of $\mathbf{R}^3$, the composite function GF is a mapping of $\mathbf{R}^3$ obtained by applying first F, then G.

1.3 Lemma   If F and G are isometries of $\mathbf{R}^3$, then the composite mapping GF is also an isometry of $\mathbf{R}^3$.

Proof.   Since G is an isometry, the distance from $G(F(\mathbf{p}))$ to $G(F(\mathbf{q}))$ is $d(F(\mathbf{p}), F(\mathbf{q}))$. But since F is an isometry, this distance equals $d(\mathbf{p}, \mathbf{q})$. Thus GF preserves distance; hence it is an isometry.   ◆

In short, a composition of isometries is again an isometry.

We also recall that if $F: \mathbf{R}^3 \to \mathbf{R}^3$ is both one-to-one and onto, then F has a unique inverse function $F^{-1}: \mathbf{R}^3 \to \mathbf{R}^3$, which sends each point $F(\mathbf{p})$ back to $\mathbf{p}$. The relationship between F and F^{-1} is best described by the formulas

$$FF^{-1} = I, \quad F^{-1}F = I,$$

where I is the *identity mapping* of $\mathbf{R}^3$, that is, the mapping such that $I(\mathbf{p}) = \mathbf{p}$ for all $\mathbf{p}$.

Translations of $\mathbf{R}^3$ (as defined in Example 1.2) are the simplest type of isometry.

1.4 Lemma   (1) If S and T are translations, then $ST = TS$ is also a translation.

(2) If T is translation by $\mathbf{a}$, then T has an inverse T^{-1}, which is translation by $-\mathbf{a}$.

(3) Given any two points $\mathbf{p}$ and $\mathbf{q}$ of $\mathbf{R}^3$, there exists a unique translation T such that $T(\mathbf{p}) = \mathbf{q}$.

Proof.   To prove (3), for example, note that translation by $\mathbf{q} - \mathbf{p}$ certainly carries $\mathbf{p}$ to $\mathbf{q}$. This is the only possibility, since if T is translation by $\mathbf{a}$ and $T(\mathbf{p}) = \mathbf{q}$, then $\mathbf{p} + \mathbf{a} = \mathbf{q}$; hence $\mathbf{a} = \mathbf{q} - \mathbf{p}$.   ◆

A useful special case of (3) is that if T is a translation such that for some one point $T(\mathbf{p}) = \mathbf{p}$, then $T = I$.

The rotation in Example 1.2 is an example of an *orthogonal transformation* of $\mathbf{R}^3$, that is, a linear transformation $C: \mathbf{R}^3 \to \mathbf{R}^3$ that preserves dot products in the sense that

$$C(\mathbf{p}) \cdot C(\mathbf{q}) = \mathbf{p} \cdot \mathbf{q} \quad \text{for all } \mathbf{p}, \mathbf{q}.$$

1.5 Lemma   If $C: \mathbf{R}^3 \to \mathbf{R}^3$ is an orthogonal transformation, then C is an isometry of $\mathbf{R}^3$.

Proof.   First we show that C preserves norms. By definition, $\|\mathbf{p}\|^2 = \mathbf{p} \cdot \mathbf{p}$; hence

$$\|C(\mathbf{p})\|^2 = C(\mathbf{p}) \cdot C(\mathbf{p}) = \mathbf{p} \cdot \mathbf{p} = \|\mathbf{p}\|^2.$$

Thus $\| C(\mathbf{p}) \| = \| \mathbf{p} \|$ for all points $\mathbf{p}$. Since C is linear, it follows easily that C is an isometry:

$$d(C(\mathbf{p}), C(\mathbf{q})) = \|C(\mathbf{p}) - C(\mathbf{q})\| = \|C(\mathbf{p} - \mathbf{q})\| = \|\mathbf{p} - \mathbf{q}\|$$

◆

$$= d(\mathbf{p}, \mathbf{q}) \quad \text{for all } \mathbf{p}, \mathbf{q}.$$

Our goal now is Theorem 1.7, which asserts that every isometry can be expressed as an orthogonal transformation followed by a translation. The main part of the proof is the following converse of Lemma 1.5.

1.6 Lemma If F is an isometry of $\mathbf{R}^3$ such that $F(\mathbf{0}) = \mathbf{0}$, then F is an orthogonal transformation.

Proof. First we show that F preserves dot products; then we show that F is a linear transformation. Note that by definition of Euclidean distance, the norm $\| \mathbf{p} \|$ of a point $\mathbf{p}$ is just the Euclidean distance $d(\mathbf{0}, \mathbf{p})$ from the origin to $\mathbf{p}$. By hypothesis, F preserves Euclidean distance, and $F(\mathbf{0}) = \mathbf{0}$; hence

$$\|F(\mathbf{p})\| = d(\mathbf{0}, F(\mathbf{p})) = d(F(\mathbf{0}), F(\mathbf{p})) = d(\mathbf{0}, \mathbf{p}) = \|\mathbf{p}\|.$$

Thus F preserves norms. Now by a standard trick ("polarization"), we shall deduce that it also preserves dot products. Since F is an isometry,

$$d(F(\mathbf{p}), F(\mathbf{q})) = d(\mathbf{p}, \mathbf{q})$$

for any pair of points. Hence

$$\|F(\mathbf{p}) - F(\mathbf{q})\| = \|\mathbf{p} - \mathbf{q}\|.$$

By the definition of norm, this implies

$$(F(\mathbf{p}) - F(\mathbf{q})) \bullet (F(\mathbf{p}) - F(\mathbf{q})) = (\mathbf{p} - \mathbf{q}) \bullet (\mathbf{p} - \mathbf{q}).$$

Hence

$$\|F(\mathbf{p})\|^2 - 2F(\mathbf{p}) \bullet F(\mathbf{q}) + \|F(\mathbf{q})\|^2 = \|\mathbf{p}\|^2 - 2\mathbf{p} \bullet \mathbf{q} + \|\mathbf{q}\|^2.$$

The norm terms here cancel, since F preserves norms, and we find

$$F(\mathbf{p}) \bullet F(\mathbf{q}) = \mathbf{p} \bullet \mathbf{q},$$

as required.

It remains to prove that F is linear. Let $\mathbf{u}_1, \mathbf{u}_2, \mathbf{u}_3$ be the unit points $(1, 0, 0)$, $(0, 1, 0)$, $(0, 0, 1)$, respectively. Then we have the identity

$$\mathbf{p} = (p_1, p_2, p_3) = \sum p_i \mathbf{u}_i.$$

Also, the points $\mathbf{u}_1$, $\mathbf{u}_2$, $\mathbf{u}_3$ are orthonormal; that is, $\mathbf{u}_i \cdot \mathbf{u}_j = \delta_{ij}$.

We know that F preserves dot products, so $F(\mathbf{u}_1)$, $F(\mathbf{u}_2)$, $F(\mathbf{u}_3)$ must also be orthonormal. Thus orthonormal expansion gives

$$F(\mathbf{p}) = \sum F(\mathbf{p}) \cdot F(\mathbf{u}_i)\, F(\mathbf{u}_i).$$

But

$$F(\mathbf{p}) \cdot F(\mathbf{u}_i) = \mathbf{p} \cdot \mathbf{u}_i = p_i,$$

so

$$F(\mathbf{p}) = \sum p_i F(\mathbf{u}_i).$$

Using this identity, it is a simple matter to check the linearity condition

$$F(a\mathbf{p} + b\mathbf{q}) = aF(\mathbf{p}) + bF(\mathbf{q}). \qquad \blacklozenge$$

We now give a concrete description of an arbitrary isometry.

1.7 Theorem   If F is an isometry of $\mathbf{R}^3$, then there exist a unique translation T and a unique orthogonal transformation C such that

$$F = TC.$$

Proof.   Let T be translation by $F(\mathbf{0})$. Then Lemma 1.4 shows that T^{-1} is translation by $-F(\mathbf{0})$. But $T^{-1}\,F$ is an isometry, by Lemma 1.3, and furthermore,

$$(T^{-1}F)(\mathbf{0}) = T^{-1}(F(\mathbf{0})) = F(\mathbf{0}) - F(\mathbf{0}) = \mathbf{0}.$$

Thus by Lemma 1.6, $T^{-1}\,F$ is an orthogonal transformation, say $T^{-1}F = C$. Applying T on the left, we get $F = TC$.

To prove the required uniqueness, we suppose that F can also be expressed as $\overline{T}\,\overline{C}$, where $\overline{T}$ is a translation and $\overline{C}$ an orthogonal transformation. We must prove $\overline{T} = T$ and $\overline{C} = C$. Now $TC = \overline{T}\,\overline{C}$; hence $C = T^{-1}\overline{T}\,\overline{C}$. Since C and $\overline{C}$ are linear transformations, they of course send the origin to itself. It follows that $(T^{-1}\overline{T})(\mathbf{0}) = \mathbf{0}$. But since $T^{-1}\overline{T}$ is a translation, we conclude that $T^{-1}\overline{T} = I$; hence $\overline{T} = T$. Then the equation $TC = \overline{T}\,\overline{C}$ becomes $TC = T\overline{C}$. Applying T^{-1} gives $C = \overline{C}$ $\qquad \blacklozenge$

Thus every isometry of $\mathbf{R}^3$ can be uniquely *described as an orthogonal transformation followed by a translation.* When $F = TC$ as in Theorem 1.7, we call

C the *orthogonal part of F,* and T the *translation part of F.* Note that CT is generally not the same as TC (Exercise 1).

This decomposition theorem is the decisive fact about isometries of $\mathbf{R}^3$ (and its proof holds for $\mathbf{R}^n$ as well). We will use it to find an explicit formula for an arbitrary isometry.

First, recall from linear algebra that if $C: \mathbf{R}^3 \to \mathbf{R}^3$ is *any* linear transformation, its *matrix* (relative to the natural basis of $\mathbf{R}^3$) is the 3×3 matrix $\{c_{ij}\}$ such that

$$C(p_1, p_2, p_3) = \left(\sum c_{1j}p_j, \sum c_{2j}p_j, \sum c_{3j}p_j \right).$$

Thus, using the *column-vector* conventions, $\mathbf{q} = C(\mathbf{p})$ can be written as

$$\begin{pmatrix} q_1 \\ q_2 \\ q_3 \end{pmatrix} = \begin{pmatrix} c_{11} & c_{12} & c_{13} \\ c_{21} & c_{22} & c_{23} \\ c_{31} & c_{32} & c_{33} \end{pmatrix} \begin{pmatrix} p_1 \\ p_2 \\ p_3 \end{pmatrix}.$$

By a standard result of linear algebra, *a linear transformation of $\mathbf{R}^3$ is orthogonal* (preserves dot products) *if and only if its matrix is orthogonal* (transpose equals inverse).

Returning to the decomposition $F = TC$ in Theorem 1.7, if T is translation by $\mathbf{a} = (a_1, a_2, a_3)$, then

$$F(\mathbf{p}) = TC(\mathbf{p}) = \mathbf{a} + C(\mathbf{p}).$$

Using the above formula for $C(\mathbf{p})$, we get

$$F(\mathbf{p}) = F(p_1, p_2, p_3) = \left(a_1 + \sum c_{1j}p_j, a_2 + \sum c_{2j}p_j, a_3 + \sum c_{3j}p_j \right).$$

Alternatively, using the column-vector conventions, $\mathbf{q} = F(\mathbf{p})$ means

$$\begin{pmatrix} q_1 \\ q_2 \\ q_3 \end{pmatrix} = \begin{pmatrix} a_1 \\ a_2 \\ a_3 \end{pmatrix} + \begin{pmatrix} c_{11} & c_{12} & c_{13} \\ c_{21} & c_{22} & c_{23} \\ c_{31} & c_{32} & c_{33} \end{pmatrix} \begin{pmatrix} p_1 \\ p_2 \\ p_3 \end{pmatrix}.$$

Exercises

Throughout these exercises, A, B, and C denote orthogonal transformations (or their matrices), and T_a is translation by $\mathbf{a}$.

1. Prove that $CT_a = T_{C(a)}C$.

2. Given isometries $F = T_aA$ and $G = T_bB$, find the translation and orthogonal part of FG and GF.

3. Show that an isometry $F = T_a C$ has an inverse mapping F^{-1}, which is also an isometry. Find the translation and orthogonal parts of F^{-1}.

4. If

$$C = \begin{pmatrix} -2/3 & 2/3 & -1/3 \\ 2/3 & 1/3 & -2/3 \\ 1/3 & 2/3 & 2/3 \end{pmatrix} \quad \text{and} \quad \begin{cases} \mathbf{p} = (3, 1, -6), \\ \mathbf{q} = (1, 0, 3), \end{cases}$$

show that C is orthogonal; then compute $C(\mathbf{p})$ and $C(\mathbf{q})$, and check that $C(\mathbf{p}) \bullet C(\mathbf{q}) = \mathbf{p} \bullet \mathbf{q}$.

5. Let $F = T_a C$, where $\mathbf{a} = (1, 3, -1)$ and

$$C = \begin{pmatrix} 1/\sqrt{2} & 0 & -1/\sqrt{2} \\ 0 & 1 & 0 \\ 1/\sqrt{2} & 0 & 1/\sqrt{2} \end{pmatrix}.$$

If $\mathbf{p} = (2, -2, 8)$, find the coordinates of the point $\mathbf{q}$ for which
(a) $\mathbf{q} = F(\mathbf{p})$.   (b) $\mathbf{q} = F^{-1}(\mathbf{p})$.
(c) $\mathbf{q} = (CT_a)(\mathbf{p})$.

6. In each case decide whether F is an isometry of $\mathbf{R}^3$. If so, find its translation and orthogonal parts.
(a) $F(\mathbf{p}) = -\mathbf{p}$.   (b) $F(\mathbf{p}) = (\mathbf{p} \bullet \mathbf{a})\,\mathbf{a}$, where $\| \mathbf{a} \| = 1$.
(c) $F(\mathbf{p}) = (p_3 - 1, p_2 - 2, p_1 - 3)$.   (d) $F(\mathbf{p}) = (p_1, p_2, 1)$.

A *group* G is a set furnished with an *operation* that assigns to each pair g_1, g_2 of elements of G an element $g_1 g_2$, subject to these rules: (1) associative law: $(g_1 g_2)g_3 = g_1(g_2 g_3)$, (2) there is a unique *identity element* e such that $eg = ge = g$ for all g in G, and (3) inverses: For each g in G there is an element g^{-1} in G such that $gg^{-1} = g^{-1}g = e$.

Groups occur naturally in many parts of geometry, and we shall mention a few in subsequent exercises. Basic properties of groups may be found in a variety of elementary textbooks.

7. Prove that the set $\mathscr{E}(3)$ of all isometries of $\mathbf{R}^3$ forms a group—with composition of functions as the operation. $\mathscr{E}(3)$ is called the *Euclidean group* of order 3.

A subset H of a group G is a *subgroup* of G provided (1) if g_1 and g_2 are in H, then so is $g_1 g_2$, (2) is g is in H, so is g^{-1}, and hence (3) the identity element e of G is in H. A subgroup H of G is automatically a group.

8. Prove that the set $\mathcal{T}(3)$ of all translations of $\mathbf{R}^3$ and the set $O(3)$ of all orthogonal transformations of $\mathbf{R}^3$ are each subgroups of the Euclidean group $\mathcal{E}(3)$. $O(3)$ is called the *orthogonal group* of order 3. Which isometries of $\mathbf{R}^3$ are in both these subgroups?

It is easy to check that the results of this section, though stated for $\mathbf{R}^3$, remain valid for Euclidean spaces $\mathbf{R}^n$ of any dimension.

9. (a) Give an explicit description of an arbitrary 2×2 orthogonal matrix C. (*Hint:* Use an angle and a sign.)
   (b) Give a formula for an arbitrary isometry F of $\mathbf{R} = \mathbf{R}^1$.

3.2   The Tangent Map of an Isometry

In Chapter 1 we showed that an arbitrary mapping $F: \mathbf{R}^3 \to \mathbf{R}^3$ has a tangent map F_* that carries each tangent vector $\mathbf{v}$ at $\mathbf{p}$ to a tangent vector $F_*(\mathbf{v})$ at $F(\mathbf{p})$. If F is an isometry, its tangent map is remarkably simple. (Since the distinction between tangent vector and point is crucial here, we temporarily restore the point of application to the notation.)

2.1 Theorem   Let F be an isometry of $\mathbf{R}^3$ with orthogonal part C. Then

$$F_*(\mathbf{v}_p) = C(\mathbf{v})_{F(p)}$$

for all tangent vectors $\mathbf{v}_p$ to $\mathbf{R}^3$.

Verbally: To get $F_*(\mathbf{v}_p)$, first shift the tangent vector $\mathbf{v}_p$ to the canonically corresponding point $\mathbf{v}$ of $\mathbf{R}^3$, then apply the orthogonal part C of F, and finally shift this point $C(\mathbf{v})$ to the canonically corresponding tangent vector at $F(\mathbf{p})$ (Fig. 3.2). *Thus all tangent vectors at all points* $\mathbf{p}$ *of* $\mathbf{R}^3$ *are "rotated" in exactly the same way by* F_* — *only the new point of application* $F(\mathbf{p})$ *depends on* $\mathbf{p}$.

Proof.   Write $F = TC$ as in Theorem 1.7. Let T be translation by $\mathbf{a}$, so $F(\mathbf{p}) = \mathbf{a} + C(\mathbf{p})$. If $\mathbf{v}_p$ is a tangent vector to $\mathbf{R}^3$, then by Definition 7.4 of Chapter 1, $F_*(\mathbf{v}_p)$ is the initial velocity of the curve $t \to F(\mathbf{p} + t\mathbf{v})$. But using the linearity of C, we obtain

$$F(\mathbf{p} + t\mathbf{v}) = TC(\mathbf{p} + t\mathbf{v}) = T(C(\mathbf{p}) + tC(\mathbf{v})) = \mathbf{a} + C(\mathbf{p}) + tC(\mathbf{v})$$

$$= F(\mathbf{p}) + tC(\mathbf{v}).$$

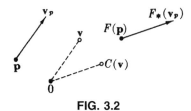

FIG. 3.2

Thus $F_*(\mathbf{v}_p)$ is the initial velocity of the curve $t \to F(\mathbf{p}) + tC(\mathbf{v})$, which is precisely the tangent vector $C(\mathbf{v})_{F(p)}$.   ◆

Expressed in terms of Euclidean coordinates, this result becomes

$$F_*\!\left(\sum_j v_j U_j\right) = \sum_{i,j} c_{ij} v_j \overline{U}_i,$$

where $C = (c_{ij})$ is the orthogonal part of the isometry F, and if U_i is evaluated at $\mathbf{p}$, then $\overline{U}_i$ is evaluated at $F(\mathbf{p})$.

2.2 Corollary   Isometries preserve dot products of tangent vectors. That is, if $\mathbf{v}_p$ and $\mathbf{w}_p$ are tangent vectors to $\mathbf{R}^3$ at the same point, and F is an isometry, then

$$F_*(\mathbf{v}_p) \bullet F_*(\mathbf{w}_p) = \mathbf{v}_p \bullet \mathbf{w}_p.$$

Proof.   Let C be the orthogonal part of F, and recall that C, being an orthogonal transformation, preserves dot products in $\mathbf{R}^3$. By Theorem 2.1,

$$F_*(\mathbf{v}_p) \bullet F_*(\mathbf{w}_p) = C(\mathbf{v})_{F(p)} \bullet C(\mathbf{w})_{F(p)} = C(\mathbf{v}) \bullet C(\mathbf{w})$$

$$= \mathbf{v} \bullet \mathbf{w} = \mathbf{v}_p \bullet \mathbf{w}_p$$

where we have twice used Definition 1.3 of Chapter 2 (dot products of tangent vectors).   ◆

Since dot products are preserved, it follows automatically that derived concepts such as norm and orthogonality are preserved. Explicitly, if F is an isometry, then $\| F_*(\mathbf{v}) \| = \| \mathbf{v} \|$, and if $\mathbf{v}$ and $\mathbf{w}$ are orthogonal, so are $F_*(\mathbf{v})$ and $F_*(\mathbf{w})$. Thus frames are also preserved: if $\mathbf{e}_1, \mathbf{e}_2, \mathbf{e}_3$ is a frame at some point $\mathbf{p}$ of $\mathbf{R}^3$ and F is an isometry, then $F_*(\mathbf{e}_1), F_*(\mathbf{e}_2), F_*(\mathbf{e}_3)$ is a frame at $F(\mathbf{p})$. (A direct proof is easy: $\mathbf{e}_i \bullet \mathbf{e}_j = \delta_{ij}$, so by Corollary 2.2, $F_*(\mathbf{e}_i) \bullet F_*(\mathbf{e}_j) = \mathbf{e}_i \bullet \mathbf{e}_j = \delta_{ij}$.)

Assertion (3) of Lemma 1.4 shows how two *points* uniquely determine a translation. We now show that two *frames* uniquely determine an isometry.

2.3 Theorem Given any two frames on $\mathbf{R}^3$, say $\mathbf{e}_1$, $\mathbf{e}_2$, $\mathbf{e}_3$ at the point $\mathbf{p}$ and $\mathbf{f}_1$, $\mathbf{f}_2$, $\mathbf{f}_3$ at the point $\mathbf{q}$, there exists a unique isometry F of $\mathbf{R}^3$ such that $F_*(\mathbf{e}_i) = \mathbf{f}_i$ for $1 \leq i \leq 3$.

Proof. First we show that there is such an isometry. Let $\hat{\mathbf{e}}_1$, $\hat{\mathbf{e}}_2$, $\hat{\mathbf{e}}_3$, and $\hat{\mathbf{f}}_1$, $\hat{\mathbf{f}}_2$, $\hat{\mathbf{f}}_3$ be the points of $\mathbf{R}^3$ canonically corresponding to the vectors in the two frames. Let C be the unique linear transformation of $\mathbf{R}^3$ such that $C(\hat{\mathbf{e}}_i) = \hat{\mathbf{f}}_i$ for $1 \leq i \leq 3$. It is easy to check that C is orthogonal. Then let T be a translation by the point $\mathbf{q} - C(\mathbf{p})$. Now we assert that the isometry $F = TC$ carries the $\mathbf{e}$ frame to the $\mathbf{f}$ frame. First note that

$$F(\mathbf{p}) = T(C(\mathbf{p})) = \mathbf{q} - C(\mathbf{p}) + C(\mathbf{p}) = \mathbf{q}.$$

Then using Theorem 2.1 we get

$$F_*(\mathbf{e}_i) = C(\hat{\mathbf{e}}_i)_{F(p)} = (\hat{\mathbf{f}}_i)_{F(p)} = (\hat{\mathbf{f}}_i)_q = \mathbf{f}_i$$

for $1 \leq i \leq 3$.

To prove uniqueness, we observe that by Theorem 2.1 this choice of C is the *only* possibility for the orthogonal part of the required isometry. The translation part is then completely determined also, since it must carry $C(\mathbf{p})$ to $\mathbf{q}$. Thus the isometry $F = TC$ is uniquely determined. ◆

To compute the isometry in the theorem, recall that the attitude matrix A of the $\mathbf{e}$ frame has the Euclidean coordinates of $\mathbf{e}_i$ as its ith row: a_{i1}, a_{i2}, a_{i3}. The attitude matrix B of the $\mathbf{f}$ frame is similar. We claim that C in the theorem (or strictly speaking, its matrix) is ${}^tBA$. To verify this it suffices to check that ${}^tBA(\mathbf{e}_i) = \mathbf{f}_i$, since this uniquely characterizes C. For $i = 1$ we find, using the column-vector conventions,

$$
{}^tBA \begin{pmatrix} a_{11} \\ a_{12} \\ a_{13} \end{pmatrix} = {}^tB \begin{pmatrix} a_{11} & a_{12} & a_{13} \\ a_{21} & a_{22} & a_{23} \\ a_{31} & a_{32} & a_{33} \end{pmatrix} \begin{pmatrix} a_{11} \\ a_{12} \\ a_{13} \end{pmatrix}
$$

$$
= {}^tB \begin{pmatrix} 1 \\ 0 \\ 0 \end{pmatrix} = \begin{pmatrix} b_{11} & b_{21} & b_{31} \\ b_{12} & b_{22} & b_{32} \\ b_{13} & b_{23} & b_{33} \end{pmatrix} \begin{pmatrix} 1 \\ 0 \\ 0 \end{pmatrix} = \begin{pmatrix} b_{11} \\ b_{12} \\ b_{13} \end{pmatrix},
$$

that is, ${}^tBA(\mathbf{e}_1) = \mathbf{f}_1$. The cases $i = 2, 3$ are similar; hence $C = {}^tBA$. As noted above, T is then necessarily translated by $\mathbf{q} - C(\mathbf{p})$.

Exercises

1. If T is a translation, show that for every tangent vector **v** the vector $T(\mathbf{v})$ is parallel to **v** (same Euclidean coordinates).

2. Prove the general formulas $(GF)_* = G_*F_*$ and $(F^{-1})_* = (F_*)^{-1}$ in the special case where F and G are isometries of $\mathbf{R}^3$.

3. Given the frame

$$\mathbf{e}_1 = (2, 2, 1)/3, \qquad \mathbf{e}_2 = (-2, 1, 2)/3, \qquad \mathbf{e}_3 = (1, -2, 2)/3$$

at $\mathbf{p} = (0, 1, 0)$ and the frame

$$\mathbf{f}_1 = (1, 0, 1)/\sqrt{2}, \qquad \mathbf{f}_2 = (0, 1, 0), \qquad \mathbf{f}_3 = (1, 0, -1)/\sqrt{2}$$

at $\mathbf{q} = (3, -1, 1)$, find a and C such that the isometry $F = T_aC$ carries the **e** frame to the **f** frame.

4. (a) Prove that an isometry $F = TC$ carries the plane through **p** orthogonal to $\mathbf{q} \neq 0$ to the plane through $F(\mathbf{p})$ orthogonal to $C(\mathbf{q})$.
(b) If P is the plane through $(1/2, -1, 0)$ orthogonal to $(0, 1, 0)$ find an isometry $F = TC$ such that $F(P)$ is the plane through $(1, -2, 1)$ orthogonal to $(1, 0, -1)$.

5. (*Computer.*)
(a) Verify that both sets of vectors in Exercise 3 form frames by showing that $A'A = I$ for their attitude matrices.
(b) Find the matrix C that carries each $\mathbf{e}_i$ to $\mathbf{f}_i$, and check this for $i = 1, 2, 3$.

3.3 Orientation

We now come to one of the most interesting and elusive ideas in geometry. Intuitively, it is *orientation* that distinguishes between a right-handed glove and a left-handed glove in ordinary space. To handle this concept mathematically, we replace gloves by frames and separate all the frames on $\mathbf{R}^3$ into two classes as follows. Recall that associated with each frame $\mathbf{e}_1$, $\mathbf{e}_2$, $\mathbf{e}_3$ at a point of $\mathbf{R}^3$ is its attitude matrix A. According to the exercises for Section 1 of Chapter 2,

$$\mathbf{e}_1 \bullet \mathbf{e}_2 \times \mathbf{e}_3 = \det A = \pm 1.$$

When this number is +1, we shall say that the frame e_1, e_2, e_3 is *positively oriented* (or right-handed); when it is -1, the frame is *negatively oriented* (or left-handed).

We omit the easy proof of the following facts.

3.1 Remark   (1) At each point of $\mathbf{R}^3$ the frame assigned by the natural frame field U_1, U_2, U_3 is positively oriented.

(2) A frame e_1, e_2, e_3 is positively oriented if and only if $e_1 \times e_2 = e_3$. Thus the orientation of a frame can be determined, for practical purposes, by the "right-hand rule" given at the end of Section 1 of Chapter 2. Pictorially, the frame (P) in Fig. 3.3 is positively oriented, whereas the frame (N) is negatively oriented. In particular, *Frenet frames are always positively oriented*, since by definition, $B = T \times N$.

(3) For a positively oriented frame e_1, e_2, e_3, the cross products are

$$e_1 = e_2 \times e_3 = -e_3 \times e_2,$$

$$e_2 = e_3 \times e_1 = -e_1 \times e_3,$$

$$e_3 = e_1 \times e_2 = -e_2 \times e_1.$$

For a negatively oriented frame, reverse the vectors in each cross product. (One need not memorize these formulas—the right-hand rule will give them all correctly.)

Having attached a sign to each frame on $\mathbf{R}^3$, we next attach a sign to each isometry F of $\mathbf{R}^3$. In Chapter 2 we proved the well-known fact that the determinant of an orthogonal matrix is either +1 or -1. Thus if C is the orthogonal part of the isometry F, we define the *sign* of F to be the determinant of C, with notation

$$\operatorname{sgn} F = \det C.$$

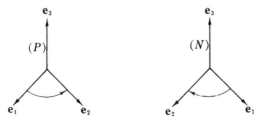

FIG. 3.3

We know that the tangent map of an isometry carries frames to frames. The following result tells what happens to their orientations.

3.2 Lemma If e_1, e_2, e_3 is a frame at some point of $\mathbf{R}^3$ and F is an isometry, then

$$F_*(e_1) \bullet F_*(e_2) \times F_*(e_3) = (\operatorname{sgn} F)\, e_1 \bullet e_2 \times e_3.$$

Proof. If $e_j = \sum a_{jk} U_k$, then by the coordinate form of Theorem 2.1 we have

$$F_*(e_j) = \sum_{i,k} c_{ik} a_{jk} \overline{U}_i,$$

where $C = (c_{ij})$ is the orthogonal part of F. Thus the attitude matrix of the frame $F_*(e_1)$, $F_*(e_2)$, $F_*(e_3)$ is the matrix

$$\left(\sum_k c_{ik} a_{jk} \right) = \left(\sum_k c_{ik}^{\,t} a_{kj} \right) = C^{\,t}A.$$

But the triple scalar product of a frame is the determinant of its attitude matrix, and by definition, $\operatorname{sgn} F = \det C$. Consequently,

$$F_*(e_1) \bullet F_*(e_2) \times F_*(e_3) = \det\left(C^{\,t}A \right)$$

$$= \det C \cdot \det {}^{t}A = \det C \cdot \det A$$

$$= (\operatorname{sgn} F)e_1 \cdot e_2 \times e_3. \qquad \blacklozenge$$

This lemma shows that if $\operatorname{sgn} F = +1$, then F_* carries positively oriented frames to positively oriented frames and carries negatively oriented frames to negatively oriented frames. On the other hand, if $\operatorname{sgn} F = -1$, positive goes to negative and negative to positive.

3.3 Definition An isometry F of $\mathbf{R}^3$ is said to be

$$orientation\text{-}preserving \text{ if } \operatorname{sgn} F = \det C = +1,$$

$$orientation\text{-}reversing \text{ if } \operatorname{sgn} F = \det C = -1,$$

where C is the orthogonal part of F.

3.4 Example (1) *Translations.* All translations are orientation-preserving. Geometrically this is clear, and in fact the orthogonal part of a translation T is just the identity mapping I, so $\operatorname{sgn} T = \det I = +1$.

(2) *Rotations.* Consider the orthogonal transformation C given in Example 1.2, which rotates $\mathbf{R}^3$ through angle θ around the z axis. Its matrix is

$$\begin{pmatrix} \cos\theta & -\sin\theta & 0 \\ \sin\theta & \cos\theta & 0 \\ 0 & 0 & 1 \end{pmatrix}.$$

Hence $\operatorname{sgn} C = \det C = +1$, so C is orientation-preserving (see Exercise 4).

(3) *Reflections.* One can (literally) see reversal of orientation by using a mirror. Suppose the yz plane of $\mathbf{R}^3$ is the mirror. If one looks toward that plane, the point $\mathbf{p} = (p_1, p_2, p_3)$ appears to be located at the point

$$R(\mathbf{p}) = (-p_1, p_2, p_3)$$

(Fig. 3.4). The mapping R so defined is called *reflection* in the yz plane. Evidently it is an orthogonal transformation, with matrix

$$\begin{pmatrix} -1 & 0 & 0 \\ 0 & 1 & 0 \\ 0 & 0 & 1 \end{pmatrix}.$$

Thus R is an orientation-reversing isometry, as confirmed by the experimental fact that the mirror image of a right hand is a left hand.

Both dot and cross product were originally defined in terms of *Euclidean* coordinates. We have seen that the dot product is given by the same formula,

$$\mathbf{v} \cdot \mathbf{w} = \left(\sum v_i \mathbf{e}_i\right) \cdot \left(\sum w_i \mathbf{e}_i\right) = \sum v_i w_i,$$

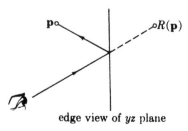

edge view of yz plane

FIG. 3.4

no matter what frame e_1, e_2, e_3 is used to get coordinates for $\mathbf{v}$ and $\mathbf{w}$. Almost the same result holds for cross products, but orientation is now involved.

3.5 Lemma Let e_1, e_2, e_3 be a frame at a point of $\mathbf{R}^3$. If $\mathbf{v} = \sum v_i e_i$ and $\mathbf{w} = \sum w_i e_i$, then

$$\mathbf{v} \times \mathbf{w} = \varepsilon \begin{vmatrix} e_1 & e_2 & e_3 \\ v_1 & v_2 & v_3 \\ w_1 & w_2 & w_3 \end{vmatrix},$$

where $\varepsilon = e_1 \cdot e_2 \times e_3 = \pm 1$.

Proof. It suffices merely to expand the cross product

$$\mathbf{v} \times \mathbf{w} = (v_1 e_1 + v_2 e_2 + v_3 e_3) \times (w_1 e_1 + w_2 e_2 + w_3 e_3)$$

using the formulas (3) of Remark 3.1. For example, if the frame is positively oriented, for the e_1 component of $\mathbf{v} \times \mathbf{w}$ we get

$$v_2 e_2 \times w_3 e_3 + v_3 e_3 \times w_2 e_2 = (v_2 w_3 - v_3 w_2) e_1.$$

Since $\varepsilon = 1$ in this case, we get the same result by expanding the determinant in the statement of this lemma. ◆

It follows immediately that the effect of an isometry on cross products also involves orientation.

3.6 Theorem Let $\mathbf{v}$ and $\mathbf{w}$ be tangent vectors to $\mathbf{R}^3$ at $\mathbf{p}$. If F is an isometry of $\mathbf{R}^3$, then

$$F_*(\mathbf{v} \times \mathbf{w}) = (\text{sgn } F) F_*(\mathbf{v}) \times F_*(\mathbf{w}).$$

Proof. Write $\mathbf{v} = \sum v_i U_i(\mathbf{p})$ *and* $\mathbf{w} = \sum w_i U_i(\mathbf{p})$. Now let

$$e_i = F_*(U_i(\mathbf{p})).$$

Since F_* is linear,

$$F_*(\mathbf{v}) = \sum v_i e_i \quad \text{and} \quad F_*(\mathbf{w}) = \sum w_i e_i.$$

A straightforward computation using Lemma 3.5 shows that

$$F_*(\mathbf{v}) \times F_*(\mathbf{w}) = \varepsilon F_*(\mathbf{v} \times \mathbf{w}),$$

where

$$\varepsilon = e_1 \cdot e_2 \times e_3 = F_*(U_1(\mathbf{p})) \cdot F_*(U_2(\mathbf{p})) \times F_*(U_3(\mathbf{p})).$$

But U_1, U_2, U_3 is positively oriented, so by Lemma 3.2, $\varepsilon = \operatorname{sgn} F$. ◆

Exercises

1. Prove

$$\operatorname{sgn}(FG) = \operatorname{sgn} F \cdot \operatorname{sgn} G = \operatorname{sgn}(GF).$$

Deduce that $\operatorname{sgn} F = \operatorname{sgn} (F^{-1})$.

2. If H_0 is an orientation-reversing isometry of $\mathbf{R}^3$, show that *every* orientation-reversing isometry has a unique expression $H_0 F$, where F is orientation-preserving.

3. Let $\mathbf{v} = (3, 1, -1)$ and $\mathbf{w} = (-3, -3, 1)$ be tangent vectors at some point. If C is the orthogonal transformation given in Exercise 4 of Section 1, check the formula

$$C_*(\mathbf{v} \times \mathbf{w}) = (\operatorname{sgn} C)C_*(\mathbf{v}) \times C_*(\mathbf{w}).$$

4. A *rotation* is an orthogonal transformation C such that $\det C = +1$. Prove that C does, in fact, rotate $\mathbf{R}^3$ around an axis. Explicitly, given a rotation C, show that there exists a number ϑ and points e_1, e_2, e_3 with $e_i \cdot e_j = \delta_{ij}$ such that (Fig. 3.5)

$$C(e_1) = \cos \vartheta \; e_1 + \sin \vartheta \; e_2,$$

$$C(e_2) = -\sin \vartheta \; e_1 + \cos \vartheta \; e_2,$$

$$C(e_3) = e_3.$$

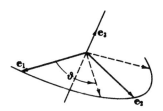

FIG. 3.5

(*Hint:* The fact that the dimension of $\mathbf{R}^3$ is odd means that C has an eigenvalue $+1$, so there is a point $\mathbf{p} \neq \mathbf{0}$ such that $C(\mathbf{p}) = \mathbf{p}$.)

5. Let $\mathbf{a}$ be a point of $\mathbf{R}^3$ such that $\| \mathbf{a} \| = 1$. Prove that the formula

$$C(\mathbf{p}) = \mathbf{a} \times \mathbf{p} + (\mathbf{p} \cdot \mathbf{a})\, \mathbf{a}$$

defines an orthogonal transformation. Describe its general effect on $\mathbf{R}^3$.

6. Prove
(a) The set $O^+(3)$ of all rotations of $\mathbf{R}^3$ is a subgroup of the orthogonal group $O(3)$ (see Ex. 8 of Sec. 3.1).
(b) The set $\mathscr{E}^+(3)$ of all orientation-preserving isometries of $\mathbf{R}^3$ is a subgroup of the Euclidean group $\mathscr{E}(3)$.

3.4 Euclidean Geometry

In the discussion at the beginning of this chapter, we recalled a fundamental feature of plane geometry: If there is an isometry carrying one triangle onto another, then the two (congruent) triangles have exactly the same geometric properties. A close examination of this statement will show that it does not admit a proof—it is, in fact, just the definition of "geometric property of a triangle." More generally, *Euclidean geometry* can be defined as the totality of concepts that are preserved by isometries of Euclidean space. For example, Corollary 2.2 shows that the notion of dot product on tangent vectors belongs to Euclidean geometry. Similarly, Theorem 3.6 shows that the cross product is preserved by isometries (except possibly for sign).

This famous definition of Euclidean geometry is somewhat generous, however. In practice, the label "Euclidean geometry" is usually attached only to those concepts that are preserved by isometries, but *not* by arbitrary mappings, or even the more restrictive class of mappings (diffeomorphisms) that possess inverse mappings. An example should make this distinction clearer. If $\alpha = (\alpha_1, \alpha_2, \alpha_3)$ is a curve in $\mathbf{R}^3$, then the various derivatives

$$\alpha' = \left(\frac{d\alpha_1}{dt}, \frac{d\alpha_2}{dt}, \frac{d\alpha_3}{dt} \right), \qquad \alpha'' = \left(\frac{d^2\alpha_1}{dt^2}, \frac{d^2\alpha_2}{dt^2}, \frac{d^2\alpha_3}{dt^2} \right), \ \cdots$$

look pretty much alike. Now, Theorem 7.8 of Chapter 1 asserts that *velocity is preserved by arbitrary mappings* $F: \mathbf{R}^3 \to \mathbf{R}^3$, that is, if $\beta = F(\alpha)$, then $\beta' = F_*(\alpha')$. But it is easy to see that *acceleration is not preserved by arbitrary mappings*. For example, if $\alpha(t) = (t, 0, 0)$ and $F = (x^2, y, z)$, then $\alpha'' = 0$; hence $F_*(\alpha'') = 0$. But $\beta = F(\alpha)$ has the formula $\beta(t) = (t^2, 0, 0)$, so $\beta'' = 2U_1$. Thus in this case, $\beta = F(\alpha)$, but $\beta'' \neq F_*(\alpha'')$. We shall see in a moment, however, that acceleration is preserved by *isometries*.

For this reason, the notion of velocity belongs to the *calculus* of Euclidean space, while the notion of acceleration belongs to Euclidean *geometry*. In this section we examine some of the concepts introduced in Chapter 2 and prove that they are, in fact, preserved by isometries. (We leave largely to the reader the easier task of showing that they are not preserved by diffeomorphisms.)

Recall the notion of vector field on a curve (Definition 2.2 of Chapter 2). If Y is a vector field on $\alpha: I \to \mathbf{R}^3$ and $F: \mathbf{R}^3 \to \mathbf{R}^3$ is any mapping, then $\overline{Y} = F_*(Y)$ is a vector field on the image curve $\overline{\alpha} = F(\alpha)$. In fact, for each t in I, $Y(t)$ is a tangent vector to $\mathbf{R}^3$ at the point $\alpha(t)$. But then $\overline{Y}(t) = F_*(Y(t))$ is a tangent vector to $\mathbf{R}^3$ at the point $F(\alpha(t)) = \overline{\alpha}(t)$.

(These relationships are illustrated in Fig. 3.6.) Isometries preserve the *derivatives* of such vector fields.

4.1 Corollary   Let Y be a vector field on a curve α in $\mathbf{R}^3$, and let F be an isometry of $\mathbf{R}^3$. Then $\overline{Y} = F_*(Y)$ is a vector field on $\overline{\alpha} = F(\alpha)$, and

$$\overline{Y}' = F_*(Y').$$

Proof.   To differentiate a vector field $Y = \sum y_j U_j$, one simply differentiates its Euclidean coordinate functions, so

$$Y' = \sum \frac{dy_j}{dt} U_j.$$

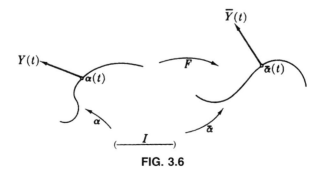

FIG. 3.6

Thus by the coordinate version of Theorem 2.1, we get

$$F_*(Y') = \sum c_{ij} \frac{dy_j}{dt} \overline{U}_i.$$

On the other hand,

$$\overline{Y} = F_*(Y) = \sum c_{ij} y_j \overline{U}_i.$$

But each c_{ij} is constant, being by definition an entry in the matrix of the orthogonal part of the isometry F. Hence

$$\overline{Y}' = \sum \frac{d}{dt}(c_{ij} y_j) \overline{U}_i = \sum c_{ij} \frac{dy_j}{dt} \overline{U}_i.$$

Thus the vector fields $F_*(Y')$ and $\overline{Y}'$ are the same.   ◆

We claimed earlier that isometries preserve acceleration: If $\overline{\alpha} = F(\alpha)$, where F is an isometry, then $\overline{\alpha}'' = F_*(\alpha'')$. This is an immediate consequence of the preceding result, for if we set $Y = \alpha'$, then by Theorem 7.8 of Chapter 1, $\overline{Y} = \overline{\alpha}'$; hence

$$\overline{\alpha}'' = \overline{Y}' = F_*(Y') = F_*(\alpha'').$$

Now we show that the Frenet apparatus of a curve is preserved by isometries. This is certainly to be expected on intuitive grounds, since a rigid motion ought to carry one curve into another that turns and twists in exactly the same way. And this is what happens *when the isometry is orientation-preserving.*

4.2 Theorem Let β be a unit-speed curve in $\mathbf{R}^3$ with positive curvature, and let $\overline{\beta} = F(\beta)$ be the image curve of β under an isometry F of $\mathbf{R}^3$. Then

$$\overline{\kappa} = \kappa, \qquad \overline{T} = F_*(T),$$
$$\overline{\tau} = (\text{sgn } F)\tau, \qquad \overline{N} = F_*(N),$$
$$\overline{B} = (\text{sgn } F)F_*(B),$$

where $\text{sgn } F = \pm 1$ is the sign of the isometry F.

Proof. Note that $\overline{\beta}$ is also a unit-speed curve, since

$$\|\overline{\beta}'\| = \|F_*(\beta')\| = \|\beta'\| = 1.$$

Thus the definitions in Section 3 of Chapter 2 apply to both β and $\bar{\beta}$, so

$$\bar{T} = \bar{\beta}' = F_*(\beta') = F_*(T).$$

Since F_* preserves both acceleration and norms, it follows from the definition of curvature that

$$\bar{\kappa} = \|\bar{\beta}''\| = \|F_*(\beta'')\| = \|\beta''\| = \kappa.$$

To get the full Frenet frame, we now use the hypothesis $\kappa > 0$ (which implies $\bar{\kappa} > 0$, since $\bar{\kappa} = \kappa$). By definition, $N = \beta''/\kappa$; hence using preceding facts, we find

$$\bar{N} = \frac{\bar{\beta}''}{\bar{\kappa}} = \frac{F_*(\beta'')}{\kappa} = F_*\left(\frac{\beta''}{\kappa}\right) = F_*(N).$$

It remains only to prove the interesting cases B and τ. Since the definition $B = T \times N$ involves a cross product, we use Theorem 3.6 to get

$$\bar{B} = \bar{T} \times \bar{N} = F_*(T) \times F_*(N) = (\operatorname{sgn} F)F_*(T \times N) = (\operatorname{sgn} F)F_*(B).$$

The definition of torsion is essentially $\tau = -B' \cdot N = B \cdot N'$. Thus, using the results above for B and N, we get

$$\bar{\tau} = \bar{B} \cdot \bar{N}' = (\operatorname{sgn} F)F_*(B) \cdot F_*(N') = (\operatorname{sgn} F)B \cdot N' = (\operatorname{sgn} F)\tau. \quad \blacklozenge$$

The presence of sgn F in the formula for the torsion of $F(\beta)$ shows that the torsion of a curve gives a more subtle description of the curve than has been apparent so far. *The sign of τ measures the orientation of the twisting of the curve.* If F is orientation-reversing, the formula $\bar{\tau} = -\tau$ proves that the twisting of the image of curve $F(\beta)$ is exactly opposite to that of β itself.

A simple example will illustrate this reversal.

4.3 Example   Let β be the unit-speed helix

$$\beta(s) = \left(\cos\frac{s}{c}, \sin\frac{s}{c}, \frac{s}{c}\right),$$

gotten from Example 3.3 of Chapter 2 by setting $a = b = 1$; hence $c = \sqrt{2}$. We know from the general formulas for helices that $\kappa = \tau = 1/2$. Now let R be reflection in the xy plane, so R is the isometry $R(x, y, z) = (x, y, -z)$. Thus the image curve $\bar{\beta} = \mathbf{R}(\beta)$ is the mirror image

$$\bar{\beta}(s) = \left(\cos\frac{s}{c}, \sin\frac{s}{c}, -\frac{s}{c}\right)$$

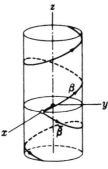

FIG. 3.7

of the original curve. One can see in Fig. 3.7 that the mirror has its usual effect: β and $\bar{\beta}$ twist in opposite ways—if β is "right-handed," then $\bar{\beta}$ is "left-handed." (The fact that β is going up and $\bar{\beta}$ down is, in itself, irrelevant.) Formally: The reflection R is orientation-reversing; hence the theorem predicts $\bar{\kappa} = \kappa = \frac{1}{2}$ and $\bar{\tau} = -\tau = -\frac{1}{2}$. Since $\bar{\beta}$ is just the helix gotten in Example 3.3 of Chapter 2 by taking $a = 1$ and $b = -1$, this may be checked by the general formulas there.

Exercises

1. Let $F = TC$ be an isometry of $\mathbf{R}^3$, β a unit speed curve in $\mathbf{R}^3$. Prove
(a) If β is a cylindrical helix, then $F(\beta)$ is a cylindrical helix.
(b) If β has spherical image σ, then $F(\beta)$ has spherical image $C(\sigma)$.

2. Let $Y = (t, 1 - t^2, 1 + t^2)$ be a vector field on the helix

$$\alpha(t) = (\cos t, \sin t, 2t),$$

and let C be the orthogonal transformation

$$C = \begin{pmatrix} -1 & 0 & 0 \\ 0 & 1/\sqrt{2} & -1/\sqrt{2} \\ 0 & 1/\sqrt{2} & 1/\sqrt{2} \end{pmatrix}.$$

Compute $\bar{\alpha} = C(\alpha)$ and $\bar{Y} = C_*(Y)$, and check that

$$C_*(Y') = \bar{Y}', \quad C_*(\alpha'') = \bar{\alpha}'', \quad Y' \cdot \alpha'' = \bar{Y}' \cdot \bar{\alpha}''.$$

3. Sketch the triangles in $\mathbf{R}^2$ that have vertices

$$\Delta_1: \ (3,1), \ (7,1), \ (7,4), \quad \Delta_2: \ (2,0), \ (2,5), \ (-2/5, 16/5).$$

Show that these triangles are congruent by exhibiting an isometry $F = TC$ that carries Δ_1 to Δ_2. (*Hint*: the orthogonal part C is not altered if the triangles are translated.)

4. If $F: \mathbf{R}^3 \to \mathbf{R}^3$ is a diffeomorphism such that F_* preserves dot products, show that F is an isometry. (*Hint:* Show that F preserves lengths of curve segments and deduce that F^{-1} does also.)

5. Let F be an isometry of $\mathbf{R}^3$. For each vector field V let $\overline{V}$ be the vector field such that $F_*(V(\mathbf{p})) = \overline{V}(F(\mathbf{p}))$ for all $\mathbf{p}$. Prove that isometries preserve covariant derivatives; that is, show $\overline{\nabla_V W} = \nabla_{\overline{V}} \overline{W}$.

3.5 Congruence of Curves

In the case of curves in $\mathbf{R}^3$, the general notion of congruence takes the following form.

5.1 Definition   Two curves α, $\beta: I \to \mathbf{E}^3$ are *congruent* provided there exists an isometry F of $\mathbf{R}^3$ such that $\beta = F(\alpha)$; that is, $\beta(t) = F(\alpha(t))$ for all t in I.

Intuitively speaking, congruent curves are the same except for position in space. They represent *trips at the same speed along routes of the same shape*. For example, the helix $\alpha(t) = (\cos t, \sin t, t)$ spirals around the z axis in exactly the same way the helix $\beta(t) = (t, \cos t, \sin t)$ spirals around the x axis. Evidently these two curves are congruent, since if F is the isometry such that

$$F(p_1, p_2, p_3) = (p_3, p_1, p_2),$$

then $F(\alpha) = \beta$.

To decide whether given curves α and β are congruent, it is hardly practical to try all the isometries of $\mathbf{R}^3$ to see whether there is one that carries α to β. What we want is a description of the shape of a unit-speed curve so accurate that if α and β have the same description, then they must be congruent. The proper description, as the reader will doubtless suspect, is given by curvature and torsion. To prove this we need one preliminary result.

Curves whose congruence is established by a translation are said to be *parallel*. Thus, curves α, $\beta: I \to \mathbf{E}^3$ are parallel if and only if there is a point

p in $\mathbf{R}^3$ such that $\beta(s) = \alpha(s) + \mathbf{p}$ for all s in I, or, in functional notation, $\beta = \alpha + \mathbf{p}$.

5.2 Lemma Two curves α, β: $I \rightarrow \mathbf{R}^3$ are parallel if their velocity vectors $\alpha'(s)$ and $\beta'(s)$ are parallel for each s in I. In this case, if $\alpha(s_0) = \beta(s_0)$ for some one s_0 in I, then $\alpha = \beta$.

Proof. By definition, if $\alpha'(s)$ and $\beta'(s)$ are parallel, they have the same Euclidean coordinates. Thus

$$\frac{d\alpha_i}{ds}(s) = \frac{d\beta_i}{ds}(s) \quad \text{for } 1 \le i \le 3,$$

where α_i and β_i are the Euclidean coordinate functions of α and β. But by elementary calculus, the equation $d\alpha_i/ds = d\beta_i/ds$ implies that there is a constant p_i such that $\beta_i = \alpha_i + p_i$. Hence $\beta = \alpha + \mathbf{p}$. Furthermore, if $\alpha(s_0) = \beta(s_0)$, we deduce that $\mathbf{p} = \mathbf{0}$; hence $\alpha = \beta$. ◆

5.3 Theorem If α, β: $I \rightarrow \mathbf{R}^3$ are unit-speed curves such that $\kappa_\alpha = \kappa_\beta$ and $\tau_\alpha = \pm\tau_\beta$, then α and β are congruent.

Proof. There are two main steps:
(1) Replace α by a suitably chosen congruent curve $F(\alpha)$.
(2) Show that $F(\alpha) = \beta$ (Fig. 3.8).
  Our guide for the choice in (1) is Theorem 4.2. Fix a number, say 0, in the interval I. If $\tau_\alpha = \tau_\beta$, then let F be the (orientation-preserving) isometry that carries the Frenet frame $T_\alpha(0)$, $N_\alpha(0)$, $B_\alpha(0)$ of α at $\alpha(0)$ to the

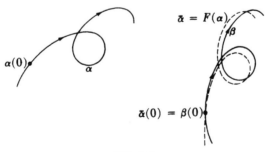

FIG. 3.8

Frenet frame $T_\beta(0)$, $N_\beta(0)$, $B_\beta(0)$, of β at $\beta(0)$. (The existence of this isometry is guaranteed by Theorem 2.3.) Denote the Frenet apparatus of $\bar\alpha = F(\alpha)$ by $\bar\kappa, \bar\tau, \bar T, \bar N, \bar B$; then it follows immediately from Theorem 4.2 and the information above that

$$\bar\alpha(0) = \beta(0), \quad \bar T(0) = T_\beta(0),$$

$$\bar\kappa = \kappa_\beta, \quad \bar N(0) = N_\beta(0), \tag{$\ddagger$}$$

$$\bar\tau = \tau_\beta, \quad \bar B(0) = B_\beta(0).$$

On the other hand, if $\tau_\alpha = -\tau_\beta$, we choose F to be the (orientation-reversing) isometry that carries $T_\alpha(0)$, $N_\alpha(0)$, $B_\alpha(0)$ at $\alpha(0)$ to the frame $T_\beta(0)$, $N_\beta(0)$, $B_\beta(0)$ at $\beta(0)$. (Frenet frames are positively oriented; hence this last frame is negatively oriented: This is why F is orientation-reversing.) Then it follows from Theorem 4.2 that the equations ($\ddagger$) hold also for $\bar\alpha = F(\alpha)$ and β. For example,

$$\bar B(0) = -F_*(B_\alpha(0)) = B_\beta(0).$$

For step (2) of the proof, we shall show $\bar T = T_\beta$; that is, the unit tangents of $\bar\alpha = F(\alpha)$ and β are parallel at each point. Since $\bar\alpha(0) = \beta(0)$, it will follow from Lemma 5.2 that $F(\alpha) = \beta$. On the interval I, consider the real-valued function $f = \bar T \cdot T_\beta + \bar N \cdot N_\beta + \bar B \cdot B_\beta$. Since these are *unit* vector fields, the Schwarz inequality (Sec. 1, Ch. 2) shows that

$$\bar T \cdot T_\beta \leqq 1;$$

furthermore, $\bar T \cdot T_\beta = 1$ if and only if $\bar T = T_\beta$. Similar remarks hold for the other two terms in f. Thus it *suffices to show that f has constant value* 3. By ($\ddagger$), $f(0) = 3$. Now consider

$$f' = \bar T' \cdot T_\beta + \bar T \cdot T'_\beta + \bar N' \cdot N_\beta + \bar N \cdot N'_\beta + \bar B' \cdot B_\beta + \bar B \cdot B'_\beta$$

A simple computation completes the proof. Substitute the Frenet formulas in this expression and use the equations $\bar\kappa = \kappa_\beta$, $\bar\tau = \tau_\beta$ from ($\ddagger$). The resulting eight terms cancel in pairs, so $f' = 0$, and f has, indeed, constant value 3. ◆

Thus, *a unit-speed curve is determined but for position in* $\mathbf{R}^3$ *by its curvature and torsion.*

Actually the proof of Theorem 5.3 does more than establish that α and β are congruent; it shows how to compute *explicitly* an isometry carrying α to β. We illustrate this in a special case.

5.4 Example   Consider the unit-speed curves α, β: $\mathbf{R} \to \mathbf{R}^3$ such that

$$\alpha(s) = \left(\cos\frac{s}{c}, \sin\frac{s}{c}, \frac{s}{c} \right),$$

$$\beta(s) = \left(\cos\frac{s}{c}, \sin\frac{s}{c}, -\frac{s}{c} \right),$$

where $c = \sqrt{2}$. Obviously, these curves are congruent by means of a reflection—they are the helices considered in Example 4.3—but we shall ignore this in order to describe a general method for computing the required isometry. According to Example 3.3 of Chapter 2, α and β have the same curvature, $\kappa_\alpha = 1/2 = \kappa_\beta$; but torsions of opposite sign, $\tau_\alpha = 1/2 = -\tau_\beta$. Thus the theorem predicts congruence by means of an orientation-reversing isometry F. From its proof we see that F must carry the Frenet frame

$$T_\alpha(0) = (0, a, a),$$

$$N_\alpha(0) = (-1, 0, 0),$$

$$B_\alpha(0) = (0, -a, a),$$

where $a = 1/\sqrt{2}$, to the frame

$$T_\beta(0) = (0, a, -a),$$

$$N_\beta(0) = (-1, 0, 0),$$

$$-B_\beta(0) = (0, -a, -a),$$

where the minus sign will produce orientation reversal. (These explicit formulas also come from Example 3.3 of Chapter 2.) By the remark following Theorem 2.3, the isometry F has orthogonal part $C = {}^tBA$, where A and B are the attitude matrices of the two frames above. Thus

$$C = \begin{pmatrix} 0 & -1 & 0 \\ a & 0 & -a \\ -a & 0 & -a \end{pmatrix}\begin{pmatrix} 0 & a & a \\ -1 & 0 & 0 \\ 0 & -a & a \end{pmatrix} = \begin{pmatrix} 1 & 0 & 0 \\ 0 & 1 & 0 \\ 0 & 0 & -1 \end{pmatrix}$$

since $a = 1/\sqrt{2}$. These two frames have the same point of application $\alpha(0) = \beta(0) = (1, 0, 0)$. But C does not move this point, so the translation part of F is just the identity map. Thus we have (correctly) found that the reflection $F = C$ carries α to β.

From the viewpoint of Euclidean geometry, two curves in $\mathbf{R}^3$ are "the same" if they differ only by an isometry of $\mathbf{R}^3$. What, for example, is a helix?

It is not just a curve that spirals around the z axis as in Example 3.3 of Chapter 2, but any curve congruent to one of these special helices. One can give general formulas, but the best characterization follows.

5.5 Corollary   Let α be a unit speed curve in $\mathbf{R}^3$. Then α is a helix if and only if both its curvature and torsion are nonzero constants.

> **Proof.**   For any numbers $a > 0$ and $b \neq 0$, let $\beta_{a,b}$ be the special helix given in Example 3.3 of Chapter 2. If α is congruent to $\beta_{a,b}$, then (changing the sign of b if necessary) we can assume the isometry is orientation-preserving. Thus, α has curvature and torsion
>
> $$\kappa = \frac{a}{a^2 + b^2}, \quad \tau = \frac{b}{a^2 + b^2}.$$
>
> Conversely, suppose α has constant nonzero κ and τ. Solving the preceding equations, we get
>
> $$a = \frac{\kappa}{\kappa^2 + \tau^2}, \quad b = \frac{\tau}{\kappa^2 + \tau^2}.$$
>
> Thus α and $\beta_{a,b}$ have the same curvature and torsion; hence they are congruent.   ◆

Our results so far demand unit speed, but it is easy to weaken this restriction.

5.6 Corollary   Let $\alpha, \beta \colon I \to \mathbf{R}^3$ arbitrary-speed curves. If

$$v_\alpha = v_\beta > 0, \quad \kappa_\alpha = \kappa_\beta > 0, \quad \text{and} \quad \tau_\alpha = \pm\tau_\beta,$$

then the curves α and β are congruent.

The proof is immediate, for the data ensures that the unit speed parametrizations of α and β have the same curvature and torsion—hence they are congruent. But then the original curves are congruent under the same isometry since their speeds are the same.

The theory of curves we have presented applies only to regular curves with positive curvature $\kappa > 0$, because only for such curves is it possible to define the Frenet frame field. However, an arbitrary curve α in $\mathbf{R}^3$ can be studied by means of an *arbitrary* frame field on α, that is, three unit-vector fields E_1, E_2, E_3 on α that are orthogonal at each point.

At a critical point later on, we will need this generalization of the congruence theorem (5.3):

5.7 Theorem   Let α, β: $I \to \mathbf{R}^3$ be curves defined on the same interval. Let E_1, E_2, E_3 be a frame field on α, and F_1, F_2, F_3 a frame field on β. If
(1) $\alpha' \cdot E_i = \beta' \cdot F_i$ $(1 \leq i \leq 3)$,
(2) $E_i' \cdot E_j = F_i' \cdot F_j$ $(1 \leq i, j \leq 3)$,

then α and β are congruent.

Explicitly, for any t_0 in I, if $\mathbf{F}$ is the unique Euclidean isometry that sends each $E_i(t_0)$ to $F_i(t_0)$, then $\mathbf{F}(\alpha) = \beta$.

Proof.   Let $\mathbf{F}$ be the specified isometry. Since $\mathbf{F}_*$ preserves dot products, it follows that the vector fields $\overline{E}_i = \mathbf{F}_*(E_i)$ for $1 \leq i \leq 3$ form a frame field on $\overline{\alpha} = \mathbf{F}(\alpha)$. And since $\mathbf{F}_*$ preserves velocities of curves and derivatives of vector fields, by using condition (1) in the theorem, we find

$$\overline{\alpha}(t_0) = \beta(t_0) \quad \text{and} \quad \overline{\alpha}' \cdot \overline{E}_i = \beta' \cdot F_i \quad \text{for } 1 \leq i \leq 3. \tag{$*$}$$

Similarly, from condition (2), we get

$$\overline{E}_i(t_0) = F_i(t_0) \quad \text{and} \quad \overline{E}_i' \cdot \overline{E}_j = F_i' \cdot F_j \quad \text{for } 1 \leq i, j \leq 3. \tag{$**$}$$

In view of this last equation, orthonormal expansion yields

$$\overline{E}_i' = \sum_j a_{ij} \overline{E}_j \quad \text{and} \quad F_i' = \sum_j a_{ij} F_j,$$

with the *same* coefficient functions a_{ij}. Note that $a_{ij} + a_{ji} = 0$; hence $a_{ii} = 0$. (*Proof:* Differentiate $\overline{E}_i \cdot \overline{E}_j = \delta_{ij}$.)

Now let $f = \sum \overline{E}_j \cdot F_i$. We prove $f = 3$ as before: $f(t_0) = 3$, and

$$f' = \sum (\overline{E}_i' \cdot F_i + \overline{E}_i \cdot F_i') = \sum_{i,j} (a_{ij} + a_{ji}) \overline{E}_j \cdot F_i = 0.$$

Thus each $\overline{E}_i \cdot F_i = 1$, that is, $\overline{E}_i$ and F_i are parallel at each point. By $(*)$ the same is true for

$$\overline{\alpha}' = \sum (\overline{\alpha}' \cdot \overline{E}_i) \overline{E}_i \quad \text{and} \quad \beta' = \sum (\beta' \cdot F_i) F_i.$$

Since $\alpha(t_0) = \beta(t_0)$, Lemma 5.2 gives the required result, $\mathbf{F}(\alpha) = \overline{\alpha} = \beta$. $\blacklozenge$

5.8 Remark   Existence theorem for curves in $\mathbf{R}^3$. Curvature and torsion tell whether two unit-speed curves are isometric, but they do more than that:

Given any two continuous functions $\kappa > 0$ and τ on an interval I, there exists a unit-speed curve $\alpha: I \to \mathbf{R}^3$ that has these functions as its curvature and torsion. (As we know, any two such curves are congruent.) Thus the natural description of curves in $\mathbf{R}^3$ is devoid of geometry, consisting of a pair of real-valued functions.

The proof of the existence theorem requires advanced methods, so we have preferred to illustrate it by the corresponding result for plane curves (Exercises 7–10). Though simpler, this 2-dimensional version has the advantage that plane curvature $\tilde{\kappa}$ is not required to be positive.

Exercises

1. Given a curve $\alpha = (\alpha_1, \alpha_2, \alpha_3): I \to \mathbf{R}^3$, prove that $\beta: I \to \mathbf{R}^3$ is congruent to α if and only if β can be written as

$$\beta(t) = \mathbf{p} + \alpha_1(t)\mathbf{e}_1 + \alpha_2(t)\mathbf{e}_2 + \alpha_3(t)\mathbf{e}_3,$$

where $\mathbf{e}_i \cdot \mathbf{e}_j = \delta_{ij}$.

2. Let E_1, E_2, E_3, be a frame field on $\mathbf{R}^3$ with dual forms θ_i and connection forms ω_{ij}. Prove that two curves $\alpha, \beta: I \to \mathbf{R}^3$ are congruent if $\theta_i(\alpha') = \theta_i(\beta')$ and $\omega_{ij}(\alpha') = \omega_{ij}(\beta')$ for $1 \le i, j \le 3$ (*Hint:* Use Thm. 5.7.)

3. Show that the curve

$$\beta(t) = \left(t + \sqrt{3} \sin t, 2 \cos t, \sqrt{3}t - \sin t\right)$$

is a helix by finding its curvature and torsion. Find a helix of the form $\alpha(t) = (a\cos t, a\sin t, bt)$ and an isometry F such that $F(\alpha) = \beta$.

4. (*Computer; see Appendix.*) (a) Show that the curves

$$\alpha(t) = \left(t + t^2, t - t^2, 1 + \sqrt{2}t^3\right), \quad \beta(t) = \left(t^2 + t^3, 1 - \sqrt{2}t, t^2 - t^3\right),$$

defined on the entire real line, have the same speed, curvature, and torsion. (b) Find formulas for T and C such that the isometry $F = TC$ carries α to β and verify explicitly that $F(\alpha) = \beta$. (*Hint:* Use Ex. 5 of Sec. 2.)

5. (*Computer optional.*) Is the following curve a helix? Prove your answer.

$$c(t) = (-2 \cos t + 2 \sin t + 2t, 2 \cos t + \sin t + 4t, \cos t + 2 \sin t - 4t).$$

6. Congruence of curves.
(a) Prove that curves $\alpha, \beta: I \to \mathbf{R}^2$ are congruent if $\tilde{\kappa}_\alpha = \tilde{\kappa}_\beta$ and they have the same speed.

(b) Show that the space curves

$$\alpha(t) = \left(\sqrt{2}t, t^2, 0\right) \quad \text{and} \quad \beta(t) = \left(-t, t, t^2\right)$$

are congruent. Find an isometry that carries α to β.

7. Given a continuous function f on an interval I, prove—using ordinary integration of functions—that there exists a unit-speed curve $\beta(s)$ in $\mathbf{R}^2$ for which $f(s)$ is the plane curvature. (*Hint:* Reverse the logic in Ex. 8 of Sec. 2.3.)

8. Show that $\beta(s) = (x(s), y(s))$ in the preceding exercise is given by the solutions of the differential equations

$$x'(s) = \cos \varphi(s), \quad y'(s) = \sin \varphi(s), \quad \varphi'(s) = f(s),$$

with initial conditions $x(0) = y(0) = \varphi(0) = 0$. (These initial conditions suffice, since any other β differs at most by a Euclidean isometry and a reparametrization $s \to s + c$.)

Explicit integration is rarely possible; the following exercises use numerical integration.

9. (*Numerical integration, computer graphics.*) Write computer commands that (a) given $f(s)$, produce a numerical description of the solution curve $\beta(s)$ in the preceding exercise, and (b) given $f(s)$, plot the solution curve.

10. (*Continuation.*) Plot unit-speed plane curves with the given plane curvature function f on at least the given interval.
 (a) $f(s) = 1 + e^s$,   on $-6 \le s \le 3$.
 (b) $f(s) = 2 + 3 \cos 3s$,   on $0 \le s \le 2\pi$.
 (c) $f(s) = 3 - 2s^2 + s^3$,   on $-2.5 \le s \le 3.5$.

Adjust scales on axes as needed.

3.6 Summary

The basic result of this chapter is that an arbitrary isometry of Euclidean space can be uniquely expressed as an orthogonal transformation followed by a translation. A consequence is that the tangent map of an isometry F is, at every point, essentially just the orthogonal part of F. Then it is a routine matter to test the concepts introduced earlier to see which belong to *Euclidean geometry*, that is, which are preserved by isometries of Euclidean space.

Finally, we proved an analogue for curves of the various criteria for congruence of triangles in plane geometry; namely, we showed that a necessary and sufficient condition for two curves in $\mathbf{R}^3$ to be congruent is that they have the same curvature and torsion (and speed). Furthermore, the sufficiency proof shows how to find the required isometry explicitly.

Chapter 4
Calculus on a Surface

This chapter begins with the definition of a surface in $\mathbf{R}^3$ and with some standard ways to construct surfaces. Although this concept is a more-or-less familiar one, it is not as widely known as it should be that each surface has a differential and integral calculus strictly comparable with the usual calculus on the Euclidean plane $\mathbf{R}^2$. The elements of this calculus—functions, vector fields, differential forms, mappings—belong strictly to the surface and not to the Euclidean space $\mathbf{R}^3$ in which the surface is located. Indeed, we shall see in the final section that this calculus survives undamaged when $\mathbf{R}^3$ is removed, leaving just the surface and nothing more.

4.1 Surfaces in R³

A surface in $\mathbf{R}^3$ is, to begin with, a *subset* of $\mathbf{R}^3$, that is, a certain collection of points of $\mathbf{R}^3$. Of course, not all subsets are surfaces: We must certainly require that a surface be smooth and two-dimensional. These requirements will be expressed in mathematical terms by the next two definitions.

1.1 Definition A *coordinate patch* $\mathbf{x}$: $D \to \mathbf{R}^3$ is a one-to-one regular mapping of an open set D of $\mathbf{R}^2$ into $\mathbf{R}^3$.

The image $\mathbf{x}(D)$ of a coordinate patch $\mathbf{x}$—that is, the set of all values of $\mathbf{x}$—is a smooth two-dimensional subset of $\mathbf{R}^3$ (Fig. 4.1). Regularity (Definition 7.9 of Chapter 1), for a patch as for a curve, is a basic smoothness condition; the one-to-one requirement is included to prevent $\mathbf{x}(D)$ from cutting across itself. Initially, in order to avoid certain technical difficulties (Example

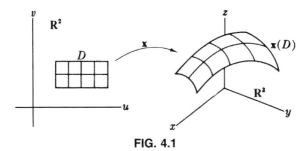

FIG. 4.1

1.6), we must use *proper* patches, those for which the inverse function $\mathbf{x}^{-1}$: $\mathbf{x}(D) \to D$ is continuous (that is, has continuous coordinate functions). If we think of D as a thin sheet of rubber, then $\mathbf{x}(D)$ is gotten by bending and stretching D in a not too violent fashion.

To construct a suitable definition of surface we start from the rough idea that *any small enough region in a surface M resembles a region in the plane* $\mathbf{R}^2$. The discussion above shows that this can be stated somewhat more precisely as, *near each of its points, M can be expressed as the image of a proper patch.* (When the image of a patch $\mathbf{x}$ is contained in M, we say that $\mathbf{x}$ is a patch *in M.*) To get the final form of the definition, it remains only to define a *neighborhood $\mathcal{N}$ of* $\mathbf{p}$ *in M* to consist of all points of M whose Euclidean distance from $\mathbf{p}$ is less than some number $\varepsilon > 0$.

1.2 Definition   A *surface in* $\mathbf{R}^3$ is a subset M of $\mathbf{R}^3$ such that for each point $\mathbf{p}$ of M there exists a proper patch in M whose image contains a neighborhood of $\mathbf{p}$ in M (Fig. 4.2).

The familiar surfaces used in elementary calculus satisfy this definition; for example, let us verify that the unit sphere Σ in $\mathbf{R}^3$ is a surface. By definition,

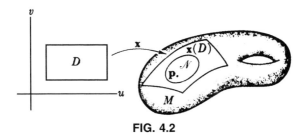

FIG. 4.2

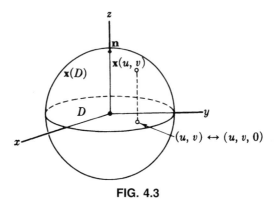

FIG. 4.3

Σ consists of all points at unit distance from the origin—that is, all points **p** such that

$$\| \mathbf{p} \| = \left(p_1^2 + p_2^2 + p_3^2\right)^{1/2} = 1.$$

To check the definition above, we start by finding a proper patch in Σ covering a neighborhood of the north pole $(0, 0, 1)$. Note that by dropping each point (q_1, q_2, q_3) of the northern hemisphere of Σ onto the xy plane at $(q_1, q_2, 0)$ we get a one-to-one correspondence of this hemisphere with a disk D of radius 1 in the xy plane (see Fig. 4.3). If this plane is identified with $\mathbf{R}^2$ by means of the natural association $(q_1, q_2, 0) \leftrightarrow (q_1, q_2)$, then D becomes the disk in $\mathbf{R}^2$ consisting of all points (u, v) such that $u^2 + v^2 < 1$. Expressing this correspondence as a function on D yields the formula

$$\mathbf{x}(u, v) = \left(u, v, \sqrt{1 - u^2 - v^2}\right).$$

Thus **x** is a one-to-one function from D onto the northern hemisphere of Σ. We claim that **x** is a proper patch. The coordinate functions of **x** are differentiable on D, so **x** is a mapping. To show that **x** is regular, we compute its Jacobian matrix (or transpose)

$$\begin{pmatrix} \dfrac{\partial u}{\partial u} & \dfrac{\partial v}{\partial u} & \dfrac{\partial f}{\partial u} \\[2mm] \dfrac{\partial u}{\partial v} & \dfrac{\partial v}{\partial v} & \dfrac{\partial f}{\partial v} \end{pmatrix} = \begin{pmatrix} 1 & 0 & \dfrac{\partial f}{\partial u} \\[2mm] 0 & 1 & \dfrac{\partial f}{\partial v} \end{pmatrix}.$$

where $f = \sqrt{1 - u^2 - v^2}$. Evidently the rows of this matrix are always linearly independent, so its rank at each point is 2. Thus, by the criterion following Definition 7.9 of Chapter 1, **x** is regular and hence is a patch. Furthermore, **x** is proper, since its inverse function $\mathbf{x}^{-1} : \mathbf{x}(D) \to D$ is given by the formula

$$\mathbf{x}^{-1}(p_1, p_2, p_3) = (p_1, p_2)$$

and hence is certainly continuous. Finally, we observe that the patch $\mathbf{x}$ covers a neighborhood of $(0, 0, 1)$ in Σ. Indeed, it covers a neighborhood of every point $\mathbf{q}$ in the northern hemisphere of Σ.

In a strictly analogous way, we can find a proper patch covering each of the other five coordinate hemispheres of Σ, and thus verify, by Definition 1.2, that Σ is a surface. Our real purpose here has been to illustrate Definition 1.2—we soon find a much quicker way to prove (in particular) that spheres are surfaces.

The argument above shows that if f is *any* differentiable real-valued function on an open set D in $\mathbf{R}^2$, then the function $\mathbf{x}: D \rightarrow \mathbf{R}^3$ such that

$$\mathbf{x}(u, v) = (u, v, f(u, v))$$

is a proper patch. We shall call patches of this type *Monge patches*.

We turn now to some standard methods of constructing surfaces. Note that the image $M = \mathbf{x}(D)$ of just one proper patch automatically satisfies 1.2; M is then called a *simple* surface. (Thus Definition 1.2 says that any surface in $\mathbf{R}^3$ can be constructed by gluing together simple surfaces.)

1.3 Example   The surface $M: z = f(x, y)$. Every differentiable real-valued function f on $\mathbf{R}^2$ determines a surface M in $\mathbf{R}^3$: the *graph* of f, that is, the set of all points of $\mathbf{R}^3$ whose coordinates satisfy the equation $z = f(x, y)$. Evidently M is the image of the Monge patch

$$\mathbf{x}(u, v) = (u, v, f(u, v));$$

hence by the remarks above, M is a simple surface.

If g is a real-valued function on $\mathbf{R}^3$ and c is a number, denote by $M: g = c$ the set of all points $\mathbf{p}$ such that $g(\mathbf{p}) = c$. For example, if g is a temperature distribution in space, then $M: g = c$ consists of all points of temperature c. There is a simple condition that tells when such a subset of $\mathbf{R}^3$ is a surface.

1.4 Theorem   Let g be a differentiable real-valued function on $\mathbf{R}^3$, and c a number. The subset $M: g(x, y, z) = c$ of $\mathbf{R}^3$ is a surface if the differential dg is not zero at any point of M.

(In Definition 1.2 and in this theorem we are tacitly assuming that M has some points in it; thus the equation $x^2 + y^2 + z^2 = -1$, for example, does not define a surface.)

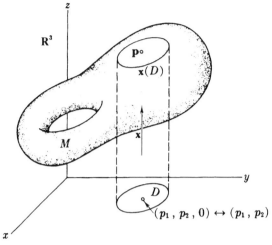

FIG. 4.4

Proof. All we do is give geometric content to a famous result of advanced calculus—the implicit function theorem. If **p** is a point of M, we must find a proper patch covering a neighborhood of **p** in M (Fig. 4.4). Since

$$dg = \frac{\partial g}{\partial x} dx + \frac{\partial g}{\partial y} dy + \frac{\partial g}{\partial z} dz,$$

the hypothesis on dg is equivalent to assuming that at least one of these partial derivatives is not zero at **p**, say $(\partial g/\partial z)(\mathbf{p}) \neq 0$. In this case, the implicit function theorem says that near **p** the equation $g(x, y, z) = c$ can be solved for z. More precisely, it asserts that there is a differentiable real-valued function h defined on a neighborhood D of (p_1, p_2) such that

(1) For each point (u, v) in D, the point $(u, v, h(u, v))$ lies in M; that is, $g(u, v, h(u, v)) = c$.

(2) Points of the form $(u, v, h(u, v))$, with (u, v) in D, fill a neighborhood of **p** in M.

It follows immediately that the Monge patch $\mathbf{x}: D \to \mathbf{R}^3$ such that

$$\mathbf{x}(u, v) = (u, v, h(u, v))$$

satisfies the requirements in Definition 1.2. Since **p** was an arbitrary point of M, we conclude that M is a surface. ◆

From now on we use the notation $M: g = c$ only when $dg \neq 0$ on M. Then M is a surface said to be defined *implicitly* by the equation $g = c$. It is now easy to prove that spheres are surfaces. The sphere Σ in $\mathbf{R}^3$ of radius $r > 0$ and center $\mathbf{c} = (c_1, c_2, c_3)$ is the set of all points at distance r from c. If $g = \sum(x_i - c_i)^2$ then Σ is defined implicitly by the equation $g = r^2$. Now,

$$dg = 2\sum(x_i - c_i)dx_i.$$

Hence dg is zero only at the point $\mathbf{c}$, which is not in Σ. Thus Σ is a surface.

An important class of surfaces is gotten by rotating curves.

1.5 Example   Surfaces of revolution. Let C be a curve in a plane $P \subset \mathbf{R}^3$, and let A be a line in P that does not meet C. When this *profile curve C* is revolved around the *axis A*, it sweeps out a *surface of revolution M* in $\mathbf{R}^3$.

Let us check that M really is a surface. For simplicity, suppose that P is a coordinate plane and A is a coordinate axis—say, the xy plane and x axis, respectively. Since C must not meet A, we put it in the upper half, $y > 0$, of the xy plane. As C is revolved, each of its points $(q_1, q_2, 0)$ gives rise to a whole circle of points

$$(q_1, q_2 \cos v, q_2 \sin v) \quad \text{for } 0 \leqq v \leqq 2\pi.$$

Thus *a point $\mathbf{p} = (p_1, p_2, p_3)$ is in M if and only if the point*

$$\bar{\mathbf{p}} = \left(p_1, \sqrt{p_2^2 + p_3^2}, 0\right)$$

is in C (Fig. 4.5).

If the profile curve is $C: f(x, y) = c$, we define a function g on $\mathbf{R}^3$ by

$$g(x, y, z) = f\left(x, \sqrt{y^2 + z^2}\right).$$

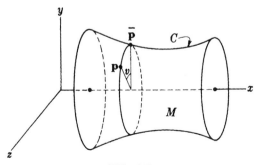

FIG. 4.5

FIG. 4.6

Then the argument above shows that the resulting surface of revolution is exactly M: $g(x, y, z) = c$. Using the chain rule, it is not hard to show that dg is never zero on M, so M is a surface.

The circles in M generated under revolution by each point of C are called the *parallels* of M; the different positions of C as it is rotated are called the *meridians* of M. This terminology derives from the geography of the sphere; however, a sphere is *not* a surface of revolution as defined above. Its profile curve must twice meet the axis of revolution, so two "parallels" reduce to single points. To simplify the statements of later theorems, we use a slightly different terminology in this case; see Exercise 12.

The necessity of the properness condition on the patches in Definition 1.2 is shown by the following example.

1.6 Example Suppose that a rectangular strip of tin is bent into a figure 8, as in Fig. 4.6. The configuration M that results does not satisfy our intuitive picture of what a surface should be, for along the axis A, M is not like the plane $\mathbf{R}^2$ but is instead like *two* intersecting planes. To express this construction in mathematical terms, let D be the rectangle $-\pi < u < \pi, 0 < v < 1$ in $\mathbf{R}^2$ and define $\mathbf{x}: D \to \mathbf{R}^3$ by $\mathbf{x}(u, v) = (\sin u, \sin 2u, v)$. It is easy to check that $\mathbf{x}$ is a patch, but its image $M = \mathbf{x}(D)$ is not a surface: $\mathbf{x}$ is not a *proper* patch. Continuity fails for $\mathbf{x}^{-1}: M \to D$ since, roughly speaking, to restore M to D, $\mathbf{x}^{-1}$ must tear M along the axis A (the z axis of $\mathbf{R}^3$).

By Example 1.5, the familiar *torus of revolution* T is a surface (Fig. 4.16). With somewhat more work, one could construct *double toruses* of various shapes, as in Fig. 4.7. By adding "handles" and "tubes" to existing surfaces one can—in principle, at least—construct surfaces of any desired degree of complexity (Fig. 4.8).

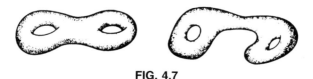

FIG. 4.7

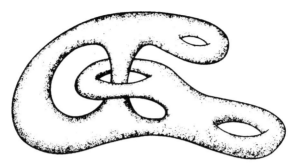

FIG. 4.8

Exercises

1. None of the following subsets M of $\mathbf{R}^3$ are surfaces. At which points $\mathbf{p}$ is it impossible to find a proper patch in M that will cover a neighborhood of $\mathbf{p}$ in M? (Sketch M—formal proofs not required.)
(a) Cone M: $z^2 = x^2 + y^2$
(b) Closed disk M: $x^2 + y^2 \leq 1, z = 0$.
(c) Folded plane M: $xy = 0, x \geq 0, y \geq 0$.

2. A *plane* in $\mathbf{R}^3$ is a surface M: $ax + by + cz = d$, where the numbers a, b, c are necessarily not all zero. Prove that every plane in $\mathbf{R}^3$ may be described by a vector equation as on page 62.

3. Sketch the general shape of the surface M: $z = ax^2 + by^2$ in each of the following cases:
(a) $a > b > 0$.   (b) $a > 0 > b$.
(c) $a > b = 0$.   (d) $a = b = 0$.

4. In which of the following cases is the mapping $\mathbf{x}$: $\mathbf{R}^2 \to \mathbf{R}^3$ a patch?
(a) $\mathbf{x}(u, v) = (u, uv, v)$.   (b) $\mathbf{x}(u, v) = (u^2, u^3, v)$.
(c) $\mathbf{x}(u, v) = (u, u^2, v + v^3)$.   (d) $\mathbf{x}(u, v) = (\cos 2\pi u, \sin 2\pi u, v)$.
(Recall that $\mathbf{x}$ is one-to-one if and only if $\mathbf{x}(u, v) = \mathbf{x}(u_1, v_1)$ implies $(u, v) = (u_1, v_1)$.)

5. (a) Prove that M: $(x^2 + y^2)^2 + 3z^2 = 1$ is a surface.
(b) For which values of c is M: $z(z - 2) + xy = c$ a surface?

6. Determine the intersection $z = 0$ of the *monkey saddle*

$$M: z = f(x, y), \quad f(x, y) = y^3 - 3yx^2,$$

with the xy plane. On which regions of the plane is $f > 0$? $f < 0$? How does this surface get its name? (*Hint:* see Fig. 5.19.)

7. Let $\mathbf{x}: D \to \mathbf{R}^3$ be a mapping, with

$$\mathbf{x}(u, v) = (x_1(u, v), x_2(u, v), x_3(u, v)).$$

(a) Prove that a point $\mathbf{p} = (p_1, p_2, p_3)$ of $\mathbf{R}^3$ is in the image $\mathbf{x}(D)$ if and only if the equations

$$p_1 = x_1(u, v), \quad p_2 = x_2(u, v), \quad p_3 = x_3(u, v)$$

can be solved for u and v, with (u, v) in D.

(b) If for every point $\mathbf{p}$ in $\mathbf{x}(D)$ these equations have the *unique* solution $u = f_1(p_1, p_2, p_3)$, $v = f_2(p_1, p_2, p_3)$, with (u, v) in D, prove that $\mathbf{x}$ is one-to-one and that $\mathbf{x}^{-1}: \mathbf{x}(D) \to D$ is given by the formula

$$\mathbf{x}^{-1}(\mathbf{p}) = (f_1(\mathbf{p}), f_2(\mathbf{p})).$$

8. Let $\mathbf{x}: D \to \mathbf{R}^3$ be the function given by

$$\mathbf{x}(u, v) = (u^2, uv, v^2)$$

on the first quadrant $D: u > 0$, $v > 0$. Show that $\mathbf{x}$ is one-to-one and find a formula for its inverse function $\mathbf{x}^{-1}: \mathbf{x}(D) \to D$. Then prove that $\mathbf{x}$ is a proper patch.

9. Let $\mathbf{x}: \mathbf{R}^2 \to \mathbf{R}^3$ be the mapping

$$\mathbf{x}(u, v) = (u + v, u - v, uv).$$

Show that $\mathbf{x}$ is a proper patch and that the image of $\mathbf{x}$ is the entire surface $M: z = (x^2 - y^2)/4$.

10. If $F: \mathbf{R}^3 \to \mathbf{R}^3$ is a diffeomorphism and M is a surface in $\mathbf{R}^3$, prove that the image $F(M)$ is also a surface in $\mathbf{R}^3$. (*Hint:* If $\mathbf{x}$ is a patch in M, then the composite function $F(\mathbf{x})$ is regular, since $F(\mathbf{x})_* = F_*\mathbf{x}_*$ by Ex. 9 of Sec. 1.7.)

11. Prove this special case of Exercise 10: If F is a diffeomorphism of $\mathbf{R}^3$, then the image of the surface $M: g = c$ is $\overline{M}: \overline{g} = c$, where $\overline{g} = g(F^{-1})$ and $\overline{M}$ is a surface. (*Hint:* If $dg(\mathbf{v}) \neq 0$ at $\mathbf{p}$ in M, show by using Ex. 7 of Sec. 1.7 that $d\overline{g}(F_*\mathbf{v}) \neq 0$ at $F(\mathbf{p})$.)

12. Let C be a Curve in the xy plane that is symmetric about the x axis. Assume C crosses the x axis and always does so orthogonally. Explain why there can be only one or two crossings. Thus C is either an arc or is closed (Fig. 4.9). Revolving C about the x axis gives a surface M, called an *augmented surface of revolution*. Explain how to define patches in M at the crossing points.

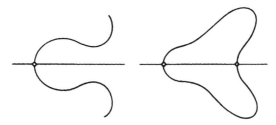

FIG. 4.9

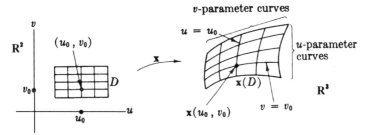

FIG. 4.10

4.2 Patch Computations

In Section 1, coordinate patches were used to define a surface; now we consider some properties of patches that will be useful in studying surfaces.

Let $\mathbf{x}: D \to \mathbf{R}^3$ be a coordinate patch. Holding u or v constant in the function $(u, v) \to \mathbf{x}(u, v)$ produces curves. Explicitly, for each point (u_0, v_0) in D the curve

$$u \to \mathbf{x}(u, v_0)$$

is called the *u-parameter curve*, $v = v_0$, of $\mathbf{x}$; and the curve

$$v \to \mathbf{x}(u_0, v)$$

is the *v-parameter curve*, $u = u_0$ (Fig. 4.10).

Thus, the image $\mathbf{x}(D)$ is covered by these two families of curves, which are the images under $\mathbf{x}$ of the horizontal and vertical lines in D, and one curve from each family goes through each point of $\mathbf{x}(D)$.

2.1 Definition   If $\mathbf{x}: D \to \mathbf{R}^3$ is a patch, for each point (u_0, v_0) in D:

(1) The velocity vector at u_0 of the *u-parameter curve*, $v = v_0$, is denoted by $\mathbf{x}_u(u_0, v_0)$.

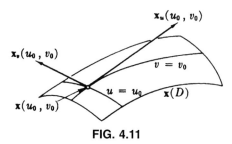

FIG. 4.11

(2) The velocity vector at v_0 of the v-parameter curve, $u = u_0$, is denoted by $\mathbf{x}_v(u_0, v_0)$.

The vectors $\mathbf{x}_u(u_0, v_0)$ and $\mathbf{x}_v(u_0, v_0)$ are called the *partial velocities* of $\mathbf{x}$ at (u_0, v_0) (Fig. 4.11).

Thus $\mathbf{x}_u$ and $\mathbf{x}_v$ are functions on D whose values at each point (u_0, v_0) are tangent vectors to $\mathbf{R}^3$ at $\mathbf{x}(u_0, v_0)$. The subscripts u and v are intended to suggest partial differentiation. Indeed if the patch is given in terms of its Euclidean coordinate functions by a formula

$$\mathbf{x}(u, v) = (x_1(u, v), x_2(u, v), x_3(u, v)),$$

then it follows from the definition above that the partial velocity functions are given by

$$\mathbf{x}_u = \left(\frac{\partial x_1}{\partial u}, \frac{\partial x_2}{\partial u}, \frac{\partial x_3}{\partial u} \right)_x,$$

$$\mathbf{x}_v = \left(\frac{\partial x_1}{\partial v}, \frac{\partial x_2}{\partial v}, \frac{\partial x_3}{\partial v} \right)_x.$$

The subscript $\mathbf{x}$ (frequently omitted) is a reminder that $\mathbf{x}_u(u, v)$ and $\mathbf{x}_v(u, v)$ have point of application $\mathbf{x}(u, v)$.

2.2 Example The *geographical patch* in the sphere. Let Σ be the sphere of radius $r > 0$ centered at the origin of $\mathbf{R}^3$. Longitude and latitude on the earth suggest a patch in Σ quite different from the Monge patch used on Σ in Section 1. The point $\mathbf{x}(u, v)$ of Σ with longitude u $(-\pi < u < \pi)$ and latitude v $(-\pi/2 < v < \pi/2)$ has Euclidean coordinates (Fig. 4.12).

$$\mathbf{x}(u, v) = (r \cos v \cos u, r \cos v \sin u, r \sin v).$$

With the domain D of $\mathbf{x}$ defined by these inequalities, the image $\mathbf{x}(D)$ of $\mathbf{x}$ is all of Σ except one semicircle from north pole to south pole. The u-

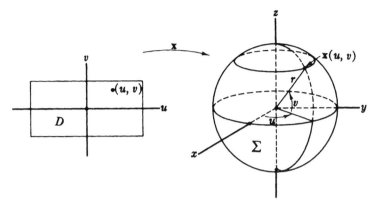

FIG. 4.12

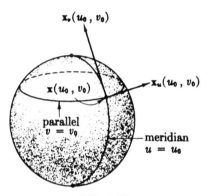

FIG. 4.13

parameter curve, $v = v_0$, is a circle—the parallel of latitude v_0. The v-parameter curve, $u = u_0$, is a semicircle—the meridian of longitude u_0.

We compute the partial velocities of $\mathbf{x}$ to be

$$\mathbf{x}_u(u, v) = r \, (-\cos v \sin u, \, \cos v \cos u, \, 0),$$

$$\mathbf{x}_v(u, v) = r \, (-\sin v \cos u, \, -\sin v \sin u, \, \cos v).$$

where r denotes a scalar multiplication. Evidently $\mathbf{x}_u$ always points due east, and $\mathbf{x}_v$ due north (Fig. 4.13). In a moment we shall give a formal proof that $\mathbf{x}$ is a patch in Σ.

To test whether a given subset M of $\mathbf{R}^3$ is a surface, Definition 1.2 demands proper patches (and Example 1.6 shows why). But once we know that M is a surface, the properness condition need no longer concern us (Exercise 14 of Section 3). Furthermore, in many situations the one-to-one restriction on patches can also be dropped.

2.3 Definition A regular mapping $\mathbf{x}: D \to \mathbf{R}^3$ whose image lies in a surface M is called a *parametrization* of the region $\mathbf{x}(D)$ in M.

(Thus a patch is merely a one-to-one parametrization.) In favorable cases this image $\mathbf{x}(D)$ may be the whole surface M, and we then have the analogue of the more familiar notion of parametrization of a Curve (see end of Section 1.4). Parametrizations will be of first importance in practical computations with surfaces, so we consider some ways of determining whether a mapping $\mathbf{x}: D \to \mathbf{R}^3$ is a parametrization of (part of) a given surface M.

The image of $\mathbf{x}$ must, of course, lie in M. Note that if the surface is given in the implicit form $M: g = c$, this means that the composite function $g(\mathbf{x})$ must have constant value c.

To test whether $\mathbf{x}$ is regular, note first that parameter curves and partial velocities $\mathbf{x}_u$ and $\mathbf{x}_v$ are well-defined for an arbitrary differentiable mapping $\mathbf{x}: D \to \mathbf{R}^3$. Also, the last two rows of the cross product

$$\mathbf{x}_u \times \mathbf{x}_v = \begin{vmatrix} U_1 & U_2 & U_3 \\ \dfrac{\partial x_1}{\partial u} & \dfrac{\partial x_2}{\partial u} & \dfrac{\partial x_3}{\partial u} \\ \dfrac{\partial x_1}{\partial v} & \dfrac{\partial x_2}{\partial v} & \dfrac{\partial x_3}{\partial v} \end{vmatrix}$$

give the (transposed) Jacobian matrix of $\mathbf{x}$ at each point. Thus the regularity of $\mathbf{x}$ is equivalent to the condition that $\mathbf{x}_u \times \mathbf{x}_v$ *is never zero*, or, by properties of the cross product, that at each point (u,v) of D *the partial velocity vectors of $\mathbf{x}$ are linearly independent.*

Let us try out these methods on the mapping $\mathbf{x}$ given in Example 2.2. Since the sphere is defined implicitly by $g = x^2 + y^2 + z^2 = r^2$, we must show that $g(\mathbf{x}) = r^2$. Substituting the coordinate functions of $\mathbf{x}$ for x, y, and z gives

$$r^{-2}g(\mathbf{x}) = \left(\cos v \cos u\right)^2 + \left(\cos v \sin u\right)^2 + \sin^2 v$$

$$= \cos^2 v + \sin^2 v = 1.$$

A short computation using the formulas for $\mathbf{x}_u$ and $\mathbf{x}_v$, given in Example 2.2, yields

$$r^{-2}\mathbf{x}_u \times \mathbf{x}_v = \cos u \cos^2 v U_1 + \sin u \cos^2 v U_2 + \cos v \sin v U_3.$$

Since $-\pi/2 < v < \pi/2$ in the domain D of $\mathbf{x}$, $\cos v$ is never zero there; but $\sin u$ and $\cos u$ are never zero simultaneously, so $\mathbf{x}_u \times \mathbf{x}_v$, is never zero on D. Thus $\mathbf{x}$ is regular—and hence is a parametrization. In fact, it remains a parametrization if the condition $-\pi < u < \pi$ is dropped, thus replacing D by the infinite strip $-\pi/2 < v < \pi/2$. In this case the u-parameter curves are periodic parametrizations of the meridians, and $\mathbf{x}$ covers the entire sphere except for the poles $(0, 0, \pm 1)$.

To show that $\mathbf{x}$ on the original domain D is a patch, it remains only to show that it is one-to-one on D, that is,

$$\mathbf{x}(u, v) = \mathbf{x}(u_1, v_1) \Rightarrow (u, v) = (u_1, v_1).$$

In view of the definition of $\mathbf{x}$, the vector equation here gives the three scalar equations

$$r \cos v \cos u = r \cos v_1 \cos u_1,$$

$$r \cos v \sin u = r \cos v_1 \sin u_1,$$

$$r \sin v = r \sin v_1.$$

Since $-\pi/2 < v < \pi/2$ in D, the last equation implies $v = v_1$. Thus $r \cos v = r \cos v_1 > 0$ can be canceled from the first two equations, and we conclude that $u = u_1$ as well.

The geographical definition of $\mathbf{x}$ in Example 2.2 makes the preceding results seem almost obvious, but the methods used will serve in more difficult cases.

2.4 Example Parametrization of a surface of revolution. Suppose that M is obtained, as in Example 1.5, by revolving a curve C in the upper half of the xy plane about the x axis. Now let

$$\alpha(u) = (g(u), h(u), 0)$$

be a parametrization of C (note that $h > 0$). As we observed in Example 1.5, when the point $(g(u), h(u), 0)$ on the profile curve C has been rotated through an angle v, it reaches a point $\mathbf{x}(u, v)$ with the same x coordinate $g(u)$, but new y and z coordinates $h(u) \cos v$ and $h(u) \sin v$, respectively (Fig. 4.14). Thus

$$\mathbf{x}(u, v) = (g(u), h(u) \cos v, h(u) \sin v).$$

Evidently this formula defines a mapping into M whose image is all of M. A short computation shows that $\mathbf{x}_u$ and $\mathbf{x}_v$ are always linearly independent, so $\mathbf{x}$ is a parametrization of M. The domain D of $\mathbf{x}$ consists of all points (u, v) for which u is in the domain of α. The u-parameter curves of $\mathbf{x}$ parametrize

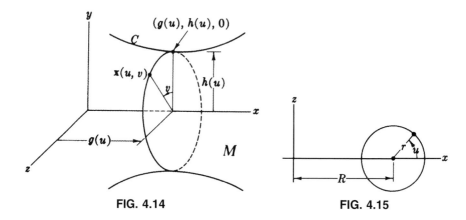

FIG. 4.14   **FIG. 4.15**

the meridians of M, the v-parameter curves the parallels. (Thus the parametrization $\mathbf{x}\colon D \to M$ is never one-to-one.)

Obviously we are not limited to rotating curves in the xy plane about the x axis. But with other choices of coordinates, we maintain the same geometric meaning for the functions g and h: g measures distance *along* the axis of revolution, while h measures distance *from* the axis of revolution.

Actually, the geographical patch in the sphere is one instance of Example 2.4 (with u and v reversed); here is another.

2.5 Example *Torus of revolution T.* This is the surface of revolution obtained when the profile curve C is a circle. Suppose that C is the circle in the xz plane with radius $r > 0$ and center $(R, 0, 0)$. We shall rotate about the z axis; hence we must require $R > r$ to keep C from meeting the axis of revolution. A natural parametrization (Fig. 4.15) for C is

$$\alpha(u) = (R + r \cos u, r \sin u).$$

Thus by the remarks above we must have $g(u) = r \sin u$ (distance *along* the z axis) and $h(u) = R + r\cos u$ (distance *from* the z axis). The general argument in Example 2.4—with coordinate axes permuted—then yields the parametrization

$$\mathbf{x}(u, v) = (h(u) \cos v, h(u) \sin v, g(u))$$

$$= ((R + r \cos u) \cos v, (R + r \cos u) \sin v, r \sin u).$$

We call $\mathbf{x}$ the *usual parametrization* of the torus (Fig. 4.16). Its domain is the whole plane $\mathbf{R}^2$, and it is periodic in both u and v:

$$\mathbf{x}(u + 2\pi, v) = \mathbf{x}(u, v + 2\pi) = \mathbf{x}(u, v) \quad \text{for all } (u, v).$$

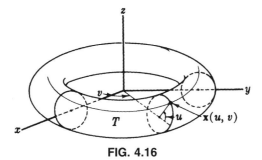

FIG. 4.16

2.6 Definition   A *ruled surface* is a surface swept out by a straight line L moving along a curve β. The various positions of the generating line L are called the *rulings* of the surface. Such a surface always has a *ruled parametrization*,

$$\mathbf{x}(u, v) = \beta(u) + v\delta(u).$$

We call β the *base curve* and δ the *director curve*, although δ is usually pictured as a vector field on β pointing along the line L.

Several examples of ruled surfaces are given in the following exercises. It is usually necessary to put restrictions on β and δ to ensure that $\mathbf{x}$ is a parametrization.

There are infinitely many different parametrizations and patches in any surface. Those we have discussed occur frequently and are fitted in a natural way to their surfaces.

Exercises

1.   Find a parametrization of the entire surface obtained by revolving:
(a) $C: y = \cosh x$ around the x axis (catenoid).
(b) $C: (x - 2)^2 + y^2 = 1$ around the y axis (torus).
(c) $C: z = x^2$ around the z axis (paraboloid).

2.   Partial velocities $\mathbf{x}_u$ and $\mathbf{x}_v$ are defined for an arbitrary mapping $\mathbf{x}: D \to \mathbf{R}^3$, so we can consider the real-valued functions

$$E = \mathbf{x}_u \cdot \mathbf{x}_u, \quad F = \mathbf{x}_u \cdot \mathbf{x}_v, \quad G = \mathbf{x}_v \cdot \mathbf{x}_v$$

on D. Prove

$$\left\| \mathbf{x}_u \times \mathbf{x}_v \right\|^2 = EG - F^2.$$

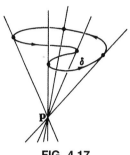

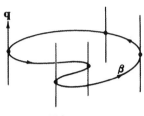

FIG. 4.17   FIG. 4.18

Deduce that $\mathbf{x}$ is a regular mapping if and only if $EG - F^2$ is never zero. (This is often the easiest way to check regularity. We will see, beginning in the next chapter, that the functions E, F, G are fundamental to the geometry of surfaces.)

3. A *generalized cone* is a ruled surface with a parametrization of the form

$$\mathbf{x}(u, v) = \mathbf{p} + v\delta(u).$$

Thus all rulings pass through the vertex $\mathbf{p}$ (Fig. 4.17). Show that $\mathbf{x}$ is regular if and only if v and $\delta \times \delta'$ are never zero. (Thus the vertex is never part of the cone. Unless the term *generalized* is used, we assume that δ is a closed curve and require either $v > 0$ or $v < 0$.)

4. A *generalized cylinder* is a ruled surface for which the rulings are all Euclidean parallel (Fig. 4.18). Thus there is always a parametrization of the form

$$\mathbf{x}(u, v) = \beta(u) + v\mathbf{q} \quad (\mathbf{q} \in \mathbf{R}^3).$$

Prove: (a) Regularity of $\mathbf{x}$ is equivalent to $\beta' \times \mathbf{q}$ never zero.
(b) If $C: f(x, y) = a$ is a Curve in the plane, show that in $\mathbf{R}^3$ the same equation defines a surface $\tilde{C}$. If $t \to (x(t), y(t))$ is a parametrization of C, find a parametrization of $\tilde{C}$ that shows it is a generalized cylinder.
Generalized cylinders are a rather broad category—including Euclidean planes when β is a straight line—so unless the term *generalized* is used, we assume that cylinders are over *closed* curves β.

5. A line L is attached orthogonally to an axis A (Fig. 4.19). If L moves steadily along A, rotating at constant speed, then L sweeps out a *helicoid H*.
When A is the z axis, H is the image of the mapping $\mathbf{x}: \mathbf{R}^2 \to \mathbf{R}^3$ such that

$$\mathbf{x}(u, v) = (u \cos v, u \sin v, bv) \quad (b \neq 0).$$

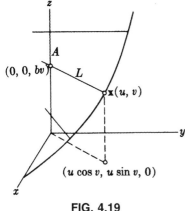

$(0, 0, bv)$

A

L

$x(u, v)$

z

y

$(u \cos v, u \sin v, 0)$

x

FIG. 4.19

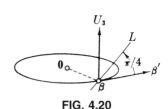

U_3

L

0

$\pi/4$

β

β'

FIG. 4.20

(a) Prove that x is a patch.

(b) Describe its parameter curves.

(c) Express the helicoid in the implicit form $g = c$.

(d) (*Computer graphics.*) Plot one full turn ($0 \leq v \leq 2\pi$) of a helicoid with $b = 1/2$. Restrict the rulings to $-1 \leq u \leq 1$.

6. (a) Show that the *saddle surface* M: $z = xy$ is doubly ruled: Find two ruled parametrizations with different rulings.

(b) (*Computer graphics.*) Plot a representative portion of M, using a patch for which the parameter curves are rulings.

7. Let β be a unit-speed parametrization of the unit circle in the xy plane. Construct a ruled surface as follows: Move a line L along β in such a way that L is always orthogonal to the radius of the circle and makes constant angle $\pi/4$ with β' (Fig. 4.20).

(a) Derive this parametrization of the resulting ruled surface M:

$$x(u, v) = \beta(u) + v(\beta'(u) + U_3).$$

(b) Express x explicitly in terms of v and coordinate functions for β.

(c) Deduce that M is given implicitly by the equation

$$x^2 + y^2 - z^2 = 1.$$

(d) Show that if the angle $\pi/4$ above is changed to $-\pi/4$, the same surface M results. Thus M is doubly ruled.

(e) Sketch this surface M showing the two rulings through each of the points $(1, 0, 0)$ and $(2, 1, 2)$.

8. Let M be the surface of revolution gotten by revolving the curve $t \rightarrow (g(t), h(t), 0)$ about the x axis ($h > 0$). Show that:

(a) If g' is never zero, then M has a parametrization of the form

$$\mathbf{x}(u, v) = (u, f(u) \cos v, f(u) \sin v).$$

(b) If h' is never zero, then M has a parametrization of the form

$$\mathbf{x}(u, v) = (f(u), u \cos v, u \sin v).$$

A *quadric surface* is a surface $M: g = 0$ in $\mathbf{R}^3$ such that g contains at most quadratic terms in x_1, x_2, x_3, that is,

$$g = \sum_{i,j} a_{ij} x_i x_j + \sum_i b_i x_i + c.$$

Trivial cases excepted, every quadric surface is congruent to one of the five types described in the next two exercises. (Use of computers is optional in these exercises.)

9. In each case, (i) show that M is a surface, and sketch its general shape when $a = 3$, $b = 2$, $c = 1$; (ii) show that $\mathbf{x}$ is a parametrization in M and describe what part of M it covers.

(a) *Ellipsoid.* $M: \dfrac{x^2}{a^2} + \dfrac{y^2}{b^2} + \dfrac{z^2}{c^2} = 1,$

$\mathbf{x}(u, v) = (a \cos u \cos v, b \cos u \sin v, c \sin u)$ on $D: -\pi/2 < u < \pi/2$.

(b) *Hyperboloid of one sheet* (Fig. 4.21).

$M: \dfrac{x^2}{a^2} + \dfrac{y^2}{b^2} - \dfrac{z^2}{c^2} = 1, \mathbf{x}(u, v) = (a \cosh u \cos v, b \cosh u \sin v, c \sinh u)$

on $\mathbf{R}^2$.

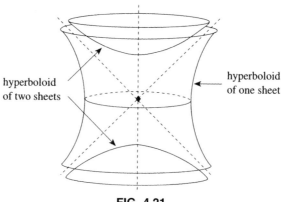

hyperboloid of two sheets

hyperboloid of one sheet

FIG. 4.21

(c) *Hyperboloid of two sheets* (Fig. 4.21).

$$M: \frac{x^2}{a^2} + \frac{y^2}{b^2} - \frac{z^2}{c^2} = -1, \; \mathbf{x}(u, v) = (a \sinh u \cos v, \, b \sinh u \sin v, \, c \cosh u)$$

on D: $u \neq 0$.

10. Sketch the following surfaces (graphs of functions) for $a = 2$, $b = 1$:

(a) *Elliptic paraboloid.* M: $z = \dfrac{x^2}{a^2} + \dfrac{y^2}{b^2}$. Show that

$$\mathbf{x}(u, v) = (au \cos v, \, bu \sin v, \, u^2), \quad u > 0,$$

is a parametrization that omits only one point of M.

(b) *Hyperbolic paraboloid.* M: $z = \dfrac{x^2}{a^2} - \dfrac{y^2}{b^2}$.

Show that M is covered by the single patch

$$\mathbf{x}(u, v) = (a(u + v), \, b(u - v), \, 4uv) \text{ on } \mathbf{R}^2.$$

11. *Doubly ruled quadrics.*

(a) Show that the hyperbolic paraboloid M in the preceding exercise is doubly ruled.

(b) (*Computer graphics.*) For $a = 2$, $b = 1$ use the patch in (b) of Exercise 10 to plot a portion of M. (Keep the same scale on all axes; the parameter curves will be the rulings.)

(c) Find two different ruled parametrizations of the hyperboloid of one sheet by using the scheme in the special case, Exercise 7.

(d) (*Computer graphics.*) Plot a portion of each of these parametrizations, taking $a = 1.5$, $b = 1$, $c = 2$.

4.3 Differentiable Functions and Tangent Vectors

We now begin an exposition of the calculus on a surface M in $\mathbf{R}^3$. The space $\mathbf{R}^3$ will gradually fade out of the picture, since our ultimate goal is a calculus for M alone. Generally speaking, we shall follow the order of topics in Chapter 1, making such changes as are necessary to adapt the calculus of the plane $\mathbf{R}^2$ to a surface M.

Suppose that f a is real-valued function defined on a surface M. If $\mathbf{x} \colon D \to M$ is a coordinate patch in M, then the composite function $f(\mathbf{x})$ is called a *coordinate expression* for f; it is just an ordinary real-valued function $(u, v) \to f(\mathbf{x}(u, v))$. We define f to be *differentiable* provided all its coordinate expressions are differentiable in the usual Euclidean sense (Definition 1.3 of Chapter 1).

For a function $F: \mathbf{R}^n \to M$, each patch $\mathbf{x}$ in M gives a *coordinate expression* $\mathbf{x}^{-1}(F)$ for F. Evidently this composite function is defined only on the set $\mathcal{O}$ of all points $\mathbf{p}$ of $\mathbf{R}^n$ such that $F(\mathbf{p})$ is in $\mathbf{x}(D)$. Again we define F to be *differentiable* provided all its coordinate expressions are differentiable in the usual Euclidean sense. We must understand that this includes the requirement that $\mathcal{O}$ be an open set of $\mathbf{R}^n$, so that the differentiability of $\mathbf{x}^{-1}(F): \mathcal{O} \to \mathbf{R}^2$ is well-defined, as in Section 7 of Chapter 1

In particular, a *curve* $\alpha: I \to M$ in a surface M is, as before, a differentiable function from an open interval I into M.

To see how this definition works out in practice, we examine an important special case.

3.1 Lemma If α is a curve $\alpha: I \to M$ whose route lies in the image $\mathbf{x}(D)$ of a single patch $\mathbf{x}$, then there exist unique differentiable functions a_1, a_2 on I such that

$$\alpha(t) = \mathbf{x}(a_1(t), a_2(t)) \quad \text{for all } t,$$

or in functional notation, $\alpha = \mathbf{x}(a_1, a_2)$. (See Fig. 4.22.)

Proof. By definition, the coordinate expression $\mathbf{x}^{-1}\alpha: I \to D$ is differentiable—it is just a curve in $\mathbf{R}^2$ whose route lies in the domain D of $\mathbf{x}$. If a_1, a_2 are the Euclidean coordinate functions of $\mathbf{x}^{-1}\alpha$, then

$$\alpha = \mathbf{x}\mathbf{x}^{-1}\alpha = \mathbf{x}(a_1, a_2).$$

These are the only such functions, for if $\alpha = \mathbf{x}(b_1, b_2)$, then

$$(a_1, a_2) = \mathbf{x}^{-1}\alpha = \mathbf{x}^{-1}\mathbf{x}(b_1, b_2) = (b_1, b_2). \quad \blacklozenge$$

These functions a_1, a_2 are called the *coordinate functions* of the curve α with respect to the patch $\mathbf{x}$.

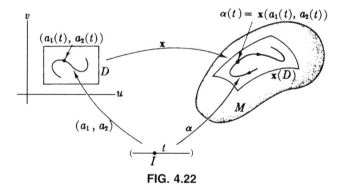

FIG. 4.22

For an arbitrary patch $\mathbf{x}: D \to M$, it is natural to think of the domain D as a *map* of the region $\mathbf{x}(D)$ in M. The functions $\mathbf{x}$ and $\mathbf{x}^{-1}$ establish a one-to-one correspondence between objects in $\mathbf{x}(D)$ and objects in D. If a curve α in $\mathbf{x}(D)$ represents the voyage of a ship, the coordinate curve (a_1, a_2) plots its position on the map D.

A rigorous proof of the following technical fact requires the methods of advanced calculus, and we shall not attempt to give a proof here.

3.2 Theorem   Let M be a surface in $\mathbf{R}^3$. If $F: \mathbf{R}^n \to \mathbf{R}^3$ is a (differentiable) mapping whose image lies in M, then considered as a function $F. \mathbf{R}^n \to M$ into M, F is differentiable (as defined above).

This theorem links the calculus of M tightly to the calculus of $\mathbf{R}^3$. For example, it implies the "obvious" result that a curve in $\mathbf{R}^3$ that lies in M is a curve of M.

Since a patch is a differentiable function from an open set of $\mathbf{R}^2$ into $\mathbf{R}^3$, it follows that a patch is a differentiable function into M. Hence its coordinate expressions are all differentiable, so *patches overlap smoothly.*

3.3 Corollary   If $\mathbf{x}$ and $\mathbf{y}$ are patches in a surface M in $\mathbf{R}^3$ whose images overlap, then the composite functions $\mathbf{x}^{-1}\mathbf{y}$ and $\mathbf{y}^{-1}\mathbf{x}$ are (differentiable) mappings defined on open sets of $\mathbf{R}^2$.

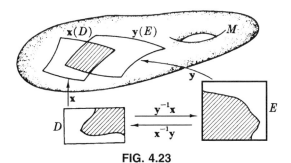

FIG. 4.23

The function $\mathbf{y}^{-1}\mathbf{x}$, for example, is defined only for those points (u, v) in D such that $\mathbf{x}(u, v)$ lies in the image $\mathbf{y}(E)$ of $\mathbf{y}$ (Fig. 4.23).

By an argument like that for Lemma 3.1, Corollary 3.3 can be rewritten:

3.4 Corollary   If $\mathbf{x}$ and $\mathbf{y}$ are overlapping patches in M, then there exist unique differentiable functions $\bar{u}$ and $\bar{v}$ such that

$$\mathbf{y}(u, v) = \mathbf{x}(\bar{u}(u, v), \bar{v}(u, v))$$

for all (u, v) in the domain of $\mathbf{x}^{-1}\mathbf{y}$. In functional notation: $\mathbf{y} = \mathbf{x}(\bar{u}, \bar{v})$.

There are, of course, symmetrical equations expressing $\mathbf{x}$ in terms of $\mathbf{y}$.

Corollary 3.3 makes it much easier to prove differentiability. For example, if f is a real-valued function on M, instead of verifying that *all* coordinate expressions $f(\mathbf{x})$ are Euclidean differentiable, we need only do so for enough patches $\mathbf{x}$ to cover all of M (so a single patch will often be enough). The proof is an exercise in checking domains of composite functions: For an *arbitrary* patch $\mathbf{y}$, $f\mathbf{x}$ and $\mathbf{x}^{-1}\mathbf{y}$ differentiable imply $f\mathbf{x}\mathbf{x}^{-1}\mathbf{y}$ differentiable. This function is in general not $f\mathbf{y}$, because its domain is too small. But since there are enough $\mathbf{x}$'s to cover M, such functions constitute all of $f\mathbf{y}$, and thus prove that it is differentiable.

It is intuitively clear what it means for a vector to be tangent to a surface M in $\mathbf{R}^3$. A formal definition can be based on the idea that a curve in M must have all its velocity vectors tangent to M.

3.5 Definition   Let $\mathbf{p}$ be a point of a surface M in $\mathbf{R}^3$. A tangent vector $\mathbf{v}$ to $\mathbf{R}^3$ at $\mathbf{p}$ is *tangent to M at $\mathbf{p}$* provided $\mathbf{v}$ is a velocity of some curve in M (Fig. 4.24).

The set of all tangent vectors to M at $\mathbf{p}$ is called the *tangent plane of M at* $\mathbf{p}$ and is denoted by $T_p(M)$. The following result shows, in particular, that at each point $\mathbf{p}$ of M the tangent plane $T_p(M)$ is actually a 2-dimensional vector subspace of the tangent space $T_p(\mathbf{R}^3)$.

3.6 Lemma   Let $\mathbf{p}$ be a point of a surface M in $\mathbf{R}^3$, and let $\mathbf{x}$ be a patch in M such that $\mathbf{x}(u_0, v_0) = \mathbf{p}$. A tangent vector $\mathbf{v}$ to $\mathbf{R}^3$ at $\mathbf{p}$ is tangent to M if and only if $\mathbf{v}$ can be written as a linear combination of $\mathbf{x}_u(u_0, v_0)$ and $\mathbf{x}_v(u_0, v_0)$.

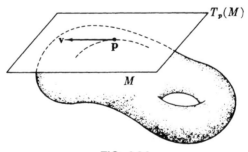

FIG. 4.24

Since partial velocities are always linearly independent, we deduce that they provided a *basis* for the tangent plane of M at each point of $\mathbf{x}(D)$.

Proof. Note that the parameter curves of $\mathbf{x}$ are curves in M, so $\mathbf{x}_u$ and $\mathbf{x}_v$ are always tangent to M at $\mathbf{p}$.

First suppose that $\mathbf{v}$ is tangent to M at $\mathbf{p}$; thus there is a curve α in M such that $\alpha(0) = \mathbf{p}$ and $\alpha'(0) = \mathbf{v}$. Now by Lemma 3.1, α may be written

$$\alpha = \mathbf{x}(a_1, a_2);$$

hence by the chain rule,

$$\alpha' = \mathbf{x}_u(a_1, a_2)\frac{da_1}{dt} + \mathbf{x}_v(a_1, a_2)\frac{da_2}{dt}.$$

But since $\alpha(0) = \mathbf{p} = \mathbf{x}(u_0, v_0)$, we have $a_1(0) = u_0$, $a_2(0) = v_0$. Hence evaluation at $t = 0$ yields

$$\mathbf{v} = \alpha'(0) = \frac{da_1}{dt}(0)\mathbf{x}_u(u_0, v_0) + \frac{da_2}{dt}(0)\mathbf{x}_v(u_0, v_0).$$

Conversely, suppose that a tangent vector $\mathbf{v}$ to $\mathbf{R}^3$ can be written

$$\mathbf{v} = c_1\mathbf{x}_u(u_0, v_0) + c_2\mathbf{x}_v(u_0, v_0).$$

By computations as above, $\mathbf{v}$ is the velocity vector at $t = 0$ of the curve

$$t \to \mathbf{x}(u_0 + tc_1, v_0 + tc_2).$$

Thus $\mathbf{v}$ is tangent to M at $\mathbf{p}$. ◆

A reasonable deduction, based on the general properties of derivatives, is that the tangent plane $T_p(M)$ is the linear approximation of the surface M near $\mathbf{p}$.

3.7 Definition A Euclidean vector field Z on a surface M in $\mathbf{R}^3$ is a function that assigns to each point $\mathbf{p}$ of M a tangent vector $Z(\mathbf{p})$ to $\mathbf{R}^3$ at $\mathbf{p}$.

A Euclidean vector field V for which each vector $V(\mathbf{p})$ is tangent to M at $\mathbf{p}$ is called a *tangent vector field* on M (Fig. 4.25). Frequently these vector fields are defined, not on all of M, but only on some region in M. As usual, we always assume differentiability.

A Euclidean vector $\mathbf{z}$ at a point $\mathbf{p}$ of M is *normal* to M if it is orthogonal to the tangent plane $T_p(M)$—that is, to every tangent vector to M at $\mathbf{p}$. And a Euclidean vector field Z on M is a *normal vector field* on M provided each vector $Z(\mathbf{p})$ is normal to M.

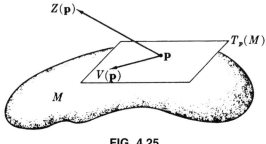

FIG. 4.25

Because $T_p(M)$ is a two-dimensional subspace of $T_p(\mathbf{R}^3)$, there is only one direction normal to M at $\mathbf{p}$: All normal vectors $\mathbf{z}$ at $\mathbf{p}$ are collinear.

Thus if $\mathbf{z}$ is not zero, it follows that $T_p(M)$ *consists of precisely those vectors* in $T_p(\mathbf{R}^3)$ *that are orthogonal* to $\mathbf{z}$.

It is particularly easy to deal with tangent and normal vector fields on a surface given in implicit form.

3.8 Lemma   If $M: g = c$ is a surface in $\mathbf{R}^3$, then the *gradient* vector field $\nabla g = \sum \partial g / \partial x_i\, U_i$ (considered only at points of M) is a nonvanishing normal vector field on the entire surface M.

Proof.   The gradient is nonvanishing (that is, never zero) on M since in the implicit case we require that the partial derivatives $\partial g / \partial x_i$ cannot simultaneously be zero at any point of M.

We must show that $(\nabla g)(\mathbf{p}) \cdot \mathbf{v} = 0$ for every tangent vector $\mathbf{v}$ to M at $\mathbf{p}$. First note that if α is a curve in M, then $g(\alpha) = g(\alpha_1, \alpha_2, \alpha_3)$ has constant value c. Thus by the chain rule,

$$\sum \frac{\partial g}{\partial x_i}(\alpha)\frac{d\alpha_i}{dt} = 0.$$

Now choose α to have initial velocity

$$\alpha'(0) = \mathbf{v} = (v_1, v_2, v_3)$$

at $\alpha(0) = \mathbf{p}$. Then

$$0 = \sum \frac{\partial g}{\partial x_i}(\alpha(0))\frac{d\alpha_i}{dt}(0) = \sum \frac{\partial g}{\partial x_i}(\mathbf{p})v_i = (\nabla g)(\mathbf{p}) \cdot \mathbf{v}.$$

$\blacklozenge$

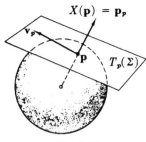

FIG. 4.26

3.9 Example Vector fields on the sphere Σ: $g = \sum x_i^2 = r^2$. The lemma shows that

$$X = \frac{1}{2}\nabla g = \sum x_i U_i$$

is a normal vector field on Σ (Fig. 4.26). This is geometrically evident, since $X(\mathbf{p}) = \sum p_i U_i(\mathbf{p})$ is the vector $\mathbf{p}$ with point of application $\mathbf{p}$! It follows by a remark above that $\mathbf{v}_p$ is tangent to Σ if and only if the dot product $\mathbf{v}_p \cdot \mathbf{p}_p = \mathbf{v} \cdot \mathbf{p}$ is zero. Similarly, a vector field V on Σ is a *tangent* vector field if and only if $V \cdot X = 0$. For example, $V(\mathbf{p}) = (-p_2, p_1, 0)$ defines a tangent vector field on Σ that points "due east" and vanishes at the north and south poles $(0, 0, \pm r)$.

We must emphsasize that only the *tangent* vector fields on M belong to the calculus of M itself, since they derive ultimately from curves in M (Definition 3.5). This is certainly not the case with *normal* vector fields. However, as we shall see in the next chapter, normal vector fields are quite useful in studying M from the viewpoint of an observer in $\mathbf{R}^3$.

Finally, we shall adapt the notion of directional derivatives to a surface. Definition 3.1 of Chapter 1 uses straight lines in $\mathbf{R}^3$; thus we must use the less restrictive formulation based on Lemma 4.6 of Chapter 1.

3.10 Definition Let $\mathbf{v}$ be a tangent vector to M at $\mathbf{p}$, and let f be a differentiable real-valued function on M. The *derivative* $\mathbf{v}[f]$ *of f with respect to $\mathbf{v}$* is the common value of $(d/dt)(f\alpha)(0)$ for all curves α in M with initial velocity $\mathbf{v}$.

Directional derivatives on a surface have exactly the same linear and Leibnizian properties as in the Euclidean case (Theorem 3.3 of Chapter 1).

Exercises

1. Let $\mathbf{x}$ be the geographical patch in the sphere Σ (Ex. 2.2). Find the coordinate expression $f(\mathbf{x})$ for the following functions on Σ:
(a) $f(\mathbf{p}) = p_1^2 + p_2^2$.   (b) $f(\mathbf{p}) = (p_1 - p_2)^2 + p_3^2$.

2. Let $\mathbf{x}$ be the usual parametrization of the torus (Ex. 2.5).
(a) Find the Euclidean coordinates α_1, α_2, α_3 of the curve $\alpha(t) = \mathbf{x}(t, t)$.
(b) Show that α is periodic, and find its period $p > 0$, the smallest number such that $\alpha(t + p) = \alpha(t)$ for all t.

3. (a) Prove Corollary 3.4.
(b) Derive the chain rule

$$ \mathbf{y}_u = \frac{\partial \bar{u}}{\partial u}\mathbf{x}_u + \frac{\partial \bar{v}}{\partial u}\mathbf{x}_v, \quad \mathbf{y}_v = \frac{\partial \bar{u}}{\partial v}\mathbf{x}_u + \frac{\partial \bar{v}}{\partial v}\mathbf{x}_v, $$

where $\mathbf{x}_u$ and $\mathbf{x}_v$ are evaluated on $(\bar{u}, \bar{v})$.
(c) Deduce that $\mathbf{y}_u \times \mathbf{y}_v = J\mathbf{x}_u \times \mathbf{x}_v$, where J is the Jacobian of the mapping $\mathbf{x}^{-1}\mathbf{y} = (\bar{u}, \bar{v})$: $D \to \mathbf{R}^2$.

4. Let $\mathbf{x}$ be a patch in M.
(a) If $\mathbf{x}_*$ is the tangent map of $\mathbf{x}$ (Sec. 7 of Ch. 1), show that

$$ \mathbf{x}_*(U_1) = \mathbf{x}_u, \quad \mathbf{x}_*(U_2) = \mathbf{x}_v, $$

where U_1, U_2 is the natural frame field on $\mathbf{R}^2$.
(b) If f is a differentiable function on M, prove

$$ \mathbf{x}_u[f] = \frac{\partial}{\partial u}(f(\mathbf{x})), \quad \mathbf{x}_v[f] = \frac{\partial}{\partial v}(f(\mathbf{x})). $$

5. Prove that:
(a) $\mathbf{v} = (v_1, v_2, v_3)$ is tangent to $M: z = f(x, y)$ at a point $\mathbf{p}$ of M if and only if

$$ v_3 = \frac{\partial f}{\partial x}(p_1, p_2)v_1 + \frac{\partial f}{\partial y}(p_1, p_2)v_2. $$

(b) if $\mathbf{x}$ is a patch in an arbitrary surface M, then $\mathbf{v}$ is tangent to M at $\mathbf{x}(u, v)$ if and only if

$$ \mathbf{v} \cdot \mathbf{x}_u(u, v) \times \mathbf{x}_v(u, v) = 0. $$

6. Let $\mathbf{x}$ and $\mathbf{y}$ be the patches in the unit sphere Σ that are defined on the unit disk $D: u^2 + v^2 < 1$ by

$$ \mathbf{x}(u, v) = (u, v, f(u, v)), \quad \mathbf{y}(u, v) = (v, f(u, v), u), $$

where $f = \sqrt{1 - u^2 - v^2}$.

(a) On a sketch of Σ indicate the images $\mathbf{x}(D)$ and $\mathbf{y}(D)$, and the region on which they overlap.

(b) At which points of D is $\mathbf{y}^{-1}\mathbf{x}$ defined? Find a formula for this function.

(c) At which points of D is $\mathbf{x}^{-1}\mathbf{y}$ defined? Find a formula for this function.

7. Find a nonvanishing normal vector field on $M: z = xy$ and two tangent vector fields that are linearly independent at each point.

8. Let C be the circular cone parametrized by

$$\mathbf{x}(u, v) = v(\cos u, \sin u, 1).$$

If α is the curve $\mathbf{x}(\sqrt{2}t, e^t)$:

(a) Express α' in terms of $\mathbf{x}_u$ and $\mathbf{x}_v$.

(b) Show that at each point of α, the velocity α' bisects the angle between $\mathbf{x}_u$ and $\mathbf{x}_v$. (*Hint:* Verify that

$$\alpha' \cdot \left(\frac{\mathbf{x}_u}{\|\mathbf{x}_u\|} \right) = \alpha' \cdot \left(\frac{\mathbf{x}_v}{\|\mathbf{x}_v\|} \right),$$

where $\mathbf{x}_u$ and $\mathbf{x}_v$ are evaluated on $(\sqrt{2}t, e^t)$.)

(c) Make a sketch of the cone C showing the curve α.

9. If $\mathbf{z}$ is a nonzero vector normal to M at $\mathbf{p}$, let $\overline{T}_p(M)$ be the Euclidean plane through $\mathbf{p}$ orthogonal to $\mathbf{z}$. Prove:

(a) If each tangent vector $\mathbf{v}_p$ to M at $\mathbf{p}$ is replaced by its *tip* $\mathbf{p} + \mathbf{v}$, then $T_p(M)$ becomes $\overline{T}_p(M)$. Thus $\overline{T}_p(M)$ gives a concrete representation of $T_p(M)$ in $\mathbf{R}^3$. It is called the *Euclidean tangent plane to M at p*.

(b) If $\mathbf{x}$ is a patch in M, then $\overline{T}_{x(u,v)}(M)$ consists of all points $\mathbf{r}$ in $\mathbf{R}^3$ such that $(\mathbf{r} - \mathbf{x}(u, v)) \cdot \mathbf{x}_u(u, v) \times \mathbf{x}_v(u, v) = 0$.

(c) If M is given implicitly by $g = c$, then $\overline{T}_P(M)$ consists of all points $\mathbf{r}$ in $\mathbf{R}^3$ such that $(\mathbf{r} - \mathbf{p}) \cdot (\nabla g)(\mathbf{p}) = 0$.

10. In each case below find an equation of the form $ax + by + cz = d$ for the plane $\overline{T}_p(M)$.

(a) $\mathbf{p} = (0, 0, 0)$ and M is the sphere

$$x^2 + y^2 + (z - 1)^2 = 1.$$

(b) $\mathbf{p} = (1, -2, 3)$ and M is the ellipsoid

$$\frac{x^2}{4} + \frac{y^2}{16} + \frac{z^2}{18} = 1.$$

(c) $\mathbf{p} = \mathbf{x}(2, \pi/4)$, where M is the helicoid parametrized by

$$\mathbf{x}(u, v) = (u \cos v, u \sin v, 2v).$$

11. (*Continuation of Ex.* 2.) With **x** the usual parametrization of the torus of revolution T, consider the curve α: $\mathbf{R} \to T$ such that $\alpha(t) = \mathbf{x}(at, bt)$.
(a) If a/b is a rational number, show that α is a simple closed curve in T, that is, periodic with no self-crossings.
(b) If a/b is irrational, show α is one-to-one. Such a curve is called a *winding line* on the torus. It is *dense* in T in the sense that given any $\varepsilon > 0$, α comes within distance ε of every point of T.
(c) (*Computer graphics.*) For reference, plot the torus T with $R = 3, r = 1$ (see Ex. 2.5). Then plot the following curves in T:
(i) $\alpha(t) = \mathbf{x}(3t, 5t)$ on intervals $0 \leq t \leq b$, for $b = \pi, 2\pi$, and larger values. Estimate the period of α, in this case the smallest number $T > 0$ such that $\alpha(T) = \alpha(0)$.
(ii) $\alpha(t) = \mathbf{x}(\pi t, 5t)$ on intervals $0 \leq t \leq b$, for increasing values of b. (Keep the curve reasonably smooth.)

12. A Euclidean vector field $Z = \sum z_i U_i$ on M is *differentiable* provided its coordinate functions z_1, z_2, z_3 (on M) are differentiable. If V is a tangent vector field on M, show that
(a) For every patch **x**: $D \to M$, V can be written as

$$V(\mathbf{x}(u, v)) = f(u, v)\mathbf{x}_u(u, v) + g(u, v)\mathbf{x}_v(u, v).$$

(b) V is differentiable if and only if the functions f and g (on D) are differentiable.
The following exercises deal with *open sets* in a surface M in $\mathbf{R}^3$, that is, sets $\mathcal{U}$ in M that contain a neighborhood in M of each of their points.

13. Prove that if **y**: $E \to M$ is a proper patch, then **y** carries open sets in E to open sets in M. Deduce that if **x**: $D \to M$ is an arbitrary patch, then the image $\mathbf{x}(D)$ is an open set in M. (*Hint:* To prove the latter assertion, use Cor. 3.3.)

14. Prove that *every* patch **x**: $D \to M$ in a surface M in $\mathbf{R}^3$ is proper. (*Hint:* Use Ex. 13. Note that $(\mathbf{x}^{-1}\mathbf{y})\mathbf{y}^{-1}$ is continuous and agrees with $\mathbf{x}^{-1}$ on an open set in $\mathbf{x}(D)$.)

15. If $\mathcal{U}$ is a subset of a surface M in $\mathbf{R}^3$, prove that $\mathcal{U}$ is itself a surface in $\mathbf{R}^3$ if and only if $\mathcal{U}$ is an open set of M.

4.4 Differential Forms on a Surface

In Chapter 1 we discussed differential forms on $\mathbf{R}^3$ only in sufficient detail to take care of the Cartan structural equations (Theorem 8.3 of Chapter 2). In the next three sections we shall give a rather complete treatment of forms *on a surface.*

Forms are just what we will need to describe the geometry of a surface, but this is only one example of their usefulness. Surfaces and Euclidean spaces are merely special cases of the general notion of *manifold* (Section 8). Every manifold has a differential and integral calculus—expressed in terms of functions, vector fields, and forms—that generalizes the usual elementary calculus on the real line. Thus forms are fundamental to all the many branches of mathematics and its applications that are based on calculus. In the special case of a surface, the calculus of forms is rather easy, but it still gives a remarkably accurate picture of the most general case.

Just as for $\mathbf{R}^3$, a 0-*form f* on a surface M is simply a (differentiable) real-valued function on M, and a 1-*form* ϕ on M is a real-valued function on tangent vectors to M that is linear at each point (Definition 5.1 of Chapter 1). We did not give a precise definition of 2-forms in Chapter 1, but we shall do so now. A 2-form will be a two-dimensional analogue of a 1-form: a real-valued function, not on single tangent vectors, but on *pairs* of tangent vectors. (In this context the term "pair" will always imply that the tangent vectors have the same point of application.)

4.1 Definition A 2-*form* η on a surface M is a real-valued function on all ordered pairs of tangent vectors $\mathbf{v}$, $\mathbf{w}$ to M such that

(1) $\eta(\mathbf{v}, \mathbf{w})$ is linear in $\mathbf{v}$ and in $\mathbf{w}$;
(2) $\eta(\mathbf{v}, \mathbf{w}) = -\eta(\mathbf{w}, \mathbf{v})$.

Since a surface is two-dimensional, *all p-forms with p > 2 are zero*, by definition. This fact considerably simplifies the theory of differential forms on a surface.

At the end of this section we will show that our new definitions are consistent with the informal exposition given in Chapter 1, Section 6.

Forms are added in the usual pointwise fashion; we add only forms of the same *degree p* = 0, 1, 2. Just as a 1-form ϕ is evaluated on a vector field V, now a 2-form η is evaluated on a pair of vector fields V, W to give a real-valued function $\eta(V, W)$ on the surface M. Of course, we shall always assume that the forms we deal with are differentiable—that is, convert differentiable vector fields into differentiable functions.

Note that the alternation rule (2) in Definition 4.1 implies that

$$\eta(\mathbf{v}, \mathbf{v}) = 0$$

for any tangent vector $\mathbf{v}$. This rule also shows that 2-forms are related to determinants.

4.2 Lemma   Let η be a 2-form on a surface M, and let $\mathbf{v}$ and $\mathbf{w}$ be (linearly independent) tangent vectors at some point of M. Then

$$\eta(a\mathbf{v} + b\mathbf{w},\, c\mathbf{v} + d\mathbf{w}) = \begin{vmatrix} a & b \\ c & d \end{vmatrix} \eta(\mathbf{v},\, \mathbf{w}).$$

Proof.   Since η is linear in its first variable, its value on the pair of tangent vectors $a\mathbf{v} + b\mathbf{w}$, $c\mathbf{v} + d\mathbf{w}$ is $a\eta(\mathbf{v},\, c\mathbf{v} + d\mathbf{w}) + b\eta(\mathbf{w},\, c\mathbf{v} + d\mathbf{w})$. Using the linearity of η in its second variable, we get

$$ac\,\eta(\mathbf{v},\, \mathbf{v}) + ad\,\eta(\mathbf{v},\, \mathbf{w}) + bc\,\eta(\mathbf{w},\, \mathbf{v}) + bd\,\eta(\mathbf{w},\, \mathbf{w}).$$

Then the alternation rule (2) gives

$$\eta(a\mathbf{v} + b\mathbf{w},\, c\mathbf{v} + d\mathbf{w}) = (ad - bc)\,\eta(\mathbf{v},\, \mathbf{w}). \qquad \blacklozenge$$

Thus the values of a 2-form on *all* pairs of tangent vectors at a point are completely determined by its value on any *one* linearly independent pair. This remark is used frequently in later work.

Wherever they appear, differential forms satisfy certain general properties, established (at least partially) in Chapter 1 for forms on $\mathbf{R}^3$. To begin with, *the wedge product of a p-form and a q-form is always a $(p + q)$-form.* If p or q is zero, the wedge product is just the usual multiplication by a function. On a surface, the wedge product is always zero if $p + q > 2$. So we need a definition only for the case $p = q = 1$.

4.3 Definition   If ϕ and ψ are 1-forms on a surface M, the *wedge product* $\phi \wedge \psi$ is the 2-form on M such that

$$(\phi \wedge \psi)(\mathbf{v},\, \mathbf{w}) = \phi(\mathbf{v})\psi(\mathbf{w}) - \phi(\mathbf{w})\psi(\mathbf{v})$$

for all pairs $\mathbf{v}$, $\mathbf{w}$ of tangent vectors to M.

Note that $\phi \wedge \psi$ really is a 2-form on M, since it is a real-valued function on all pairs of tangent vectors and satisfies the conditions in Definition 4.1. The wedge product has all the usual algebraic properties except commutativity; in general, *if ξ is a p-form and η is a q-form, then*

$$\xi \wedge \eta = (-1)^{pq}\eta \wedge \xi.$$

On a surface the only minus sign occurs in the multiplication of 1-forms, where just as in Chapter 1, we have $\phi \wedge \psi = -\psi \wedge \phi$, and hence $\phi \wedge \phi = 0$.

The differential calculus of forms is based on the exterior derivative d. For a 0-form (function) f on a surface, the exterior derivative is, as before, the

1-form df such that $df(\mathbf{v}) = \mathbf{v}[f]$. Wherever forms appear, the exterior derivative of a p-form is a $(p + 1)$-form. Thus, for surfaces the only new definition we need is that of the exterior derivative $d\phi$ of a 1-form ϕ.

4.4 Definition Let ϕ be a 1-form on a surface M. Then the *exterior derivative $d\phi$* of ϕ is the 2-form such that for any patch $\mathbf{x}$ in M,

$$d\phi(\mathbf{x}_u, \mathbf{x}_v) = \frac{\partial}{\partial u}(\phi(\mathbf{x}_v)) - \frac{\partial}{\partial v}(\phi(\mathbf{x}_u)).$$

As it stands, this is not yet a valid definition; there is a problem of consistency. What we have actually defined is a form $d_\mathbf{x}\phi$ on the image of each patch $\mathbf{x}$ in M. So what we must prove is that on the region where two patches overlap, the forms $d_\mathbf{x}\phi$ and $d_\mathbf{y}\phi$ are equal. Only then will we have obtained from ϕ a single form $d\phi$ on M.

4.5 Lemma Let ϕ be a 1-form on M. If $\mathbf{x}$ and $\mathbf{y}$ are patches in M, then $d_\mathbf{x}\phi = d_\mathbf{y}\phi$ on the overlap of $\mathbf{x}(D)$ and $\mathbf{y}(E)$.

Proof. Because $\mathbf{y}_u$ and $\mathbf{y}_v$ are linearly independent at each point, it suffices by Lemma 4.2 to show that

$$(d_\mathbf{y}\phi)(\mathbf{y}_u, \mathbf{y}_v) = (d_\mathbf{x}\phi)(\mathbf{y}_u, \mathbf{y}_v).$$

Now, as in Corollary 3.4, we write $\mathbf{y} = \mathbf{x}(\bar{u}, \bar{v})$ and deduce by the chain rule that

$$\mathbf{y}_u = \frac{\partial \bar{u}}{\partial u}\mathbf{x}_u + \frac{\partial \bar{v}}{\partial u}\mathbf{x}_v,$$

$$\mathbf{y}_v = \frac{\partial \bar{u}}{\partial v}\mathbf{x}_u + \frac{\partial \bar{v}}{\partial v}\mathbf{x}_v, \tag{1}$$

where $\mathbf{x}_u$ and $\mathbf{x}_v$ are henceforth *evaluated on* $(\bar{u}, \bar{v})$. Then by Lemma 4.2,

$$(d_\mathbf{x}\phi)(\mathbf{y}_u, \mathbf{y}_v) = J(d_\mathbf{x}\phi)(\mathbf{x}_u, \mathbf{x}_v), \tag{2}$$

where J is the Jacobian $(\partial\bar{u}/\partial u)(\partial\bar{v}/\partial v) - (\partial\bar{u}/\partial v)(\partial\bar{v}/\partial u)$. Thus it is clear from Definition 4.4 that to prove $(d_\mathbf{y}\phi)(\mathbf{y}_u, \mathbf{y}_v) = (d_\mathbf{x}\phi)(\mathbf{y}_u, \mathbf{y}_v)$, all we need is the equation.

$$\frac{\partial}{\partial u}(\phi(\mathbf{y}_v)) - \frac{\partial}{\partial v}(\phi(\mathbf{y}_v)) = J\left\{\frac{\partial}{\partial\bar{u}}(\phi(\mathbf{x}_v)) - \frac{\partial}{\partial\bar{v}}(\phi(\mathbf{x}_v))\right\}. \tag{3}$$

It suffices to operate on $(\partial/\partial u)(\phi(\mathbf{y}_v))$, for merely reversing u and v will then yield $(\partial/\partial v)(\phi(\mathbf{y}_u))$. Since (3) requires us to *subtract* these two derivatives, we can *discard any terms that will cancel* when u and v are everywhere reversed.

Applying ϕ to the second equation in (1) yields

$$\phi(\mathbf{y}_v) = \phi(\mathbf{x}_u)\frac{\partial \bar{u}}{\partial v} + \phi(\mathbf{x}_v)\frac{\partial \bar{v}}{\partial v}.$$

Hence by the chain rule,

$$\frac{\partial}{\partial u}(\phi(\mathbf{y}_v)) = \frac{\partial}{\partial u}(\phi(\mathbf{x}_u))\frac{\partial \bar{u}}{\partial v} + \frac{\partial}{\partial u}(\phi(\mathbf{x}_v))\frac{\partial \bar{v}}{\partial v} + \cdots, \tag{4}$$

where in accordance with the remark above we have discarded two symmetric terms. Next we use the chain rule—and the same remark—to get

$$\frac{\partial}{\partial u}(\phi(\mathbf{y}_v)) = \left(\frac{\partial}{\partial v}(\phi(\mathbf{x}_u))\frac{\partial \bar{v}}{\partial u} + \cdots\right)\frac{\partial \bar{u}}{\partial v} + \left(\frac{\partial}{\partial u}(\phi(\mathbf{x}_v))\frac{\partial \bar{u}}{\partial u} + \cdots\right)\frac{\partial \bar{v}}{\partial v}. \tag{5}$$

Now reverse u and v in (5) (and also $\bar{u}$ and $\bar{v}$) and subtract from (5) itself. The result is precisely equation (3).   ◆

It is difficult to exaggerate the importance of the exterior derivative. We have already seen in Chapter 1 that it generalizes the familiar notion of differential of a function, and that it contains the three fundamental derivative operations of classical vector analysis (Exercise 8 in Section 1.6). In Chapter 2 it is essential to the Cartan structural equations (Theorem 8.3). Perhaps the clearest statement of its meaning will come in Stokes's theorem (6.5), which could actually be used to define the exterior derivative of a 1-form.

On a surface, the exterior derivative of a wedge product displays the same linear and Leibnizian properties (Theorem 6.4 of Chapter 2) as in $\mathbf{R}^3$; see Exercise 3. For practical computations these properties are apt to be more efficient than a direct appeal to the definition. Examples of this technique appear in subsequent exercises.

The most striking property of this notion of derivative is that there are no *second* exterior derivatives: Wherever forms appear, *the exterior derivative applied twice always gives zero.* For a surface, we need only prove this for 0-forms, since even for a 1-form ϕ, the second derivative $d(d\phi)$ is a 3-form, and hence is automatically zero.

4.6 Theorem   If f is a real-valued function on M, then $d(df) = 0$.

Proof.   Let $\psi = df$, so we must show $d\psi = 0$. It suffices by Lemma 4.2 to prove that for any patch $\mathbf{x}$ in M we have $(d\psi)(\mathbf{x}_u, \mathbf{x}_v) = 0$. Now using Exercise 4 of Section 3, we get

$$\psi(\mathbf{x}_u) = df(\mathbf{x}_u) = \mathbf{x}_u[f] = \frac{\partial}{\partial u}(f(\mathbf{x}))$$

and similarly

$$\psi(\mathbf{x}_v) = \frac{\partial}{\partial v}(f(\mathbf{x})).$$

Hence

$$d\psi(\mathbf{x}_u, \mathbf{x}_v) = \frac{\partial}{\partial u}(\psi(\mathbf{x}_v)) - \frac{\partial}{\partial v}(\psi(\mathbf{x}_u)) = \frac{\partial^2(f(\mathbf{x}))}{\partial u \partial v} - \frac{\partial^2(f(\mathbf{x}))}{\partial v \partial u} = 0. \qquad \blacklozenge$$

Many computations and proofs reduce to the problem of showing that two forms are equal. As we have seen, to do so it is not necessary to check that the forms have the same value on *all* tangent vectors. In particular, if $\mathbf{x}$ is a coordinate patch, then

(1) for 1-forms on $\mathbf{x}(D)$: $\phi = \psi$ if and only if $\phi(\mathbf{x}_u) = \psi(\mathbf{x}_u)$ and $\phi(\mathbf{x}_v) = \psi(\mathbf{x}_v)$;
(2) for 2-forms on $\mathbf{x}(D)$: $\mu = v$ if and only if $\mu(\mathbf{x}_u, \mathbf{x}_v) = v(\mathbf{x}_u, \mathbf{x}_v)$.

(To prove these criteria, we express arbitrary tangent vectors as linear combinations of $\mathbf{x}_u$ and $\mathbf{x}_v$.) More generally, $\mathbf{x}_u$ and $\mathbf{x}_v$ may be replaced by any two vector fields that are linearly independent at each point.

Let us now check that the rigorous results proved in this section are consistent with the rules of operation stated in Chapter 1, Section 6.

4.7 Example Differential forms on the plane $\mathbf{R}^2$. Let $u_1 = u$ and $u_2 = v$ be the natural coordinate functions, and U_1, U_2 the natural frame field on $\mathbf{R}^2$. The differential calculus of forms on $\mathbf{R}^2$ is expressed in terms of u_1 and u_2 as follows:

If f is a function, ϕ a 1-form, and η a 2-form, then

(1) $\phi = f_1 du_1 + f_2 du_2$, where $f_i = \phi(U_i)$.
(2) $\eta = g du_1 du_2$, where $g = \eta(U_1, U_2)$.
(3) for $\psi = g_1 du_1 + g_2 du_2$ and ϕ as above,

$$\phi \wedge \psi = (f_1 g_2 - f_2 g_1) du_1 \, du_2.$$

(4) $df = \dfrac{\partial f}{\partial u_1} du_1 + \dfrac{\partial f}{\partial u_2} du_2.$

(5) $d\phi = \left(\dfrac{\partial f_2}{\partial u_1} - \dfrac{\partial f_1}{\partial u_2} \right) du_1 \, du_2$ (ϕ as above).

For the proof of these formulas, see Exercise 4.

Similar definitions and coordinate expressions may be established on any Euclidean space. In the case of the real line $\mathbf{R}^1$, the natural frame field reduces to the single vector field U_1 for which $U_1 = [f] = df/dt$. All p-forms with $p > 0$ are zero, and for a 1-form, $\phi = \phi(U_1)\, dt$.

Wherever differential forms are used, the following conditions are fundamental.

4.8 Definition   A differential form ϕ is *closed* if its exterior derivative is zero, $d\phi = 0$; and ϕ is *exact* if it is the exterior derivative of some form, $\phi = d\xi$.

Since d applied twice is always 0, *every exact form is closed.* In the case of a surface, since d increases degrees by 1, every 2-form is closed and no 0-form (i.e., function) is exact. Thus 1-forms are the important case, and for a 1-form ϕ exactness always means that there is a *function f* such that $df = \phi$. The analytical and topological consequences of these definitions run deep.

Exercises

1.   Prove the Leibnizian formulas

$$d(fg) = g\, df + f\, dg, \quad d(f\phi) = df \wedge \phi + f\, d\phi,$$

where f and g are functions on M and ϕ is a 1-form.
(*Hint:* By definition, $(f\phi)(v_p) = f(\mathbf{p})\phi(v_p)$; hence $f\phi$ evaluated on $\mathbf{x}_u$ is $f(\mathbf{x})\phi(\mathbf{x}_u)$.)

2.   (a) Prove formulas (1) and (2) in Example 4.7 using the remark preceding Example 4.7. (*Hint:* Show $(du_1 du_2)\,(U_1, U_2) = 1$.)
(b) Derive the remaining formulas using the properties of d and the wedge product.

3.   If f is a real-valued function on a surface, and g is a function on the real line, show that

$$v_p[g(f)] = g'(f)v_p[f].$$

Deduce that

$$d(g(f)) = g'(f)df.$$

4.   If f, g, and h are functions on a surface M, and ϕ is a 1-form, prove:
(a) $d(fgh) = ghdf + fhdg + fgdh$,
(b) $d(\phi f) = fd\phi - \phi \wedge df, \quad (\phi f = f\phi)$,
(c) $(df \wedge dg)\,(\mathbf{v}, \mathbf{w}) = \mathbf{v}[f]\mathbf{w}[g] - \mathbf{v}[g]\mathbf{w}[f]$.

5.   Suppose that M is covered by open sets $\mathcal{U}_1, \ldots, \mathcal{U}_k$, and on each $\mathcal{U}_i$ there is defined a function f_i such that $f_i - f_j$ is constant on the overlap of $\mathcal{U}_i$ and $\mathcal{U}_j$. Show that there is a 1-form ϕ on M such that $\phi = df_i$ on each $\mathcal{U}_i$. Generalize to the case of 1-forms ϕ_i such that $\phi_i - \phi_j$ is closed.

6.   Let $\mathbf{y}: E \to M$ be an arbitrary mapping of an open set of $\mathbf{R}^2$ into a surface M. If ϕ is a 1-form on M, show that the formula

$$d\phi(\mathbf{y}_u, \mathbf{y}_v) = \frac{\partial}{\partial u}(\phi(\mathbf{y}_v)) - \frac{\partial}{\partial v}(\phi(\mathbf{y}_u))$$

is still valid even when $\mathbf{y}$ is not regular or one-to-one.
   (*Hint:* In the proof of Lem. 4.5, check that equation (3) is still valid in this case.)

   A patch $\mathbf{x}$ in M establishes a one-to-one correspondence between an open set D of $\mathbf{R}^2$ and an open set $\mathbf{x}(D)$ of M. Although we have emphasized the function $\mathbf{x}: D \to \mathbf{x}(D)$, there are some advantages to emphasizing instead the inverse function $\mathbf{x}^{-1}: \mathbf{x}(D) \to D$.

7.   If $\mathbf{x}: D \to M$ is a patch in M, let $\tilde{u}$ and $\tilde{v}$ be the coordinate functions of $\mathbf{x}^{-1}$, so $\mathbf{x}^{-1}(\mathbf{p}) = (\tilde{u}(\mathbf{p}), \tilde{v}(\mathbf{p}))$ for all $\mathbf{p}$ in $\mathbf{x}(D)$. Show that
   (a) $\tilde{u}$ and $\tilde{v}$ are differentiable functions on $\mathbf{x}(D)$ such that:

$$\tilde{u}(\mathbf{x}(u, v)) = u, \quad \tilde{v}(\mathbf{x}(u, v)) = v.$$

These functions constitute the *coordinate system* associated with $\mathbf{x}$.

   (b) $d\tilde{u}(\mathbf{x}_u) = 1, \quad d\tilde{u}(\mathbf{x}_v) = 0,$

   $d\tilde{v}(\mathbf{x}_u) = 0, \quad d\tilde{v}(\mathbf{x}_v) = 1.$

   (c) If ϕ is a 1-form and η is a 2-form, then

$$\phi = f \, d\tilde{u} + g \, d\tilde{v}, \quad \text{where } f(\mathbf{x}) = \phi(\mathbf{x}_u), \, g(\mathbf{x}) = \phi(\mathbf{x}_v);$$
$$\eta = h \, d\tilde{u} \, d\tilde{v}, \quad \text{where } h(\mathbf{x}) = \eta(\mathbf{x}_u, \mathbf{x}_v).$$

(*Hint:* for (b) use Ex. 4(b) of Sec. 3.)

8.   Identify (or describe) the associated coordinate system $\tilde{u}$, $\tilde{v}$ of
   (a) The polar coordinate patch $\mathbf{x}(u, v) = (u\cos v, u\sin v)$ defined on the domain $D: u > 0, 0 < v < 2\pi$.
   (b) The identity patch $\mathbf{x}(u, v) = (u, v)$ in $\mathbf{R}^2$.
   (c) The geographical patch $\mathbf{x}$ in the sphere.

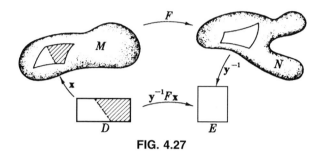

FIG. 4.27

4.5 Mappings of Surfaces

To define differentiability of a function *from* a surface *to* a surface, we follow
the same general scheme used in Section 3 and require that all its coordinate
expressions be differentiable.

5.1 Definition   A function $F: M \to N$ from one surface to another is
differentiable provided that for each patch $\mathbf{x}$ in M and $\mathbf{y}$ in N the composite
function $\mathbf{y}^{-1}F\mathbf{x}$ is Euclidean differentiable (and defined on an open set of $\mathbf{R}^2$).
F is then called a *mapping of surfaces*.

   Evidently the function $\mathbf{y}^{-1}F\mathbf{x}$ is defined at all points (u, v) of D such that
$F(\mathbf{x}(u, v))$ lies in the image of $\mathbf{y}$ (Fig. 4.27). As in Section 3 we deduce from
Corollary 3.3 that in applying this definition, it suffices to check enough
patches to cover both M and N.

5.2 Example   (1) Let Σ be the unit sphere in $\mathbf{R}^3$ (center at $\mathbf{0}$) but with
north and south poles removed, and let C be the cylinder based on the unit
circle in the xy plane. So C is in contact with the sphere along the equator.
We define a mapping $F: \Sigma \to C$ as follows: If $\mathbf{p}$ is a point of Σ, draw the line
orthogonally out from the z azis through $\mathbf{p}$, and let $F(\mathbf{p})$ be the point at which
this line first meets C, as in Fig. 4.28. To prove that F is a mapping, we use
the geographical patch $\mathbf{x}$ in Σ (Example 2.2), and for C the patch $\mathbf{y}(u, v) =$
$(\cos u, \sin u, v)$. Now $\mathbf{x}(u, v) = (\cos v \cos u, \cos v \sin u, \sin v)$, so from the definition of F we get

$$F(\mathbf{x}(u, v)) = (\cos u, \sin u, \sin v).$$

But this point of C is $\mathbf{y}(u, \sin v)$; hence

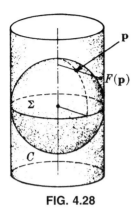

FIG. 4.28

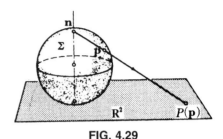

FIG. 4.29

$$F(\mathbf{x}(u, v)) = \mathbf{y}(u, \sin v).$$

Applying $\mathbf{y}^{-1}$ to both sides of this equation gives

$$(\mathbf{y}^{-1} F \mathbf{x})(u, v) = (u, \sin v),$$

so $\mathbf{y}^{-1} F \mathbf{x}$ is certainly differentiable. (Actually, $\mathbf{x}$ does not entirely cover Σ, but the missing semicircle can be covered by a patch like $\mathbf{x}$.) We conclude that F is a mapping.

(2) *Stereographic projection of the punctured sphere* Σ *onto the plane.* Let Σ be a unit sphere resting on the xy plane at the origin, so the center of Σ is at $(0, 0, 1)$. *Delete the north pole* $\mathbf{n} = (0, 0, 2)$ from Σ. Now imagine that there is a light source at the north pole, and for each point $\mathbf{p}$ of Σ, let $P(\mathbf{p})$ be the shadow of $\mathbf{p}$ in the xy plane (Fig. 4.29). As usual, we identify the xy plane with $\mathbf{R}^2$ by $(p_1, p_2, 0) \leftrightarrow (p_1, p_2)$. Thus we have defined a function P from Σ onto $\mathbf{R}^2$. Evidently P has the form

$$P(p_1, p_2, p_3) = \left(\frac{Rp_1}{r}, \frac{Rp_2}{r} \right),$$

where r and R are the distances from $\mathbf{p}$ and $P(\mathbf{p})$, respectively, to the z axis. But from the similar triangles in Fig. 4.30, we see that $R/2 = r/(2 - p_3)$; hence

$$P(p_1, p_2, p_3) = \left(\frac{2p_1}{2 - p_3}, \frac{2p_2}{2 - p_3} \right).$$

Now if $\mathbf{x}$ is any patch in Σ, the composite function $P\mathbf{x}$ is Euclidean differentiable, so $P: \Sigma \to \mathbf{R}^2$ is a mapping.

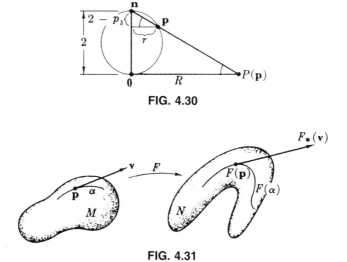

FIG. 4.30

FIG. 4.31

Just as for mappings of Euclidean space, each mapping of surfaces has a tangent map.

5.3 Definition Let $F: M \to N$ be a mapping of surfaces. The *tangent map* F_* of F assigns to each tangent vector $\mathbf{v}$ to M the tangent vector $F_*(\mathbf{v})$ to N such that if $\mathbf{v}$ is the initial velocity of a curve α in M, then $F_*(\mathbf{v})$ is the initial velocity of the image curve $F(\alpha)$ in N (Fig. 4.31).

Furthermore, at each point $\mathbf{p}$, the tangent map F_* is a linear transformation from the tangent plane $T_p(M)$ to the tangent plane $T_{F(p)}(N)$ (see Exercise 9). It follows immediately from the definition that F_* preserves velocities of curves: If $\bar{\alpha} = F(\alpha)$ is the image in N of a curve α in M, then $F_*(\alpha') = \bar{\alpha}'$. As in the Euclidean case, we deduce the convenient property that the tangent map of a composition is the composition of the tangent maps (Exercise 10).

The tangent map of a mapping $F: M \to N$ may be computed in terms of partial velocities as follows. If $\mathbf{x}: D \to M$ is a parametrization in M, let $\mathbf{y}$ be the composite mapping $F(\mathbf{x}): D \to N$ (which need not be a parametrization). Obviously, F carries the parameter curves of $\mathbf{x}$ to the corresponding parameter curves of $\mathbf{y}$. Since F_* preserves velocities of curves, it follows at once that

$$F_*(\mathbf{x}_u) = \mathbf{y}_u, \quad F_*(\mathbf{x}_v) = \mathbf{y}_v.$$

Since $\mathbf{x}_u$ and $\mathbf{x}_v$ give a basis for the tangent space of M at each point of $\mathbf{x}(D)$, these readily computable formulas completely determine F_*.

The discussion of regular mappings in Section 7 of Chapter 1 translates easily to the case of a mapping of surfaces $F: M \to N$. F is *regular* provided all of its derivative maps $F*_p: T_p(M) \to T_{F(p)}(N)$ are one-to-one. Since these tangent planes have the same dimension, we know from linear algebra that the one-to-one requirement is equivalent to $F*$ being a linear isomorphism. A mapping $F: M \to N$ that has an inverse mapping $F^{-1}: N \to M$ is called a *diffeomorphism*. We may think of a diffeomorphism F as smoothly distorting M to produce N. By applying the Euclidean formulation of the inverse function theorem to a coordinate expression $\mathbf{y}^{-1}F\mathbf{x}$ for F, we can deduce this extension of the inverse function theorem (7.10 of Chapter 1).

5.4 Theorem   Let $F: M \to N$ be a mapping of surfaces, and suppose that $F*_p: T_p(M) \to T_{F(p)}(N)$ is a linear isomorphism at some one point $\mathbf{p}$ of M. Then there exists a neighborhood $\mathscr{U}$ of $\mathbf{p}$ in M such that the restriction of F to $\mathscr{U}$ is a diffeomorphism onto a neighborhood $\mathscr{V}$ of $F(\mathbf{p})$ in N.

An immediate consequence is this useful result: *A one-to-one regular mapping F of M onto N is a diffeomorphism.* For since F is one-to-one and onto, it has a unique inverse function F^{-1}, and F^{-1} is a differentiable mapping since on each neighborhood $\mathscr{V}$ as above, it coincides with the inverse of the diffeomorphism $\mathscr{U} \to \mathscr{V}$. Surfaces M and N are said to be *diffeomorphic* if there exists a diffeomorphism from M to N.

Diffeomorphisms have little respect for size or shape; here are some examples.

5.5 Example   (1) Any open rectangle in the plane $\mathbf{R}^2$ is diffeomorphic to the entire plane. Take $R: -\pi/2 < u, v < \pi/2$ for simplicity. Then $F(u,v) = (\tan u, \tan v)$ is a mapping of R onto $\mathbf{R}^2$. Using a branch of the inverse tangent function, the mapping $F^{-1}(u_1, v_1) = (\tan^{-1}(u_1), \tan^{-1}(v_1))$ is a differentiable inverse of F, so F is a diffeomorphism.

(2) The sphere Σ minus one point is also diffeomorphic to the entire plane. Stereographic projection P, as in Example 5.2(2), is a one-to-one mapping of the punctured sphere Σ_0 onto $\mathbf{R}^2$. A variant

$$\mathbf{x}(u, v) = (\cos v \cos u, \cos v \sin u, 1 + \sin v)$$

of the usual geographical parametrization is a parametrization of $\Sigma_0 - \mathbf{0}$. The formula for P in Example 5.2 gives

$$\mathbf{y}(u, v) = P(\mathbf{x}(u, v)) = \frac{2 \cos v}{1 - \sin v}(\cos u, \sin u).$$

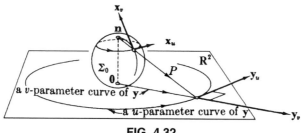

FIG. 4.32

Since $P_*(\mathbf{x}_u) = \mathbf{y}_u$ and $P_*(\mathbf{x}_v) = \mathbf{y}_v$, the regularity of F can be checked by computing $\mathbf{y}_u$ and $\mathbf{y}_v$. These turn out to be orthogonal and nonzero, hence linearly independent, as suggested by Fig. 4.32.

(3) A cylinder C over a closed curve is diffeomorphic to the plane minus one point. For simplicity, take $C: x^2 + y^2 = 1$, and define a mapping $F: C \to \mathbf{R}^2$ by $F(x, y, z) = e^z(x, y)$. Since e^z takes on all values $r > 0$, F maps C onto $\mathbf{R}^2 - \mathbf{0}$.

For the inverse of F, experimentation suggests

$$G(u, v) = \left(\frac{u}{\sqrt{u^2 + v^2}}, \frac{v}{\sqrt{u^2 + v^2}}, \log \sqrt{u^2 + v^2} \right).$$

To prove $G = F^{-1}$, compute $G(F(x, y, z)) = (x, y, z)$ and $F(G(u, v)) = (u, v)$.

Differential forms have the remarkable property that they can be moved from one surface to another by means of an arbitrary mapping.† Let us experiment with a 0-form, that is, a real-valued function. If $F: M \to N$ is a mapping of surfaces and f is a function on M, there is simply no reasonable general way to move f over to a function on N. But if instead, f is a function on N, the problem is easy; we pull f back to the composite function $f(F)$ on M. The corresponding pull-back for 1-forms and 2-forms is accomplished as follows.

5.6 Definition   Let $F: M \to N$ be a mapping of surfaces.
   (1) If ϕ is a 1-form on N, let $F^*\phi$ be the 1-form on M such that

$$(F^*\phi)(\mathbf{v}) = \phi(F_*\mathbf{v})$$

for all tangent vectors $\mathbf{v}$ to M.
   (2) If η is a 2-form on N, let $F^*\eta$ be the 2-form on M such that

---

† This is not true for vector fields.

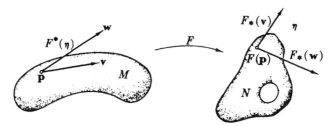

FIG. 4.33

$$(F^*\eta)(\mathbf{v}, \mathbf{w}) = \eta(F_*\mathbf{v}, F_*\mathbf{w})$$

for all pairs of tangent vectors $\mathbf{v}$, $\mathbf{w}$ on M (Fig. 4.33).

When we are dealing with a function f in its role as a 0-form, we shall sometimes write F^*f instead of $f(F)$, in accordance with the notation for the pullback of 1-forms and 2-forms.

The essential operations on forms are sum, wedge product, and exterior derivative; all are preserved by mappings.

5.7 Theorem Let $F: M \to N$ be a mapping of surfaces, and let ξ and η be forms on N. Then
(1) $F^*(\xi + \eta) = F^*\xi + F^*\eta$,
(2) $F^*(\xi \wedge \eta) = F^*\xi \wedge F^*\eta$,
(3) $F^*(d\xi) = d(F^*\xi)$.

Proof. In (1), ξ and η are both assumed to be p-forms (degree $p = 0, 1,$ 2) and the proof is a routine computation. In (2), we must allow ξ and η to have different degrees. When, say, ξ is a function f, the given formula means simply $F^*(f\eta) = f(F)F^*(\eta)$. In any case, the proof of (2) is also a straightforward computation. But (3) is more interesting. The easier case when ξ is a function is left as an exercise, and we address ourselves to the case where ξ is a 1-form.

It suffices to show that for every patch $\mathbf{x}: D \to M$,

$$(d(F^*\xi))(\mathbf{x}_u, \mathbf{x}_v) = (F^*(d\xi))(\mathbf{x}_u, \mathbf{x}_v).$$

Let $\mathbf{y} = F(\mathbf{x})$, and recall that $F_*(\mathbf{x}_u) = \mathbf{y}_u$ and $F_*(\mathbf{x}_v) = \mathbf{y}_v$. Thus, using the definitions of d and F^*, we get

$$d(F^* \xi)(\mathbf{x}_u, \mathbf{x}_v) = \frac{\partial}{\partial u}\{(F^* \xi)(\mathbf{x}_v)\} - \frac{\partial}{\partial v}\{(F^* \xi)(\mathbf{x}_u)\}$$

$$= \frac{\partial}{\partial u}\{\xi(F_* \mathbf{x}_v)\} - \frac{\partial}{\partial v}\{\xi(F_* \mathbf{x}_u)\}$$

$$= \frac{\partial}{\partial u}\{\xi(\mathbf{y}_v)\} - \frac{\partial}{\partial v}\{\xi(\mathbf{y}_u)\}.$$

Even if $\mathbf{y}$ is not a patch, Exercise 6 of Section 4 shows that this last expression is still equal to $d\xi(\mathbf{y}_u, \mathbf{y}_v)$. But

$$d\xi(\mathbf{y}_u, \mathbf{y}_v) = d\xi(F_* \mathbf{x}_u, F_* \mathbf{x}_v) = (F^* (d\xi))(\mathbf{x}_u, \mathbf{x}_v).$$

Thus we conclude that $d(F^*\xi)$ and $F^*(d\xi)$ have the same value on $(\mathbf{x}_u, \mathbf{x}_v)$. ◆

The elegant formulas in Theorem 5.7 are the key to the deeper study of mappings. In Chapter 6 we shall apply them to the connection forms of frame fields to get fundamental information about the geometry of mappings of surfaces.

Exercises

1. Let M and N be surfaces in $\mathbf{R}^3$. If $F: \mathbf{R}^3 \to \mathbf{R}^3$ is a mapping such that the image $F(M)$ of M is contained in N, then the restriction of F to M is a function $F|M: M \to N$. Prove that $F|M$ is a mapping of surfaces. (*Hint:* Use Thm. 3.2.)

2. Let Σ be the sphere of radius r with center at the origin of $\mathbf{R}^3$. Describe the effect of the following mappings $F: \Sigma \to \Sigma$ on the meridians and parallels of Σ.
(a) $F(\mathbf{p}) = -\mathbf{p}$.
(b) $F(p_1, p_2, p_3) = (p_3, p_1, p_2)$.
(c) $F(p_1, p_2, p_3) = \left(\dfrac{p_1 + p_2}{\sqrt{2}}, \dfrac{p_1 - p_2}{\sqrt{2}}, -p_3 \right).$

3. Let M be a *simple surface*, that is, one that is the image of a single proper patch $\mathbf{x}: D \to \mathbf{R}^3$. If $\mathbf{y}: D \to N$ is any mapping into a surface N, show that the function $F: M \to N$ such that

$$F(\mathbf{x}(u, v)) = \mathbf{y}(u, v) \quad \text{for all } (u, v) \text{ in } D$$

is a mapping of surfaces. (*Hint:* Write $F = \mathbf{y}\mathbf{x}^{-1}$, and use Cor. 3.3.)

4. Use the preceding exercise to construct a mapping of the helicoid H (Ex. 2.5) onto the torus T (Ex. 2.5) such that the rulings of H are carried to the meridians of T.

5. If Σ is the sphere $\| \mathbf{p} \| = r$, the mapping $A: \Sigma \to \Sigma$ such that $A(\mathbf{p}) = -\mathbf{p}$ is called the *antipodal map* of Σ. Prove that A is a diffeomorphism and that $A*(\mathbf{v}_p) = (-\mathbf{v})_{-p}$.

6. A regular mapping $F: M \to N$ of surfaces is often called a *local diffeomorphism*. For such a mapping F, prove that, in fact, every point $\mathbf{p}$ of M has a neighborhood $\mathcal{U}$ such that $F|\mathcal{U}$ is a diffeomorphism of $\mathcal{U}$ onto a neighborhood of $F(\mathbf{p})$ in N.

7. If $\mathbf{x}: D \to M$ is a parametrization, prove that the restriction of $\mathbf{x}$ to a sufficiently small neighborhood of a point (u_0, v_0) in D is a patch in M. (Thus any parametrization can be cut into patches.)

8. Let $F: M \to N$ be a mapping. If $\mathbf{x}$ is a patch in M, then as in the text, let $\mathbf{y} = F(\mathbf{x})$. (Although $\mathbf{y}$ maps into N, it is not necessarily a patch.) For a curve

$$\alpha(t) = \mathbf{x}(a_1(t), a_2(t))$$

in M, show that the image curve $\bar{\alpha} = F(\alpha)$ in N, has velocity

$$\bar{\alpha}' = \frac{da_1}{dt} \mathbf{y}_u(a_1, a_2) + \frac{da_2}{dt} \mathbf{y}_v(a_1, a_2).$$

9. Prove: (a) The invariance property needed to justify the definition (5.3) of the tangent map.
(b) Tangent maps $F_*: T_p(M) \to T_{F(p)}(N)$ are linear transformations.

10. Given mappings $M \xrightarrow{F} N \xrightarrow{G} P$, let $GF: M \to P$ be the composite mapping. Show that
(a) GF is differentiable,   (b) $(GF)_* = G_* F_*$,
(c) $(GF)^* = F^* G^*$,
that is, for any form ξ on P, $(GF)^*(\xi) = F^*(G^*(\xi))$. (Note the reversal of factors, caused by the fact that forms travel in the opposite direction from points and tangent vectors.)

11. Prove that every surface of revolution is diffeomorphic to either a torus or a cylinder. (*Hint:* Parametrize profile curves on the same interval.) (As Fig. 4.9 suggests, every *augmented* surface of revolution is diffeomorphic to either a plane or a sphere.)

12. (a) Show that the inverse mapping P^{-1} of the stereographic projection $P: \Sigma_0 \to \mathbf{R}^2$ is given by

$$P^{-1}(u, v) = \frac{(4u, 4v, 2f)}{f + 4}, \quad \text{where } f = u^2 + v^2.$$

(Check that both PP^{-1} and $P^{-1}P$ are identity maps.)

(b) Deduce that the entire sphere Σ can be covered by only two patches. (The scheme in Section 1 requires six.)

13. (*Consistent formulas.*) If $G: \tilde{M} \to M$ is a regular mapping onto M, and $\tilde{F}: \tilde{M} \to N$ is an arbitrary mapping, we say that the formula $F(G(\mathbf{q})) = \tilde{F}(\mathbf{q})$ is *consistent* provided

$$G(\mathbf{q}_1) = G(\mathbf{q}_2) \Rightarrow \tilde{F}(\mathbf{q}_1) = \tilde{F}(\mathbf{q}_2)$$

for $\mathbf{q}_1$, $\mathbf{q}_2$ in $\tilde{M}$. Prove:

(a) In this case, F is a well-defined differentiable mapping from M to N.

(b) Furthermore, if the reverse implication

$$\tilde{F}(\mathbf{q}_1) = \tilde{F}(\mathbf{q}_2) \Rightarrow G(\mathbf{q}_1) = G(\mathbf{q}_2)$$

also holds, then F is one-to-one.

This result is helpful in constructing maps $F: M \to N$ with specified properties. Often G will be a parametrization of M.

4.6   Integration of Forms

Differential forms are no less important in integral calculus than in differential calculus. Indeed, they are just what is needed to establish integration theory on an arbitrary surface. In a sense, integration takes place only on Euclidean space, so a form on a surface is integrated by first pulling it back to Euclidean space.

Consider the one-dimensional case. Let $\alpha: [a, b] \to M$ be a curve segment on a surface M. The pullback $\alpha^*\phi$ of a 1-form ϕ on M to the interval $[a, b]$ has the expression $f(t)dt$, where by the remarks following Example 4.7,

$$f(t) = (\alpha^* \phi)(U_1(t)) = \phi(\alpha_*(U_1(t))) = \phi(\alpha'(t)).$$

Thus the scheme mentioned above yields the following result:

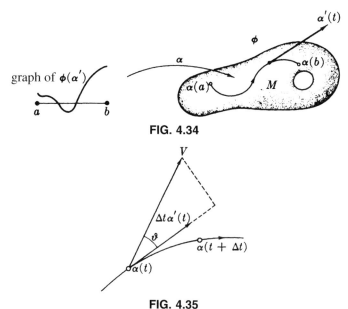

FIG. 4.34

FIG. 4.35

6.1 Definition   Let ϕ be a 1-form on M, and let $\alpha\colon [a, b] \to M$ be a curve segment in M (Fig. 4.34). Then the *integral of ϕ over α* is

$$\int_\alpha \phi = \int_{[a,b]} \alpha^* \phi = \int_a^b \phi(\alpha'(t))dt.$$

The integral $\int_\alpha \phi$, often called a *line integral*, has a wide variety of uses in science and engineering. For example, let us consider a vector field V on a surface M as a *force field*, and a curve $\alpha\colon [a, b] \to M$ as a description of a moving particle, with $\alpha(t)$ its position at time t. What is the total amount of *work* W done by the force on the particle as it moves from $\mathbf{p} = \alpha(a)$ to $\mathbf{q} = \alpha(b)$? The discussion of velocity in Chapter 1, Section 4, shows that for Δt small, the subsegment of α from $\alpha(t)$ to $\alpha(t + \Delta t)$ is approximated by the straight line segment $\Delta t\, \alpha'(t)$. Work is done on the particle only by the component of force *tangent* to α, that is,

$$V(\alpha) \cdot \frac{\alpha'}{\|\alpha'\|} = \|V(\alpha)\| \cos \vartheta,$$

where ϑ is the angle between $V(\alpha(t))$ and $\alpha'(t)$ (Fig. 4.35). Thus the work done by the force during time Δt is approximately the *force* (as above) times the *distance* $\|\alpha'(t)\|\, \Delta t$. Adding these contributions over the whole time interval $[a, b]$ and taking the usual limit, we get

$$W = \int_a^b V(\alpha(t)) \cdot \alpha'(t)dt.$$

To express this more simply, we introduce the 1-form ϕ *dual* to the vector field V; its value on a tangent vector $\mathbf{w}$ at $\mathbf{p}$ is $\mathbf{w} \cdot V(\mathbf{p})$. Then by Definition 6.1, the total work is just

$$W = \int_a \phi.$$

We emphasize that this notion of line integral—like everything we do with forms—applies without essential change if the surface M is replaced by a Euclidean space or, indeed, by any *manifold* (Section 8).

When the 1-form being integrated is the differential of a function, we get the following generalization of the fundamental theorem of calculus.

6.2 Theorem   Let f be a function on M, and let $\alpha: [a, b] \to M$ be a curve segment in M from $\mathbf{p} = \alpha(a)$ to $\mathbf{q} = \alpha(b)$. Then

$$\int_\alpha df = f(\mathbf{q}) - f(\mathbf{p}).$$

Proof.   By definition,

$$\int_\alpha df = \int_a^b df(\alpha')dt.$$

But

$$df(\alpha') = \alpha'[f] = \frac{d}{dt}(f(\alpha)).$$

Hence, by the fundamental theorem of calculus,

$$\int_a df = \int_a^b \frac{d}{dt}(f(\alpha))\,dt = f(\alpha(b)) - f(\alpha(a)) = f(\mathbf{q}) - f(\mathbf{p}). \qquad \blacklozenge$$

Thus the integral $\int_\alpha df$ is *path independent:* its value is the same for all curves from $\mathbf{p}$ to $\mathbf{q}$. Hence it is zero for all closed curves, $\alpha(a) = \alpha(b)$.

The preceding theorem can be interpreted roughly as follows: The "boundary" of the curve segment α from $\mathbf{p}$ to $\mathbf{q}$ is $\mathbf{q} - \mathbf{p}$, where the purely formal minus sign indicates that $\mathbf{p}$ is the starting point of α. Then the integral of df over α equals the "integral" of f over the boundary of α, namely, $f(\mathbf{q}) - f(\mathbf{p})$. This interpretation will be justified by the analogous theorem (6.5) in dimension 2.

If we consider a closed rectangle $R: a \leq u \leq b, c \leq v \leq d$ in $\mathbf{R}^2$ as a 2-dimensional interval, then a *2-segment* is a differentiable map $\mathbf{x}: R \to M$ of R into M (Fig. 4.36). As for a 1-segment, differentiability means that $\mathbf{x}$ can be extended over a larger open set containing R. Although we use the patch notation $\mathbf{x}$, we do not assume that $\mathbf{x}$ is either regular or one-to-one—but the partial velocities $\mathbf{x}_u$ and $\mathbf{x}_v$ are still well defined.

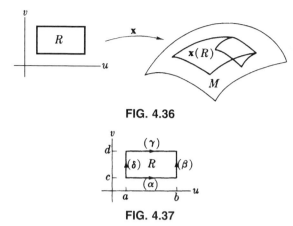

FIG. 4.36

FIG. 4.37

If η is a 2-form on M, then the pullback $\mathbf{x}^*\eta$ of η has, using Example 4.7, the expression $h\,du\,dv$, where

$$h = (\mathbf{x}^*\eta)(U_1, U_2) = \eta(\mathbf{x}_*U_1, \mathbf{x}_*U_2) = \eta(\mathbf{x}_u, \mathbf{x}_v).$$

Then strict analogy with Definition 6.1 yields:

6.3 Definition   Let η be a 2-form on M, and let $\mathbf{x}: R \to M$ be a 2-segment in M. The *integral of η over* $\mathbf{x}$ is

$$\iint_{\mathbf{x}} \eta = \iint_R \mathbf{x}^*\eta = \int_a^b \int_c^d \eta(\mathbf{x}_u, \mathbf{x}_v)\,du\,dv.$$

The physical applications of this integral are no less rich than those of Definition 6.1; however, we proceed directly toward the analogue of Theorem 6.2.

6.4 Definition   Let $\mathbf{x}: R \to M$ be a 2-segment in M, with R the closed rectangle $a \leqq u \leqq b$, $c \leqq v \leqq d$ (Fig. 4.37). The *edge curves* of $\mathbf{x}$ are the curve segments $\alpha, \beta, \gamma, \delta$ such that

$$\alpha(u) = \mathbf{x}(u, c),$$

$$\beta(v) = \mathbf{x}(b, v),$$

$$\gamma(u) = \mathbf{x}(u, d),$$

$$\delta(v) = \mathbf{x}(a, v).$$

Then the *boundary* $\partial\mathbf{x}$ of the 2-segment $\mathbf{x}$ is the formal expression

$$\partial\mathbf{x} = \alpha + \beta - \gamma - \delta.$$

FIG. 4.38

These four curve segments are gotten by restricting the function $\mathbf{x}: R \to M$ to the four line segments that make up the boundary of the rectangle R (Fig. 4.37). The formal minus signs before γ and δ signal that the parametrizations of γ and δ would have to be reversed to give a consistent trip around the rim of $\mathbf{x}(R)$ (Fig. 4.38).

Then if ϕ is a 1-form on M, the *integral of ϕ over the boundary $\partial \mathbf{x}$ of $\mathbf{x}$* is defined to be

$$\int_{\partial \mathbf{x}} \phi = \int_{\alpha} \phi + \int_{\beta} \phi - \int_{\gamma} \phi - \int_{\delta} \phi.$$

The 2-dimensional analogue of Theorem 6.2 is

6.5 Theorem (Stokes' theorem) If ϕ is a 1-form on M, and $\mathbf{x}: R \to M$ is a 2-segment, then

$$\iint_{\mathbf{x}} d\phi = \int_{\partial \mathbf{x}} \phi.$$

Proof. We work on the double integral and show that it turns into the integral of ϕ over the boundary of $\mathbf{x}$. Combining Definitions 6.3 and 4.4 gives

$$\iint_{R} d\phi(\mathbf{x}_u, \mathbf{x}_v) \, du \, dv = \iint_{R} \left(\frac{\partial}{\partial u}(\phi(\mathbf{x}_v)) - \frac{\partial}{\partial v}(\phi(\mathbf{x}_u)) \right) du \, dv.$$

Let $f = \phi(\mathbf{x}_u)$ and $g = \phi(\mathbf{x}_v)$. Then the equation above becomes

$$\iint_{\mathbf{x}} d\phi = \iint_{R} \frac{\partial g}{\partial u} du \, dv - \iint_{R} \frac{\partial f}{\partial v} du \, dv. \tag{1}$$

Now we treat these double integrals as iterated integrals. Suppose the rectangle R is given, as usual, by the inequalities $a \leqq u \leqq b$, $c \leqq v \leqq d$. Then integrating first with respect to u, we find

$$\iint_{R} \frac{\partial g}{\partial u} du \, dv = \int_{c}^{d} I(v) \, dv, \quad \text{where } I(v) = \int_{a}^{b} \frac{\partial g}{\partial u}(u, v) \, du.$$

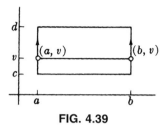

FIG. 4.39

In the partial integral defining $I(v)$, v is constant, so the integrand is just the ordinary derivative with respect to u. Thus, the fundamental theorem of calculus applies to give $I(v) = g(b, v) - g(a, v)$ (Fig. 4.39). Hence

$$\iint_R \frac{\partial g}{\partial u} du\, dv = \int_c^d g(b, v)\, dv - \int_c^d g(a, v)\, dv. \tag{2}$$

Consider the first integral on the right. By definition, $g(b, v) = \phi(\mathbf{x}_v(b, v))$. But $\mathbf{x}_v(b, v)$ is exactly the velocity $\beta'(v)$ of the "right side" curve β in $\partial \mathbf{x}$. Hence by Definition 6.1,

$$\int_c^d g(b, v)\, dv = \int_c^d \phi(\beta'(v))\, dv = \int_\beta \phi.$$

A similar argument shows that the other integral in (2) is $\int_\delta \phi$. Thus

$$\iint_R \frac{\partial g}{\partial u}\, du\, dv = \int_\beta \phi - \int_\delta \phi. \tag{3}$$

In the same way—but integrating first with respect to v—we find

$$\iint_R \frac{\partial f}{\partial v}\, du\, dv = \int_\gamma \phi - \int_\alpha \phi. \tag{4}$$

Assembling the information in (1), (3), and (4) gives the required result,

$$\iint_\mathbf{x} d\phi = \left(\int_\beta \phi - \int_\delta \phi \right) - \left(\int_\gamma \phi - \int_\alpha \phi \right) = \int_{\partial \mathbf{x}} \phi. \qquad \blacklozenge$$

Stokes' theorem ranks as one of the most useful results in all mathematics. Alternative formulations and extensive applications can be found in texts on advanced calculus and applied mathematics. We will use it to study the geometry and topology of surfaces.

The line integral $\int_\alpha \phi$ is not particularly sensitive to reparametrization of the curve segment α. All that matters is the overall direction in which the route of α is traversed, as indicated by what the reparametrization does to end points.

6.6 Lemma   Let $\alpha(h)$: $[a, b] \to M$ be a reparametrization of a curve segment α: $[c, d] \to M$ by h: $[a, b] \to [c, d]$. For any 1-form ϕ on M,

(1) If h is *orientation-preserving*, that is, if $h(a) = c$ and $h(b) = d$, then

$$\int_{\alpha(h)} \phi = \int_\alpha \phi.$$

(2) If h is *orientation-reversing*, that is, if $h(a) = d$ and $h(b) = c$, then

$$\int_{\alpha(h)} \phi = -\int_\alpha \phi.$$

Proof.   The velocity of $\alpha(h)$ is $\alpha(h)' = dh/dt \; \alpha'(h)$, so

$$\int_{\alpha(h)} \phi = \int_a^b \phi\left(\alpha(h)'\right) dt = \int_a^b \phi(\alpha'(h)) \frac{dh}{dt} dt.$$

Now we apply the theorem on change of variables in an integral to the last integral above. If h is orientation-preserving, then

$$\int_{\alpha(h)} \phi = \int_c^d \phi(\alpha') \, du = \int_\alpha \phi,$$

while in the orientation-reversing case,

$$\int_{\alpha(h)} \phi = \int_d^c \phi(\alpha') \, du = -\int_c^d \phi(\alpha') \, du = -\int_\alpha \phi. \qquad \blacklozenge$$

This lemma provides a concrete interpretation to the formal minus signs in the boundary $\partial \mathbf{x} = \alpha + \beta - \gamma - \delta$ of a 2-segment. For any curve

$$\xi: [t_0, t_1] \to M,$$

let $-\xi$ be any orientation-*reversing* reparametrization of ξ, for instance, $(-\xi)(t) = \xi(t_0 + t_1 - t)$. Then by the lemma,

$$\int_{-\xi} \phi = -\int_\xi \phi.$$

Thus the formula for $\int_{\partial \mathbf{x}} \phi$ just before Theorem 6.5 can be rewritten as

$$\int_{\partial \mathbf{x}} \phi = \int_\alpha \phi + \int_\beta \phi + \int_{-\gamma} \phi + \int_{-\delta} \phi.$$

Exercises

1.   If α is a curve in $\mathbf{R}^2$ and ϕ is a 1-form, prove this computational rule for finding $\phi(\alpha')dt$: Substitute $u = \alpha_1$ and $v = \alpha_2$ into the coordinate expression $\phi = f(u, v) \, du + g(u, v) \, dv$.

2.   Let α: $[-1, 1] \to \mathbf{R}^2$ be the curve segment given by $\alpha(t) = (t, t^2)$.

(a) If $\phi = v^2 \, du + 2uv \, dv$, compute $\int_\alpha \phi$.

(b) Find a function f such that $df = \phi$ and check Theorem 6.2 in this case.

3. Let ϕ be a 1-form on a surface M. Show:

(a) If ϕ is closed, then $\int_{\mathbf{x}} \phi = 0$ for every 2-segment $\mathbf{x}$ in M.

(b) If ϕ is exact, then more generally,

$$\int_{\alpha} \phi = \sum_i \int_{\alpha_i} \phi = 0$$

for any closed, piecewise smooth curve whose smooth segments are $\alpha_1, \ldots, \alpha_k$ (hence α_k ends at the start of α_1).

4. The 1-form

$$\psi = \frac{u\, dv - v\, du}{u^2 + v^2}$$

is well-defined on the plane $\mathbf{R}^2$ with the origin $\mathbf{0}$ removed. Show:

(a) ψ is closed but not exact on $\mathbf{R}^2 - \mathbf{0}$. (*Hint:* Integrate around the unit circle and use Ex. 3.)

(b) The restriction of ψ to, say, the right half-plane $u > 0$ is exact.

5. (a) Show that every curve α in $\mathbf{R}^2$ that does not pass through the origin has an (orientation-preserving) reparametrization in the polar form

$$\alpha(t) = (r(t) \cos \vartheta(t), r(t) \sin \vartheta(t)).$$

(*Hint:* Use Ex. 12 of Sec. 2.1.)

If the curve $\alpha\colon [a, b] \to \mathbf{R}^2 - \mathbf{0}$ is closed, prove:

(b) $\operatorname{wind}(\alpha) = \dfrac{\vartheta(b) - \vartheta(a)}{2\pi}$ is an integer.

This integer, called the *winding number of α about* $\mathbf{0}$, represents the total algebraic number of times α has gone around the origin in the counterclockwise direction. (Note that $\operatorname{wind}(\alpha) = \operatorname{wind}(\alpha/\|\alpha\|)$.)

(c) If ψ is the 1-form in Exercise 4, then $\operatorname{wind}(\alpha) = \dfrac{1}{2\pi} \int_{\alpha} \psi$.

(d) If $\alpha = (f, g)$, then

$$\operatorname{wind}(\alpha) = \frac{1}{2\pi} \int_a^b \frac{fg' - gf'}{f^2 + g^2}\, dt = \frac{1}{2\pi} \int_a^b \frac{\det(\alpha(t), \alpha'(t))}{\alpha(t) \cdot \alpha(t)}\, dt.$$

(The determinant is of the 2×2 matrix whose rows are $\alpha(t)$ and $\alpha'(t)$.)

6. (*Continuation, by computer.*) For a point $\mathbf{p} \in \mathbf{R}^2$ not on a closed curve α, the *winding number of α about* $\mathbf{p}$ is defined to be $\operatorname{wind}(\alpha - \mathbf{p})$.

(a) Write commands that given a closed curve α in $\mathbf{R}^2$ and a point $\mathbf{p}$ not on α, return the winding number of α about $\mathbf{p}$. (*Hint:* Use either integral in (d) of the preceding exercise.)

(b) In each case, plot the curve α (observing its orientation) and estimate the winding numbers about the indicated points **p**. Then calculate these numbers, using the integral in (d) above. (Numerical integration is efficient here, since the result is known to be an integer.)

(i) lemniscate, $\alpha(t) = (2\sin t, \sin 2t)$; **p** $= (1, 0), (0, 1), (-1, 0)$.

(ii) limaçon, $\beta(t) = (3\sin t + 1) (\cos t, \sin t)$; **p** $= (0, 1), (0, 3), (0, 5)$.

7. Let $F: M \rightarrow N$ be a mapping. Prove:

(a) If α is a curve segment in M, and ϕ is a 1-form on N, then
$$\int_\alpha F^*\phi = \int_{F(\alpha)} \phi.$$

(b) If **x** is 2-segment in M, and η is a 2-form on N, then $\iint_\mathbf{x} F^*\eta = \iint_{F(\mathbf{x})} \eta.$

8. Let **x** be a patch in a surface M. For a curve segment
$$\alpha(t) = \mathbf{x}(a_1(t), a_2(t)), \quad a \leq t \leq b,$$

in $\mathbf{x}(R)$, show that

$$\int_\alpha \phi = \int_a^b \left(\phi(\mathbf{x}_u) \frac{da_1}{dt} + \phi(\mathbf{x}_v) \frac{da_2}{dt} \right) dt,$$

where $\mathbf{x}_u$ and $\mathbf{x}_v$ are evaluated on (a_1, a_2). (This generalizes Ex. 1, which is recovered by using the identity patch $\mathbf{x}(u, v) = (u, v)$ in $\mathbf{R}^2$.)

9. Let **x** be the usual parametrization of the torus T (Ex. 2.5). For integers m and n, let α be the closed curve

$$\alpha(t) = \mathbf{x}(mt, nt) \quad (0 \leq t \leq 2\pi).$$

Find:

(a) $\int_\alpha \xi$, where ξ is the 1-form such that $\xi(\mathbf{x}_u) = 1$ and $\xi(\mathbf{x}_v) = 0$.

(b) $\int_\alpha \eta$, where η is the 1-form such that $\eta(\mathbf{x}_u) = 0$ and $\eta(\mathbf{x}_v) = 1$.

For an arbitrary closed curve γ in T, $\int_\gamma \xi/(2\pi)$ is an integer that counts the total (algebraic) number of times γ goes around the torus in the general direction of the parallels, and $\int_\gamma \eta/(2\pi)$ gives a similar count for the meridians. (This suggests the informal notation $\xi = d\vartheta, \eta = d\varphi$, but see Ex. 7 of Sec. 7.)

10. Let $\mathbf{x}: R \rightarrow M$ be a 2-segment defined on the unit square $R: 0 \leq u$, $v \leq 1$. If ϕ is the 1-form on M such that

$$\phi(\mathbf{x}_u) = u + v \quad \text{and} \quad \phi(\mathbf{x}_v) = uv,$$

compute $\iint_{\mathbf{x}} d\phi$ and $\int_{\partial \mathbf{x}} \phi$ separately, and check the results by Stokes' theorem. (*Hint:* $\mathbf{x}^*d\phi = d(\mathbf{x}^*\phi)$.)

11. Same as Exercise 10, except that $R: 0 \leq u \leq \pi/2,\ 0 \leq v \leq \pi$, and $\phi(\mathbf{x}_u) = u\cos v,\ \phi(\mathbf{x}_v) = v\sin u$.
The following exercise is a 2-dimensional analogue of Lemma 6.6. However, with future applications in mind, we generalize 2-segments $\mathbf{x}: R \to M$ by replacing the rectangle R by any compact region $\mathscr{R}$ in $\mathbf{R}^2$ whose boundary consists of smooth curve segments. (Compactness ensures that integrals over $\mathscr{R}$ will be finite.)

12. (*Effect of change of variables.*) Let $\mathbf{x}: \mathscr{S} \to M$ be a differentiable mapping and let $(U, V): \mathscr{R} \to \mathscr{S}$ be a one-to-one regular map whose Jacobian determinant

$$J(u, v) = \det \begin{bmatrix} \dfrac{\partial U}{\partial u} & \dfrac{\partial U}{\partial v} \\[2ex] \dfrac{\partial V}{\partial u} & \dfrac{\partial V}{\partial v} \end{bmatrix}$$

is always positive (*orientation-preserving* case) or always negative (*orientation-reversing* case). Then let $\mathbf{y}: \mathscr{R} \to M$ be given by $\mathbf{y}(u, v) = \mathbf{x}(U(u, v), V(u, v))$.
  (a) For a 2-form η on M, use

$$\mathbf{x}_u = \mathbf{y}_U \frac{\partial U}{\partial u} + \mathbf{y}_V \frac{\partial V}{\partial u}, \quad \mathbf{x}_v = \mathbf{y}_U \frac{\partial U}{\partial v} + \mathbf{y}_V \frac{\partial V}{\partial v}$$

to prove

$$\eta(\mathbf{x}_u, \mathbf{x}_v) = J(u, v)\, \eta(\mathbf{y}_U(U, V), \mathbf{y}_V(U, V)).$$

  (b) Deduce that $\iint_{\mathbf{x}} \eta = \iint_{\mathbf{y}} \eta$ in the orientation-preserving case, and minus this in the orientation-reversing case.
  (*Hint:* The formula for change of variables in a double integral involves the absolute value of a Jacobian determinant.)

13. The *classical Stokes' theorem* asserts, typically, that if $\mathbf{x}: D \to \mathbf{R}^3$ is a 2-segment and V is a vector field on $\mathbf{R}^3$, then

$$\int_{\partial \mathbf{x}} V \cdot ds = \iint_{\mathbf{x}} U \cdot (\nabla \times V)\, dA,$$

where $\nabla \times V = \operatorname{curl} V$. Interpret this as a special case of Theorem 6.5. (*Hint:* Assume $dA \approx \sqrt{EG - F^2}\ du\, dv$†, and use Ex. 8 of Sec. 1.6.)

---

† This geometric result will be clarified later on. Note that Theorem 6.5 does not involve geometry.

4.7   Topological Properties of Surfaces

Topological properties are the most basic a surface can have. In this section we discuss four such properties, phrasing the definitions in terms most efficient for geometry.

7.1 Definition   A surface is *connected* provided that for any two points **p** and **q** of M there is a curve segment in M from **p** to **q**. (See also Exercise 9.)

Thus a connected surface M is all in one piece, since one can travel from any point in M to any other without leaving M. Most of the surfaces we have met so far have been connected; the surface M: $z^2 - x^2 - y^2 = 1$ (a hyperboloid of two sheets) is not. Connectedness is a mild and natural condition that is sometimes included in the definition of surface.

The general definition of *compactness* is expressed in terms of open coverings. An *open covering* of a set A is a collection of open sets that *covers A* in the sense that each point of A is in at least one of the sets.

A subset A in a space S (for us, either a Euclidean space or a surface) is *compact* provided that given any open covering of A some finite number of the sets already covers A. In elementary calculus it is usually proved that a closed interval I: $a \leqq t \leqq b$ in **R** is compact, and this result extends to higher dimensions. In particular, any closed rectangle R: $a \leqq u \leqq b$, $c \leqq v \leqq d$ in $\mathbf{R}^2$ is compact.

We will need this abstract definition at a few crucial points, but in surface theory the following concrete criterion is more useful.

7.2 Lemma   A surface M is compact if and only if it can be covered by the images of a finite number of 2-segments in M.

Proof.   Suppose M is compact. For each point **p** in M, by using a coordinate patch containing **p**, we can construct a 2-segment whose image contains a neighborhood of **p**. The definition of compactness shows that a finite number of these neighborhoods already covers M; hence the corresponding 2-segments cover M.

The converse is an exercise in finiteness. First we show that the image **x**(R) of a single 2-segment is compact. Recall that the definition of differentiability for 2-segments allows us to assume that **x** has been smoothly extended over an open set containing the (closed) rectangle R.

Let $\{\mathcal{U}_\alpha\}$ be an open covering of M. For each point **r** in R, one of these sets, say $\mathcal{U}_\mathbf{r}$, contains **x**(**r**). Being differentiable, **x** is also continuous, so **r** has a neighborhood $\mathcal{N}_r$ that is carried into $\mathcal{U}_r$ by $\mathbf{x}$. For all $\mathbf{r}$, these neighborhoods $\mathcal{N}_r$ form an open covering of R. As mentioned above, the rectangle R is compact, so some finite number of these neighborhoods cover R. But this means that the corresponding original sets $\mathcal{U}_r$ (finite in number!) cover $\mathbf{x}(R)$; hence it is compact.

Now suppose that M is covered by the images of a finite number of 2-segments $\mathbf{x}_1, \ldots, \mathbf{x}_k$. If $\{\mathcal{U}_\alpha\}$ is an open covering of M, then we have just shown that a finite number of these sets suffice to cover the image of each $\mathbf{x}_i$. Collecting these sets for $i = 1, 2, \ldots, k$ produces a finite number of sets $\mathcal{U}_a$ that cover M. Thus M is compact. ◆

It follows at once from this proof that *a region R in M is compact if it is composed of the images of finitely many 2-segments in M*. For example, spheres are compact, since if the formula for the geographical patch (Example 2.2) is applied on the closed rectangle,

$$R: -\pi \leqq u \leqq \pi, \quad -\pi/2 \leqq v \leqq \pi/2,$$

then this single 2-segment covers the entire sphere†. Similarly, the torus (Example 2.5) is compact, as is every surface of revolution whose profile curve is closed.

The following lemma generalizes this fundamental fact: A continuous real-valued function on a closed rectangle R in the plane takes on a maximum at some point of R.

7.3 Lemma   A continuous function f on a compact region $\mathcal{R}$ in a surface M takes on a maximum at some point of M.

Proof.   We show this in the only case we need: where $\mathcal{R}$ consists of the images $\mathbf{x}_1(R_1), \ldots, \mathbf{x}_k(R_k)$ of a finite number of 2-segments (for example, where $\mathcal{R}$ is an entire surface).

Since each $\mathbf{x}_i$ is continuous, the composite functions $f(\mathbf{x}_i)$: $R_i \to \mathbf{R}$ are all continuous. By the remark above, for each index i, there is a point (u_i, v_i) in R_i where $f(\mathbf{x}_i)$ is a maximum. Let $f(\mathbf{x}_j(u_j, v_j))$ be the largest of these finite number k of maximum values; then evidently $f(\mathbf{p}) \leqq f(\mathbf{x}_j(u_j, v_j))$ for all $\mathbf{p}$ in M. ◆

---

† Amazingly, *every* compact surface can be expressed as the image of a single 2-segment. See Ch. 1 of [Ma].

This lemma is useful in proving noncompactness. For example, a cylinder C such as $x^2 + y^2 = r^2$ is not compact since the coordinate function z is unbounded on C.

Finite size alone does not produce compactness. For example, the open disk $\mathscr{D}$: $x^2 + y^2 < 1$ in the xy plane is itself a surface. Although $\mathscr{D}$ is bounded and has finite area π, it is not compact since the function $f = (1 - x^2 - y^2)^{-1}$ is continuous on $\mathscr{D}$ and does not have a maximum.

In general, a compact surface cannot have open edges, as $\mathscr{D}$ does, but must be smoothly closed up everywhere and finite in size—like a sphere or torus.

Roughly speaking, an orientable surface is one that is not twisted. Of the many equivalent definitions of orientability for surfaces, the following is perhaps the simplest.

7.4 Definition   A surface M is *orientable* if there exists a differentiable (or merely continuous) 2-form μ on M that is nonzero at every point of M.

Recall that a 2-form is zero at a point $\mathbf{p}$ if it is zero on every pair of tangent vectors at $\mathbf{p}$—or equivalently, on one linearly independent pair. Thus the plane $\mathbf{R}^2$ is orientable since $du\, dv$ is a nonvanishing 2-form. This definition of orientability is somewhat mysterious, so for a surface M in $\mathbf{R}^3$ we give a more intuitive description in terms of Euclidean geometry. A *unit normal* U on M is a differentiable Euclidean vector field on M that has unit length and is everywhere normal to M.

7.5 Proposition   A surface $M \subset \mathbf{R}^3$ is orientable if and only if there exists a unit normal vector field on M. If M is connected as well as orientable, there are exactly two unit normals, $\pm U$.

Proof.   We use the cross product of $\mathbf{R}^3$ to convert normal vector fields into 2-forms, and vice versa.

Let U be a unit normal on M. If $\mathbf{v}$ and $\mathbf{w}$ are tangent vectors to M at $\mathbf{p}$, define

$$\mu(\mathbf{v}, \mathbf{w}) = U(\mathbf{p}) \bullet \mathbf{v} \times \mathbf{w}.$$

Standard properties of the cross product show that μ is a 2-form on M. When $\mathbf{v}$ and $\mathbf{w}$ are linearly independent, so are all three vectors, so $\mu(\mathbf{v}, \mathbf{w}) \neq 0$. Thus μ is nonvanishing, which proves that M is orientable.

Conversely, suppose M is orientable, with μ a nonvanishing 2-form. Again, nonvanishing implies that if $\mathbf{v}, \mathbf{w}$ are linearly independent vectors at a point $\mathbf{p}$ of M, then $\mu(\mathbf{v}, \mathbf{w}) \neq 0$. Now define

$$Z(\mathbf{p}) = \frac{\mathbf{v} \times \mathbf{w}}{\mu(\mathbf{v}, \mathbf{w})}.$$

This formula is independent of the choice of $\mathbf{v}$, $\mathbf{w}$. Explicitly, for any other such pair $\mathbf{v}'$, $\mathbf{w}'$, it follows from Lemma 4.2 and the analogous formula for cross products that

$$\frac{\mathbf{v}' \times \mathbf{w}'}{\mu(\mathbf{v}', \mathbf{w}')} = \frac{\mathbf{v} \times \mathbf{w}}{\mu(\mathbf{v}, \mathbf{w})}.$$

Properties of the cross product show that $Z(p)$ is nonzero and normal to M. The formula for cross product shows that Z is differentiable. Thus $U = Z/\|Z\|$ is the required unit normal.

If U is a unit normal on M, then so is $-U$. To show that there are no others, let V also be a unit normal. At each point these (differentiable) unit vector fields are collinear, so the only values for the dot product $V \cdot U$ are $+1$ and -1. On a connected surface, a nonvanishing differentiable function cannot change sign (Exercise 4), hence *either* $V \cdot U = +1$ everywhere, so $V = U$, *or* $V \cdot U = -1$ everywhere, so $V = -U$.   ◆

For example, all spheres, cylinders, surfaces of revolution, and quadric surfaces are orientable. It follows from Lemma 3.8 that every surface in $\mathbf{R}^3$ that can be defined implicitly is orientable.

However, nonorientable surfaces do exist in $\mathbf{R}^3$. The simplest example is the famous Möbius band B, made from a strip of paper by giving it a half twist, then gluing its ends together (Fig. 4.40). B is nonorientable since it cannot have a (differentiable) unit normal. To see this let γ be a closed curve, as in Fig. 4.40, that runs once around the band with $\gamma(0) = \gamma(1) = \mathbf{p}$. Now suppose a unit normal vector U at $\gamma(0)$ moves continuously around γ. As the figure shows, the twist in B forces the contradiction

$$U(\mathbf{p}) = U(\gamma(1)) = -U(\gamma(0)) = -U(\mathbf{p}).$$

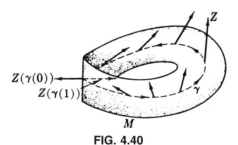

FIG. 4.40

The last of the topological properties we consider will let us express formally the intuitive idea that a plane is simpler than a cylinder, and a sphere is simpler than a torus. The key is that in the cylinder or a torus there are closed curves that cannot be continuously shrunk down to a point.

7.6 Definition   A closed curve α in M is *homotopic to a constant* provided there is a 2-segment $\mathbf{x}: R \to M$ (called a *homotopy*) defined on

$$R: a \leqq u \leqq b, 0 \leqq v \leqq 1$$

such that α is the base curve of $\mathbf{x}$ and the other three edge curves are constant at $\mathbf{p} = \alpha(a) = \alpha(b)$.

A curve such as α for which $\alpha(a) = \alpha(b)$ holds but not necessarily $\alpha'(a) = \alpha'(b)$ is often called a *loop at* $\mathbf{p}$. Since the sides β and δ of $\mathbf{x}$ are constant at $\mathbf{p}$, for every $0 \leq v \leq 1$ the u-parameter curve $\alpha_v(u) = \mathbf{x}(u, v)$ is also a loop at $\mathbf{p}$. As v varies from 0 to 1, the loop α_v varies continuously from $\alpha_0 = \alpha$ to the curve $\alpha_1 = \gamma$, which is constant at $\mathbf{p}$.

It is easy to show that in the plane every loop is homotopic to a constant. Given a loop $\alpha: [a, b] \to \mathbf{R}^2$ at $\mathbf{p}$ in $\mathbf{R}^2$, use scalar multiplication in $\mathbf{R}^2$ to define $\mathbf{x}(u, v) = v\alpha(a) + (1 - v)\alpha(u)$. Then

$$\mathbf{x}(u, 0) = \alpha(u) \text{ and } \mathbf{x}(u, 1) = \mathbf{p} \quad \text{for all } a \leqq u \leqq b, \text{ and}$$

$$\mathbf{x}(a, v) = \mathbf{x}(b, v) = \mathbf{p} \qquad \text{for all } 0 \leqq v \leqq 1.$$

Hence $\mathbf{x}$ is a homotopy from α to a constant.

7.7 Definition   A surface M is *simply connected* provided it is connected and every loop in M is homotopic to a constant.

(This definition is valid for any manifold—and more generally.) The preceding homotopy shows that the plane $\mathbf{R}^2$ is simply connected, and the same formula works for any Euclidean space.

The 2-sphere Σ is also simply connected. Consider the following scheme of proof. Let α be a loop in Σ at, say, the north pole of Σ. Pick a point $\mathbf{q}$ not on α. For simplicity, suppose $\mathbf{q}$ is the south pole. Now let $\mathbf{x}$ be the homotopy under which each point of α moves due north along a great circle, reaching $\mathbf{p}$ in unit time. This $\mathbf{x}$ is a homotopy of α to a constant, as required.

But there is a difficulty here: finding the point $\mathbf{q}$. In our usual case, where α is differentiable, techniques from advanced calculus will show that there is always a point $\mathbf{q}$ not on α. However, if α is merely continuous, it may actually fill the entire sphere. In this case, topological methods can be used to deform α slightly, making it no longer space-filling; then the scheme above is valid.

To show that a given surface is simply connected, we can always try to construct the necessary homotopies; however, to show that a surface is *not* simply connected, indirect means are usually required. One of the most effective derives from integration. Recall that a differential form ϕ is *closed* if $d\phi = 0$.

7.8 Lemma   Let ϕ be a closed 1-form on a surface M. If a loop α in M is homotopic to a constant, then $\int_\alpha \phi = 0$.

Proof.   Suppose $\mathbf{x}$ is a homotopy showing that a loop α is homotopic to a constant, say $\mathbf{p}$. Now we apply Stokes' theorem (Theorem 6.5). The integral over a constant curve is zero, and $d\phi = 0$, hence

$$0 = \iint_{\mathbf{x}} d\phi = \int_\alpha \phi. \qquad \blacklozenge$$

Now suppose we remove a single point, say the origin $\mathbf{0}$, from the (simply connected) plane $\mathbf{R}^2$. The loop $\alpha: [0, 2\pi] \to C$ given by $\alpha(t) = (\cos t, \sin t)$ circles once around the missing point. It seems obvious that α cannot be shrunk down to a point in the punctured plane $P = \mathbf{R}^2 - \mathbf{0}$. The preceding lemma provides an easy way to prove it.

Exercise 4 of the preceding section shows that the 1-form

$$\psi = \frac{x\,dy - y\,dx}{x^2 + y^2}$$

on P is closed and that its integral around α is 2π (these are easy computations). By the lemma, α is not homotopic to a constant; hence P is not simply connected.

As noted earlier, since $d^2 = 0$, exact forms are always closed. However, closed forms need not be exact. For example, if the closed 1-form ψ were exact, it would follow from Stokes' theorem (6.5) that $\int_\alpha \psi = 0$. However, in an important special case, closed 1-forms *are* exact.

7.9 Lemma   (Poincaré) On a simply connected surface, every closed 1-form is exact.

Proof.   First we show that *the integral of a closed 1-form is path independent*, that is, if α and β are curve segments from $\mathbf{p}$ to $\mathbf{q}$, then $\int_\alpha \phi = \int_\beta \phi$. In fact, if $-\beta$ is an orientation-reversing reparametrization of β, then $\alpha + (-\beta) = \alpha - \beta$ is a loop. By simple connectedness, it is homotopic to a constant. Thus, using Lemma 6.6,

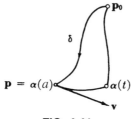

FIG. 4.41

$$0 = \int_{\alpha - \beta} \phi = \int_{\alpha} \phi + \int_{-\beta} \phi = \int_{\alpha} \phi - \int_{\beta} \phi.$$

Now suppose ϕ is a closed 1-form on a simply connected surface M. Pick a point $\mathbf{p}_0$ and define $f(\mathbf{p}) = \int_{\delta} \phi$ for any curve segment δ from $\mathbf{p}_0$ to $\mathbf{p}$. Path independence makes f a well-defined function on M.

To show that $df = \phi$, we must show that $df(\mathbf{v}) = \phi(\mathbf{v})$ for every tangent vector $\mathbf{v}$ at a point $\mathbf{p}$. This is equivalent to $\mathbf{v}[f] = \phi(\mathbf{v})$.

Let $\alpha: [a, b] \to M$ be a curve with initial velocity $\alpha'(a) = \mathbf{v}$. Then $\delta + \alpha|[a, t]$ is a curve from $\mathbf{p}_0$ to $\alpha(t)$ (Fig. 4.41), so by the definition of f,

$$f(\alpha(t)) = \int_{\delta + \alpha|[a, t]} \phi = f(\mathbf{p}) + \int_{a}^{t} \phi(\alpha'(u)) \, du.$$

Taking the derivative with respect to t gives

$$\alpha'(t)[f] = (f(\alpha))'(t) = \phi(\alpha'(t)).$$

When $t = 0$ this becomes $\mathbf{v}[f] = \phi(\mathbf{v})$, as required.   ◆

Among the four properties we have discussed there are two direct implications—both yielding orientability.

7.10 Theorem   A compact surface in $\mathbf{R}^3$ is orientable.

This is an easy consequence of the following nontrivial topological theorem, a 2-dimensional version of the Jordan Curve Theorem. *If M is a compact surface in $\mathbf{R}^3$, then M separates $\mathbf{R}^3$ into two nonempty open sets: an* exterior *(the points that can escape to infinity) and an* interior *(the points trapped inside M).* So we need only pick the unit normal vector at each $\mathbf{p}$ in M that points into, say, the exterior and apply Proposition 7.5. Thus orientation is at stake when in elementary calculus the "outward unit normal" is assigned to a surface in $\mathbf{R}^3$.

7.11 Theorem   A simply connected surface is orientable.

We defer the proof until Section 4 of Chapter 8. (Notice, for future reference, that this theorem does not mention $\mathbf{R}^3$.)

A final note: Because the properties discussed in this section are topological, they can be defined solely in terms of open sets and continuous functions. However, differentiable versions are usually more practical for use in geometry.

Exercises

1. Which of the following surfaces are compact and which are connected?
(a) A sphere with one point removed.
(b) The region $z > 0$ in M: $z = xy$.
(c) A torus with the curve $\alpha(t) = \mathbf{x}(t, t)$ removed. (See Ex. 2 of Sec. 4.3.)
(d) The surface in Fig. 4.8.
(e) M: $x^2 + y^4 + z^6 = 1$.

2. Let F be a mapping of a surface M *onto* a surface N. Prove:
(a) If M is connected, then N is connected.
(b) If M is compact, then N is compact. (Try both the covering definition and the criterion in Lem. 7.2.)

3. Let $F: M \rightarrow N$ be a regular mapping. Prove that if N is orientable, then M is orientable.

4. Let f be a differentiable real-valued function on a connected surface. Prove:
(a) If $df = 0$, then f is constant.
(b) If f is never zero then either $f > 0$ or $f < 0$.

5. Of the four basic types of surfaces of revolution (see Ex. 11 of Sec. 5)—plane, sphere, cylinder, torus—which are,
(a) connected?   (b) compact?
(c) orientable?   (d) simply connected?

A closed curve α in M is *freely* homotopic to a constant if the conditions on $\mathbf{x}$ in Definition 7.6 are weakened to $\beta = \delta$ with only γ required to be constant (Fig. 4.42). Then the v constant curves α_v are loops that move along the curve $\beta = \delta$ as they shrink to $\mathbf{p}$.

6. (a) If a loop α is freely homotopic to a constant via $\mathbf{x}$, show that for any 1-form ϕ,

FIG. 4.42

$$\int_\alpha \phi = \iint_x d\phi.$$

(b) If closed curves α and β in $\mathbf{R}^2 - \mathbf{p}$ are freely homotopic in $\mathbf{R}^2 - \mathbf{p}$, show that they have the same winding number about $\mathbf{p}$.

(c) A *smooth disk* $\mathscr{D}$ in a surface is the image of the unit disk $x^2 + y^2 \le 1$ in $\mathbf{R}^2$ under a one-to-one regular map F. Show that the 2-segment

$$\mathbf{x}(u, v) = F(u \cos v, u \sin v), \quad (0 \le u \le 1, 0 \le v \le 2\pi),$$

fills $\mathscr{D}$ and is a free homotopy of the (closed) boundary curve $v \to \mathbf{x}(1, v)$ to a constant.

7. Let ϕ be a closed 1-form and α a closed curve.

(a) Show that $\int_\alpha \phi = 0$ if either ϕ is exact or α is freely homotopic to a constant.

(b) Deduce that on a torus T the meridians and parallels are not freely homotopic to constants, and the closed 1-forms ξ and η of Exercise 9 of Section 6 are not exact.

8. (*Counterexamples.*) Give examples to show that the following are false:

(a) Converses of (a) and (b) of Exercise 2.

(b) Exercise 3 with F not regular.

(c) Converse of Exercise 3.

9. (a) If $\mathbf{p}$ is a point of a surface S, show that the set of all points of S that can be connected to $\mathbf{p}$ by a (piecewise smooth) curve in S is an open set of S. (*Hint:* Each point of a surface has a neighborhood that is connected in the sense of Def. 7.1.)

(b) Same as (a) but with *can* replaced by *cannot*.

(c) For a surface M, show that Definition 7.1 ("path-connectedness") is equivalent to the general topological definition of connectedness, namely: If $\mathscr{U}$ and $M - \mathscr{U}$ are open sets of M, and $\mathscr{U}$ contains at least one point, then $M = \mathscr{U}$. (Use the corresponding property for a closed interval in $\mathbf{R}$, a standard result of analysis.)

10. The Hausdorff axiom asserts that distinct points $\mathbf{p} \ne \mathbf{q}$ have disjoint neighborhoods. Prove:

(a) $\mathbf{R}^3$ obeys the Hausdorff axiom. (The same proof works for all $\mathbf{R}^n$.)

(b) A surface M in $\mathbf{R}^3$ obeys the Hausdorff axiom.

11. If $\mathscr{R}$ is a compact region in a surface M, prove that $\mathscr{R}$ is a closed set of M, that is, $M - \mathscr{R}$ is an open set. (*Hint*: To show that $M - \mathscr{R}$ is open, use the preceding exercise and the fact that a finite intersection of neighborhoods of $\mathbf{p}$ is again a neighborhood of $\mathbf{p}$.)

12. Let M and N be surfaces in $\mathbf{R}^3$ such that M is contained in N.

(a) If M is compact and N is connected, prove that $M = N$. (*Hint:* Show that M is both closed and open in N.)

(b) Give examples to show that (a) fails if either M is not compact or N is not connected.

(c) Deduce from (a) that if $F: M \to N$ is a local diffeomorphism with M compact and N connected, then $F(M) = N$.

4.8 Manifolds

Surfaces in $\mathbf{R}^3$ are a matter of everyday experience, so it is reasonable to investigate them mathematically. But examining this concept with a critical eye, we may well ask whether there could not be surfaces in $\mathbf{R}^4$, or in $\mathbf{R}^n$—or even surfaces that are not contained in any Euclidean space. To devise a definition for such a surface, we must rely not on our direct experience of the real world, but on our mathematical experience of surfaces in $\mathbf{R}^3$. Thus we shall strip away from Definition 1.2 every feature that involves $\mathbf{R}^3$. What is left will be just a surface.

To begin with, a surface will be a set M: a collection of any objects whatsoever. We call these objects the *points* of M, but as examples below will show, they definitely need not be the usual points of some Euclidean space. An *abstract patch in* M will now be just a one-to-one function $\mathbf{x}: D \to M$ from an open set D of $\mathbf{R}^2$ into the set M.

To get a workable definition of surface we must find a way to define what it means for functions involving M to be differentiable. The key to this problem turns out to be the smooth overlap condition in Corollary 3.3. To *prove* this condition is now a logical impossibility since $\mathbf{R}^3$ is gone, so in the usual fashion of mathematics, we make it an axiom.

8.1 Definition A *surface* is a set M furnished with a collection $\mathscr{P}$ of abstract patches in M satisfying

(1) *The covering axiom:* The images of the patches in the collection $\mathscr{P}$ cover M.

(2) *The smooth overlap axiom:* For any patches **x**, **y** in $\mathscr{P}$, the composite functions $\mathbf{y}^{-1}\mathbf{x}$ and $\mathbf{x}^{-1}\mathbf{y}$ are Euclidean differentiable—and defined on open sets of $\mathbf{R}^2$.

This definition generalizes Definition 1.2: A surface in $\mathbf{R}^3$ *is* a surface in this sense. However, there is a technical gap in this definition that requires attention. First, for any patch **x**: $D \to M$ in a surface, define a set $\mathbf{x}(\mathscr{U})$ to be open provided $\mathscr{U}$ is open in $D \subset \mathbf{R}^2$. Then the *open sets* of M are all unions of such sets. (This is consistent with the case $M \subset \mathbf{R}^3$, since there **x** and $\mathbf{x}^{-1}$ are continuous.)

Examples like that in Exercise 11 show that for the open sets to behave properly we must add another axiom to the definition of surface.

(3) *The Hausdorff axiom:* For any points $\mathbf{p} \neq \mathbf{q}$ in M there exist disjoint (that is, nonoverlapping) patches **x** and **y** with **p** in $\mathbf{x}(D)$ and **q** in $\mathbf{y}(E)$.

Here is an example of an important surface that, as we will soon see, cannot be found in $\mathbf{R}^3$.

8.2 Example   *The projective plane P.* Starting from the unit sphere Σ in $\mathbf{R}^3$ we construct P by identifying *antipodal points* in Σ, that is, by considering **p** and $-\mathbf{p}$ to be the same point of P (Fig. 4.43). Formally, this means that the set P consists of all antipodal pairs $\{\mathbf{p}, -\mathbf{p}\}$ of points in Σ. Order is not relevant here; that is, $\{\mathbf{p}, -\mathbf{p}\} = \{-\mathbf{p}, \mathbf{p}\}$. (Working with the projective plane often involves looking back and forth between antipodal points.)

There are two important mappings associated with P: the *antipodal map* $A(\mathbf{p}) = -\mathbf{p}$ on Σ and the *projection* $F(\mathbf{p}) = \{\mathbf{p}, -\mathbf{p}\}$ of Σ onto P. Note that $FA = F$.

Call a patch **x** in Σ "small" if it is contained in a single open hemisphere. Then the composite function $F\mathbf{x}$ is one-to-one, and is thus an abstract patch. The collection of all such abstract patches makes P a surface. In fact, the covering condition (1) is clear, and the Hausdorff axiom derives from the corresponding property for Euclidean spaces. The smooth overlap axiom (2) can be checked as follows.

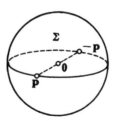

FIG. 4.43

Suppose $F\mathbf{x}$ and $F\mathbf{y}$ overlap in Σ; that is, their images have points in common. If $\mathbf{x}$ and $\mathbf{y}$ overlap in Σ, then $(F\mathbf{y})^{-1}F\mathbf{x} = \mathbf{y}^{-1}\mathbf{x}$, which by Corollary 3.3 is differentiable and defined on an open set. On the other hand, if $\mathbf{x}$ and $\mathbf{y}$ do not overlap, replace $\mathbf{y}$ by $A\mathbf{y}$. Then $\mathbf{x}$ and $A\mathbf{y}$ *do* overlap, so the previous argument applies.

To emphasize the distinction between a surface in $\mathbf{R}^3$ and the general notion of surface defined above, we sometimes call the latter an *abstract surface*.

To get as many patches as possible in an abstract surface M we always understand that its patch collection $\mathscr{P}$ has been enlarged to include all the abstract patches in M that overlap smoothly with those originally in $\mathscr{P}$. We emphasize that abstract surfaces M_1 and M_2 with the *same* set of points are nevertheless different surfaces if their (enlarged) patch collections $\mathscr{P}_1$ and $\mathscr{P}_2$ are different.

There is essentially only one problem to solve in establishing calculus on an abstract surface M, and that is to define the *velocity* of a curve in M. In the old definition a velocity vector was a tangent vector to $\mathbf{R}^3$, so something new is needed. For everything else—differentiable functions, curves themselves, tangent vectors, vector fields, differentiable forms, and so on—the *definitions* and *theorems* given for surfaces in $\mathbf{R}^3$ apply without change. It is necessary to tinker with a few *proofs*, but no serious problems arise.

It makes little difference what we define velocity to be—provided the new definition produces the same essential behavior as before. The most efficient choice is based on the directional derivative property in Lemma 4.6, Chapter 1.

8.3 Definition   Let $\alpha\colon I \to M$ be a curve in an abstract surface M. For each t in I the *velocity vector* $\alpha'(t)$ is the function such that

$$\alpha'(t)[f] = \frac{d(f\alpha)}{dt}(t)$$

for every differentiable real-valued function f on M.

Thus $\alpha'(t)$ is a real-valued function whose domain is the set $\mathscr{F}$ of all real-valued functions on M. This is all we need to generalize the calculus on surfaces in $\mathbf{R}^3$ to the case of an abstract surface.

We now have a calculus for $\mathbf{R}^n$ (Chapter 1) and another one for surfaces. These are strictly analogous, but analogies in mathematics, though useful initially, can be annoying in the long run. What we need is a single calculus of which these two will be special cases. The most general object on which calculus can be conducted is called a *manifold*. It is simply an abstract surface of arbitrary dimension n.

8.4 Definition   An *n-dimensional manifold M* is a set furnished with a collection $\mathscr{P}$ of *abstract patches* (one-to-one functions $\mathbf{x}: D \to M$, D an open set in $\mathbf{R}^n$) satisfying

(1) The covering property: The images of the patches in the collection $\mathscr{P}$ cover M.

(2) The smooth overlap property: For any patches $\mathbf{x}$, $\mathbf{y}$ in $\mathscr{P}$, the composite functions $\mathbf{y}^{-1}\mathbf{x}$ and $\mathbf{x}^{-1}\mathbf{y}$ are Euclidean differentiable—and defined on open sets of $\mathbf{R}^n$.

(3) The Hausdorff property: For any points $\mathbf{p} \neq \mathbf{q}$ in M there are disjoint patches $\mathbf{x}$ and $\mathbf{y}$ with $\mathbf{p}$ in $\mathbf{x}(D)$ and $\mathbf{q}$ in $\mathbf{y}(E)$.

Thus a surface (Definition 8.1) is just a 2-dimensional manifold. As before, Euclidean n-space $\mathbf{R}^n$ is an n-dimensional manifold whose (initial) patch collection consists only of the identity map.

To keep this definition as close as possible to that of a surface in $\mathbf{R}^3$, we have deviated somewhat from the standard definition of manifold in which it is the inverse functions $\mathbf{x}^{-1}: \mathbf{x}(D) \to D$ that are axiomatized.

The calculus of an arbitrary n-dimensional manifold is defined in the same way as for $n = 2$. Usually we need only replace $i = 1, 2$ by $i = 1, 2, \ldots, n$. Differential forms on an n-dimensional manifold have the same general properties as in the case $n = 2$, which we have explored in Sections 4, 5, and 6. But there are p-forms for $0 \leqq p \leqq n$, so when n is large, the algebra becomes more complicated.

Wherever calculus appears in mathematics and its applications, manifolds will be found, and higher dimensional manifolds turn out to be important in problems—both pure and applied—that initially seem to involve only dimensions 2 or 3. For example, here is a 4-dimensional manifold that has already appeared, implicitly at least, in this chapter.

8.5 Example   *The tangent bundle of a surface.* For a surface M, let $T(M)$ be the set of all tangent vectors to M at all points of M. (For simplicity we assume M is a surface in $\mathbf{R}^3$, but it could just as well be an abstract surface or, indeed, a manifold of any dimension.) Since M has dimension 2 and each tangent space $T_p(M)$ has dimension 2, we anticipate that $T(M)$ will have dimension 4.

To get a natural patch collection for $T(M)$ we derive from each patch $\mathbf{x}$ in M an abstract patch $\tilde{x}$ in $T(M)$. Given $\mathbf{x}: D \to M$, let $\tilde{D}$ be the open set in $\mathbf{R}^4$

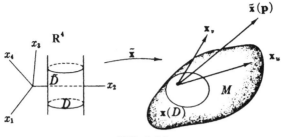

FIG. 4.44

consisting of all points (p_1, p_2, p_3, p_4) for which (p_1, p_2) is in D. Then let $\tilde{\mathbf{x}}$ be the function $\tilde{D} \to T(M)$ given by

$$\tilde{\mathbf{x}}(p_1, p_2, p_3, p_4) = p_3 \mathbf{x}_u(p_1, p_2) + p_4 \mathbf{x}_v(p_1, p_2).$$

(In Fig. 4.44 we identify $\mathbf{R}^2$ with the $x_1 x_2$ plane of $\mathbf{R}^4$ and deal as best we can with dimension 4.)

Using Exercise 3 of Section 3 and the proof of Lemma 3.6, it is not difficult to check that each such function $\tilde{\mathbf{x}}$ is one-to-one and that the collection of all such patches satisfies the conditions in Definition 8.4. Thus $T(M)$ is a 4-dimensional manifold, called the *tangent bundle* of M.

Exercises

1. Prove that a surface M is nonorientable if there is a smoothly closed curve α: $[a, b] \to M$ and a tangent vector field Y on α such that
   (i) Y and α' are linearly independent at every point, and
   (ii) $Y(b) = -Y(a)$.
(*Hint:* Assume M is orientable and deduce a contradiction.)

2. Establish the following properties of the projective plane P.
   (a) If $F: \Sigma \to P$ is the projection, then each tangent vector to P is the image under F_* of exactly two tangent vectors to Σ, these of the form $\mathbf{v}_p$ and $-\mathbf{v}_{-p}$.
   (b) P is compact, connected, and nonorientable—hence P is not diffeomorphic to any surface in $\mathbf{R}^3$.

3. (a) Prove that the tangent bundle $T(M)$ of a surface is a manifold. (If $\mathbf{x}$ and $\mathbf{y}$ are overlapping patches in M, find an explicit formula for $\tilde{\mathbf{y}}^{-1}\tilde{\mathbf{x}}$.)
   (b) If M is the image of a single patch $\mathbf{x}: \mathbf{R}^2 \to M$, show that the tangent bundle of M is diffeomorphic to $\mathbf{R}^4$.

4. A surface M in $\mathbf{R}^3$ is *closed* if it is a closed set of $\mathbf{R}^3$, that is, $\mathbf{R}^3 - M$ is an open set. (Confusingly, closed surface has sometimes been defined by analogy with closed curve to mean *compact surface.*)

(a) Prove that every surface given in the implicit form $M: g = c$ is closed.

(b) Prove that a compact surface in $\mathbf{R}^3$ is closed. (*Hint:* Use the scheme of Ex. 11 of Sec. 7.)

(c) Give an example of a closed surface in $\mathbf{R}^3$ that is not compact.

5. A surface M in $\mathbf{R}^3$ is *bounded* provided there is a number $R > 0$ such that $\|\mathbf{p}\| \leqq R$ for all $\mathbf{p} \in M$.

(a) Prove that a compact surface $M \subset \mathbf{R}^3$ is bounded.

(b) Give an example of a surface in $\mathbf{R}^3$ that is bounded but not compact. It follows from this exercise and Exercise 4 that if $M \subset \mathbf{R}^3$ is compact, it is closed and bounded. The converse is true but is more difficult to prove (see [Mu]).

6. Let $\hat{M}$ be the set of all the unit normal vectors on a surface M in $\mathbf{R}^3$. (So there are two points of $\hat{M}$ for each point of M.) For each patch $\mathbf{x}$ in M we define two patches in $\hat{M}$, namely,

$$\mathbf{x}_\pm(u, v) = \pm \frac{\mathbf{x}_u \times \mathbf{x}_v}{\|\mathbf{x}_u \times \mathbf{x}_v\|}.$$

(a) Show that the set of all such patches makes $\hat{M}$ a surface, called the *orientation covering surface* of M.

(b) Describe the orientation covering surface of the unit sphere $\Sigma \subset \mathbf{R}^3$ and of the torus $T \subset \mathbf{R}^3$.

(c) If a connected surface M in $\mathbf{R}^3$ is orientable, show that $\hat{M}$ consists of two diffeomorphic copies of M. (*Hint:* M has smooth unit normals $\pm U$.)

(d) The natural map $\pm U \to \mathbf{p}$ of $\hat{M}$ onto M is regular.

7. (*Continuation.*) If a connected surface M in $\mathbf{R}^3$ is nonorientable, show that $\hat{M}$ is (i) connected and (ii) orientable. (Thus nonorientability can be cured by doubling. For an example, see Ex. 10.)

(*Hints:* (i) If $\alpha: [a, b] \to M$ is a curve, then any unit normal vector at $\alpha(a)$ can be moved differentiably along α as a unit normal $U_\alpha(t)$—thus giving a curve in $\hat{M}$. Assume this fact: if M is nonorientable, there exists a loop α in M with $U_{\alpha(b)} = -U_{\alpha(a)}$.

(ii) Any patch $\mathbf{z}$ determines a unit vector field $U_\mathbf{z} = \mathbf{z}_u \times \mathbf{z}_v / \|\mathbf{z}_u \times \mathbf{z}_v\|$. If patches $\mathbf{x}$ and $\mathbf{y}$ in M meet, but $U_\mathbf{y} = -U_\mathbf{x}$ on the overlap, then $\hat{\mathbf{x}}$ and $\hat{\mathbf{y}}$ do not meet.)

8. A Möbius band B can be constructed as a ruled surface by

$$\mathbf{x}(u, v) = \beta(u) + v\delta(u), \quad \text{with, say, } -1/3 < v < 1/3,$$

where $\beta(u) = (\cos u,\ \sin u,\ 0)$ and

$$\delta(u) = \left(\cos \frac{u}{2} \cos u,\ \cos \frac{u}{2} \sin u,\ \sin \frac{u}{2} \right).$$

(The ruling makes only a half turn as it traverses the circle β.)
  Show that $\mathbf{x}$ is one-to-one and regular, with unit normal

$$U(u,\ v) = \frac{\mathbf{x}_u(u,\ v) \times \delta(u)}{\|\mathbf{x}_u(u,\ v)\|}.$$

9. (*Continuation, computer graphics.*)
  (a) Plot the Möbius band B.
  (b) Plot a surface that represents B with its central circle β removed (for clarity, remove a small band around β). Is this surface connected? orientable?

Although a point of $\hat{M}$ is a *vector* in $\mathbf{R}^3$, not a point of $\mathbf{R}^3$, in favorable cases $\hat{M}$ can be turned into a surface in $\mathbf{R}^3$ by mapping each U_p in $\hat{M}$ to the point $\mathbf{p} + \varepsilon U_p$ (Euclidean coordinates) in $\mathbf{R}^3$ for some small $\varepsilon > 0$. (For $\varepsilon = 1$, the point would be the tip of the "arrow" U_p.)

10. (*Continuation.*)
  (a) Using the scheme above, plot the orientation covering surface $\hat{B}$ of the Möbius band B. (Take $\varepsilon = 1/4$.)
  (b) By inspection, is $\hat{B}$ connected? orientable? How is $\hat{B}$ related to the surface in (b) of the preceding exercise?

11. (*Plane with two origins.*) Let Z consist of all ordered pairs of real numbers and one additional point $\mathbf{0}^*$. Let $\mathbf{x}$ and $\mathbf{y}$ be the functions from $\mathbf{R}^2$ to Z such that

$$\mathbf{x}(u,\ v) = \mathbf{y}(u,\ v) = (u,\ v) \quad \text{if} \quad (u,\ v) \neq (0,\ 0),$$

but

$$\mathbf{x}(0,\ 0) = \mathbf{0} = (0,\ 0) \quad \text{and} \quad \mathbf{y}(0,\ 0) = \mathbf{0}^*.$$

(a) Show that the abstract patches $\mathbf{x}$ and $\mathbf{y}$ constitute a patch collection that satisfies the first two conditions in Definition 8.1, but not the Hausdorff axiom. Without the Hausdorff axiom, strange things can happen. For example, prove:
(b) A convergent sequence in Z can have two limits.
(c) The function $F: Z \to Z$ that reverses $\mathbf{0}$ and $\mathbf{0}^*$, leaving all other points fixed, is a differentiable mapping.

12. (a) Given a one-to-one function H from a manifold M onto an arbitrary set A, prove there is a unique way to make A a manifold so that H becomes a diffeomorphism. (*Hint*: Diffeomorphisms move patches to patches.)

(b) In each of the following cases, find natural choices of H and M that make the set a manifold.

(i) The set of all 2×2 real symmetric matrices.

(ii) The set of all circles in $\mathbf{R}^2$.

(iii) The set of all great circles on a sphere Σ.

(iv) The set of all (finite) closed intervals in $\mathbf{R}$.

13. (*Integral curves.*) A curve α in M is an *integral curve* of a vector field V on M provided $\alpha'(t) = V(\alpha(t))$ for all t. Thus an integral curve has at each point the velocity prescribed by V. If $\alpha(0) = \mathbf{p}$, we say that α *starts at* $\mathbf{p}$.

(a) In $\mathbf{R}^2$, show that the curve $\alpha(t) = (u(t), v(t))$ is an integral curve of $V = f_1 U_1 + f_2 U_2$ starting at $(a, b) \in \mathbf{R}^2$ if and only if

$$u' = f_1(u, v), \quad u(0) = a,$$
$$v' = f_2(u, v), \quad v(0) = b.$$

The theory of differential equations guarantees that there is a unique solution for α.

(b) Find an explicit formula for the integral curve of $V = -u^2 U_1 + uv U_2$ on $\mathbf{R}^2$ that starts at the point $(1, -1)$. (The differential equations involved can be solved by elementary methods since one of them is particularly simple.)

(c) Sketch (by hand or by computer) the integral curve α on suitable intervals $A < t < -1$ and $-1 < t < B$.

14. (*Continuation.*) Show that every vector field V on a surface M has an integral curve β starting at any given point $\mathbf{p}$. Specifically, if $\mathbf{x}: D \rightarrow M$ is a patch with $\mathbf{x}(a, b) = \mathbf{p}$, and $\overline{V}$ is the vector field on D such that $\mathbf{x}_*(\overline{V}) = V$, show that $\beta(t) = \mathbf{x}(u(t), v(t))$, where $(u(t), v(t))$ is the integral curve of $\overline{V}$ starting at (a, b).

15. (*Cartesian products.*) For any sets A and B the Cartesian product $A \times B$ consists of all ordered pairs (a, b), with a in A and b in B. If $\mathbf{x}: D \rightarrow M$ and $\mathbf{y}: E \rightarrow N$ are patches in surfaces M and N, define $\mathbf{x} \times \mathbf{y}: D \times E \rightarrow M \times N$ by

$$(\mathbf{x} \times \mathbf{y})(u, v, u', v') = (\mathbf{x}(u, v), \mathbf{y}(u', v')).$$

Show that $\mathbf{x} \times \mathbf{y}$ is an abstract patch and that the collection $\mathscr{P}$ of all such patches makes $M \times N$ a 4-dimensional manifold. $M \times N$ is called the *Cartesian product* of M and N.

The same scheme works for any two manifolds. It derives from the way the x axis and y axis produce the xy plane; indeed, $\mathbf{R} \times \mathbf{R}$ is precisely $\mathbf{R}^2$.

16. If M is an abstract surface, a *proper imbedding* of M into $\mathbf{R}^3$ is a one-to-one regular mapping $F: M \to \mathbf{R}^3$ such that the inverse function $F^{-1}: F(M) \to M$ is continuous. Prove that the image $F(M)$ of a proper imbedding is a surface in $\mathbf{R}^3$ (Def. 1.2) and that it is diffeomorphic to M.

If $F: M \to \mathbf{R}^3$ is merely regular, then F is an *immersion* of M into $\mathbf{R}^3$, and the image $F(M)$ is often called an "immersed surface," even though it can cut across itself and hence not satisfy Definition 1.2.

4.9 Summary

The discovery of calculus made it possible to study arbitrary curved surfaces M in $\mathbf{R}^3$. Initially this was done mostly in terms of the natural Euclidean coordinates $\{x, y, z\}$ of $\mathbf{R}^3$. However, it gradually became clear that in many contexts, coordinates $\{u, v\}$ *in the surface itself* were more efficient. Thus a two-dimensional calculus was developed for surfaces, one that remains valid even if the surface is not contained in $\mathbf{R}^3$.

Along with the Euclidean spaces, such surfaces are prime examples of the general notion of *manifold*. The calculus of any manifold involves differentiable functions, vector fields, differential forms, mappings—and various operations of differentiation and integration. These features are all preserved in a suitable sense by diffeomorphisms—indeed, this criterion gives a formal definition of *manifold theory*.

Chapter 5
Shape Operators

In Chapter 2 we measured the shape of a curve in $\mathbf{R}^3$ by its curvature and torsion functions. Now we consider the analogous measurement problem for surfaces. It turns out that the shape of a surface M in $\mathbf{R}^3$ is described infinitesimally by a certain linear operator S defined on each of the tangent planes of M. As with curves, to say that two surfaces in $\mathbf{R}^3$ have the same shape means simply that they are congruent. And just as with curves, we shall justify our infinitesimal measurements by proving that two surfaces with "the same" shape operators are, in fact, congruent. The *algebraic* invariants (determinant, trace, ...) of its shape operators thus have *geometric* meaning for the surface M. We investigate this matter in detail and find efficient ways to compute these invariants, which we test on a number of geometrically interesting surfaces.

From now on, the notation $M \subset \mathbf{R}^3$ means a connected surface M in $\mathbf{R}^3$ as defined in Chapter 4.

5.1   The Shape Operator of $M \subset \mathbf{R}^3$

Suppose that Z is a Euclidean vector field (Definition 3.7 of Chapter 4) on a surface M in $\mathbf{R}^3$. Although Z is defined only at points of M, the covariant derivative $\nabla_\mathbf{v} Z$ (Chapter 2, Section 5) still makes sense *as long as* $\mathbf{v}$ *is tangent to* M. As usual, $\nabla_\mathbf{v} Z$ is the rate of change of Z in the $\mathbf{v}$ direction, and there are two main ways to compute it.

Method 1.   Let α be a curve in M that has initial velocity $\alpha'(0) = \mathbf{v}$. Let Z_α be the restriction of Z to α, that is, the vector field $t \rightarrow Z(\alpha(t))$ on α. Then

$$\nabla_\mathbf{v} Z = (Z_\alpha)'(0),$$

where the derivative is that of Chapter 2, Section 2.

Method 2. Express Z in terms of the natural frame field of $\mathbf{R}^3$ by

$$Z = \sum z_i U_i.$$

Then

$$\nabla_v Z = \sum v[z_i] U_i,$$

where the directional derivative is that of Definition 3.10 in Chapter 4.

It is easy to show that these two methods give the same result. In fact, since $Z = \sum z_i U_i$,

$$(Z_\alpha)'(0) = \sum z_i(\alpha)'(0) U_i = \sum v[z_i] U_i.$$

(Compare Lemma 5.2 of Chapter 2.) Note that even if Z is a tangent vector field, the covariant derivative $\nabla_v Z$ need not be tangent to M.

If M is an *orientable* surface, Proposition 7.5 of Chapter 4 shows that there is always a (differentiable) unit normal vector field U on the entire surface, and in fact—since M is now assumed connected—there are exactly two, $\pm U$. Even if M is not orientable, unit normals $\pm U$ are available *locally*, since a small region around any point is orientable. In fact, we will find explicit formulas for U on the image $x(D) \subset M$ of any patch.

We are now in a position to find a mathematical measurement of the shape of a surface in $\mathbf{R}^3$.

1.1 Definition If $\mathbf{p}$ is a point of M, then for each tangent vector $\mathbf{v}$ to M at $\mathbf{p}$, let

$$S_p(\mathbf{v}) = -\nabla_v U,$$

where U is a unit normal vector field on a neighborhood of $\mathbf{p}$ in M. S_p is called the *shape operator* of M at $\mathbf{p}$ derived from U.† (Fig. 5.1.)

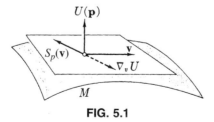

FIG. 5.1

† The minus sign artificially introduced in this definition will sharply reduce the total number of minus signs needed later on.

The tangent plane of M at any point $\mathbf{q}$ consists of all Euclidean vectors orthogonal to $U(\mathbf{q})$. Thus the rate of change $\nabla_v U$ of U in the $\mathbf{v}$ direction tells how the tangent planes of M are varying in the $\mathbf{v}$ direction—and this gives an infinitesimal description of the way M itself is curving in $\mathbf{R}^3$.

Note that if U is replaced by $-U$, then S_p changes to $-S_p$.

1.2 Lemma   For each point $\mathbf{p}$ of $M \subset \mathbf{R}^3$, the shape operator is a linear operator

$$S_p: T_p(M) \to T_p(M)$$

on the tangent plane of M at $\mathbf{p}$.

Proof.   In Definition 1.1, U is a unit vector field, so $U \cdot U = 1$. Thus by a Leibnizian property of covariant derivatives,

$$0 = \mathbf{v}[U \cdot U] = 2\nabla_v U \cdot U(\mathbf{p}) = -2S_p(\mathbf{v}) \cdot U(\mathbf{p}),$$

where $\mathbf{v}$ is tangent to M at $\mathbf{p}$. Since U is also a *normal* vector field, it follows that $S_p(\mathbf{v})$ is tangent to M at $\mathbf{p}$. Thus S_p is a function from $T_p(M)$ to $T_p(M)$. (It is to emphasize this that we use the term "operator" instead of "transformation.")

The linearity of S_p is a consequence of a linearity property of covariant derivatives:

$$S_p(a\mathbf{v} + b\mathbf{w}) = -\nabla_{av+bw}U = -(a\nabla_v U + b\nabla_w U)$$
$$= aS_p(\mathbf{v}) + bS_p(\mathbf{w}). \qquad \blacklozenge$$

At each point $\mathbf{p}$ of $M \subset \mathbf{R}^3$ there are actually two shape operators, $\pm S_p$, derived from the two unit normals $\pm U$ near $\mathbf{p}$. We shall refer to all of these, collectively, as *the shape operator S of M*. Thus if a choice of unit normal is not specified, there is a relatively harmless ambiguity of sign.

1.3 Example   Shape operators of some surfaces in $\mathbf{R}^3$.

(1) Let Σ be the sphere of radius r consisting of all points $\mathbf{p}$ of $\mathbf{R}^3$ with $\| \mathbf{p} \| = r$. Let U be the outward normal on Σ. Now as U moves away from any point $\mathbf{p}$ in the direction $\mathbf{v}$, evidently U topples forward in the exact direction of $\mathbf{v}$ itself (Fig. 5.2). Thus $S(\mathbf{v})$ must have the form $-c\mathbf{v}$.

In fact, using *gradients* as in Example 3.9 of Chapter 4, we find

$$U = \frac{1}{r}\sum x_i U_i.$$

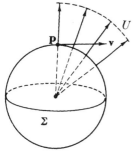

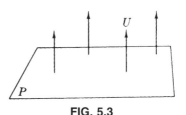

FIG. 5.3

FIG. 5.2

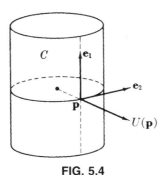

FIG. 5.4

But then

$$\nabla_v U = \frac{1}{r}\sum v[x_i]U_i(\mathbf{p}) = -\frac{\mathbf{v}}{r}.$$

Thus $S(\mathbf{v}) = -\mathbf{v}/r$ for all $\mathbf{v}$. So the shape operator S is merely scalar multiplication by $-1/r$. This uniformity in S reflects the roundness of spheres: They bend the same way in all directions at all points.

(2) Let P be a plane in $\mathbf{R}^3$. A unit normal vector field U on P is evidently *parallel* in $\mathbf{R}^3$ (constant Euclidean coordinates) (Fig. 5.3). Hence

$$S(\mathbf{v}) = -\nabla_v U = 0$$

for all tangent vectors $\mathbf{v}$ to P. Thus the shape operator is identically zero, which is to be expected, since planes do not bend at all.

(3) Let C be the circular cylinder $x^2 + y^2 = r^2$ in $\mathbf{R}^3$. At any point $\mathbf{p}$ of C, let $\mathbf{e}_1$ and $\mathbf{e}_2$ be unit tangent vectors, with $\mathbf{e}_1$ tangent to the ruling of the cylinder through $\mathbf{p}$, and $\mathbf{e}_2$ tangent to the cross-sectional circle. Use the outward normal U as indicated in Fig. 5.4.

Now, when U moves from $\mathbf{p}$ in the $\mathbf{e}_1$ direction, it stays parallel to itself just as on a plane; hence $S(\mathbf{e}_1) = 0$. When U moves in the $\mathbf{e}_2$ direction, it topples forward exactly as on a sphere of radius r; hence $S(\mathbf{e}_2) = -\mathbf{e}_2/r$. In this way S describes the "half-flat, half-round" shape of a cylinder.

(4) The *saddle surface M: z = xy*. For the moment we investigate S only at $\mathbf{p} = (0, 0, 0)$ in M. Since the x and y axes of $\mathbf{R}^3$ lie in M, the vectors

$$\mathbf{u}_1 = (1, 0, 0) \text{ and } \mathbf{u}_2 = (0, 1, 0)$$

are tangent to M at $\mathbf{p}$. We use the "upward" unit normal U, which at $\mathbf{p}$ is $(0, 0, 1)$. Along the x axis, U stays orthogonal to the x axis, and as it proceeds in the $\mathbf{u}_1$ direction, U swings from left to right (Fig. 5.5).

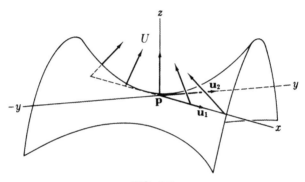

FIG. 5.5

In fact, a routine computation (Exercise 3) shows that $\nabla_{\mathbf{u}_1} U = -\mathbf{u}_2$. Similarly, we find $\nabla_{\mathbf{u}_2} U = -\mathbf{u}_1$. Thus the shape operator of M at $\mathbf{p}$ is given by the formula

$$S(a\mathbf{u}_1 + b\mathbf{u}_2) = b\mathbf{u}_1 + a\mathbf{u}_2.$$

These examples clarify the analogy between the shape operator of a surface and the curvature and torsion of a curve. In the case of a curve, there is only one direction to move, and κ and τ measure the rate of change of the unit vector fields T and B (hence N). For a surface, only one unit vector field is intrinsically determined—the unit normal U. Furthermore, at each point, there is now a whole plane of directions in which U can move, so that rates of change of U are measured, not numerically, but by the linear operators S.

1.4 Lemma For each point $\mathbf{p}$ of $M \subset \mathbf{R}^3$, the shape operator

$$S: T_p(M) \rightarrow T_p(M)$$

is a *symmetric* linear operator; that is,

$$S(\mathbf{v}) \bullet \mathbf{w} = S(\mathbf{w}) \bullet \mathbf{v}$$

for any pair of tangent vectors to M at $\mathbf{p}$.

We postpone the proof of this crucial fact to Section 4, where it occurs naturally in the course of general computations.

From the viewpoint of linear algebra, a symmetric linear operator on a two-dimensional vector space is a very simple object indeed. For a shape operator, its eigenvalues and eigenvectors, its trace and determinant, all turn out to have geometric meaning of first importance for the surface $M \subset \mathbf{R}^3$.

Exercises

1. Let α be a curve in $M \subset \mathbf{R}^3$. If U is a unit normal of M restricted to the curve α, show that $S(\alpha') = -U'$.

2. Consider the surface $M: z = f(x, y)$, where

$$f(0, 0) = f_x(0, 0) = f_y(0, 0) = 0.$$

(The subscripts indicate partial derivatives.) Show that
(a) The vectors $\mathbf{u}_1 = U_1(\mathbf{0})$ and $\mathbf{u}_2 = U_2(\mathbf{0})$ are tangent to M at the origin $\mathbf{0}$, and

$$U = \frac{-f_x U_1 - f_y U_2 + U_3}{\sqrt{1 + f_x^2 + f_y^2}}$$

is a unit normal vector field on M.
(b) $S(\mathbf{u}_1) = f_{xx}(0, 0)\mathbf{u}_1 + f_{xy}(0, 0)\mathbf{u}_2$,
$S(\mathbf{u}_2) = f_{yx}(0, 0)\mathbf{u}_1 + f_{yy}(0, 0)\mathbf{u}_2$.

(*Note:* The square root in the denominator is no real problem here because of the special character of f at $(0, 0)$. In general, direct computation of S is awkward, and in Section 4 we shall establish indirect ways of getting at it.)

3. (*Continuation.*) In each case, express $S(a\mathbf{u}_1 + b\mathbf{u}_2)$ in terms of $\mathbf{u}_1$ and $\mathbf{u}_2$, and determine the rank of S at $\mathbf{0}$ (rank S is dimension of image S: 0, 1, or 2).
(a) $z = xy$.   (b) $z = 2x^2 + y^2$.
(c) $z = (x + y)^2$.   (d) $z = xy^2$.

4. Let M be a surface in $\mathbf{R}^3$ oriented by a unit normal vector field

$$U = g_1 U_1 + g_2 U_2 + g_3 U_3.$$

Then the *Gauss map* $G: M \to \Sigma$ of M sends each point $\mathbf{p}$ of M to the point $(g_1(\mathbf{p}), g_2(\mathbf{p}), g_3(\mathbf{p}))$ of the unit sphere Σ. Pictorially: Move $U(\mathbf{p})$ to the origin by parallel motion; there it points to $G(\mathbf{p})$ (Fig. 5.6).

Thus G completely describes the turning of U as it traverses M.

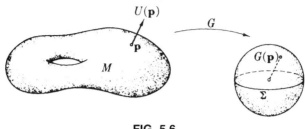

FIG. 5.6

For each of the following surfaces, describe the image $G(M)$ of the Gauss map in the sphere Σ (use either normal):

(a) Cylinder, $x^2 + y^2 = r^2$.

(b) Cone, $z = \sqrt{x^2 + y^2}$.

(c) Plane, $x + y + z = 0$.

(d) Sphere, $(x - 1)^2 + y^2 + (z + 2)^2 = 1$.

5. Let $G: T \to \Sigma$ be the Gauss map of the torus T (Ex. 2.5 of Ch. 4) derived from its outward unit normal U. What are the image curves under G of the meridians and parallels of T? Which points of Σ are the image of exactly two points of T?

6. Let $G: M \to \Sigma$ be the Gauss map of the saddle surface $M: z = xy$ derived from the unit normal U obtained as in Exercise 2. What is the image under G of one of the straight lines, $y = $ constant, in M? How much of the sphere is covered by the entire image $G(M)$?

7. Show that the shape operator of M is (minus) the tangent map of its Gauss map: If S and $G: M \to \Sigma$ are both derived from U, then $S(\mathbf{v})$ and $-G_*(\mathbf{v})$ are parallel for every tangent vector $\mathbf{v}$ to M.

8. An orientable surface has two Gauss maps derived from its two unit normals. Show that they differ only by the antipodal mapping of Σ (Ex. 8.2 of Ch. 4). Define a Gauss-type mapping for a nonorientable surface in $\mathbf{R}^3$.

9. If V is a tangent vector field on M (with unit normal U), then $S(V)$ is the tangent vector field on M whose value at each point $\mathbf{p}$ is $S_p(V(\mathbf{p}))$. Show that if W is also tangent to M, then

$$S(V) \bullet W = \nabla_V W \bullet U.$$

Deduce that the symmetry of S is equivalent to the assertion that the *bracket*

$$[V, W] = \nabla_V W - \nabla_W V$$

of two tangent vector fields is again a tangent vector field.

5.2 Normal Curvature

Throughout this section we shall work in a region of $M \subset \mathbf{R}^3$ that has been *oriented* by the choice of a unit normal vector field U, and we use the shape operator S derived from U.

The shape of a surface in $\mathbf{R}^3$ influences the shape of the curves in M.

2.1 Lemma If α is a curve in $M \subset \mathbf{R}^3$, then

$$\alpha'' \cdot U = S(\alpha') \cdot \alpha'.$$

Proof. Since α is in M, its velocity α' is always tangent to M. Thus

$$\alpha' \cdot U = 0,$$

where U is restricted to the curve α. Differentiation yields

$$\alpha'' \cdot U + \alpha' \cdot U' = 0.$$

But from Section 1, we know that $S(\alpha') = -U'$. Hence

$$\alpha'' \cdot U = -U' \cdot \alpha' = S(\alpha') \cdot \alpha' \qquad \blacklozenge$$

Geometric interpretation: at each point, $\alpha'' \cdot U$ is the component of acceleration α'' normal to the surface M (Fig. 5.7). The lemma shows that this component depends only on the velocity α' and the shape operator of M. Thus *all curves in M with a given velocity $\mathbf{v}$ at point $\mathbf{p}$ will have the same normal component of acceleration at $\mathbf{p}$, namely, $S(\mathbf{v}) \cdot \mathbf{v}$*. This is the component of acceleration that *the bending of M in $\mathbf{R}^3$ forces them to have*.

Thus if $\mathbf{v}$ is standardized by reducing it to a unit vector $\mathbf{u}$, we get a measurement of the way M is bent in the $\mathbf{u}$ direction.

2.2 Definition Let $\mathbf{u}$ be a unit vector tangent to $M \subset \mathbf{R}^3$ at a point $\mathbf{p}$. Then the number $k(\mathbf{u}) = S(\mathbf{u}) \cdot \mathbf{u}$ is called the *normal curvature of M in the $\mathbf{u}$ direction*.

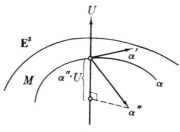

FIG. 5.7

In this context, we define a *tangent direction to M at* **p** to be a one-dimensional subspace L of $T_p(M)$, that is, a line through the zero vector (located for intuitive purposes at **p**). Any nonzero tangent vector at **p** determines a direction L, but we prefer to use one of the two unit vectors $\pm\mathbf{u}$ in L. Note that

$$k(\mathbf{u}) = S(\mathbf{u}) \cdot \mathbf{u} = S(-\mathbf{u}) \cdot (-\mathbf{u}) = k(-\mathbf{u}).$$

Thus, although we evaluate k on unit vectors, it is, in effect, a real-valued function on the set of all tangent directions to M.

Given a unit tangent vector **u** to M at **p**, let α be a unit-speed curve in M with initial velocity $\alpha'(0) = \mathbf{u}$. Using the Frenet apparatus of α, the preceding lemma gives

$$k(\mathbf{u}) = S(\mathbf{u}) \cdot \mathbf{u} = \alpha''(0) \cdot U(\mathbf{p}) = \kappa(0)N(0) \cdot U(\mathbf{p})$$
$$= \kappa(0)\cos\vartheta.$$

Thus the normal curvature of M in the **u** direction is $\kappa(0)\cos\vartheta$, where $\kappa(0)$ is the curvature of α at $\alpha(0) = \mathbf{p}$, and ϑ is the angle between the principal normal $N(0)$ and the surface normal $U(\mathbf{p})$, as in Fig. 5.8.

Given **u**, there is a natural way to choose the curve so that ϑ is 0 or π. In fact, if P is the plane determined by **u** and $U(\mathbf{p})$, then P cuts from M (near **p**) a curve σ called the *normal section* of M in the **u** direction. If we give σ unit-speed parametrization with $\sigma'(0) = \mathbf{u}$, then $N(0) = \pm U(p)$, since

$$\sigma''(0) = \kappa(0)N(0)$$

is orthogonal to $\sigma'(0) = \mathbf{u}$ and tangent to P. So for a normal section in the **u** direction (Fig. 5.9),

$$k(\mathbf{u}) = \kappa_\sigma(0)N(0) \cdot U(\mathbf{p}) = \pm\kappa_\sigma(0).$$

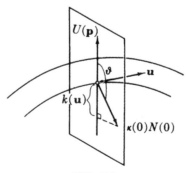

FIG. 5.8

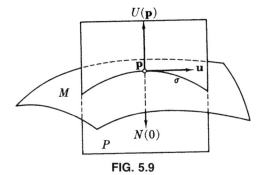

FIG. 5.9

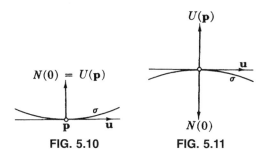

FIG. 5.10   **FIG. 5.11**

Thus it is possible to make a reasonable estimate of the normal curvatures in various directions on a surface $M \subset \mathbf{R}^3$ by picturing what the corresponding normal sections would look like. We know that the principal normal N of a curve tells in which direction it is turning. Thus the preceding discussion gives geometric meaning to the sign of the normal curvature $k(\mathbf{u})$ (relative to our fixed choice of U).

(1) If $k(\mathbf{u}) > 0$, then $N(0) = U(\mathbf{p})$, so the normal section σ is bending toward $U(\mathbf{p})$ at $\mathbf{p}$ (Fig. 5.10). Thus in the $\mathbf{u}$ direction the surface M is bending *toward* $U(\mathbf{p})$.

(2) If $k(\mathbf{u}) < 0$, then $N(0) = -U(\mathbf{p})$, so the normal section σ is bending *away from* $U(\mathbf{p})$ at $\mathbf{p}$. Thus in the $\mathbf{u}$ direction M is bending away from $U(\mathbf{p})$ (Fig. 5.11).

(3) If $k(\mathbf{u}) = 0$, then $k_\sigma(0) = 0$ and $N(0)$ is undefined. Here the normal section σ is not turning at $\sigma(0) = \mathbf{p}$. We cannot conclude that in the $\mathbf{u}$ direction M is not bending at all, since κ might be zero only at $\sigma(0) = \mathbf{p}$. But we can conclude that its rate of bending is unusually small.

In different directions at a fixed point $\mathbf{p}$, the surface may bend in quite different ways. For example, consider the saddle surface $z = xy$ in Example 1.3. If we identify the tangent plane of M at $\mathbf{p} = (0, 0, 0)$ with the xy plane of

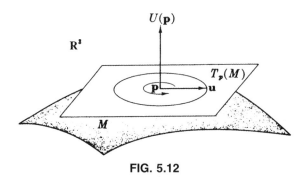

FIG. 5.12

$\mathbf{R}^3$, then clearly the normal curvature in the direction of the x and y axes is zero, since the normal sections are straight lines. However, Fig. 5.5 shows that in the tangent direction given by the line $y = x$, the normal curvature is positive, for the normal section is a parabola bending upward. ($U(\mathbf{p}) = (0, 0, 1)$ is "upward.") But in the direction of the line $y = -x$, normal curvature is negative, since this parabola bends downward.

Let us now fix a point $\mathbf{p}$ of $M \subset \mathbf{R}^3$ and imagine that a unit tangent vector $\mathbf{u}$ at $\mathbf{p}$ revolves, sweeping out the unit circle in the tangent plane $T_p(M)$. From the corresponding normal sections, we get a moving picture of the way M is bending in *every* direction at $\mathbf{p}$ (Fig. 5.12).

2.3 Definition   Let $\mathbf{p}$ be a point of $M \subset \mathbf{R}^3$. The maximum and minimum values of the normal curvature $k(\mathbf{u})$ of M at $\mathbf{p}$ are called the *principal curvatures* of M at $\mathbf{p}$, and are denoted by k_1 and k_2. The directions in which these extreme values occur are called *principal directions* of M at $\mathbf{p}$. Unit vectors in these directions are called *principal vectors* of M at $\mathbf{p}$.

Using the normal-section scheme discussed above, it is often fairly easy to pick out the directions of maximum and minimum bending. For example, if we use the outward normal (U) on a circular cylinder C as in Fig. 5.4, then the normal sections of C all bend away from U, so $k(\mathbf{u}) \leqq 0$. Furthermore, it is reasonably clear that the maximum value $k_1 = 0$ occurs only in the direction $\mathbf{e}_1$ of a ruling; minimum value $k_2 < 0$ occurs only in the direction $\mathbf{e}_2$ tangent to a cross-section.

An interesting special case occurs at points $\mathbf{p}$ for which $k_1 = k_2$. The maximum and minimum normal curvature being equal, it follows that $k(\mathbf{u})$ is constant: *M bends the same amount in all directions at* $\mathbf{p}$ (so all directions are principal).

2.4 Definition   A point $\mathbf{p}$ of $M \subset \mathbf{R}^3$ is *umbilic* provided the normal curvature $k(\mathbf{u})$ is constant on all unit tangent vectors $\mathbf{u}$ at $\mathbf{p}$.

For example, what we found in (1) of Example 1.3 was that every point of the sphere Σ is umbilic, with $k_1 = k_2 = -1/r$.

2.5 Theorem (1) If $\mathbf{p}$ is an umbilic point of $M \subset \mathbf{R}^3$, then the shape operator S at $\mathbf{p}$ is just scalar multiplication by $k = k_1 = k_2$.
(2) If $\mathbf{p}$ is a nonumbilic point, $k_1 \neq k_2$, then there are exactly two principal directions, and these are orthogonal. Furthermore, if $\mathbf{e}_1$ and $\mathbf{e}_2$ are principal vectors in these directions, then

$$S(\mathbf{e}_1) = k_1 \mathbf{e}_1, \quad S(\mathbf{e}_2) = k_2 \mathbf{e}_2.$$

In short, the principal curvatures of M at $\mathbf{p}$ are the *eigenvalues* of S, and the principal vectors of M at $\mathbf{p}$ are the *eigenvectors* of S.

Proof. Suppose that k takes on its maximum value k_1 at $\mathbf{e}_1$, so

$$k_1 = k(\mathbf{e}_1) = S(\mathbf{e}_1) \cdot \mathbf{e}_1.$$

Let $\mathbf{e}_2$ be merely a unit tangent vector orthogonal to $\mathbf{e}_1$ (presently we shall show that it is also a principal vector).
If $\mathbf{u}$ is any unit tangent vector at $\mathbf{p}$, we write

$$\mathbf{u} = \mathbf{u}(\vartheta) = c\mathbf{e}_1 + s\mathbf{e}_2,$$

where $c = \cos\vartheta$, $s = \sin\vartheta$ (Fig. 5.13). Thus normal curvature k at $\mathbf{p}$ becomes a function on the real line: $k(\vartheta) = k(\mathbf{u}(\vartheta))$.
For $1 \leq i, j \leq 2$, let S_{ij} be the number $S(\mathbf{e}_i) \cdot \mathbf{e}_j$. Note that $S_{11} = k_1$, and by the symmetry of the shape operator, $S_{12} = S_{21}$. We compute

$$k(\vartheta) = S(c\mathbf{e}_1 + s\mathbf{e}_2) \cdot (c\mathbf{e}_1 + s\mathbf{e}_2)$$
$$= c^2 S_{11} + 2sc S_{12} + s^2 S_{22}. \tag{1}$$

Hence

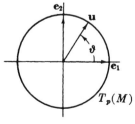

FIG. 5.13

$$\frac{dk}{d\vartheta}(\vartheta) = 2sc(S_{22} - S_{11}) + 2(c^2 - S^2)S_{12}. \tag{2}$$

If $\vartheta = 0$, then $c = 1$ and $s = 0$, so $\mathbf{u}(0) = \mathbf{e}_1$. Thus, by assumption, $k(\vartheta)$ is a maximum at $\vartheta = 0$, so $(dk/d\vartheta)(0) = 0$. It follows immediately from (2) that $S_{12} = 0$.

Since $\mathbf{e}_1$, $\mathbf{e}_2$ is an orthonormal basis for $T_p(M)$, we deduce by orthonormal expansion that

$$S(\mathbf{e}_1) = S_{11}\mathbf{e}_1, \quad S(\mathbf{e}_2) = S_{22}\mathbf{e}_2. \tag{3}$$

Now if $\mathbf{p}$ is umbilic, then $S_{22} = k(\mathbf{e}_2)$ is the same as $S_{11} = k(\mathbf{e}_1) = k_1$, so (3) shows that S is scalar multiplication by $k_1 = k_2$.

If $\mathbf{p}$ is *not* umbilic, we look back at (1), which has become

$$k(\vartheta) = c^2 k_1 + s^2 S_{22}. \tag{4}$$

Since k_1 is the maximum value of $k(\vartheta)$, and $k(\vartheta)$ is now nonconstant, it follows that $k_1 > S_{22}$. But then (4) shows: (a) the maximum value k_1 is taken on *only* when $c = \pm 1$, $s = 0$, that is, in the $\mathbf{e}_1$ direction; and (b) the minimum value k_2 is S_{22}, and is taken on *only* when $c = 0$, $s = \pm 1$ that is, in the $\mathbf{e}_2$ direction. This proves the second assertion in the theorem, since (3) now reads:

$$S(\mathbf{e}_1) = k_1\mathbf{e}_1, \quad S(\mathbf{e}_2) = k_2\mathbf{e}_2. \qquad \blacklozenge$$

Contained in the preceding proof is Euler's formula for the normal curvature of M in *all* directions at $\mathbf{p}$.

2.6 Corollary Let k_1, k_2 and $\mathbf{e}_1$, $\mathbf{e}_2$ be the principal curvatures and vectors of $M \subset \mathbf{R}^3$ at $\mathbf{p}$. Then if $\mathbf{u} = \cos\vartheta\mathbf{e}_1 + \sin\vartheta\mathbf{e}_2$, the normal curvature of M in the $\mathbf{u}$ direction is (Fig. 5.13)

$$k(\mathbf{u}) = k_1 \cos^2\vartheta + k_2 \sin^2\vartheta.$$

Here is another way to show how the principal curvatures k_1 and k_2 control the shape of M near an arbitrary point $\mathbf{p}$. Since the position of M in $\mathbf{R}^3$ is of no importance, we can assume that (1) $\mathbf{p}$ is at the origin of $\mathbf{R}^3$, (2) the tangent plane $T_p(M)$ is the xy plane of $\mathbf{R}^3$, and (3) the x and y axes are the principal directions. *Near* $\mathbf{p}$, M can be expressed as $M: z = f(x, y)$, as shown in Fig. 5.14, and the idea is to construct an *approximation* of M near $\mathbf{p}$ by using only terms up to quadratic in the Taylor expansion of the function f. Now (1) and (2) imply $f^0 = f_x^0 = f_y^0 = 0$, where the subscripts indicate partial derivatives and the superscript zero denotes evaluation at $x = 0$, $y = 0$. Thus the quadratic approximation of f near $(0, 0)$ reduces to

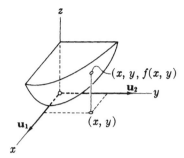

FIG. 5.14

$$f(x, y) \sim \frac{1}{2}\left(f_{xx}^0 x^2 + 2 f_{xy}^0 xy + f_{yy}^0 y^2\right).$$

In Exercise 2 of Section 1 we found that for the tangent vectors

$$\mathbf{u}_1 = (1, 0, 0) \quad \text{and} \quad \mathbf{u}_2 = (0, 1, 0)$$

at $\mathbf{p} = 0$,

$$S(\mathbf{u}_1) = -\nabla_{\mathbf{u}_1} U = f_{xx}^0 \mathbf{u}_1 + f_{xy}^0 \mathbf{u}_2,$$

$$S(\mathbf{u}_2) = -\nabla_{\mathbf{u}_2} U = f_{xy}^0 \mathbf{u}_1 + f_{yy}^0 \mathbf{u}_2.$$

By condition (3) above, $\mathbf{u}_1$ and $\mathbf{u}_2$ are *principal* vectors, so it follows from Theorem 2.5 that $k_1 = f_{xx}^0$, $k_2 = f_{yy}^0$, and $f_{xy}^0 = 0$.

Substituting these values in the quadratic approximation of f, we conclude that *the shape of M near* $\mathbf{p}$ *is approximately the same as that of the surface*

$$M': z = \frac{1}{2}\left(k_1 x^2 + k_2 y^2\right)$$

near $\mathbf{0}$. M' is called the *quadratic approximation* of M near $\mathbf{p}$. It is an analogue for surfaces of a Frenet approximation of a curve.

From Definition 2.2 through Corollary 2.6 we have been concerned with the geometry of $M \subset \mathbf{R}^3$ near one of its points $\mathbf{p}$. These results thus apply simultaneously to all the points of the oriented region $\mathcal{O}$ on which, by our initial assumption, the unit normal U is defined. In particular then, we have actually defined principal curvature *functions* k_1 and k_2 on $\mathcal{O}$, where at each point $\mathbf{p}$ of $\mathcal{O}$, $k_1(\mathbf{p})$ and $k_2(\mathbf{p})$ are the principal curvatures of M at $\mathbf{p}$. Note that these functions are only defined "modulo sign": If U is replaced by $-U$, they become $-k_1$ and $-k_2$.

Exercises

1. Use the results of Example 1.3 to find the principal curvatures and principal vectors of
  (a) The cylinder, at every point.
  (b) The saddle surface, at the origin.

2. If $\mathbf{v} \neq \mathbf{0}$ is a tangent vector (not necessarily of unit length), show that the normal curvature of M in the direction of $\mathbf{v}$ is $k(\mathbf{v}) = S(\mathbf{v}) \cdot \mathbf{v}/\mathbf{v} \cdot \mathbf{v}$.

3. For each integer $n \geq 2$, let α_n be the curve $t \to (r\cos t, r\sin t, \pm t^n)$ in the cylinder $M: x^2 + y^2 = r^2$. These curves all have the same velocity at $t = 0$; test Lemma 2.1 by showing that they all have the same normal component of acceleration at $t = 0$.

4. For each of the following surfaces, find the quadratic approximation near the origin:
  (a) $z = \exp(x^2 + y^2) - 1$.   (b) $z = \log\cos x - \log\cos y$.
  (c) $z = (x + 3y)^3$.

5.3 Gaussian Curvature

The preceding section found the geometrical meaning of the eigenvalues and eigenvectors of the shape operator. Now we examine the determinant and trace of S.

3.1 Definition   The *Gaussian curvature* of $M \subset \mathbf{R}^3$ is the real-valued function $K = \det S$ on M. Explicitly, for each point $\mathbf{p}$ of M, the Gaussian curvature $K(\mathbf{p})$ of M at $\mathbf{p}$ is the determinant of the shape operator S of M at $\mathbf{p}$.

The *mean curvature* of $M \subset \mathbf{R}^3$ is the function $H = 1/2$ trace S. Gaussian and mean curvature are expressed in terms of principal curvature by

3.2 Lemma $K = k_1 k_2, \quad H = \dfrac{1}{2}(k_1 + k_2).$

Proof.   The determinant (and trace) of a linear operator may be defined as the common value of the determinant (and trace) of all its matrices. If e_1 and e_2 are principal vectors at a point $\mathbf{p}$, then by Theorem 2.5, we have $S(e_1) = k_1(\mathbf{p})e_1$ and $S(e_2) = k_2(\mathbf{p})e_2$. Thus the matrix of S at $\mathbf{p}$ with respect to e_1, e_2 is

$$\begin{pmatrix} k_1(\mathbf{p}) & 0 \\ 0 & k_2(\mathbf{p}) \end{pmatrix}.$$

This immediately gives the required result. ◆

A significant fact about the Gaussian curvature: It is independent of the choice of the unit normal U. If U is changed to $-U$, then the signs of *both* k_1 and k_2 change, so $K = k_1 k_2$ is unaffected. This is obviously not the case with mean curvature $H = (k_1 + k_2)/2$, which has the same ambiguity of sign as the principal curvatures themselves.

The normal section method in Section 2 lets us tell, by inspection, approximately what the principal curvatures of M are at each point. Thus we get a reasonable idea of what the Gaussian curvature $K = k_1 k_2$ is at each point $\mathbf{p}$ by merely *looking* at the surface M. In particular, we can usually tell what the sign of $K(\mathbf{p})$ is—and this sign has an important geometric meaning, which we now illustrate.

3.3 Remark *The sign of Gaussian curvature at a point* $\mathbf{p}$.

(1) *Positive.* If $K(\mathbf{p}) > 0$, then by Lemma 3.2, the principal curvatures $k_1(\mathbf{p})$ and $k_2(\mathbf{p})$ have the same sign. By Corollary 2.6, either $k(\mathbf{u}) > 0$ for all unit vectors $\mathbf{u}$ at $\mathbf{p}$ or $k(\mathbf{u}) < 0$. Thus *M is bending away from its tangent plane* $T_p(M)$ *in all tangent directions at* $\mathbf{p}$ (Fig. 5.15)

The quadratic approximation of M near $\mathbf{p}$ is the paraboloid

$$z = \frac{1}{2} k_1(\mathbf{p})x^2 + \frac{1}{2} k_2(\mathbf{p})y^2.$$

(2) *Negative.* If $K(\mathbf{p}) < 0$, then by Lemma 3.2, the principal curvatures $k_1(\mathbf{p})$ and $k_2(\mathbf{p})$ have opposite signs. Thus the quadratic approximation of M near $\mathbf{p}$ is a hyperboloid, so M is also saddle-shaped *near* $\mathbf{p}$ (Fig. 5.16).

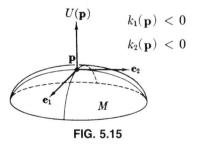

FIG. 5.15   **FIG. 5.16**

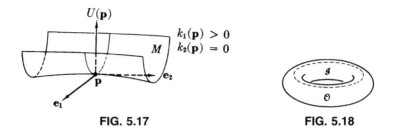

FIG. 5.17   FIG. 5.18

(3) *Zero.* If $K(\mathbf{p}) = 0$, then by Lemma 3.2 there are two cases:
   (a) If only one principal curvature is zero, say

$$k_1(\mathbf{p}) \neq 0, \quad k_2(\mathbf{p}) = 0.$$

   (b) If both principal curvatures are zero, say

$$k_1(\mathbf{p}) = k_2(\mathbf{p}) = 0.$$

In case (a) the quadratic approximation is the cylinder $2z = k_1(\mathbf{p})x^2$, so M is trough-shaped near $\mathbf{p}$ (Fig. 5.17).

In case (b), the quadratic approximation reduces simply to the plane $z = 0$, so we get no information about the shape of M near $\mathbf{p}$.

A torus of revolution T provides a good example of these different cases. At points on the outer half $\mathcal{O}$ of T, the torus bends away from its tangent plane as one can see from Fig. 5.18; hence $K > 0$ on $\mathcal{O}$. But near each point $\mathbf{p}$ of the inner half $\mathcal{I}$, T is saddle-shaped and cuts through $T_p(M)$. Hence $K < 0$ on $\mathcal{I}$.

Near each point on the two circles (top and bottom) that separate $\mathcal{O}$ and $\mathcal{I}$, the torus is trough-shaped; hence $K = 0$ there. (A quantitative check of these qualitative results is given in Section 7.)

In case 3(b) above, where both principal curvatures vanish, $\mathbf{p}$ is called a *planar* point of M. (There are no planar points on the torus.) For example, the central point $\mathbf{p}$ of a *monkey saddle*, say

$$M: z = x\left(x + \sqrt{3}y\right)\left(x - \sqrt{3}y\right),$$

is planar. Here three hills and valleys meet, as shown in Fig. 5.19. Thus $\mathbf{p}$ *must* be a planar point—the shape of M near $\mathbf{p}$ is too complicated for the other three possibilities in Remark 3.3.

We consider now some ways to compute Gaussian and mean curvature.

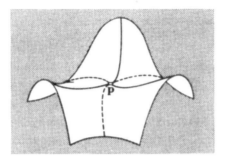

FIG. 5.19

3.4 Lemma If $\mathbf{v}$ and $\mathbf{w}$ are linearly independent tangent vectors at a point $\mathbf{p}$ of $M \subset \mathbf{R}^3$, then

$$S(\mathbf{v}) \times S(\mathbf{w}) = K(\mathbf{p})\mathbf{v} \times \mathbf{w},$$

$$S(\mathbf{v}) \times \mathbf{w} + \mathbf{v} \times S(\mathbf{w}) = 2H(\mathbf{p})\mathbf{v} \times \mathbf{w}.$$

Proof. Since $\mathbf{v}, \mathbf{w}$ is a basis for the tangent plane $T_p(M)$, we can write

$$S(\mathbf{v}) = a\mathbf{v} + b\mathbf{w},$$

$$S(\mathbf{w}) = c\mathbf{v} + d\mathbf{w}.$$

This shows that

$$\begin{pmatrix} a & b \\ c & d \end{pmatrix}$$

is the matrix of S with respect to the basis $\mathbf{v}, \mathbf{w}$. Hence

$$K(\mathbf{p}) = \det S = ad - bc, \quad H(\mathbf{p}) = \frac{1}{2}\,\text{trace}\,S = \frac{1}{2}(a + d).$$

Using standard properties of the cross product, we compute

$$S(\mathbf{v}) \times S(\mathbf{w}) = (a\mathbf{v} + b\mathbf{w}) \times (c\mathbf{v} + d\mathbf{w})$$

$$= (ad - bc)\mathbf{v} \times \mathbf{w} = K(\mathbf{p})\mathbf{v} \times \mathbf{w},$$

and a similar calculation gives the formula for $H(\mathbf{p})$. ◆

Thus if V and W are tangent vector fields that are linearly independent at each point of an oriented region, we have vector field equations

$$S(V) \times S(W) = KV \times W,$$

$$S(V) \times W + V \times S(W) = 2HV \times W.$$

These may be solved for K and H by dotting each side with the normal vector field $V \times W$, and using the Lagrange identity (Exercise 6). We then find

$$K = \frac{\begin{vmatrix} S(V) \cdot V & S(V) \cdot W \\ S(W) \cdot V & S(W) \cdot V \end{vmatrix}}{\begin{vmatrix} V \cdot V & V \cdot W \\ W \cdot V & W \cdot W \end{vmatrix}},$$

$$H = \frac{\begin{vmatrix} S(V) \cdot V & S(V) \cdot W \\ W \cdot V & W \cdot W \end{vmatrix} + \begin{vmatrix} V \cdot V & V \cdot W \\ S(W) \cdot V & S(W) \cdot W \end{vmatrix}}{2 \begin{vmatrix} V \cdot V & V \cdot W \\ W \cdot V & W \cdot W \end{vmatrix}}.$$

The denominators are never zero, since the independence of V and W is equivalent to $(V \times W) \cdot (V \times W) > 0$. In particular, the functions K and H are *differentiable*.

Once K and H are known, it is a simple matter to find k_1 and k_2.

3.5 Corollary On an oriented region $\mathcal{O}$ in M, the principal curvature functions are

$$k_1, k_2 = H \pm \sqrt{H^2 - K}.$$

Proof. To verify the formula, it suffices to substitute

$$K = k_1 k_2 \quad \text{and} \quad H = \frac{k_1 + k_2}{2}$$

and note that

$$H^2 - K = \frac{(k_1 + k_2)^2}{4} - k_1 k_2 = \frac{(k_1 - k_2)^2}{4}. \qquad \blacklozenge$$

A more enlightening derivation (Exercise 4) uses the characteristic polynomial of S.

This formula shows only that k_1 and k_2 are *continuous* functions on $\mathcal{O}$; they need not be differentiable since the square-root function is badly behaved at zero. The identity in the proof shows that $H^2 - K$ is zero only at umbilic points, however, so k_1 *and* k_2 *are differentiable on any oriented region free of umbilics.*

A natural way to single out special types of surfaces in $\mathbf{R}^3$ is by restrictions on Gaussian and mean curvature.

3.6 Definition A surface M in $\mathbf{R}^3$ is *flat* provided its Gaussian curvature is zero, and *minimal* provided its mean curvature is zero.

As expected, a plane is flat, for by Example 1.3 its shape operators are all zero, so $K = \det S = 0$. On a circular cylinder, (3) of Example 1.3 shows that S is *singular* at each point $\mathbf{p}$, that is, has rank less than the dimension of the tangent plane $T_p(M)$. Thus, although S itself is never zero, its determinant is always zero, so cylinders are also flat. This terminology seems odd at first for a surface so obviously curved, but it will be amply justified in later work.

Note that minimal surfaces have Gaussian curvature $K \leq 0$, because if

$$H = \frac{k_1 + k_2}{2} = 0,$$

then $k_1 = -k_2$, so $K = k_1 k_2 \leq 0$.

Another notable class of surfaces consists of those with *constant* Gaussian curvature. As mentioned earlier, Example 1.3 shows that a sphere of radius r has $k_1 = k_2 = -1/r$ (for U outward). Thus the sphere Σ has constant positive curvature $K = 1/r^2$: The smaller the sphere, the larger its curvature.

We shall find many examples of these various special types of surface as we proceed through this chapter.

Exercises

1. Show that there are no umbilics on a surface with $K < 0$, and that when $K \leq 0$, umbilic points are planar.

2. Let $\mathbf{u}_1$ and $\mathbf{u}_2$ be orthonormal tangent vectors at a point $\mathbf{p}$ of M. What geometric information can be deduced from each of the following conditions on S at $\mathbf{p}$?

(a) $S(\mathbf{u}_1) \cdot \mathbf{u}_2 = 0$.   (b) $S(\mathbf{u}_1) + S(\mathbf{u}_2) = 0$.
(c) $S(\mathbf{u}_1) \times S(\mathbf{u}_2) = 0$.   (d) $S(\mathbf{u}_1) \cdot S(\mathbf{u}_2) = 0$.

3. (*Mean curvature.*) Prove that
(a) the average value of the normal curvature in *any* two orthogonal directions at $\mathbf{p}$ is $H(\mathbf{p})$. (The analogue for K is false.)

(b) $H(\mathbf{p}) = (1/2)\pi \int_0^{2\pi} k(\vartheta)\, d\vartheta$,

where $k(\vartheta)$ is normal curvature.

4. The *characteristic polynomial* of an arbitrary linear operator S is

$$p(k) = \det(A - kI),$$

where A is any matrix of S.

(a) Show that the characteristic polynomial of the shape operator is

$$k^2 - 2Hk + K.$$

(b) Every linear operator satisfies its characteristic equation; that is, $p(S)$ is the zero operator when S is formally substituted in $p(k)$. Prove this in the case of the shape operator by showing that

$$S(\mathbf{v}) \bullet S(\mathbf{w}) - 2HS(\mathbf{v}) \bullet \mathbf{w} + K\mathbf{v} \bullet \mathbf{w} = 0$$

for any pair of tangent vectors to M.
The real-valued functions

$$I(\mathbf{v}, \mathbf{w}) = \mathbf{v} \bullet \mathbf{w}, \quad II(\mathbf{v}, \mathbf{w}) = S(\mathbf{v}) \bullet \mathbf{w},$$

$$III(\mathbf{v}, \mathbf{w}) = S^2(\mathbf{v}) \bullet \mathbf{w} = S(\mathbf{v}) \bullet S(\mathbf{w}),$$

defined for all pairs of tangent vectors to an oriented surface, are traditionally called the *first, second*, and *third fundamental forms* of M. They are not differential forms; in fact, they are symmetric in $\mathbf{v}$ and $\mathbf{w}$ rather than alternating. The shape operator does not appear explicitly in the classical treatment of this subject; it is replaced by the second fundamental form.

5. (*Dupin curve.*) For a point $\mathbf{p}$ of an oriented region of M, let C_0 be the intersection of M near $\mathbf{p}$ with its tangent plane $T_p(M)$; specifically, C_0 consists of those points of M *near* $\mathbf{p}$ that lie in the plane through $\mathbf{p}$ orthogonal to $U(\mathbf{p})$. C_0 may be approximated by substituting for M its quadratic approximation $\hat{M}$; thus C_0 is approximated by the curve

$$\hat{C}_0: k_1 x^2 + k_2 y^2 = 0, \quad \text{near } (0, 0).$$

(a) Describe $\hat{C}_0$ in each of the three cases $K(\mathbf{p}) > 0$, $K(\mathbf{p}) < 0$, and $K(\mathbf{p}) = 0$ (not planar).
(b) Repeat (a) with C_0 replaced by C_ε and $C_{-\varepsilon}$, where the tangent plane has been replaced by the two parallel planes at distance $\pm\varepsilon$ from it.
(c) This scheme fails for planar points since the quadratic approximation becomes $\hat{M}: z = 0$. For the monkey saddle, sketch C_0, C_ε, and $C_{-\varepsilon}$.

6. For vectors $\mathbf{x}, \mathbf{y}, \mathbf{v}, \mathbf{w}$ in $\mathbf{R}^3$, prove the *Lagrange identity*

$$(\mathbf{x} \times \mathbf{y}) \bullet (\mathbf{v} \times \mathbf{w}) = \begin{vmatrix} \mathbf{x} \bullet \mathbf{v} & \mathbf{x} \bullet \mathbf{w} \\ \mathbf{y} \bullet \mathbf{v} & \mathbf{y} \bullet \mathbf{w} \end{vmatrix}.$$

(a) By hand. (*Hint:* Since both sides are linear in each vector separately, it suffices to prove the identity when each vector is one of the unit vectors $\mathbf{u}_1, \mathbf{u}_2, \mathbf{u}_3$.)
(b) By computer. (For dot and cross products, see the Appendix.)

7. (*Parallel surfaces.*) Let M be a surface oriented by U; for a fixed number ε (positive or negative) let $F: M \to \mathbf{R}^3$ be the mapping such that

$$F(\mathbf{p}) = \mathbf{p} + \varepsilon U(\mathbf{p}).$$

(a) If $\mathbf{v}$ is tangent to M at $\mathbf{p}$, show that $\overline{\mathbf{v}} = F_*(\mathbf{v})$ is $\mathbf{v} - \varepsilon S(\mathbf{v})$. Deduce that

$$\overline{\mathbf{v}} \times \overline{\mathbf{w}} = J(\mathbf{p})\mathbf{v} \times \mathbf{w},$$

where

$$J = 1 - 2\varepsilon H + \varepsilon^2 K = (1 - \varepsilon k_1)(1 - \varepsilon k_2).$$

When the function J does not vanish on M (for example, if M is compact and $|\varepsilon|$ small), this shows that F is a regular mapping, so the image

$$\overline{M} = F(M)$$

is at least an immersed surface in $\mathbf{R}^3$ (Ex. 16 in Sec. 4.8). $\overline{M}$ is said to be *parallel* to M at distance ε (Fig. 5.20).

(b) Show that the canonical isomorphisms of $\mathbf{R}^3$ make U a unit normal on $\overline{M}$ for which $\overline{S}(\overline{\mathbf{v}}) = S(\mathbf{v})$.

(c) Derive the following formulas for the Gaussian and mean curvatures of $\overline{M}$:

$$\overline{K}(F) = \frac{K}{J}; \quad \overline{H}(F) = \frac{H - \varepsilon K}{J}.$$

8. (*Continuation.*)

(a) Check the results in (c) in the case of a sphere of radius r oriented by the outward normal U. Describe the mapping $F = F_\varepsilon$ when ε is 0, $-r$, and $-2r$.

(b) Starting from an orientable surface with constant positive Gaussian curvature, construct a surface with constant mean curvature.

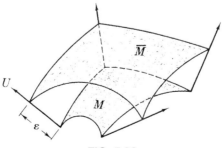

FIG. 5.20

5.4 Computational Techniques

We have defined the shape operators S of a surface M in $\mathbf{R}^3$ and found geometrical meaning for its main algebraic invariants: Gaussian curvature K, mean curvature H, principal curvatures k_1 and k_2, and (at each point) principal vectors $\mathbf{e}_1$ and $\mathbf{e}_2$. We shall now see how to express these invariants in terms of patches in M.

If $\mathbf{x}: D \to M$ is a patch in $M \subset \mathbf{R}^3$, we have already used the three real-valued functions

$$E = \mathbf{x}_u \bullet \mathbf{x}_u, \quad F = \mathbf{x}_u \bullet \mathbf{x}_v = \mathbf{x}_v \bullet \mathbf{x}_u, \quad G = \mathbf{x}_v \bullet \mathbf{x}_v$$

on D. Here $E > 0$ and $G > 0$ are the squares of the speeds of the u- and v-parameter curves of $\mathbf{x}$, and F measures the *coordinate angle* ϑ between $\mathbf{x}_u$ and $\mathbf{x}_v$, since

$$F = \mathbf{x}_u \bullet \mathbf{x}_v = \|\mathbf{x}_u\| \|\mathbf{x}_v\| \cos \vartheta = \sqrt{EG} \cos \vartheta$$

(Fig. 5.21). E, F, and G are the "warping functions" of the patch $\mathbf{x}$: They measure the way $\mathbf{x}$ distorts the flat region D in $\mathbf{R}^2$ in order to apply it to the curved region $\mathbf{x}(D)$ in M. These functions completely determine the dot product of tangent vectors at points of $\mathbf{x}(D)$, for if

$$\mathbf{v} = v_1\mathbf{x}_u + v_2\mathbf{x}_v \quad \text{and} \quad \mathbf{w} = w_1\mathbf{x}_u + w_2\mathbf{x}_u,$$

then

$$\mathbf{v} \bullet \mathbf{w} = Ev_1w_1 + F(v_1w_2 + v_2w_1) + Gv_2w_2.$$

(In such equations we understand that $\mathbf{x}_u$, $\mathbf{x}_v$, E, F, and G are evaluated at (u, v) where $\mathbf{x}(u, v)$ is the point of application of $\mathbf{v}$ and $\mathbf{w}$.)

Now $\mathbf{x}_u \times \mathbf{x}_v$ is a function on D whose value at each point (u, v) of D is a vector orthogonal to both $\mathbf{x}_u(u, v)$ and $\mathbf{x}_v(u, v)$—and hence normal to M at the point $\mathbf{x}(u, v)$. Furthermore, by Exercise 6 of Section 3,

$$\|\mathbf{x}_u \times \mathbf{x}_v\|^2 = EG - F^2.$$

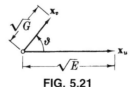

FIG. 5.21

Since $\mathbf{x}$ is, by definition, regular, this real-valued function on D is never zero. Thus we can construct the *unit normal function*

$$U = \frac{\mathbf{x}_u \times \mathbf{x}_v}{\|\mathbf{x}_u \times \mathbf{x}_v\|}$$

on D, which assigns to each (u, v) in D a unit normal vector to M at $\mathbf{x}(u, v)$. We emphasize that in this context, U, like $\mathbf{x}_u$ and $\mathbf{x}_v$, is not a vector field on $\mathbf{x}(D)$, but merely a vector-valued function on D. Nevertheless, we may regard the system $\mathbf{x}_u$, $\mathbf{x}_v$, U as a kind of defective frame field. At least U has unit length and is orthogonal to both $\mathbf{x}_u$ and $\mathbf{x}_v$, even though $\mathbf{x}_u$ and $\mathbf{x}_v$ are generally not orthonormal.

In this context, covariant derivatives are usually computed along the parameter curves of $\mathbf{x}$, where by the discussion in Section 1, they reduce to partial differentiation with respect to u and v. As in the case of $\mathbf{x}_u$ and $\mathbf{x}_v$, these partial derivatives are again denoted by subscripts u and v. If

$$\mathbf{x}(u, v) = (x_1(u, v), x_2(u, v), x_3(u, v)),$$

then just as for $\mathbf{x}_u$ and $\mathbf{x}_v$ on page 140, we have

$$\mathbf{x}_{uu} = \left(\frac{\partial^2 x_1}{\partial u^2}, \frac{\partial^2 x_2}{\partial u^2}, \frac{\partial^2 x_3}{\partial u^2} \right)_{\mathbf{x}},$$

$$\mathbf{x}_{vu} = \left(\frac{\partial^2 x_1}{\partial u\,\partial v}, \frac{\partial^2 x_2}{\partial u\,\partial v}, \frac{\partial^2 x_3}{\partial u\,\partial v} \right)_{\mathbf{x}},$$

$$\mathbf{x}_{vv} = \left(\frac{\partial^2 x_1}{\partial v^2}, \frac{\partial^2 x_2}{\partial v^2}, \frac{\partial^2 x_3}{\partial v^2} \right)_{\mathbf{x}}.$$

Evidently $\mathbf{x}_{uu}$ and $\mathbf{x}_{vv}$ give the accelerations of the u- and v-parameter curves. Since order of partial differentiation is immaterial, $\mathbf{x}_{uv} = \mathbf{x}_{vu}$, which gives both the covariant derivative of $\mathbf{x}_u$ in the $\mathbf{x}_v$ direction and of $\mathbf{x}_v$ in the $\mathbf{x}_u$ direction.

Now if S is the shape operator derived from U, we define three more real-valued functions on D:

$$\mathrm{L} = S(\mathbf{x}_u) \bullet \mathbf{x}_u,$$

$$\mathrm{M} = S(\mathbf{x}_u) \bullet \mathbf{x}_v = S(\mathbf{x}_v) \bullet \mathbf{x}_u,$$

$$\mathrm{N} = S(\mathbf{x}_v) \bullet \mathbf{x}_v.$$

Because $\mathbf{x}_u$, $\mathbf{x}_v$ gives a basis for the tangent space of M at each point of $\mathbf{x}(D)$, it is clear that these functions uniquely determine the shape operator. Since this basis is generally not orthonormal, L, M, and N do not lead to simple expressions for $S(\mathbf{x}_u)$ and $S(\mathbf{x}_v)$ in terms of $\mathbf{x}_u$ and $\mathbf{x}_v$. In the formulas preceding Corollary 3.5, however, they *do* provide simple expressions for Gaussian and mean curvature.

4.1 Corollary If $\mathbf{x}$ is a patch in $M \subset \mathbf{R}^3$, then

$$K(\mathbf{x}) = \frac{\text{LN} - \text{M}^2}{EG - F^2}, \quad H(\mathbf{x}) = \frac{G\text{L} + E\text{N} - 2F\text{M}}{2(EG - F^2)}.$$

Proof. Evaluated on $\mathbf{x}(D)$, the formulas on page 220 express K and H in terms of tangent vector fields V and W. If the latter are replaced by $\mathbf{x}_u$ and $\mathbf{x}_v$, respectively, we find the required formulas for K and H. ◆

When the patch $\mathbf{x}$ is clear from context, we shall usually abbreviate the composite functions $K(\mathbf{x})$ and $H(\mathbf{x})$ to merely K and H.

By a device like that used in Lemma 2.1, we can find a simple way to *compute* L, M, and N—and thereby K and H. For example, since $U \cdot \mathbf{x}_u = 0$, partial differentiation with respect to v—that is, ordinary differentiation along v-parameter curves—yields

$$0 = \frac{\partial}{\partial v}(U \cdot \mathbf{x}_u) = U_v \cdot \mathbf{x}_u + U \cdot \mathbf{x}_{uv}.$$

(Recall that U_v is the covariant derivative of the vector field $v \rightarrow U(u_0, v)$ on each v-parameter curve $u = u_0$.) Since $\mathbf{x}_v$ gives the velocity vectors of such curves, Exercise 1.1 shows that $U_v = -S(\mathbf{x}_v)$. Thus the preceding equation becomes

$$S(\mathbf{x}_v) \cdot \mathbf{x}_u = U \cdot \mathbf{x}_{uv}$$

(Fig. 5.22). Three similar equations may be found by replacing u by v, and v by u. In particular,

$$S(\mathbf{x}_u) \cdot \mathbf{x}_v = U \cdot \mathbf{x}_{vu} = U \cdot \mathbf{x}_{uv} = S(\mathbf{x}_v) \cdot \mathbf{x}_u.$$

Again, since $\mathbf{x}_u$ and $\mathbf{x}_v$ give a basis for the tangent space at each point, this is sufficient to prove that S *is symmetric* (Lemma 1.4).

4.2 Lemma If $\mathbf{x}$ is a patch in $M \subset \mathbf{R}^3$, then

$$\text{L} = S(\mathbf{x}_u) \cdot \mathbf{x}_u = U \cdot \mathbf{x}_{uu},$$

$$\text{M} = S(\mathbf{x}_u) \cdot \mathbf{x}_v = U \cdot \mathbf{x}_{uv},$$

$$\text{N} = S(\mathbf{x}_v) \cdot \mathbf{x}_v = U \cdot \mathbf{x}_{vv}.$$

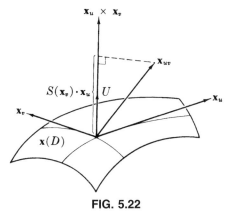

FIG. 5.22

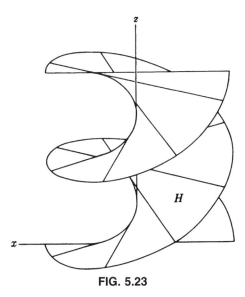

FIG. 5.23

The first equation in each case is just the definition, and *u and v* may be reversed in the formulas for M.

4.3 Example *Computation of Gaussian and mean curvature*
(1) **Helicoid** (Exercise 5 of Section 4.2). This surface *H*, shown in Fig. 5.23, is covered by a single patch

$$\mathbf{x}(u, v) = (u \cos v,\, u \sin v,\, bv), \quad b \neq 0,$$

for which

$$\mathbf{x}_u = (\cos v, \sin v, 0), \qquad E = 1,$$
$$\mathbf{x}_v = (-u \sin v, u \cos v, b), \quad F = 0,$$
$$G = b^2 + u^2.$$

Hence

$$\mathbf{x}_u \times \mathbf{x}_v = (b \sin v, -b \cos v, u).$$

Because coordinate patches are, by definition, regular mappings, we have seen in Chapter 4 that the function

$$\|\mathbf{x}_u \times \mathbf{x}_v\| = \sqrt{EG - F^2}$$

is never zero. For any patch we denote this useful function by W, that is,

$$W = \sqrt{EG - F^2}.$$

In the case at hand, $W = W(u, v) = \sqrt{b^2 + u^2}$, so the unit normal function is

$$U = \frac{\mathbf{x}_u \times \mathbf{x}_v}{W} = \frac{(b \sin v, -b \cos v, u)}{\sqrt{b^2 + u^2}}.$$

(A computation of U can always be checked by verifying that the result is a unit vector orthogonal to both $\mathbf{x}_u$ and $\mathbf{x}_v$.)

Next we find

$$\mathbf{x}_{uu} = 0,$$
$$\mathbf{x}_{uv} = (-\sin v, \cos v, 0),$$
$$\mathbf{x}_{vv} = (-u \cos v, -u \sin v, 0).$$

Here $\mathbf{x}_{uu} = 0$ is obvious, since the u-parameter curves are straight lines. The v-parameter curves are helices, and this formula for the acceleration $\mathbf{x}_{vv}$ was found already in Chapter 2. Now by Lemma 4.2,

$$L = \mathbf{x}_{uu} \cdot \frac{(\mathbf{x}_u \times \mathbf{x}_v)}{W} = 0,$$

$$M = \mathbf{x}_{uv} \cdot \frac{(\mathbf{x}_u \times \mathbf{x}_v)}{W} = -\frac{b}{W},$$

$$N = \mathbf{x}_{vv} \cdot \frac{(\mathbf{x}_u \times \mathbf{x}_v)}{W} = 0.$$

Hence by Corollary 4.1 and the results above,

$$K = \frac{\text{LN} - \text{M}^2}{EG - F^2} = \frac{-(b/W)^2}{W^2} = \frac{-b^2}{W^4} = \frac{-b^2}{(b^2 + u^2)^2},$$

$$H = \frac{\text{GL} + \text{EN} - 2\text{FM}}{2(EG - F^2)} = 0.$$

Thus the helicoid is a minimal surface with Gaussian curvature

$$-\frac{1}{b^2} \le K < 0.$$

The minimum value $K = -1/b^2$ occurs on the central axis ($u = 0$) of the helicoid, and $K \to 0$ as distance $|u|$ from the axis increases to infinity.

(2) *The saddle surface* M: $z = xy$ (Example 1.3). This time we use the Monge patch $\mathbf{x}(u, v) = (u, v, uv)$ and with the same format as above, compute

$$\mathbf{x}_u = (1, 0, v), \qquad E = 1 + v^2,$$
$$\mathbf{x}_v = (0, 1, u), \qquad F = uv,$$
$$G = 1 + u^2,$$
$$U = (-v, -u, 1)/W, \quad W = \sqrt{1 + u^2 + v^2},$$
$$\mathbf{x}_{uu} = 0, \qquad \text{L} = 0,$$
$$\mathbf{x}_{uv} = (0, 0, 1), \qquad \text{M} = 1/W,$$
$$\mathbf{x}_{vv} = 0, \qquad \text{N} = 0.$$

Hence

$$K = \frac{-1}{(1 + u^2 + v^2)^2}, \quad H = \frac{-uv}{(1 + u^2 + v^2)^{3/2}}.$$

Strictly speaking, these functions are $K(\mathbf{x})$ and $H(\mathbf{x})$ defined on the domain $\mathbf{R}^2$ of $\mathbf{x}$. In this case, it is easy to express K and H directly as functions on M by using the cylindrical coordinate functions $r = \sqrt{x^2 + y^2}$ and z. Note from Fig. 5.24 that

$$r(\mathbf{x}(u, v)) = \sqrt{u^2 + v^2}$$

and

$$z(\mathbf{x}(u, v)) = uv;$$

hence on M,

$$K = \frac{-1}{(1 + r^2)^2}, \quad H = \frac{-z}{(1 + r^2)^{3/2}}.$$

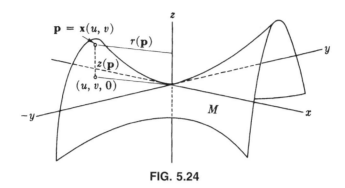

FIG. 5.24

Thus the Gaussian curvature of M depends only on distance to the z axis, rising from $K = -1$ (at the origin) toward zero as r goes to infinity, while H varies more radically.

Like all simple (that is, one-patch) surfaces, the helicoid and saddle surfaces are orientable, since computations as above provide a unit normal on the whole surface. Thus the principal curvature functions $k_1 \geq k_2$ can be defined unambiguously on each surface. These can always be found from K and H by Corollary 3.5. Since the helicoid is a minimal surface, we get the simple result

$$k_1, k_2 = \frac{\pm b}{\left(b^2 + u^2\right)}.$$

For the saddle surface,

$$k_1, k_2 = \frac{-z \pm \sqrt{1 + r^2 + z^2}}{\left(1 + r^2\right)^{3/2}}.$$

Techniques for computing principal vectors are left to the exercises.

The computational results in this section, though stated for coordinate patches, remain valid for arbitrary regular mappings $\mathbf{x}\colon D \to \mathbf{R}^3$ since the restriction of $\mathbf{x}$ to any small enough open set in D is a patch.

Exercises

1. For the geographical parametrization
$\mathbf{x}(u, v) = (r\cos v \cos u,\ r\cos v \sin u,\ r\sin v)$, of the sphere Σ of radius r, find E,

F, G, and W, and then the unit normal U, Gaussian curvature K, and mean curvature H.

2.   For a Monge patch $\mathbf{x}(u, v) = (u, v, f(u, v))$, show that,

$$E = 1 + f_u^2, \quad L = \frac{f_{uu}}{W},$$

$$F = f_u f_v, \quad M = \frac{f_{uv}}{W},$$

$$G = 1 + f_v^2, \quad N = \frac{f_{vv}}{W},$$

where

$$W = \sqrt{EG - F^2} = \left(1 + f_u^2 + f_v^2\right)^{1/2}$$

Then find formulas for K and H.

3.   (*Continuation.*) Deduce that the image of $\mathbf{x}$ is
(a) flat if and only if

$$f_{uu} f_{vv} - f_{uv}^2 = 0;$$

(b) minimal if and only if

$$\left(1 + f_u^2\right)f_{vv} - 2 f_u f_v f_{uv} + \left(1 + f_v^2\right)f_{uu} = 0.$$

4.   Let $\mathbf{x}$ be the patch

$$\mathbf{x}(u, v) = \left(u, v, \log \cos v - \log \cos u\right)$$

defined on $-\pi/2 < u$, $v < \pi/2$. Show that the image of $\mathbf{x}$ is a minimal surface with Gaussian curvature

$$K = \frac{-\sec^2 u \sec^2 v}{W^4},$$

where $W^2 = 1 + \tan^2 u + \tan^2 v$. (This patch is in Scherk's surface, Ex. 5 of Sec. 5.5.)

5.   Show that a curve segment

$$\alpha(t) = \mathbf{x}(a_1(t), a_2(t)), \quad a \le t \le b,$$

has length

$$L(\alpha) = \int_a^b \left(E a_1'^2 + 2 F a_1' a_2' + G a_2'^2\right)^{\frac{1}{2}} dt,$$

where E, F, and G are evaluated on a_1, a_2.

6. Find the Gaussian curvature of the elliptic and hyperbolic paraboloids

$$M: \ z = \frac{x^2}{a^2} + \varepsilon \frac{y^2}{b^2},$$

where $\varepsilon = \pm 1$.

7. Find the curvature of the monkey saddle $M: z = x^3 - 3xy^2$, and express it in terms of $r = \sqrt{x^2 + y^2}$.

8. A patch x in M is *orthogonal* provided $F = 0$ (so x_u and x_v are orthogonal at each point). Show that in this case

$$S(x_u) = \frac{L}{E} x_u + \frac{M}{G} x_v,$$

$$S(x_v) = \frac{M}{E} x_u + \frac{N}{G} x_v.$$

(b) A patch x in M is *principal* provided $F = M = 0$. Prove that x_u and x_v are principal vectors at each point, with corresponding principal curvatures L/E and N/G.

9. Prove that a nonzero tangent vector $v = v_1 x_u + v_2 x_v$ is a principal vector if and only if

$$\begin{vmatrix} v_2^2 & -v_1 v_2 & v_1^2 \\ E & F & G \\ L & M & N \end{vmatrix} = 0.$$

(*Hint:* v is principal if and only if $S(v) \times v = 0$.)

10. Show that on the saddle surface $z = xy$ the two vector fields

$$\left(\sqrt{1+x^2} \pm \sqrt{1+y^2}, \ y\sqrt{1+x^2} \pm x\sqrt{1+y^2} \right)$$

are principal at each point. Check that they are orthogonal and tangent to M.

11. If $v = v_1 x_u + v_2 x_v$ is tangent to M at $x(u, v)$, show that the normal curvature in the direction $u = v/\|v\|$ is

$$k(u) = \frac{L v_1^2 + 2M v_1 v_2 + N v_2^2}{E v_1^2 + 2F v_1 v_2 + G v_2^2},$$

where the various functions are evaluated at (u, v).

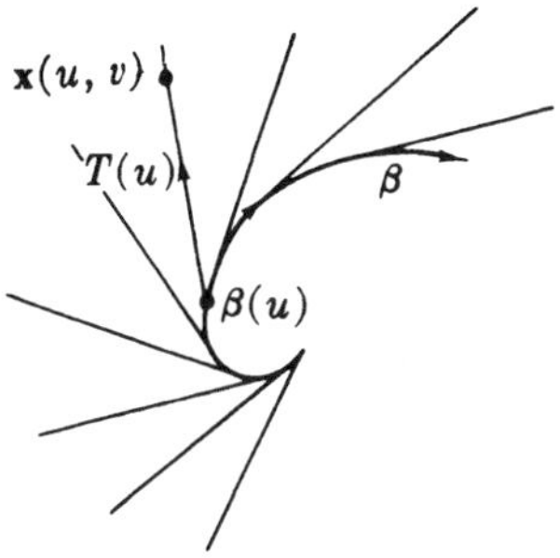

FIG. 5.25

12. Show that a ruled surface $\mathbf{x}(u, v) = \beta(u) + v\delta(u)$ has Gaussian curvature

$$K = \frac{-M^2}{EG - F^2} = \frac{-(\beta' \cdot \delta \times \delta')^2}{W^4},$$

where $W = \|\beta' \times \delta + v\delta \times \delta\|$.

13. (*Flat ruled surfaces.*)
   (a) Show that generalized cones and cylinders are flat (Exs. 3 and 4 of Sec. 4.2).
   (b) If β is a unit-speed curve in $\mathbf{R}^3$ with $\kappa > 0$, the ruled surface

$$\mathbf{x}(u, v) = \beta(u) + vT(u), \quad v \neq 0,$$

where $T(u) = \beta'(u)$, is called the *tangent surface* of β. Prove that $\mathbf{x}$ is regular and the tangent surface is flat. (The surface is separated into two pieces by the curve; Fig. 5.25 shows the $v > 0$ half.)

14. (*Patch criterion for umbilics.*)
   (a) Show that a point $\mathbf{x}(u, v)$ is umbilic if and only if there is a number k such that at (u, v),

$$L = kE, \quad M = kF, \quad N = kG.$$

(Then k is the principal curvature $k_1 = k_2$.)

15. Find the umbilic points, if any, on the following surfaces:
   (a) Monkey saddle (Ex. 7).
   (b) Elliptic paraboloid (Ex. 6), assuming $a \geqq b$.

(*Hint*: Compute the "vectors" (E, F, G) and (L, M, N) for arbitrary (u, v), discarding common factors if convenient. Then solve $(E, F, G) \times (L, M, N) = 0$ for (u, v).)

16. (*Loxodromes.*) For $a \neq 0$, let f_a: $(-\pi/2, \pi/2) \to \mathbf{R}$ be the unique function such that

$$f_a'(t) = \frac{a}{\cos t} \quad \text{and} \quad f(0) = 0.$$

If $\mathbf{x}$ is the geographical parametrization of a sphere, the curve $\lambda_a(t) = \mathbf{x}(f_a(t),t)$ is a *loxodrome*.

(a) Prove that λ_a' always makes a constant angle with the due-north vector field $\mathbf{x}_v$. Thus λ_a represents a trip with constant (idealized) compass bearing.

(b) Show that the length of λ_a from the south pole $(0, 0, -r)$ to the north pole $(0, 0, r)$ (limit values) is $\sqrt{1 + a^2}\, \pi r$.

(c) (*Computer.*) Verify that $f_a(t) = a \log\tan\left(t/2 + \pi/4\right)$, and plot λ_{10} from near the south pole to near the north pole on a unit sphere. (Require smoothness, and keep the same scale on each axis.)

17. (*Tubes.*) If β is a curve in $\mathbf{R}^3$ with $0 < \kappa \leq b$, let

$$\mathbf{x}(u, v) = \beta(u) + \varepsilon(\cos v\, N(u) + \sin v\, B(u)).$$

Thus the v-parameter curves are circles of constant radius ε in planes orthogonal to β. Show that

(a) $\mathbf{x}_u \times \mathbf{x}_v = -\varepsilon(1 - \kappa\varepsilon \cos v)(\cos vN(u) + \sin vB(u))$.

(b) If ε is small enough, $\mathbf{x}$ is regular. So $\mathbf{x}$ is at least an immersed surface, called a *tube* around β.

(c) $U = \cos v\, N(u) + \sin v\, B(u)$ is a unit normal vector on the tube.

(d) $K = \dfrac{-\kappa(u)\cos v}{\varepsilon(1 - \kappa(u)\varepsilon \cos v)}$.

(*Hint:* Use $S(\mathbf{x}_u) \times S(\mathbf{x}_v) = K\, \mathbf{x}_u \times \mathbf{x}_v$.)

The following exercises deal with use of the computer in patch computations.

18. (*Computer.*) The Appendix gives computer commands for the functions E, F, G, W, L, M, N derived from a patch.

(a) Write the computer commands, based on Corollary 4.1, that give the Gaussian curvature and mean curvature of a patch in terms of these functions.

(b) To test these commands, find E, F, G, W, L, M, N, K, H for each of the cases in Example 4.3. Compare with the text computations.

19. (*Computer.*) Make a save file containing the following patches or parametrizations. (See Appendix for "save files" and format for parameters.)

(a) the patch in Exercise 4.

(b) a single Monge patch—with parameters a, b, ε—for the hyperboloids in Exercise 6.

(c) a Monge patch for the monkey saddle (Ex. 7), in terms of (i) rectangular coordinates u, v and (ii) polar coordinates r, ϑ on $\mathbf{R}^2$.

(d) the parametrization of Enneper's surface in Exercise 15 of Section 7.

(e) the geographical parametrization of a sphere of radius r.

20. (*Computer formulas.*)

(a) For a patch $\mathbf{x}$ in $\mathbf{R}^3$, show that Gaussian curvature can be expressed directly in terms of $\mathbf{x}$ as

$$K(u, v) = \frac{\left(\mathbf{x}_{uu} \cdot \mathbf{x}_u \times \mathbf{x}_v\right)\left(\mathbf{x}_{vv} \cdot \mathbf{x}_u \times \mathbf{x}_v\right) - \left(\mathbf{x}_{uv} \cdot \mathbf{x}_u \times \mathbf{x}_v\right)^2}{\left(\left(\mathbf{x}_u \cdot \mathbf{x}_u\right)\left(\mathbf{x}_v \cdot \mathbf{x}_v\right) - \left(\mathbf{x}_u \cdot \mathbf{x}_v\right)^2\right)^2}.$$

This formula gives the fastest general computer computation of K. The Appendix has computer commands for it in the *Mathematica* and *Maple* systems.

(b) Test this command on the two cases in Example 4.3 and the patches in Exercise 19.

The derivation of the corresponding formula for *mean* curvature is rather tedious. This formula may be found in Alfred Gray's book [G].

(c) Find a computer formula for the Gaussian curvature of the graph of a function $f: \mathbf{R}^2 \to \mathbf{R}$. (*Hint:* use Ex. 2.) Test this command on the Monge patches referenced in Exercises 18 and 19.

21. (*Computer.*)

(a) Write commands that, given a curve α on some interval, plot the tube of radius r around α. (See Ex. 17.)

(b) Use part (a) to plot the tube of radius $\frac{1}{2}$ around two turns of the helix $t \to (3 \cos t, 3 \sin t, t/2)$.

(c) Plot the tube of radius $\frac{1}{2}$ around the curve τ in Exercise 19 of Sec. 2.4. (This makes it clear that τ is a trefoil knot.)

5.5 The Implicit Case

In this brief section we describe a way to compute the geometry of a surface $M \subset \mathbf{R}^3$ that has a nonvanishing normal vector field Z defined on the entire surface. The main case is a surface given in implicit form $M: g = 0$, for then, by Lemma 3.8 of Chapter 4, the gradient

$$\nabla g = \sum \frac{\partial g}{\partial x_i}$$

is such a vector field.

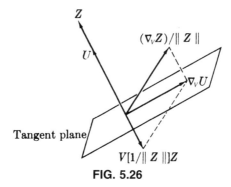

FIG. 5.26

Let S be the shape operator on M derived from the unit normal

$$U = Z/\|Z\|.$$

Write $Z = \sum z_i U_i$. Then if V is a tangent vector field on M, Method 2 in Section 1 gives

$$\nabla_V Z = \sum v[z_i] U_i.$$

Hence, using a Leibnizian property of such derivatives,

$$\nabla_V U = \nabla_V \left(\frac{Z}{\|Z\|} \right) = \frac{\nabla_V Z}{\|Z\|} + V\left[\frac{1}{\|Z\|} \right] Z$$

(Fig. 5.26). The last term here, $V[1/\|Z\|] Z$, is evidently a *normal* vector field; we do not care which one it is, so we denote it merely by $-N_V$. Thus

$$S(V) = -\nabla_V U = -\frac{\nabla_V Z}{\|Z\|} + N_V.$$

Note that if W is another tangent vector field on M, then $N_V \times N_W = 0$, while products such as $N_V \times Y$ are tangent to M for *any* Euclidean vector field Y on M. Thus it is a routine matter to deduce the following lemma from Lemma 3.4.

5.1 Lemma   Let Z be a nonvanishing normal vector field on M. If V and W are tangent vector fields such that $V \times W = Z$, then

$$K = \frac{Z \cdot \nabla_V Z \times \nabla_W Z}{\|Z\|^4},$$

$$H = -Z \cdot \frac{\nabla_V Z \times W + V \times \nabla_W Z}{2\|Z\|^3}.$$

To compute, say, the Gaussian curvature of a surface M: $g = c$ using patches, one must begin by explicitly finding enough of them to cover all of M; a complete computation of K may thus be tedious, even when g is a rather simple function. The following example shows to advantage the approach just described.

5.2 Example Curvature of the ellipsoid

$$M:\ g = \frac{x^2}{a^2} + \frac{y^2}{b^2} + \frac{z^2}{c^2} = 1.$$

We write $g = \sum \dfrac{x_i^2}{a_i^2}$, and use the (nonvanishing) normal vector field

$$Z = \frac{1}{2}\nabla g = \sum \frac{x_i}{a_i^2} U_i.$$

Then if $V = \sum v_i U_i$ is a tangent vector field on M,

$$\nabla_V Z = \sum \frac{V[x_i]}{a_i^2} U_i = \sum \frac{v_i}{a_i^2} U_i,$$

since

$$V[x_i] = dx_i(V) = v_i.$$

Similar results for another tangent vector field W yield

$$Z \bullet \nabla_V Z \times \nabla_W Z = \begin{vmatrix} \dfrac{x_1}{a_1^2} & \dfrac{x_2}{a_2^2} & \dfrac{x_3}{a_3^2} \\ \dfrac{v_1}{a_1^2} & \dfrac{v_2}{a_2^2} & \dfrac{v_3}{a_3^2} \\ \dfrac{w_1}{a_1^2} & \dfrac{w_2}{a_2^2} & \dfrac{w_3}{a_3^2} \end{vmatrix} = \frac{1}{a_1^2 a_2^2 a_3^2} X \bullet V \times W,$$

where X is the special vector field $\sum x_i U_i$ used in Example 3.9 of Chapter 4. It is always possible to choose V and W so that $V \times W = Z$. But then

$$X \bullet V \times W = X \bullet Z = \sum \frac{x_i^2}{a_i^2} = 1.$$

Thus by Lemma 5.1 we have found

$$K = \frac{1}{a_1^2 a_2^2 a_3^2 \|Z\|^4}, \quad \text{where } \|Z\|^4 = \left(\sum \frac{x_i^2}{a_i^4} \right)^2,$$

that is,

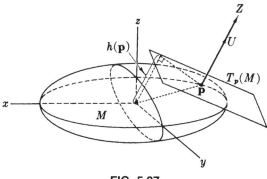

FIG. 5.27

$$K = \frac{1}{a^2b^2c^2\|Z\|^4}, \quad \text{where} \ \|Z\|^4 = \left(\frac{x^2}{a^4} + \frac{y^2}{b^4} + \frac{z^2}{c^4}\right)^2.$$

For *any* oriented surface in $\mathbf{R}^3$, its *support function* h assigns to each point $\mathbf{p}$ the orthogonal distance $h(\mathbf{p}) = \mathbf{p} \cdot U(\mathbf{p})$ from the origin to the Euclidean tangent plane $T_p(M)$, as shown in Fig. 5.27 for the ellipsoid. Using the above-mentioned vector field X (whose value at $\mathbf{p}$ is the tangent vector $\mathbf{p}_p$), we find for the ellipsoid that

$$h = X \cdot U = X \cdot \frac{Z}{\|Z\|} = \frac{1}{\|Z\|}.$$

Thus a clearer expression of the Gaussian curvature of the ellipsoid is

$$K = \frac{h^4}{a^2b^2c^2}.$$

Note that if $a = b = c = r$ (so M is a sphere), then $h = 1/\|Z\|$ has constant value r, and this formula reduces to $K = 1/r^2$.

Exercises

1. Show that the elliptic hyperboloids of one and two sheets (Ex. 2.9 of Ch. 4) have Gaussian curvatures $K = -h^4/a^2b^2c^2$ and $K = h^4/a^2b^2c^2$, respectively, where both support functions h are given by the same formula as for the ellipsoid in Example 5.2.

2. If h is the support function of an oriented surface $M \subset \mathbf{R}^3$, show that
(a) A point $\mathbf{p}$ of M is a critical point of h if and only if $\mathbf{p} \cdot S(\mathbf{v}) = 0$ for all tangent vectors to M at $\mathbf{p}$. (*Hint:* Write h as $X \cdot U$, where $X = \sum x_i U_i$.)

(b) When $K(\mathbf{p}) \neq 0$, $\mathbf{p}$ is a critical point of h if and only if $\mathbf{p}$ (considered as a vector) is orthogonal to M at $\mathbf{p}$.

3. (a) Use the preceding exercises to find the critical points of the Gaussian curvature function K on the ellipsoid and on the hyperboloids of one and two sheets (Ex. 2.9 of Ch. 4).

(b) Assuming $a \geq b \geq c$ for these surfaces, find their Gaussian curvature intervals.

4. Compute K and H for the saddle surface M: $z = xy$ by the method of this section. (*Hint*: Take V and W tangent to the rulings of M.)

5. *Scherk's minimal surface, M: $e^z \cos x = \cos y$.* Let $\mathscr{R}$ be the region in the xy plane on which $\cos x \cos y > 0$. $\mathscr{R}$ is a checkerboard pattern of open squares, with vertices $(\pi/2 + m\pi, \ \pi/2 + n\pi)$. Show that:

(a) M is a surface.

(b) For each point (u, v) in $\mathscr{R}$ there is exactly one point (u, v, w) in M. The only other points of M are entire vertical lines over each of the vertices of $\mathscr{R}$ (Fig. 5.28).

(c) M is a minimal surface with $K = -e^{2z}/(e^{2z} \sin^2 x + 1)^2$.

(*Hint:* $V = \cos x \ U_1 + \sin x \ U_3$ is a tangent vector field.)

(d) The patch in Exercise 4 of Section 4 parametrizes the part of M over a typical open square. Show that the curvature $K(u, v)$ calculated there is consistent with (c).

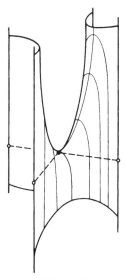

FIG. 5.28

6. Let Z be a nonvanishing normal vector field on M. Show that a tangent vector $\mathbf{v}$ to M at $\mathbf{p}$ is principal if and only if

$$\mathbf{v} \cdot Z(\mathbf{p}) \times \nabla_v Z = 0.$$

(*Hint:* Recall that $\mathbf{v}$ is principal if and only if $S(\mathbf{v}) \times \mathbf{v} = 0$.)

The preceding equation together with the tangency equation $Z(\mathbf{p}) \cdot \mathbf{v} = 0$ can be solved for the principal directions. Thus umbilics can be located using these equations, since $\mathbf{p}$ is umbilic if and only if every tangent vector at $\mathbf{p}$ is principal.

7. For the ellipsoid M: $\sum x_i^2 / a_i^2 = 1$, show that:
(a) A tangent vector $\mathbf{v}$ at $\mathbf{p}$ is principal if and only if

$$0 = p_1 v_2 v_3 (a_2^2 - a_3^2) + p_2 v_3 v_1 (a_3^2 - a_1^2) + p_3 v_1 v_2 (a_1^2 - a_2^2).$$

(b) Assuming $a_1 > a_2 > a_3$, there are exactly four umbilics on M, with coordinates

$$p_1 = \pm a_1 \left(\frac{a_1^2 - a_2^2}{a_1^2 - a_3^2} \right)^{1/2}, \quad p_2 = 0, \quad p_3 = \pm a_3 \left(\frac{a_2^2 - a_3^2}{a_1^2 - a_3^2} \right)^{1/2}.$$

5.6 Special Curves in a Surface

In this section we consider three geometrically significant types of curves in a surface $M \subset \mathbf{R}^3$.

6.1 Definition A regular curve α in $M \subset \mathbf{R}^3$ is a *principal curve* provided that the velocity α' of α always points in a principal direction.

Thus principal curves always travel in directions for which the bending of M in $\mathbf{R}^3$ takes its extreme values. Neglecting changes of parametrization, there are exactly two principal curves through each nonumbilic point of M—and these necessarily cut orthogonally across each other. (At an umbilic point $\mathbf{p}$, every direction is principal, and near $\mathbf{p}$ the pattern of principal curves can be quite complicated.)

6.2 Lemma Let α be a regular curve in $M \subset \mathbf{R}^3$, and let U be a unit normal vector field restricted to α. Then
(1) The curve α is principal if and only if U' and α' are collinear at each point.

(2) If α is a principal curve, then the principal curvature of M in the direction of α' is $\alpha'' \cdot U / \alpha' \cdot \alpha'$.

Proof. (1) Exercise 1.1 shows that $S(\alpha') = -U'$. Thus U' and α' are collinear if and only if $S(\alpha')$ and α' are collinear. But by Theorem 2.5, this amounts to saying that α' always points in a principal direction or, equivalently, that α is a principal curve.

(2) Since α is a principal curve, the vector field $\alpha' / \|\alpha'\|$ consists entirely of (unit) principal vectors belonging to, say, the principal curvature k_i. Thus

$$k_i = k(\alpha' / \|\alpha'\|) = S(\alpha' / \|\alpha'\|) \cdot \alpha' / \|\alpha'\|$$
$$= \frac{S(\alpha') \cdot \alpha'}{\alpha' \cdot \alpha'} = \frac{\alpha'' \cdot U}{\alpha' \cdot \alpha'},$$

where the last equality uses Lemma 2.1. ◆

In this lemma, (1) is a simple criterion for a curve to be principal, while (2) gives the principal curvature along a curve known to be principal.

6.3 Lemma Let α be a curve cut from a surface $M \subset \mathbf{R}^3$ by a plane P. If the angle between M and P is constant along α, then α is a principal curve of M.

Proof. Let U and V be unit normal vector fields to M and P (respectively) along the curve α, as shown in Fig. 5.29. Since P is a plane, V is parallel, that is, $V' = 0$. The constant-angle assumption means that $U \cdot V$ is constant; thus

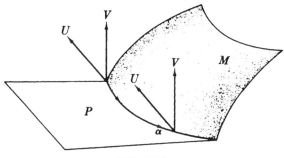

FIG. 5.29

$$0 = (U \cdot V)' = U' \cdot V.$$

Since U is a unit vector, U' is orthogonal to U as well as to V. The same is of course true of α', since α lies in both M and P. If U and V are linearly independent (as in Fig. 5.29) we conclude that U' and α' are collinear; hence by Lemma 6.2, α is principal.

However, linear independence fails only when $U = \pm V$. But then $U' = 0$, so α is (trivially) principal in this case as well.   ◆

Using this result, it is easy to see that *the meridians and parallels of a surface of revolution M are its principal curves.* Indeed, each meridian μ is sliced from M by a plane *through* the axis of revolution and hence orthogonal to M along μ, while each parallel π is sliced from M by a plane *orthogonal* to the axis, and by rotational symmetry such a plane makes a constant angle with M along π.

Directions tangent to $M \subset \mathbf{R}^3$ in which the normal curvature is zero are called *asymptotic directions.* Thus a tangent vector $\mathbf{v}$ is *asymptotic* provided $k(\mathbf{v}) = S(\mathbf{v}) \cdot \mathbf{v} = 0$, so in an asymptotic direction, M is (instantaneously, at least) not bending away from its tangent plane.

Using Corollary 2.6 we can get a complete analysis of asymptotic directions in terms of Gaussian curvature.

6.4 Lemma   Let $\mathbf{p}$ be a point of $M \subset \mathbf{R}^3$.

(1) If $K(\mathbf{p}) > 0$, then there are no asymptotic directions at $\mathbf{p}$.

(2) If $K(\mathbf{p}) < 0$, then there are exactly two asymptotic directions at $\mathbf{p}$, and these are bisected by the principal directions (Fig. 5.30) at angle ϑ such that

$$\tan^2 \vartheta = \frac{-k_1(\mathbf{p})}{k_2(\mathbf{p})}.$$

(3) If $K(\mathbf{p}) = 0$, then *every* direction is asymptotic if $\mathbf{p}$ is a planar point; otherwise there is exactly one asymptotic direction and it is also principal.

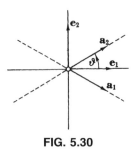

FIG. 5.30

Proof. These cases all derive from Euler's formula

$$k(\mathbf{u}) = k_1(\mathbf{p})\cos^2\vartheta + k_2(\mathbf{p})\sin^2\vartheta$$

in Corollary 2.6.

(1) Since $k_1(\mathbf{p})$ and $k_2(\mathbf{p})$ have the same sign, $k(\mathbf{u})$ is never zero.

(2) Here $k_1(\mathbf{p})$ and $k_2(\mathbf{p})$ have opposite signs, and we can solve the equation $0 = k_1(\mathbf{p})\cos^2\vartheta + k_2(\mathbf{p})\sin^2\vartheta$ to obtain the two asymptotic directions.

(3) If $\mathbf{p}$ is planar, then

$$k_1(\mathbf{p}) = k_2(\mathbf{p}) = 0;$$

hence $k(\mathbf{u})$ is identically zero. If just $k_2(\mathbf{p}) = 0$, then

$$k(\mathbf{u}) = k_1(\mathbf{p})\cos^2\vartheta$$

is zero only when $\cos\vartheta = 0$, that is, in the principal direction $\mathbf{u} = \mathbf{e}_2$. ◆

We can get an approximate idea of the asymptotic directions at a point $\mathbf{p}$ of a given surface M by picturing the intersection of the tangent plane $T_p(M)$ with M near $\mathbf{p}$. When $K(\mathbf{p})$ is negative, this intersection will consist of two curves through $\mathbf{p}$ whose tangent lines (at $\mathbf{p}$) are asymptotic directions (Exercise 5 of Section 53).

Figure 5.31 shows the two asymptotic directions A and A' at a point $\mathbf{p}$ on the inner equator of a torus.

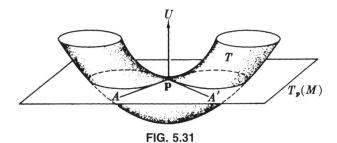

FIG. 5.31

6.5 Definition A regular curve α in $M \subset \mathbf{R}^3$ is an *asymptotic curve* provided its velocity α' always points in an asymptotic direction.

Thus α is asymptotic if and only if

$$k(\alpha') = S(\alpha') \bullet \alpha' = 0.$$

Since $S(\alpha') = -U'$, this gives a criterion, $U' \cdot \alpha' = 0$, for α to be asymptotic. Asymptotic curves are more sensitive to Gaussian curvature than are principal curves: Lemma 6.3 shows that there are none in regions where K is positive, but two cross (at an angle depending on K) at each point of a region where K is negative.

The simplest criterion for a curve in M to be asymptotic is that *its acceleration α'' always be tangent to M*. In fact, differentiation of $U \cdot \alpha' = 0$ gives

$$U' \cdot \alpha' + U \cdot \alpha'' = 0,$$

so $U' \cdot \alpha' = 0$ (α asymptotic) if and only if $U \cdot \alpha'' = 0$.

The analysis of asymptotic directions in Lemma 6.4 has consequences for both flat and minimal surfaces. First, *a surface M in $\mathbf{R}^3$ is minimal if and only if there exist two* orthogonal *asymptotic directions at each of its points.* In fact, $H(\mathbf{p}) = 0$ is equivalent to $k_1(\mathbf{p}) = -k_2(\mathbf{p})$, and an examination of the possibilities in Lemma 5.4 shows that $k_1(\mathbf{p}) = -k_2(\mathbf{p})$ if and only if either (a) $\mathbf{p}$ is planar (so the criterion holds trivially) or (b)

$$K(\mathbf{p}) < 0 \quad \text{with } \vartheta = \pm\pi/4,$$

which means that the two asymptotic directions are orthogonal.

Thus a surface is minimal if and only if through each point there are two asymptotic curves that cross *orthogonally.* This observation gives geometric meaning to the calculations in Example 4.3, which show that the helicoid is a minimal surface. In fact, the u- and v-parameter curves of the patch $\mathbf{x}$ are orthogonal since $F = 0$, and their accelerations are tangent to the surface since $\mathrm{L} = U \cdot \mathbf{x}_{uu} = 0$ and $\mathrm{N} = U \cdot \mathbf{x}_{vv} = 0$.

Recall that a *ruled surface* is swept out by a line moving through $\mathbf{R}^3$ (Definition 2.6 in Chapter 4). We have seen, for example, that the helicoid and saddle surface in Example 4.3 are ruled surfaces. Thus it is no accident that both these surfaces have K negative, since:

6.6 Lemma   A ruled surface M has Gaussian curvature $K \leqq 0$. Furthermore, $K = 0$ if and only if the unit normal U is parallel along each ruling of M (so all points $\mathbf{p}$ on a ruling have the same Euclidean tangent plane $T_p(M)$).

Proof.   A straight line $t \to \mathbf{p} + t\mathbf{q}$ is certainly *asymptotic* since its acceleration is zero and is thus trivially tangent to M. By definition a ruled surface contains a line through each of its points, so there is an asymptotic direction at each point. Hence, by Lemma 6.4, $K \leqq 0$.

Now let $\alpha(t) = \mathbf{p} + t\mathbf{q}$ be an arbitrary ruling in M. If U is parallel along α, then $S(\alpha') = -U' = 0$. Thus α is a principal curve with principal curvature $k(\alpha') = 0$, so $K = k_1 k_2 = 0$.

Conversely, if $K = 0$ we deduce from Case (3) in Lemma 6.4 that asymptotic directions (and curves) in M are also *principal*. Thus each ruling is principal ($S(\alpha') = k(\alpha')\alpha'$) as well as asymptotic ($k(\alpha') = 0$); hence

$$U' = -S(\alpha') = 0, \qquad \blacklozenge$$

and U is parallel along each ruling of M.

We come now to the last and most important of the three types of curves under discussion.

6.8 Definition A curve α in $M \subset \mathbf{R}^3$ is a *geodesic* of M provided its acceleration α'' is always normal to M.

Since α'' is normal to M, the inhabitants of M perceive no acceleration at all—for them the geodesic is a "straight line." A full study of geodesics is given in later chapters, where, in particular, we examine their character as shortest routes of travel. Geodesics are far more plentiful in a surface M than are principal or asymptotic curves. Indeed, Theorem 4.2 of Chapter 7 will show that given any tangent vector $\mathbf{v}$ to M there is a (unique) geodesic with initial velocity $\mathbf{v}$.

Because the acceleration α'' of a geodesic is orthogonal to M, it is orthogonal to the velocity α' of α. Thus *geodesics have constant speed*, since differentiation of $\|\alpha'\|^2 = \alpha' \cdot \alpha'$ gives $2\alpha' \cdot \alpha'' = 0$.

A straight line $\alpha(t) = \mathbf{p} + t\mathbf{q}$ contained in M is always a geodesic of M since its acceleration $\alpha'' = 0$ is trivially normal to M. Though they lack any geometric significance, constant curves are also geodesics, but to avoid clutter this case is often neglected.

6.9 Example *Geodesics of some surfaces in $\mathbf{R}^3$.*

(1) *Planes.* If α is a geodesic in a plane P orthogonal to $\mathbf{u}$, then $\alpha' \cdot \mathbf{u} = 0$, hence $\alpha'' \cdot \mathbf{u} = 0$. But α'' is by definition normal to P, hence collinear with $\mathbf{u}$, so $\alpha'' = 0$. Thus α is a straight line. Since as noted above, every such line is a geodesic, we conclude that *the geodesics of P are the straight lines in P.*

(2) *Spheres.* A *great circle* in a sphere $\Sigma \subset \mathbf{R}^3$ is a circle cut from Σ by a plane P through the center (Fig. 5.32). If α is a constant-speed parametrization of any circle, we know that its acceleration α'' points toward the center of the circle. In the case of a great circle that center is also the center of the sphere Σ. Thus α'' is normal to Σ, so α is a geodesic of Σ.

We can find such a geodesic with any given initial velocity $\mathbf{v}_p$ (the required plane P passes through $\mathbf{p}$ orthogonal to $\mathbf{p} \times \mathbf{v}$). Hence by the uniqueness

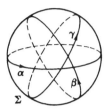

FIG. 5.32

feature mentioned earlier, this construction yields all the geodesics of Σ. Explicitly, *the geodesics of a sphere are the constant-speed parametrizations of its great circles* (Fig. 5.32).

(3) *Cylinders.* The geodesics of, say, the circular cylinder $M: x^2 + y^2 = r^2$ are all curves of the form

$$\alpha(t) = (r\cos(at + b), r\sin(at + b), ct + d).$$

To see this, write an arbitrary curve in M as

$$\alpha(t) = (r\cos \vartheta(t), r\sin \vartheta(t), h(t)).$$

A vector normal to M must have z coordinate zero. Thus if α is a geodesic, $h'' = 0$, so $h(t) = ct + d$. Since the speed of a geodesic is constant, the speed $(r^2\vartheta'^2 + h'^2)^{1/2}$ of α is constant, so ϑ' is constant. Hence $\vartheta(t) = at + b$.

When both constants a and c are nonzero, α is a helix on M. In extreme cases, α parametrizes a ruling if $a = 0$ and a cross-sectional circle if $c = 0$.◆

A *closed geodesic* is a geodesic segment $\alpha: [a, b] \to M$ that is smoothly closed ($\gamma'(b) = \gamma'(a)$) and hence extendible by periodicity over the whole real line. Thus closed geodesics and periodic geodesics are effectively the same thing. In the surfaces above, every geodesic of the sphere is closed, while on the cylinder only the cross-sectional circles are closed.

6.10 Remark Here is a simple geometric way to find examples of geodesics. If a unit-speed curve α in M lies in a plane P everywhere orthogonal to M along α, then α is a geodesic of M. *Proof.* Since α has constant speed, α'' is always orthogonal to α', but these two vectors lie in a plane orthogonal to M, and α' is always tangent to M. Hence α'' must be orthogonal to M, so α is geodesic.

Using this remark we could have found all the geodesics in the preceding example except the helices in the cylinder. It shows at once that on a surface of revolution M, *all meridians are geodesics*, since they are cut from M by planes passing through the axis of rotation and hence orthogonal to M.

The essential properties of the three types of curves we have considered can be summarized as follows:

Principal curves $k(\alpha') = k_1$ or k_2, $S(\alpha')$ collinear α',
Asymptotic curves $k(\alpha') = 0$, $S(\alpha')$ orthogonal to α', α'' tangent to M
Geodesics α'' normal to M

Exercises

1.   Prove that a curve α in M is a straight line of $\mathbf{R}^3$ if and only if α is both geodesic and asymptotic.

2.   To which of the three types—principal, asymptotic, geodesic—do the following curves belong?
   (a) The top circle α of a torus (Fig. 5.33).
   (b) The outer equator β of a torus.
   (c) The x axis in $M: z = xy$.

(Assume constant-speed parametrizations.)

3.   (*Closed geodesics.*) Show:
   (a) In a surface of revolution, a parallel through a point $\alpha(t)$ on the profile curve is a (necessarily closed) geodesic if and only if $\alpha'(t)$ is parallel to the axis of revolution.
   (b) There are at least three closed geodesics on every ellipsoid (Ex. 9 of Sec. 4.2).

4.   Let α be an asymptotic curve in $M \subset \mathbf{R}^3$ with curvature $\kappa > 0$.
   (a) Prove that the binormal B of α is normal to the surface along α, and deduce that $S(T) = \tau N$.
   (b) Show that along α the surface has Gaussian curvature $K = -\tau^2$.
   (c) Use (b) to find the Gaussian curvature of the helicoid (Ex. 4.3).

5.   Suppose that a curve α lies in two surfaces M and N that make a constant angle along α (that is, $U \cdot V$ constant). Show that α is principal in M if and only if principal in N.

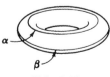

FIG. 5.33

6. If $\mathbf{x}$ is a patch in M, prove that a curve $\alpha(t) = \mathbf{x}(a_1(t), a_2(t))$ is
(a) Principal if and only if

$$\begin{vmatrix} a_2'^2 & -a_1'a_2' & a_1'^2 \\ E & F & G \\ L & M & N \end{vmatrix} = 0.$$

(b) Asymptotic if and only if $La_1'^2 + 2Ma_1'a_2' + Na_2'^2 = 0$.

7. Let α be a unit-speed curve in $M \subset \mathbf{R}^3$. Instead of the Frenet frame field on α, consider the *Darboux frame field* T, V, U—where T is the unit tangent of α, U is the surface normal restricted to α, and $V = U \times T$ (Fig. 5.34).
(a) Show that

$$\begin{aligned} T' &= & gV & + kU, \\ V' &= -gT & & + tU, \\ U' &= -kT & - tV, \end{aligned}$$

where $k = S(T) \cdot T$ is the normal curvature $k(T)$ of M in the T direction, and $t = S(T) \cdot V$.
The new function g is called the *geodesic curvature* of α.
(b) Deduce that α is

$$\text{geodesic} \Leftrightarrow g = 0,$$
$$\text{asymptotic} \Leftrightarrow k = 0,$$
$$\text{principal} \Leftrightarrow t = 0.$$

8. If α is a (unit speed) curve in M, show that
(a) α is both principal and geodesic if and only if it lies in a plane everywhere orthogonal to M along α.
(b) α is both principal and asymptotic if and only if it lies in a plane everywhere tangent to M along α.

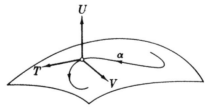

FIG. 5.34

9. On the monkey saddle M (see Fig. 5.19) find *three* asymptotic curves and *three* principal curves passing through the origin **0**. (This is possible only because **0** is a planar umbilic point.)

10. Let α be a regular curve in $M \subset \mathbf{R}^3$, and let U be the unit normal of M along α. Show that α is a principal curve of M if and only if the ruled surface $\mathbf{x}(u, v) = \alpha(u) + vU(u)$ is flat.

11. A ruled surface is *noncylindrical* if its rulings are always changing directions; thus for any director curve, $\delta \times \delta' \neq 0$. Show that:
(a) a noncylindrical ruled surface has a parametrization

$$\mathbf{x}(u, v) = \sigma(u) + v\delta(u)$$

for which $\|\delta\| = 1$ and $\sigma' \cdot \delta = 0$.
(b) for this parametrization,

$$K = \frac{-p^2(u)}{\left(p^2(u) + v^2\right)^2}, \quad \text{where } p = \frac{\sigma' \cdot \delta \times \delta'}{\delta' \cdot \delta'}.$$

The curve σ is called the *striction curve*, and the function p is the *distribution parameter.*
(c) Deduce from the behavior of K on each ruling that the route of the striction curve is independent of parametrization, and hence that the distribution parameter is essentially a function on the set of rulings.

(*Hint:* For (a), find f such that $\sigma = \alpha + f\delta$. For (b), show that $\sigma' \times \delta = p\delta'$.)

12. In each case below, find the striction curve and distribution parameter, and check the formula for K in (b) of the preceding exercise.
(a) the helicoid in Example 4.3.
(b) the tangent surface of a curve (Ex. 13 of Sec. 4).
(c) both sets of rulings of the saddle surface in Example 4.3.

(*Hint:* In the usual ruled parametrization $(u, 0, 0) + v(0, 1, u)$, this last vector must be replaced by a unit vector in order to apply Ex. 11. The curvature formula in Ex. 4.3 will then change.)

13. If $\mathbf{x}(u, v) = \alpha(u) + v\delta(u)$ parametrizes a noncylindrical ruled surface, let $L(u)$ be the ruling through $\alpha(u)$. Show that:
(a) If ϑ_ε is the smallest angle from $L(u)$ to $L(u + \varepsilon)$, and d_ε is the orthogonal distance from $L(u)$ to $L(u + \varepsilon)$, then

$$\lim_{\varepsilon \to 0} \frac{d_\varepsilon}{\vartheta_\varepsilon} = p(u).$$

Thus the distribution parameter is the rate of turning of L.

FIG. 5.35

(b) There is a unique point $\mathbf{p}_\varepsilon$ of $L(u)$ that is nearest to $L(u + \varepsilon)$, and

$$\lim_{\varepsilon \to 0} \mathbf{p}_\varepsilon = \sigma(u)$$

(Fig. 5.35). (This gives another characterization of the striction curve σ.)

(*Hint:* The common perpendicular to $L(u)$ and $L(u + \varepsilon)$ is in the direction of $\delta(u) \times \delta(u + \varepsilon) \approx \varepsilon\delta(u) \times \delta'(u)$.)

14. Let $\mathbf{x}(u, v) = \alpha(u) + v\delta(u)$, with $\|\delta\| = 1$, parametrize a flat ruled surface M. Show that:
(a) If α' is always zero, then M is a generalized cone.
(b) If δ' is always zero, then M is a generalized cylinder.
(c) If both α' and δ' are never zero, then M is the tangent surface of its striction curve. (*Hint:* Parametrize by $\sigma + v\delta$ as in Ex. 11, giving σ unit speed. Use $K = 0$ to show that $T = \sigma'$ and δ are collinear.)

These are only the extreme cases. For example, a flat piece of paper could be bent cylindrically at one end and conically at the other. Note that of the three types, only the cylinder has rulings that are entire straight lines.

15. (*Enneper's minimal surface.*) Prove:
(a) The mapping $\mathbf{x}: \mathbf{R}^2 \to \mathbf{R}^3$ given by

$$\mathbf{x}(u, v) = \left(u - \frac{u^3}{3} + uv^2, \ v - \frac{v^3}{3} + vu^2, \ u^2 - v^2 \right),$$

though not one-to-one, is regular, and hence defines an immersed surface $\mathscr{E}$.
(b) $\mathbf{x}$ is a principal parametrization of $\mathscr{E}$, that is, the u- and v-parameter curves are principal curves.
(c) $\mathscr{E}$ is a minimal surface.
(d) The asymptotic curves of $\mathscr{E}$ are $u \to \mathbf{x}(u, \pm u)$.

16. (*Continuation by computer graphics.*)
(a) Plot $\mathbf{x}(D) \subset \mathscr{E}$, for D: $-3 \le u, v \le 3$. (Note that by the preceding exercise the parameter curves are principal.)
(b) Show that the Euclidean isometry $(x, y, z) \to (-y, x, -z)$ carries $\mathscr{E}$ to itself.

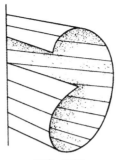

FIG. 5.36

Thus the $z < 0$ half of $\mathscr{E}$ is the mirror image of a 90° rotation of the $z > 0$ half. Further properties of $\mathscr{E}$ are developed in Exercise 10 of Section 6.8.

17. A *right conoid* is a ruled surface whose rulings all pass orthogonally through a fixed axis (Fig. 5.36). Taking this axis as the z axis of $\mathbf{R}^3$, we find the parametrization

$$\mathbf{x}(u, v) = (u\cos\vartheta(v),\ u\sin\vartheta(v),\ h(v)),$$

where the u-parameter curves are the rulings. (This reversal of u and v from earlier exercises makes it clear that the helicoid is a conoid.)

(a) Find the Gaussian and mean curvature of $\mathbf{x}$.

(b) Show that the surface is noncylindrical if ϑ' is never zero, and in this case, find the striction curve and parameter of distribution.

18. (*Computer graphics.*)

(a) A right conoid has base curve $\alpha(v) = (0, 0, \cos 2v)$ and director curve $\delta(v) = (\cos v,\ \sin v,\ 0)$. For the resulting ruled parametrization (with u-parameter curves as rulings), plot the portion with $-2.5 \leqq u \leqq 2.5$, $0 \leqq v \leqq \pi$.

(b) The axis of a right conoid is the z axis, and its rulings pass through every point of the circle $y^2 + z^2 = r^2$ in the plane $x = c$. Verify that

$$\mathbf{x}(u, v) = (uc,\ ur\cos v,\ r\sin v)$$

parametrizes this conoid, and for $r = 2$, $c = 1$ plot the portion between the axis and the circle.

(c) Same as (b) except that the circle is replaced by the curve $y = \sin z$. Find a ruled parametrization, and for $c = 4$, plot the region with $0 \leqq z \leqq 4\pi$ and rulings running from $x = -4$ to $x = +4$.

19. For curves β and δ in $\mathbf{R}^3$, let $\mathbf{x}(u, v) = \beta(u) + v\delta(u)$. Find, in simplest terms:

(a) A necessary and sufficient condition that $\mathbf{x}$ parametrize a (ruled) surface in $\mathbf{R}^3$.

(b) A formula for the Gaussian curvature K of this surface. (*Hint:* Show $N = 0$.)

5.7 Surfaces of Revolution

The geometry of a surface of revolution is rather simple, yet these surfaces exhibit a wide variety of geometric behavior; thus they offer a good field for experiment.

We apply the methods of Section 4 to study an arbitrary surface of revolution M, with the usual parametrization, given in Example 2.4 of Chapter 4 by

$$\mathbf{x}(u, v) = (g(u),\ h(u)\cos v,\ h(u)\sin v).$$

Here $h(u) > 0$ is the radius of the parallel at distance $g(u)$ along the axis of revolution of M, as shown in Fig. 4.14. This geometric significance for g and h means that our results do not depend on the particular position of M relative to the coordinate axes of $\mathbf{R}^3$.

Because g and h are functions of u alone, we can write

$$\mathbf{x}_u = (g',\ h'\cos v,\ h'\sin v),$$

$$\mathbf{x}_v = (0,\ -h\sin v,\ h\cos v),$$

and hence

$$E = g'^2 + h'^2,\quad F = 0,\quad G = h^2.$$

Here E is the square of the speed of the profile curve and hence of all the meridians (u-parameter curves), while G is the square of the speed of the parallels (v-parameter curves). Next we find, successively,

$$\mathbf{x}_u \times \mathbf{x}_v = (hh',\ -hg'\cos v,\ -hg'\sin v),$$

$$\|\mathbf{x}_u \times \mathbf{x}_v\| = \sqrt{EG - F^2} = h\sqrt{g'^2 + h'^2},$$

$$U = \frac{1}{\sqrt{g'^2 + h'^2}}(h',\ -g'\cos v,\ -g'\sin v).$$

Taking second derivatives gives

$$\mathbf{x}_{uu} = (g'',\ h''\cos v,\ h''\sin v),$$

$$\mathbf{x}_{uv} = (0,\ -h'\sin v,\ h'\cos v),$$

$$\mathbf{x}_{vv} = (0,\ -h\cos v,\ -h\sin v),$$

Hence

$$L = \frac{-g'h'' + g''h'}{\sqrt{g'^2 + h'^2}}, \quad M = 0, \quad N = \frac{g'h}{\sqrt{g'^2 + h'^2}}.$$

Since $F = M = 0$, $\mathbf{x}$ is a principal parametrization (Exercise 8 of Section 4), and for the shape operator S derived from U,

$$S(\mathbf{x}_u) = \frac{L}{E}\mathbf{x}_u, \quad S(\mathbf{x}_v) = \frac{N}{G}\mathbf{x}_v.$$

This is an analytical proof that the *meridians* and *parallels* of a surface of revolution are its principal curves. Furthermore, if the corresponding principal curvature functions are denoted by k_μ and k_π, instead of k_1 and k_2, we have

$$k_\mu = \frac{L}{E} = \frac{-\begin{vmatrix} g' & h' \\ g'' & h'' \end{vmatrix}}{\left(g'^2 + h'^2\right)^{3/2}}, \quad k_\pi = \frac{N}{G} = \frac{g'}{h\left(g'^2 + h'^2\right)^{1/2}}. \tag{1}$$

Thus the Gaussian curvature of M is

$$K = k_\mu k_\pi = \frac{-g'\begin{vmatrix} g' & h' \\ g'' & h'' \end{vmatrix}}{h\left(g'^2 + h'^2\right)^2}. \tag{2}$$

This formula defines K as a real-valued function on the domain of the profile curve

$$\alpha(u) = \left(g(u), h(u), 0\right).$$

By the conventions of Section 4, $K(u)$ is the Gaussian curvature $K(\mathbf{x}(u, v))$ of M *at every point of the parallel through* $\alpha(u)$. The same is true for the other functions above. The rotational symmetry of M about its axis of revolution means that its geometry is "constant on parallels"—completely determined by the profile curve.

In the special case where the profile curve passes at most once over each point of the axis of rotation, we can usually arrange for the function g to be simply $g(u) = u$ (Example 2.8 of Chapter 4). Then the formulas (1) and (2) above reduce to

$$k_\mu = \frac{-h''}{\left(1 + h'^2\right)^{3/2}}, \quad k_\pi = \frac{1}{h\left(1 + h'^2\right)^{1/2}},$$

$$K = \frac{-h''}{h\left(1 + h'^2\right)^2}. \tag{3}$$

7.1 Example   *Surfaces of revolution.*

(1) *Torus of revolution T.* The usual parametrization **x** in Example 2.5 of Chapter 4 has

$$g(u) = r \sin u, \quad h(u) = R + r \cos u,$$

for constants $0 < r < R$. Although the axis of revolution is now the z axis, formulas (1) and (2) above remain valid, and we compute

$$E = r^2, \quad F = 0, \quad G = (R + r \cos u)^2,$$

$$\text{L} = r, \quad \text{M} = 0, \quad \text{N} = (R + r \cos u)\cos u,$$

$$k_\mu = \frac{1}{r}, \quad k_\pi = \frac{\cos u}{R + r \cos u},$$

$$K = \frac{\cos u}{r(R + r \cos u)}.$$

This gives an analytical proof that the Gaussian curvature of the torus is positive on the outer half and negative on the inner half. K has its maximum value $1/r(R + r)$ on the outer equator ($u = 0$), its minimum value $-1/r(R - r)$ on the inner equator ($u = \pi$), and is zero on the top and bottom circles ($u = \pm\pi/2$).

(2) *Catenoid.* The curve $y = c \cosh(x/c)$ is a *catenary*; its shape is that of a chain hanging under the influence of gravity. The surface obtained by rotating this curve around the x axis is called a *catenoid* (Fig. 5.37). From the formulas (3) we find,

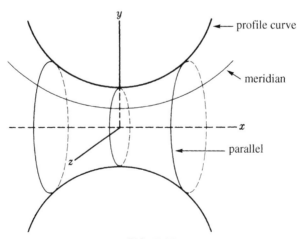

FIG. 5.37

$$-k_\mu = k_\pi = \frac{1}{c \cosh^2(x/c)},$$

and hence

$$H = 0, \quad K = \frac{-1}{c^2 \cosh^4(x/c)}.$$

Since its mean curvature H is zero, the catenoid is a minimal surface. Its Gaussian curvature interval is $-1/c^2 \leqq K < 0$, with minimum value $K = -1/c^2$ on the central circle $(u = 0)$. ◆

7.2 Theorem If a surface of revolution M is a minimal surface, then M is contained in either a plane or a catenoid.

Proof. M is parametrized as usual by

$$\mathbf{x}(u, v) = (g(u), h(u)\cos v, h(u)\sin v),$$

with u in a (possibly infinite) interval I.

Case 1. g' is identically zero. Then g is constant, so M is part of a plane orthogonal to the axis of revolution.

Case 2. g' is never zero. By Exercise 8 in Section 4.2, M has a parametrization of the form

$$\mathbf{y}(u, v) = (u, h(u)\cos v, h(u)\sin v).$$

The formulas for k_μ and k_π in (3) above then show that the minimality condition is equivalent to

$$hh'' = 1 + h'^2.$$

Because u does not appear explicitly in this differential equation, there is a standard elementary way to solve it. We merely record that the solution is

$$h(u) = a \cosh\left(\frac{u}{a} + b\right),$$

where $a \neq 0$ and b are constants. Thus M is part of a catenoid.

Case 3. g' is zero at some points, nonzero at others. This cannot happen. For definiteness, suppose that $g'(u) > 0$ for $u < u_0$ but $g'(u_0) = 0$. By Case 2, the profile curve $(g(u), h(u), 0)$ is a catenary for $u < u_0$. The shape of the catenary makes it clear that slope h'/g' cannot approach infinity as $u \to u_0$. ◆

This result shows that *catenoids are the only complete nonplanar surfaces of revolution that are minimal.* (Completeness, discussed in Chapter 8, implies that the surface cannot be part of a larger surface.)

Helicoids and catenoids are called the elementary minimal surfaces. Two others are given in the exercises for this chapter (Exercise 5 in Section 5 and Exercise 15 in Section 6). Soap-film models of an immense variety of minimal surfaces can easily be exhibited by the methods given in [dC], where the term "minimal" is explained.

The expression $\sqrt{g'^2 + h'^2}$, which appears so frequently in the formulas above, is just the speed of the profile curve $\alpha(u) = (g(u), h(u), 0)$. Thus we can radically simplify these formulas by a reparametrization that has unit speed. The surface of revolution is unchanged; it has merely been given a new parametrization, called *canonical.*

7.3 Lemma   For a canonical parametrization of a surface of revolution,

$$E = 1, \quad F = 0, \quad G = h^2,$$

and the Gaussian curvature is

$$K = \frac{-h''}{h}.$$

Proof.   Since $g'^2 + h'^2 = 1$ for a canonical parametrization, these expressions for E, F, and G follow immediately from those at the start of this section. The formula for K in (2) becomes

$$K = \frac{-g'}{h} \begin{vmatrix} g' & h' \\ g'' & h'' \end{vmatrix} = \frac{-g'^2 h'' + g' g'' h'}{h}.$$

But this can be simplified. Differentiation of $g'^2 + h'^2 = 1$ gives $g' g'' = -h' h''$, and when this is substituted above, we get $K = -h''/h$.   ◆

The effect of using a canonical parametrization is to shift the emphasis from measurements in the space *outside M* (for example, along the axis of revolution) to measurement *within M*. This important idea will be developed more fully as we proceed.

7.4 Example   *Canonical parametrization of the catenoid ($c = 1$).*
An arc length function for the catenary $\alpha(u) = (u, \cosh u)$ is $s(u) = \sinh u$. Hence a unit-speed reparametrization is

$$\beta(s) = (g(s), h(s)) = \left(\sinh^{-1} s, \sqrt{1 + s^2} \right),$$

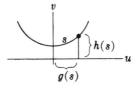

FIG. 5.38

as indicated in Fig. 5.38. The resulting canonical parametrization of the catenoid is given by

$$\bar{x}(s, v) = \left(\sinh^{-1} s, \sqrt{1+s^2} \cos v, \sqrt{1+s^2} \sin v\right).$$

Hence by the preceding lemma,

$$K(s) = -\frac{h''(s)}{h(s)} = \frac{-1}{\left(1+s^2\right)^2}.$$

This formula for K in terms of $\bar{x}$ is consistent with the formula

$$K(u) = \frac{1}{\cosh^4 u}$$

found in Example 6.1 for the parametrization $\mathbf{x}_1$. In fact, since $s(u) = \sinh u$, we have

$$K(s(u)) = \frac{-1}{\left(1+s^2(u)\right)^2} = \frac{-1}{\left(1+\sinh^2 u\right)^2} = \frac{-1}{\cosh^4 u}.$$

The simple formula for K in Lemma 7.3 suggests a way to construct surfaces of revolution with *prescribed* Gaussian curvature. Given a function

$$K = K(u) \text{ on some interval,}$$

first solve the differential equation $h'' + Kh = 0$ for h, subject to initial conditions $h(0) > 0$ and $|h'(0)| < 1$. (The first of these conditions is a convenience; the second is a necessity since we must have $g'^2 + h'^2 = 1$.)

To get a canonical parametrization, we need a function g satisfying the equation $g'^2 + h'^2 = 1$. Evidently,

$$g(u) = \int_0^u \sqrt{1 - h'^2(t)} \, dt$$

will do the job.

We conclude that for any interval around 0 on which the initial conditions

$$h > 0 \text{ and } |h'| < 1$$

both hold, revolving the profile curve $(g(u), h(u), 0)$ around the x axis produces a surface that has, by Lemma 7.3, Gaussian curvature $K = -h''/h$.

A natural use of this scheme is to look for surfaces that have *constant* curvature. Consider first the K positive case.

7.5 Example *Surfaces of revolution with constant positive curvature.*
We apply the procedure to the constant function $K = 1/c^2$. The differential equation $h'' + h/c^2 = 0$ has general solution

$$h(u) = a \cos\left(\frac{u}{c} + b\right).$$

The constant b represents only a translation of coordinates so we may as well set $b = 0$. As usual, nothing is lost by requiring $h > 0$; hence $a > 0$. Thus the functions

$$g(u) = \int_0^u \left(1 - \frac{a^2}{c^2} \sin^2 \frac{t}{c}\right)^{\frac{1}{2}} dt, \quad h(u) = a \cos \frac{u}{c}$$

give rise to a surface of revolution M_a with constant Gaussian curvature

$$K = 1/c^2.$$

As mentioned above, the conditions $h > 0$ and $|h'| < 1$ determine the largest interval I on which the procedure works. The constant c is fixed, but the constant a is at our disposal, and it distinguishes three cases.

Case 1. $a = c$. Here

$$g(u) = \int_0^u \cos \frac{t}{c} dt = c \sin \frac{u}{c}, \quad h(u) = c \cos \frac{u}{c}. \tag{4}$$

Thus the maximum interval I is $-\pi c/2 < u < \pi c/2$, and the profile curve $(g(u), h(u))$ is a semicircle. Revolution about the x axis produces a sphere Σ of radius c—except for its two points on the axis.

Case 2. $0 < a < c$. Here h is positive on the same interval as above and $|h'| < 1$ is always true, so g is well defined. The profile curve has the same length $\pi c/2$, but it now forms a shallower arch, which rests on the x axis at $\pm a^*$, where $a^* = g(\pi c/2) > a$ (Fig. 5.39). As a shrinks down from c to 0, one can check that a^* increases from c to $\pi c/2$. The resulting surface of revolution, round when $a = c$, first becomes football-shaped and then grows ever thinner, becoming, for a small, a needle of length just less than πc.

By contrast with Case 1, the intercepts $(\pm a_*, 0, 0)$ cannot be added to M now since this surface is actually pointed at each end (Fig. 5.39).

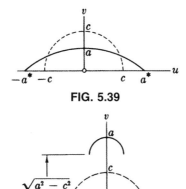

FIG. 5.39

FIG. 5.40

The differential equation $h'' + h/c^2 = 0$ has delicately adjusted the shape of M_a so that its principal curvatures are no longer equal but still give

$$K = k_\mu k_\pi = \frac{1}{c^2}.$$

Case 3. $a > c$. Here the maximum interval is shorter than in Case 1. The formula for $g(u)$ in (4) shows that the endpoints now are $\pm a_*$, where $a_* < c$ is determined by $\sin a_*/c = c/a < 1$. Thus,

$$h(a_*) = a \cos a_*/c = \sqrt{a^2 - c^2}.$$

As a increases from $a = c$, the resulting surface of revolution M_a is at first somewhat like the outer half of a torus. But when a is very large, it becomes a huge circular band (Fig. 5.40), whose very short profile curve is sharply curved (k_μ must be large since $k_\pi \approx 1/a$ and $k_\mu k_\pi = 1/c^2$).

A corresponding analysis for constant negative curvature leads to an infinite family of surfaces of revolution with $K = -1/c^2$ (Exercises 7 and 8). The simplest of these surfaces is

7.6 Example *The bugle surface B.* The profile curve of B (in the xy plane) is characterized by this geometric condition: It starts at the point $(0, c)$ and moves so that its tangent line reaches the x axis after running for distance exactly c. This curve, a *tractrix*, can be described analytically as

$$\alpha(u) = (u, h(u)), \quad u > 0,$$

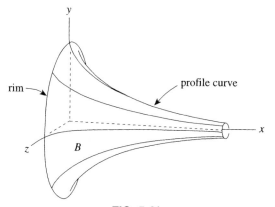

FIG. 5.41

where h is the solution of the differential equation

$$h' = \frac{-h}{\sqrt{c^2 - h^2}}$$

such that $h(u) \to c$ as $u \to 0$. The resulting surface of revolution B is called a *bugle surface* or *tractroid* (Fig. 5.41). Using the differential equation above, we deduce from the earlier formulas (3) that the principal curvatures of B are

$$k_\mu = \frac{-h'}{c}, \quad k_\pi = \frac{1}{ch'}.$$

Thus the bugle surface has constant negative curvature

$$K = -\frac{1}{c^2}.$$

This surface cannot be extended across its rim—not part of B—to form a larger surface in $\mathbf{R}^3$ since $k_\mu(u) \to \infty$ as $u \to 0$. ◆

When this surface was first discovered, it seemed to be the analogue, for K a negative constant, of the sphere; it was thus called a *pseudosphere*. However, as we shall see later on, the true analogue of the sphere is quite a different surface and cannot be found in $\mathbf{R}^3$.

Exercises

1. Find the Gaussian curvature of the surface obtained by revolving the curve $y = e^{-x^2/2}$ around the x axis. Sketch this surface and indicate the regions where $K > 0$ and $K < 0$.

2. (a) Show that when $y = f(x)$ is revolved around the x axis, the Gaussian curvature $K(x)$ has the same sign $(-, 0, +)$ as $-f''(x)$ for all x.

(b) Deduce that for a surface of revolution with arbitrary axis, the Gaussian curvature K is positive on parallels through *convex* intervals on the profile curve (where the curve bulges away from the axis) and negative on parallels through *concave* intervals (where the curve sags toward the axis).

3. Prove that a flat surface of revolution is part of a plane, cone, or cylinder.

4. (*Computer.*)

(a) Write computer commands that, given a profile curve $u \to (g(u), h(u))$, (i) plot the resulting surface of revolution for $a \leq u \leq b$, and (ii) return its Gaussian curvature $K(u)$.

(b) Test (a) on the torus and catenoid in Example 7.1.

5. If $r = \sqrt{x^2 + y^2}$ is the usual polar coordinate function on the xy plane, and f is a differentiable function, show the $M: z = f(r)$ is a surface of revolution and that its Gaussian curvature K is given by

$$K(r) = \frac{f'(r)f''(r)}{r\left(1 + f'(r)^2\right)^2}.$$

6. Find the Gaussian curvature of the surface $M: z = e^{-r^2/2}$. Sketch this surface, indicating the regions where $K > 0$ and $K < 0$.

7. (*Surfaces of revolution with negative curvature $K = -1/c^2$*) As in the corresponding positive case, there is a family of such surfaces, separated into two subfamilies by a special surface. Essentially all these surfaces are given, using canonical parametrization, by solutions of $h'' - h/c^2 = 0$ as follows:

(a) If $0 < a < c$, let M_a be the surface given by $h(u) = a \sinh(u/c)$, $u > 0$. Show that its profile curve $(g(u), h(u))$ leaves the origin with slope $a/\sqrt{c^2 - a^2}$ and rises to a maximum height of $\sqrt{c^2 - a^2}$.

(b) If $a = c$, let $\bar{B}$ be the surface given by $h(u) = ce^{u/c}$, $u < 0$. Show that its mirror image B, given by $h(u) = ce^{-u/c}$, $u > 0$, is the bugle surface in Example 7.6.

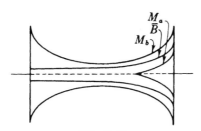

FIG. 5.42

(c) If $b > c$, let M_b be the surface given by $h(u) = b \cosh(u/c)$. Show that as $|u|$ increases from 0, its profile curve rises symmetrically from height b to

height $\sqrt{c^2 + b^2}$.

Sample profile curves of all three types are shown in Fig. 5.42, where M_a and M_b have been translated along the axis of revolution. Explicit formulas for the profile curves in (a) and (b) involve elliptic integrals (see [G]).

8. (a) Taking $c = 1$ for simplicity, show that the tractrix has a parametrization (g, h) with

$$g(u) = e^{-u}, \ h(u) = -\sqrt{1 - e^{-2u}} + \mathrm{arctanh}\sqrt{1 - e^{-2u}}.$$

(b) (*Computer graphics.*) Plot a view of the resulting bugle surface similar to that in Fig. 5.41.

9. In a *twisted* surface of revolution, as points rotate around the axis they also move evenly in the axis direction. Explicitly, if the original surface has a usual parametrization in terms of functions $g(u)$ and $h(u)$, then the twisted surface has parametrization

$$\mathbf{x}(u, v) = (g(u) + pv, \ h(u)\cos v, \ h(u)\sin v).$$

where p is a constant.

(a) Find a parametrization of the twisted bugle surface D (*Dini's surface*) with data as in the preceding exercise and $p = 1/5$.
(b) (*Computer.*) Plot the surface D in (*a*) for $0.01 \le u \le 2$ and $0 \le v \le 6\pi$. (Impose smoothness and view the surface from a point with $x < 0$.)
(c) Show that D has constant negative curvature.

5.8 Summary

The shape operator S of a surface M in $\mathbf{R}^3$ measures the rate of change of a unit normal U in any direction on M and thus describes the way the shape of M is changing in that direction. If we imagine U as the "first derivative" of M, then S is the "second derivative." But the shape operator is an algebraic object consisting of linear operators on the tangent planes of M. And it is by an algebraic analysis of S that we have been led to the main geometric invariants of a surface in $\mathbf{R}^3$: its principal curvatures and directions, and its Gaussian and mean curvatures.

Chapter 6
Geometry of Surfaces in R³

Now that we know how to measure the shape of a surface M in $\mathbf{R}^3$, the next step is to see how the shape of M is related to its other properties. Near each point of M, the Gaussian curvature has a strong influence on shape (Remark 3.3 of Chapter 5), but we are now interested in the situation *in the large*—over the whole extent of M. For example, what can be said about the shape of M if it is compact, or flat, or both?

In the early 1800s Gauss raised a question that led to a new and deeper understanding of what geometry is: How much of the geometry of a surface in $\mathbf{R}^3$ is *independent of* its shape? At first glance this seems a strange question—what can we possibly say about a sphere, for example, if we ignore the fact that it is round? To get some grip on Gauss's question, let us imagine that the surface $M \subset \mathbf{R}^3$ has inhabitants who are unaware of the space outside their surface, and thus have no conception of its shape in $\mathbf{R}^3$. Nevertheless, they will still be able to measure the distance from place to place in M and find the area of regions in M. We shall see that, in fact, they can construct an *intrinsic geometry* for M that is richer and no less interesting than the familiar Euclidean geometry of the plane $\mathbf{R}^2$.

6.1 The Fundamental Equations

To study the geometry of a surface M in $\mathbf{R}^3$ we shall apply the Cartan methods outlined in Chapter 2. As with the Frenet theory of a curve in $\mathbf{R}^3$, this requires that we put frames on M and examine their rates of change along M. Formally, a *Euclidean frame field* on $M \subset \mathbf{R}^3$ consists of three Euclidean vector fields (Definition 3.7, Chapter 4) that are orthonormal at each point. Such a frame field can be fitted to its surface as follows.

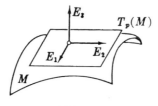

FIG. 6.1

1.1 Definition   An *adapted frame field E_1, E_2, E_3 on a region $\mathcal{O}$ in $M \subset \mathbf{R}^3$* is a Euclidean frame field such that E_3 is always normal to M (hence E_1 and E_2 are tangent to M) (Fig. 6.1).

Thus the normal vector field denoted by U in the preceding chapter now becomes E_3. For brevity we often refer to an adapted frame field "on M," but the actual domain of definition is in general only some region in M, since an adapted frame field need not exist on all of M.

1.2 Lemma   There is an adapted frame field on a region $\mathcal{O}$ in $M \subset R^3$ if and only if $\mathcal{O}$ is orientable and there exists a nonvanishing tangent vector field on $\mathcal{O}$.

Proof.   This condition is certainly necessary, since E_3 orients $\mathcal{O}$, and E_1 and E_2 are unit tangent vector fields. To show that it is sufficient, let $\mathcal{O}$ be oriented by a unit normal vector field U, and let V be a tangent vector field that does not vanish on $\mathcal{O}$. But then it is easy to see that

$$E_1 = \frac{V}{\|V\|}, \quad E_2 = U \times E_1, \quad E_3 = U$$

is an adapted frame field on $\mathcal{O}$.   ◆

1.3 Example   Adapted frame fields.
(1) Cylinder M: $x^2 + y^2 = r^2$. The gradient of $g = x^2 + y^2$ leads to the unit normal vector field $E_3 = (xU_1 + yU_2)/r$. Obviously the unit vector field U_3 is tangent to M at each point. Setting $E_2 = U_3 \times E_3$, we then get the adapted frame field

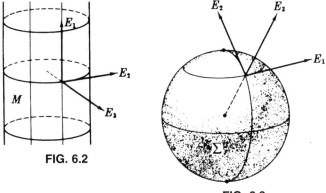

FIG. 6.2

FIG. 6.3

$$E_1 = U_3,$$

$$E_2 = \frac{-yU_1 + xU_2}{r},$$

$$E_3 = \frac{xU_1 + yU_2}{r}$$

on the whole cylinder M (Fig. 6.2).

(2) Sphere Σ: $x^2 + y^2 + z^2 = r^2$. The outward unit normal

$$E_3 = \frac{xU_1 + yU_2 + zU_3}{r}$$

is defined on all of Σ, but as we shall see in Chapter 7, every tangent vector field on Σ must vanish somewhere. For example, the "due east" vector field $V = -yU_1 + xU_2$ is zero at the the north and south poles $(0, 0, \pm r)$. Thus the adapted frame field

$$E_1 = \frac{V}{\|V\|},$$

$$E_2 = E_3 \times E_1,$$

$$E_3 = \frac{xU_1 + yU_2 + zU_3}{r}$$

(Fig. 6.3) is defined on the region $\mathcal{O}$ in Σ gotten by deleting the north and south poles.

Lemma 1.2 implies in particular that there is an adapted frame field on the image $\mathbf{x}(D)$ of any patch in M; thus such fields exist *locally* on any surface in $\mathbf{R}^3$.

Now we shall bring the connection equations (Theorem 7.2 of Chapter 2) to bear on the study of a surface M in $\mathbf{R}^3$. Let E_1, E_2, E_3 be an adapted frame field on M. By moving each frame $E_1(\mathbf{p})$, $E_2(\mathbf{p})$, $E_3(\mathbf{p})$ over a short interval on the normal line at each point $\mathbf{p}$, we can extend the given frame field to one defined on an open set in $\mathbf{R}^3$. Thus the connection equations

$$\nabla_v E_i = \sum \omega_{ij}(v) E_j(\mathbf{p})$$

are available for use. *We shall apply them only to vectors* $\mathbf{v}$ *tangent to* M. In particular, *the connection forms* ω_{ij}, *become* 1-*forms on* M in the sense of Section 7 of Chapter 2. Thus we have

1.4 Theorem If E_1, E_2, E_3 is an adapted frame field on $M \subset \mathbf{R}^3$, and $\mathbf{v}$ is tangent to M at $\mathbf{p}$, then

$$\nabla_v E_i = \sum_{j=1}^{3} \omega_{ij}(v) E_j(\mathbf{p}) \quad (1 \leq i \leq 3).$$

The usual interpretation of the connection forms may be read from these equations, and it bears repetition: $\omega_{ij}(\mathbf{v})$ *is the initial rate at which* E_i *rotates toward* E_j *as* $\mathbf{p}$ *moves in the* $\mathbf{v}$ *direction.* Since E_3 is a unit normal vector field on M, the shape operator of M can be described by connection forms.

1.5 Corollary Let S be the shape operator gotten from E_3, where E_1, E_2, E_3 is an adapted frame field on $M \subset \mathbf{R}^3$. Then for each tangent vector $\mathbf{v}$ to M at $\mathbf{p}$,

$$S(\mathbf{v}) = \omega_{13}(\mathbf{v}) E_1(\mathbf{p}) + \omega_{23}(\mathbf{v}) E_2(\mathbf{p}).$$

Proof. By definition, $S(\mathbf{v}) = -\nabla_v E_3$. The connection equation for $i = 3$ then gives the result, since the connection form $\omega = (\omega_{ij})$ is skew-symmetric: $\omega_{ij} = -\omega_{ji}$. ◆

In addition to its connection forms, the adapted frame field E_1, E_2, E_3 also has *dual* 1-*forms* θ_1, θ_2, θ_3 (Definition 8.1 of Chapter 2) that give the co-ordinates $\theta_i(\mathbf{v}) = \mathbf{v} \cdot E_i(\mathbf{p})$ of any tangent vector $\mathbf{v}_p$ with respect to the frame $E_1(\mathbf{p})$, $E_2(\mathbf{p})$, $E_3(\mathbf{p})$. As with the connection forms, the dual forms will be applied *only* to vectors tangent to M, so they become forms on M. This restriction is fatal to θ_3, for if $\mathbf{v}$ is tangent to M, it is orthogonal to E_3, so $\theta_3(\mathbf{v}) = \mathbf{v} \cdot E_3(\mathbf{p}) = 0$. Thus θ_3 is identically zero on M.

Because of the skew-symmetry of the connection form, we are left with essentially only five 1-forms:

$\theta_1,\ \theta_2$ provide a dual description of the tangent vector fields $E_1,\ E_2$;

ω_{12} gives the rate of rotation of $E_1,\ E_2$;

$\omega_{13},\ \omega_{23}$ describe the shape operator derived from E_3.

1.6 Example   *The sphere.* Consider the adapted frame field $E_1,\ E_2,\ E_3$ defined on the (doubly punctured) sphere Σ in Example 1.3. By extending this frame field to an open set of $\mathbf{R}^3$ we get the *spherical frame field* given in Example 6.2 of Chapter 2, provided the indices of the latter are shifted by $1 \to 3,\ 2 \to 1,\ 3 \to 2$. Thus, in terms of the spherical coordinate functions, Example 8.4 of Chapter 2 gives

$$\theta_1 = r \cos \varphi \, d\vartheta, \quad \omega_{12} = \sin \varphi \, d\vartheta,$$
$$\theta_2 = r \, d\varphi, \quad \omega_{13} = -\cos \varphi \, d\vartheta,$$
$$\omega_{23} = -d\varphi.$$

Because all forms (including functions) are now restricted to the surface Σ, the spherical coordinate function ρ has become a constant: the radius r of the sphere.

In general, the forms associated with an adapted frame field obey the following remarkable set of equations.

1.7 Theorem   If $E_1,\ E_2,\ E_3$ is an adapted frame field on $M \subset \mathbf{R}^3$, then its dual forms and connection forms on M satisfy:

(1) $\begin{cases} d\theta_1 = \omega_{12} \wedge \theta_2 \\ d\theta_2 = \omega_{21} \wedge \theta_1 \end{cases}$ *First structural equations*

(2) $\omega_{31} \wedge \theta_1 + \omega_{32} \wedge \theta_2 = 0$ *Symmetry equation*

(3) $d\omega_{12} = \omega_{13} \wedge \omega_{32}$ *Gauss equation*

(4) $\begin{cases} d\omega_{13} = \omega_{12} \wedge \omega_{23} \\ d\omega_{23} = \omega_{21} \wedge \omega_{13} \end{cases}$ *Codazzi equations*

Proof.   We merely apply the Cartan structural equations in Theorem 8.3 of Chapter 2. The first structural equation,

$$d\theta_i = \sum_j \omega_{ij} \wedge \theta_j$$

yields (1) and (2) above. In fact, for $i = 1$, 2, we get (1), since $\theta_3 = 0$ on the surface M. But $\theta_3 = 0$ implies $d\theta_3 = 0$, so for $i = 3$ we get (2).

Then the second structural equation yields the Gauss (3) and Codazzi (4) equations. ◆

Because connection forms are skew-symmetric and a wedge product of 1-forms satisfies $\phi \wedge \psi = -\psi \wedge \phi$, the fundamental equations above can be rewritten in a variety of equivalent ways. However we shall stick to the index pattern used above, which, on the whole, seems the easiest to remember.

We emphasize that the forms introduced in this section describe not the surface M directly, but only the particular adapted frame field E_1, E_2, E_3 from which they are derived: A different choice of the frame field will produce different forms. Nevertheless, the six fundamental equations in Theorem 1.7 contain a tremendous amount of information about the surface $M \subset \mathbf{R}^3$, and we shall call on each in turn as we come to a geometric situation that it governs. For example, since ω_{13} and ω_{23} describe the shape operator of M, the Codazzi equations (4) express the rate at which *the shape of M is changing* from one point to another.

The first of the following exercises shows how the Cartan approach automatically singles out the three types of curves considered in Chapter 5, Section 6.

Exercises

1.   Let α be a unit-speed curve in $M \subset \mathbf{R}^3$. If E_1, E_2, E_3 is an adapted frame field such that E_1 restricted to α is its unit tangent T, show that
(a) α is a geodesic of M if and only if $\omega_{12}(T) = 0$.
(b) If $E_3 = E_1 \times E_2$, then

$$g = \omega_{12}(T), \quad k = \omega_{13}(T), \quad t = \omega_{23}(T),$$

where g, k, and t are the functions defined in Exercise 7 of Section 5.6. (*Hint:* If $T = E_1$ along α, then $E_i' = \nabla_{E_1} E_i$ along α.)

2.   (*Sphere.*) For the frame field in Example 1.6:
(a) Verify the fundamental equations (Thm. 1.7).
(b) Deduce from the formulas for θ_1 and θ_2 that

$$E_1[\vartheta] = \frac{1}{r \cos \varphi}, \qquad E_1[\varphi] = 0,$$

$$E_2[\vartheta] = 0, \qquad E_2[\varphi] = \frac{1}{r}.$$

(c) Use Corollary 1.5 to find the shape operator S of the sphere.

3. Give a new proof that shape operators are symmetric by using the symmetry equation (Thm. 1.7).

6.2 Form Computations

From now on, our study of the geometry of surfaces will be carried on mostly in terms of differential forms, so the reader may wish to look back over their general properties in Sections 4 and 5 of Chapter 4. Increasingly, we shall tend to compare M with the Euclidean plane $\mathbf{R}^2$. Thus, if E_1, E_2, E_3 is an adapted frame field on $M \subset \mathbf{R}^3$, we say that E_1, E_2 constitutes a *tangent frame field* on M. Any tangent vector field V on M may be expressed in terms of E_1 and E_2 by the orthonormal expansion

$$V = (V \cdot E_1)E_1 + (V \cdot E_2)E_2.$$

To show that two forms are equal, we do not have to check their values on *all* tangent vectors, but only on the "basis" vector fields E_1, E_2. (See the remarks preceding Example 4.7 of Chapter 4). Explicitly: 1-*forms ϕ and ψ are equal if and only if*

$$\phi(E_1) = \psi(E_1) \quad \text{and} \quad \phi(E_2) = \psi(E_2);$$

2-*forms μ and v are equal if and only if*

$$\mu(E_1, E_2) = v(E_1, E_2).$$

The dual forms θ_1, θ_2 are, as we have emphasized, merely another description of the tangent frame field E_1, E_2; they are completely characterized by the equations

$$\theta_i(E_j) = \delta_{ij} \quad (1 \leq i, j \leq 2).$$

These forms provided a "basis" for the forms on M (or, strictly speaking, on the region of definition of E_1, E_2).

2.1 Lemma (The Basis Formulas) Let θ_1, θ_2 be the dual 1-forms of E_1, E_2 on M. If ϕ is a 1-form and μ a 2-form, then

(1) $\phi = \phi(E_1)\theta_1 + \phi(E_2)\theta_2,$
(2) $\mu = \mu(E_1, E_2)\theta_1 \wedge \theta_2.$

Proof. Apply the equality criteria above, observing for (2) that by definition of the wedge product,

$$(\theta_1 \wedge \theta_2)(E_1, E_2) = \theta_1(E_1)\theta_2(E_2) - \theta_1(E_2)\theta_2(E_1)$$
$$= 1 \cdot 1 - 0 \cdot 0 = 1. \qquad \blacklozenge$$

Assuming throughout that the forms θ_1, θ_2, ω_{12}, ω_{13}, ω_{23} derive as in Section 1 from an adapted frame field E_1, E_2, E_3 on a region in M, let us see what some of the concepts introduced in Chapter 5 look like when expressed in terms of forms. We begin with the analogue of Lemma 3.4 of Chapter 5.

2.2 Lemma (1) $\omega_{13} \wedge \omega_{23} = K\theta_1 \wedge \theta_2$
(2) $\omega_{13} \wedge \theta_2 + \theta_1 \wedge \omega_{23} = 2H\theta_1 \wedge \theta_2.$

Proof. To apply the definitions $K = \det S$, $2H = \operatorname{trace} S$, we shall find the matrix of S with respect to E_1 and E_2. As in Corollary 1.5, the connection equations give

$$S(E_1) = -\nabla_{E_1} E_3 = -\omega_{31}(E_1)E_1 - \omega_{32}(E_1)E_2,$$
$$S(E_2) = -\nabla_{E_2} E_3 = -\omega_{31}(E_2)E_1 - \omega_{32}(E_2)E_2.$$

Thus the matrix of S is

$$\begin{pmatrix} \omega_{13}(E_1) & \omega_{13}(E_1) \\ \omega_{23}(E_2) & \omega_{23}(E_2) \end{pmatrix}.$$

Now, using the second formula in Lemma 2.1, what we must show is that $(\omega_{13} \wedge \omega_{23})(E_1, E_2) = K$ and $(\omega_{13} \wedge \theta_2 + \theta_1 \wedge \omega_{23})(E_1, E_2) = 2H$. But

$$(\omega_{13} \wedge \omega_{23})(E_1, E_2) = \omega_{13}(E_1)\omega_{23}(E_2) - \omega_{13}(E_2)\omega_{23}(E_1)$$
$$= \text{determinant of matrix of } S = \det S = K,$$

and a similar computation gives the trace formula. $\qquad \blacklozenge$

Comparing the first formula above with the Gauss equation (3) in Theorem 1.7, we get

2.3 Corollary $d\omega_{12} = -K\theta_1 \wedge \theta_2.$

We shall call this the *second structural equation*,† and derive from it a new interpretation of Gaussian curvature: ω_{12} measures the rate of rotation of

----

† This equation will be shown to be the analogue for M of the second structural equation for R^3 (Theorem 8.3 of Chapter 2).

tangent frame field E_1, E_2—and since K determined the exterior derivative $d\omega_{12}$, it becomes a kind of "second derivative" of E_1, E_2.

For example, on a sphere of radius r, the formulas in Example 1.6 give

$$\theta_1 \wedge \theta_2 = r^2 \cos \varphi \, d\vartheta d\varphi = -r^2 \cos \varphi \, d\varphi d\vartheta.$$

But

$$d\omega_{12} = d(\sin \varphi \, d\vartheta) = d(\sin \varphi) \wedge d\vartheta = \cos \varphi \, d\varphi d\vartheta.$$

Thus the second structural equation gives the expected result, $K = 1/r^2$.

As we have emphasized, a frame field on a surface is most efficient when it is derived in some natural way from the geometry of that surface, as with the Frenet frame field in the analogous case of a curve. Here is an important example.

2.4 Definition   A *principal frame field* on $M \subset \mathbf{R}^3$ is an adapted frame field E_1, E_2, E_3 such that at each point E_1 and E_2 are principal vectors of M.

So long as its domain of definition contains no umbilics, a principal frame field is uniquely determined—except for changes of sign—by the two principal directions at each point.

Occasionally it may be possible to get a principal frame field on an entire surface. For example, on a surface of revolution, we can take E_1 tangent to meridians, E_2 tangent to parallels. In general, however, about the best we can do is as follows.

2.5 Lemma   If $\mathbf{p}$ is a nonumbilic point of $M \subset \mathbf{R}^3$, then there exists a principal frame field on some neighborhood of $\mathbf{p}$ in M.

Proof.   Let F_1, F_2, F_3 be an arbitrary adapted frame field on a neighborhood $\mathcal{N}$ of $\mathbf{p}$. Since $\mathbf{p}$ is not umbilic, we can assume (by rotating F_1, F_2 if necessary) that $F_1(\mathbf{p})$ and $F_2(\mathbf{p})$ are not principal vectors at $\mathbf{p}$. By hypothesis $k_1(\mathbf{p}) \neq k_2(\mathbf{p})$; hence by continuity, k_1 and k_2 remain distinct near $\mathbf{p}$. On a small enough neighborhood N of $\mathbf{p}$, all these conditions are thus in force.

Let S_{ij} be the matrix of S with respect to F_1, F_2. It is now just a standard problem in linear algebra to compute—*simultaneously at all points of* $\mathcal{N}$—eigenvectors of S, that is, principal vectors of M. In fact, at each point the tangent vector fields.

$$V_1 = S_{12}F_1 + (k_1 - S_{11})F_2,$$
$$V_2 = (k_2 - S_{22})F_1 + S_{12}F_2$$

give eigenvectors of S. (This can be checked by a direct computation if one does not care to appeal to linear algebra.) Furthermore, the function $S_{12} = S(F_1) \cdot F_2$ is never zero on our selected neighborhood $\mathcal{N}$, so $\|V_1\|$ and $\|V_2\|$ are never zero. Thus the vector fields

$$E_1 = \frac{V_1}{\|V_1\|}, \quad E_2 = \frac{V_2}{\|V_2\|}$$

consist only of principal vectors, so E_1, E_2, $E_3 = F_3$ is a principal frame field on $\mathcal{N}$.   ◆

If E_1, E_2, E_3 is a principal frame field on M, then the vector fields E_1 and E_2 consist of eigenvectors of the shape operator derived from E_3. Thus we can label the principal curvature functions so that $S(E_1) = k_1 E_1$ and $S(E_2) = k_2 E_2$. Comparison with Corollary 1.5 then yields

$$\omega_{13}(E_1) = k_1, \qquad \omega_{13}(E_2) = 0,$$
$$\omega_{23}(E_1) = 0, \qquad \omega_{23}(E_2) = k_2.$$

Thus the basis formula (1) in Lemma 2.1 gives

$$\omega_{13} = k_1\theta_1, \quad \omega_{23} = k_2\theta_2. \tag{$*$}$$

This leads to an interesting version of the Codazzi equations.

2.6 Theorem   If E_1, E_2, E_3 is a principal frame field on $M \subset \mathbf{R}^3$, then

$$E_1[k_2] = (k_1 - k_2)\omega_{12}(E_2),$$
$$E_2[k_1] = (k_1 - k_2)\omega_{12}(E_1).$$

Proof.   The Codazzi equations (Theorem 1.7) read

$$d\omega_{13} = \omega_{12} \wedge \omega_{23}, \quad d\omega_{23} = \omega_{21} \wedge \omega_{13}$$

The proof is now an exercise in the calculus of forms as discussed in Chapter 4, Section 4. Substituting from $(*)$ above in the first of these equations, we get

$$d(k_1\theta_1) = \omega_{12} \wedge k_2\theta_2;$$

hence

$$dk_1 \wedge \theta_1 + k_1 d\theta_1 = k_2\omega_{12} \wedge \theta_2.$$

If we substitute the structural equation $d\theta_1 = \omega_{12} \wedge \theta_2$, this becomes

$$dk_1 \wedge \theta_1 = (k_2 - k_1)\omega_{12} \wedge \theta_2.$$

Now apply these 2-forms to the pair of vector fields E_1, E_2 to obtain

$$0 - dk_1(E_2) = (k_2 - k_1)\omega_{12}(E_1) - 0;$$

hence

$$E_2[k_1] = dk_1(E_2) = (k_1 - k_2)\omega_{12}(E_1).$$

The other required equation derives in the same way from the Codazzi equation $d\omega_{23} = \omega_{21} \wedge \omega_{13}$. ◆

Note that for a principal frame field, $\omega_{12}(\mathbf{v})$ tells how the principal directions are changing in the $\mathbf{v}$ direction.

Exercises

1. (*Cylinder.*) Restricting the cylindrical frame field (Example 6.2(1) in Ch. 2) to the cylinder $x^2 + y^2 = r^2$ and reversing indices 1 and 3 gives the adapted frame field in Example 1.3(1). In Chapter 2 we found that the only nonzero connection forms of this frame field were $\omega_{23} = -d\vartheta = -\omega_{32}$ (new indices).
(a) Express the dual forms θ_1, θ_2 of this frame field in terms of cylindrical coordinates r, ϑ, z. (*Hint:* See end of Sec. 8, Ch. 2.)
(b) Verify the fundamental equations (Thm. 1.7) in this case.
(c) Express the shape operator of M in terms of cylindrical coordinates, showing that the frame field in (a) is principal.
(d) Use Lemma 2.2 to find K and H.

2. (a) If E_1, E_2 is a tangent frame field on M with connection form ω_{12}, show that

$$K = E_2[\omega_{12}(E_1)] - E_1[\omega_{12}(E_2)] - \omega_{12}(E_1)^2 - \omega_{12}(E_2)^2.$$

(*Hint:* Write $\omega_{12} = f_1\theta_1 + f_2\theta_2$, where $f_i = \omega_{12}(E_i)$, and use Cor. 2.3.)
(b) Check this formula on the sphere in Example 1.6.

6.3 Some Global Theorems

We have claimed all along that the shape operator S is the analogue for a surface in $\mathbf{R}^3$ of the curvature and torsion of a curve in $\mathbf{R}^3$. Simple hypotheses on κ and τ singled out some important types of curves. Let us see now

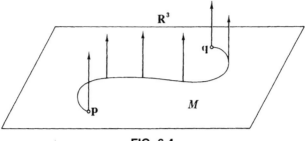

FIG. 6.4

what can be done with S in the case of surfaces. (Throughout this section, surfaces are assumed to be *connected.*)

3.1 Theorem If its shape operator is identically zero, then M is part of a plane in $\mathbf{R}^3$.

Proof. By the definition of shape operator, $S = 0$ means that any unit normal vector field E_3 on M is Euclidean parallel, and hence can be identified with a point of $\mathbf{R}^3$ (see Fig. 6.4).

Choose any point $\mathbf{p}$ in M. We will show that M lies in the plane through $\mathbf{p}$ orthogonal to E_3. If $\mathbf{q}$ is an arbitrary point of M, then since M is connected, there is a curve α in M from $\alpha(0) = \mathbf{p}$ to $\alpha(1) = \mathbf{q}$. Consider the function

$$f(t) = (\alpha(t) - \mathbf{p}) \cdot E_3.$$

But then

$$\frac{df}{dt} = \alpha' \cdot E_3 = 0 \quad \text{and} \quad f(0) = 0;$$

hence f is identically zero. In particular,

$$f(1) = (\mathbf{q} - \mathbf{p}) \cdot E_3 = 0,$$

so every point $\mathbf{q}$ of M is in the required plane (Fig. 6.4). ◆

We saw in Chapter 5, Section 3 that requiring a single point $\mathbf{p}$ of $M \subset \mathbf{R}^3$ to be planar ($k_1 = k_2 = 0$, or equivalently $S = 0$) produces no significant effect on the shape of M near $\mathbf{p}$. But the result above shows that if *every* point is planar, then M is, in fact, part of a plane.

Perhaps the next simplest hypothesis on a surface M in $\mathbf{R}^3$ is that at each point $\mathbf{p}$, the shape operator is merely scalar multiplication by some number—which a priori may depend on $\mathbf{p}$. This means that M is *all-umbilic*, that is, consists entirely of umbilic points.

3.2 Lemma If M is an all-umbilic surface in $\mathbf{R}^3$, then M has constant Gaussian curvature $K \geqq 0$.

Proof. Let E_1, E_2, E_3 be an adapted frame field on some region $\mathcal{O}$ in M. Since M is all-umbilic, the principal curvature functions on $\mathcal{O}$ are equal, $k_1 = k_2 = k$, and furthermore, E_1, E_2, E_3 is actually a principal frame field (since every direction on M is principal). Thus we can apply Theorem 2.6 to conclude that $E_1[k] = E_2[k] = 0$. Alternatively, we may write

$$dk(E_1) = dk(E_2) = 0,$$

so by Lemma 2.1, $dk = 0$ on $\mathcal{O}$. But $K = k_1 k_2 = k^2$, so $dK = 2k\,dk = 0$ on $\mathcal{O}$. Since every point of M is in such a region $\mathcal{O}$, we conclude that $dK = 0$ on all of M. It follows that K is constant (Exercise 4 of Section 4.7). ◆

3.3 Theorem If $M \subset \mathbf{R}^3$ is all-umbilic and $K > 0$, then M is part of a sphere in $\mathbf{R}^3$ of radius $1/\sqrt{K}$.

Proof. Pick at random a point $\mathbf{p}$ in M and a unit normal vector $E_3(\mathbf{p})$ to M at $\mathbf{p}$. *We shall prove that the point*

$$\mathbf{c} = \mathbf{p} + \frac{1}{k(\mathbf{p})} E_3(\mathbf{p})$$

is equidistant from every point of M. (Here $k(\mathbf{p}) = k_1(\mathbf{p}) = k_2(\mathbf{p})$ is the principal curvature corresponding to $E_3(\mathbf{p})$.)

Now let $\mathbf{q}$ be any point of M, and let α be a curve segment in M from $\alpha(0) = \mathbf{p}$ to $\alpha(1) = \mathbf{q}$. Extend $E_3(\mathbf{p})$ to a unit normal vector field E_3 on α, as shown in Fig. 6.5, and consider the curve

$$\gamma = \alpha + \frac{1}{k} E_3 \quad \text{in } \mathbf{R}^3.$$

Here we understand that the principal curvature function k derives from E_3; thus k is continuous. But $K = k^2$, and by the preceding lemma, K is constant, so k is constant. Thus

$$\gamma' = \alpha' + \frac{1}{k} E_3'.$$

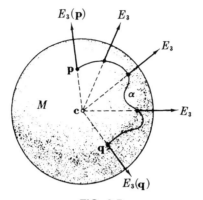

FIG. 6.5

But

$$E_3' = -S(\alpha') = -k\alpha',$$

since by the all-umbilic hypothesis, S is scalar multiplication by k. Thus

$$\gamma' = \alpha' + \frac{1}{k}(-k\alpha') = 0,$$

so the curve γ must be constant. In particular,

$$\mathbf{c} = \gamma(0) = \gamma(1) = \mathbf{q} + \frac{1}{k}E_3(\mathbf{q}),$$

so $d(\mathbf{c}, \mathbf{q}) = 1/|k|$ for every point $\mathbf{q}$ of M. Since $K = k_1 k_2 = k^2$, we have shown that M is contained in the sphere of center $\mathbf{c}$ and radius $1/\sqrt{K}$. ◆

Using all three of the preceding results, we conclude that a surface M in $\mathbf{R}^3$ is all-umbilic *if and only if M is part of a plane or a sphere.*

3.4 Corollary A compact all-umbilic surface M in $\mathbf{R}^3$ is an entire sphere.

Proof. By the preceding remark, we deduce from Exercise 12 of Section 4.7 that M must be an *entire* plane or sphere. The former is impossible, since M is compact, but planes are not. ◆

We now find a useful geometric consequence of a topological assumption about surfaces in $\mathbf{R}^3$.

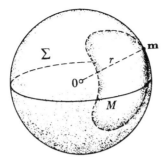

FIG. 6.6

3.5 Theorem   On every compact surface M in $\mathbf{R}^3$ there is a point at which the Gaussian curvature K is strictly positive.

Proof.   Consider the real-valued function f on M such that $f(\mathbf{p}) = \|\,\mathbf{p}\,\|^2$. Thus in terms of the natural coordinates of $\mathbf{R}^3$, $f = \sum x_i^2$. Now, f is differentiable, hence continuous, and M is compact. Thus by Lemma 7.3 of Chapter 4, f takes on its maximum at some point $\mathbf{m}$ of M. Since f measures the square of the distance to the origin, $\mathbf{m}$ is simply a point of M at maximum distance $r = \|\,\mathbf{m}\,\| > 0$ from the origin. Intuitively, it is clear that M is tangent at $\mathbf{p}$ to the sphere Σ of radius r—and that M lies inside Σ, and hence is more curved than Σ (Fig. 6.6). Thus we would expect that $K(\mathbf{m}) \geq 1/r^2 > 0$. Let us *prove* this inequality.

Given any unit tangent vector $\mathbf{u}$ to M at the maximum point $\mathbf{m}$, pick a unit-speed curve α in M such that $\alpha(0) = \mathbf{m}$, $\alpha'(0) = \mathbf{u}$. It follows from the derivation of $\mathbf{m}$ that the composite function $f(\alpha)$ also has a maximum at $t = 0$. Thus

$$\frac{d}{dt}(f\alpha)(0) = 0, \quad \frac{d^2}{dt^2}(f\alpha)(0) \leqq 0. \tag{1}$$

But $f(\alpha) = \alpha \bullet \alpha$, so $\dfrac{d(f\alpha)}{dt} = 2\alpha \bullet \alpha'$. Evaluating at $t = 0$, we find

$$0 = \frac{d(f\alpha)}{dt}(0) = 2\alpha(0) \bullet \alpha'(0) = 2\mathbf{m} \bullet \mathbf{u}.$$

Since $\mathbf{u}$ was any unit tangent vector to M at $\mathbf{m}$, this means that $\mathbf{m}$ (considered as a vector) is normal to M at $\mathbf{m}$.

Differentiating again, we get

$$\frac{d^2(f\alpha)}{dt^2} = 2\alpha' \bullet \alpha' + 2\alpha \bullet \alpha''.$$

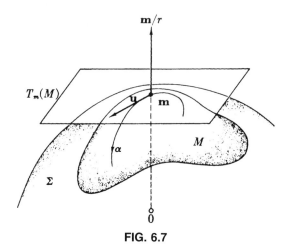

FIG. 6.7

By (1), at $t = 0$ this yields

$$0 \geqq \mathbf{u} \cdot \mathbf{u} + \mathbf{m} \cdot \alpha''(0)$$

$$= 1 + \mathbf{m} \cdot \alpha''(0). \tag{2}$$

The discussion above shows that $\mathbf{m}/r$ may be considered as a unit normal vector to M at $\mathbf{m}$, as shown in Fig. 6.7. Thus $\mathbf{m}/r \cdot \alpha''$ is precisely the normal curvature $k(\mathbf{u})$ of M in the $\mathbf{u}$ direction, and it follows from (2) that $k(\mathbf{u}) \leqq -1/r$. In particular, both principal curvatures satisfy this inequality, so

$$K(\mathbf{m}) \geqq \frac{1}{r^2} > 0. \qquad \blacklozenge$$

Thus there are no compact surfaces in $\mathbf{R}^3$ *with* $K \leqq 0$.

Maintaining the hypothesis of compactness, we consider the effect of requiring that Gaussian curvature be constant. Theorem 3.5 shows that the only possibility is $K > 0$. Spheres are obvious examples of compact surfaces in $\mathbf{R}^3$ with constant positive Gaussian curvature. It is one of the most remarkable facts of surface theory that they are the *only* such surfaces. To prove this we need a nontrivial preliminary result.

3.6 Lemma (Hilbert)   Let $\mathbf{m}$ be a point of $M \subset \mathbf{R}^3$ such that
(1) k_1 has a local maximum at $\mathbf{m}$;
(2) k_2 has a local minimum at $\mathbf{m}$;
(3) $k_1(\mathbf{m}) > k_2(\mathbf{m})$.
Then $K(\mathbf{m}) \leqq 0$.

For example, it is easy to see that these hypotheses hold at any point on the inner equator of a torus or on the minimal circle ($x = 0$) of the catenoid. And K is, in fact, negative in both these examples.

To convert hypotheses (1) and (2) into usable form in the proof that follows, we recall some facts about maxima and minima. If f is a (differentiable) function on a surface M and V is a tangent vector field, then the *first* derivative $V[f]$ is again a function on M. Thus we can apply V again to obtain the *second* derivative $V[V[f]] = VV[f]$. A straightforward computation shows that if f has a local maximum at a point $\mathbf{m}$, then the analogues of the usual conditions in elementary calculus hold, namely,

$$V[f] = 0, \quad VV[f] \leqq 0 \quad \text{at } \mathbf{m}.$$

For a local minimum, of course, the inequality is reversed.

Proof. Since $k_1(\mathbf{m}) > k_2(\mathbf{m})$, $\mathbf{m}$ is not umbilic; hence by Lemma 2.5 there exists a principal frame field E_1, E_2, E_3 on a neighborhood of $\mathbf{m}$ in M. By the remark above, the hypotheses of minimality and maximality at $\mathbf{m}$ imply in particular

$$E_1[k_2] = E_2[k_1] = 0 \quad \text{at } \mathbf{m} \tag{1}$$

and

$$E_1 E_1[k_2] \geqq 0 \quad \text{and} \quad E_2 E_2[k_1] \leqq 0 \quad \text{at } \mathbf{m}. \tag{2}$$

Now we use the Codazzi equations (Theorem 2.6). From (1) it follows that

$$\omega_{12}(E_1) = \omega_{12}(E_2) = 0 \quad \text{at } \mathbf{m}$$

since $k_1 - k_2 \neq 0$ at $\mathbf{m}$. Thus by Exercise 2(a) of Section 2,

$$K = E_2[\omega_{12}(E_1)] - E_1[\omega_{12}(E_2)] \quad \text{at } \mathbf{m}. \tag{3}$$

Applying E_1 to the first Codazzi equation in Theorem 2.6 yields

$$E_1 E_1[k_2] = (E_1[k_1] - E_1[k_2])\omega_{12}(E_2) + (k_1 - k_2)E_1[\omega_{12}(E_2)].$$

But at the special point $\mathbf{m}$, we have $\omega_{12}(E_2) = 0$ and $k_1 - k_2 > 0$; hence from (2) we deduce

$$E_1[\omega_{12}(E_2)] \geqq 0 \quad \text{at } \mathbf{m}. \tag{4}$$

A similar argument starting from the second Codazzi equation gives

$$E_2[\omega_{12}(E_1)] \leqq 0 \quad \text{at } \mathbf{m}. \tag{5}$$

Using (4) and (5) in the expression (3) for the Gaussian curvature at $\mathbf{m}$, we conclude that $K(\mathbf{m}) \leqq 0$.   ◆

3.7 Theorem (Liebmann)   If M is a compact surface in $\mathbf{R}^3$ with constant Gaussian curvature K, then M is a sphere of radius $1/\sqrt{K}$. (Theorem 3.5 implies K is positive.)

Proof.   Since $M \subset \mathbf{R}^3$ is compact, Theorem 7.10 of Chapter 4 shows it is orientable, so a unit normal exists with shape operators S that are smoothly defined on the entire surface. Thus continuous principal curvature functions $k_1 \geqq k_2 \geqq 0$ are also globally well-defined. By Lemma 7.3 of Chapter 4, k_1 has a maximum at some point $\mathbf{p}$ of M. Since $K = k_1 k_2$ is constant, k_2 has a minimum there. Now it cannot be true that $k_1(\mathbf{p}) > k_2(\mathbf{p})$, for then Hilbert's lemma would give $K \leqq 0$. Thus $k_1(\mathbf{p}) = k_2(\mathbf{p})$. By the choice of $\mathbf{p}$, it follows that M is all umbilic; hence it is a standard sphere of radius $r = 1/\sqrt{K}$.   ◆

Liebmann's theorem is false if the compactness hypothesis is omitted, for we saw in Chapter 5, Section 7 that there are many nonspherical surfaces with constant positive curvature.

Exercises

1.   If M is a flat minimal surface, prove that M is part of a plane.

2.   *Flat surfaces in* $\mathbf{R}^3$ *can be bent only along straight lines* (see Fig. 6.9). If $k_1 = 0$ but k_2 is never zero, show that the principal curves of k_1 are line segments in $\mathbf{R}^3$. (*Hint:* With $\{E_i\}$ principal and $\alpha'' = \nabla_{E_1} E_1$, use Thm. 1.4.)

3.   Let $M \subset \mathbf{R}^3$ be a compact surface with $K > 0$. If M has constant mean curvature H, show that M is a sphere of radius $1/|H|$.

4.   Prove that in a region free of umbilics there are two principal curves through each point, these crossing orthogonally. (*Hint:* Use Ex. 14 of Sec. 4.8.)

5.   If the principal curvatures of a surface $M \subset \mathbf{R}^3$ are constant, show that M is part of either a plane, a sphere, or a circular cylinder. (In the case $k_1 \neq k_2$, assume there is a principal frame field on all of M.)

6.4 Isometries and Local Isometries

We remarked earlier that the inhabitants of a surface M in $\mathbf{R}^3$, unaware of the space outside their surface, could nevertheless determine the distance *in M* between any two points of M—just as distance on the surface of the earth can be determined by its inhabitants. The mathematical formulation is as follows.

4.1 Definition If **p** and **q** are points of $M \subset \mathbf{R}^3$, consider the collection of all curve segments α *in M* from **p** to **q**. The *intrinsic distance* $\rho(\mathbf{p}, \mathbf{q})$ from **p** to **q** in M is the greatest lower bound of the lengths $L(\alpha)$ of these curve segments.

There need not be a curve α whose length is exactly $\rho(\mathbf{p}, \mathbf{q})$ (see Exercise 3). The intrinsic distance $\rho(\mathbf{p}, \mathbf{q})$ will generally be greater than the straightline Euclidean distance $d(\mathbf{p}, \mathbf{q})$, since the curves α are obliged to stay in M (Fig. 6.8).

On the surface of the earth (sphere of radius about 4000 miles) it is, of course, intrinsic distance that is of practical interest. One says, for example, that it is 12,500 miles from the north pole to the south pole, though the Euclidean distance through the center of the earth is only 8000 miles.

We saw in Chapter 3 how Euclidean geometry is based on the notion of isometry, a distance-preserving mapping. For surfaces in M we shall *prove* the distance-preserving property and use its infinitesimal form (Corollary 2.2 of Chapter 3) as the definition.

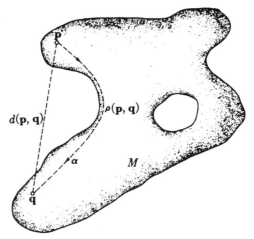

FIG. 6.8

4.2 Definition   An *isometry* $F: M \rightarrow \overline{M}$ of surfaces in $\mathbf{R}^3$ is a one-to-one mapping of M onto $\overline{M}$ that preserves dot products of tangent vectors. Explicitly, if F_* is the derivative map of F, then

$$F_*(\mathbf{v}) \bullet F_*(\mathbf{w}) = \mathbf{v} \bullet \mathbf{w}$$

for any pair of tangent vectors $\mathbf{v}$, $\mathbf{w}$ to M.

If F_* preserves dot products, then it also preserves lengths of tangent vectors. It follows that an isometry is a regular mapping (Chapter 4, Section 5), for if $F_*(\mathbf{v}) = 0$, then

$$\|\mathbf{v}\| = \|F_*(\mathbf{v})\| = 0;$$

hence $\mathbf{v} = 0$. Thus by the remarks following Theorem 5.4 of Chapter 4, an isometry $F: M \rightarrow \overline{M}$ is in particular a diffeomorphism, that is, has an inverse mapping $F^{-1}: \overline{M} \rightarrow M$. Furthermore, F^{-1} is also an isometry.

4.3 Theorem   Isometries preserve intrinsic distance: if $F: M \rightarrow \overline{M}$ is an isometry of surfaces in $\mathbf{R}^3$, then

$$\rho(\mathbf{p}, \mathbf{q}) = \overline{\rho}(F(\mathbf{p}), F(\mathbf{q}))$$

for any two points $\mathbf{p}$, $\mathbf{q}$ in M.

(Here ρ and $\overline{\rho}$ are the intrinsic distance functions of M and $\overline{M}$ respectively.)

Proof.   First note that isometries preserve the speed and length of curves. The proof is just like the Euclidean case: If α is a curve segment in M, then $\overline{\alpha} = F(\alpha)$ is a curve segment in $\overline{M}$ with velocity $\overline{\alpha}' = F_*(\alpha')$. Since F_* preserves dot products, it preserves norms, so $\| \alpha' \| = \| F_*(\alpha') \| = \| F(\alpha)' \| = \| \overline{\alpha}' \|$. Hence

$$L(\alpha) = \int_a^b \|\alpha'(t)\|\, dt = \int_a^b \|\overline{\alpha}'(t)\|\, dt = L(\overline{\alpha}).$$

Now, if α runs from $\mathbf{p}$ to $\mathbf{q}$ in M, its image $\overline{\alpha} = F(\alpha)$ runs from $F(\mathbf{p})$ to $F(\mathbf{q})$ in $\overline{M}$. Reciprocally, if β is a curve segment in $\overline{M}$ from $F(\mathbf{p})$ to $F(\mathbf{q})$ in $\overline{M}$, then $F^{-1}(\beta)$ runs from $\mathbf{p}$ to $\mathbf{q}$ in M. We have, in fact, established a one-to-one correspondence between the collection of curve segments used to define $\rho(\mathbf{p},\mathbf{q})$ and those used for $\overline{\rho}(F(\mathbf{p}),F(\mathbf{q}))$. But as was shown above, corresponding curves have the same length, so it follows at once that $\rho(\mathbf{p}, \mathbf{q}) = \overline{\rho}(F(\mathbf{p}), F(\mathbf{q}))$. ◆

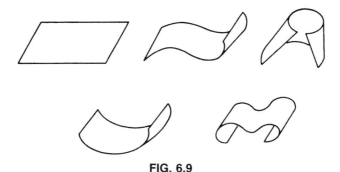

FIG. 6.9

Thus we may think of an isometry as bending a surface into a different shape without changing the intrinsic distance between any of its points. *Consequently, the inhabitants of the surface are not aware of any change at all, for their geometric measurements all remain exactly the same.*

If there exists an isometry from M to $\overline{M}$, then these two surfaces are said to be *isometric*. For example, if a piece of paper is bent into various shapes without creasing or stretching, the resulting surfaces are all isometric (Fig. 6.9).

To study isometries it is convenient to separate the geometric condition of preservation of dot products from the one-to-one and onto requirements.

4.4 Definition   A *local isometry* $F: M \to N$ of surfaces is a mapping that preserves dot products of tangent vectors (that is, F_* does).

Thus an isometry is a local isometry that is both one-to-one and onto.

If F is a local isometry, the earlier argument still shows that F is a regular mapping. Then for each point $\mathbf{p}$ of M the inverse function theorem (5.4 of Chapter 4) asserts that there is a neighborhood $\mathcal{U}$ of $\mathbf{p}$ in M that F carries diffeomorphically onto a neighborhood $\mathcal{V}$ of $F(\mathbf{p})$ in N. Now, $\mathcal{U}$ and $\mathcal{V}$ are themselves surfaces in $\mathbf{R}^3$, and thus the mapping $F|_{\mathcal{U}}: \mathcal{U} \to \mathcal{V}$ is an isometry. In this sense a local isometry is, indeed, *locally an isometry*.

There is a simple patch criterion for local isometries using the functions E, F, and G defined in Section 4 of Chapter 5.

4.5 Lemma   Let $F: M \to N$ be a mapping. For each patch $\mathbf{x}: D \to M$, consider the composite mapping

$$\overline{\mathbf{x}} = F(\mathbf{x}): D \to N.$$

Then F is a local isometry if and only if for each patch $\mathbf{x}$ we have

$$E = \overline{E}, \quad F = \overline{F}, \quad G = \overline{G}.$$

(Here $\overline{\mathbf{x}}$ need not be a patch, but $\overline{E}$, $\overline{F}$ and $\overline{G}$ are defined for it as usual.)

Proof. Suppose the criterion holds—and only for enough patches to cover all of M. Then by one of the equivalences in Exercise 1, to show that F_* preserves dot products we need only prove that

$$\|\mathbf{x}_u\| = \|F_*(\mathbf{x}_u)\|, \quad \mathbf{x}_u \cdot \mathbf{x}_v = F_*(\mathbf{x}_u) \cdot F_*(\mathbf{x}_v), \quad \|\mathbf{x}_v\| = \|F_*(\mathbf{x}_v)\|.$$

But as we saw in Chapter 4, it follows immediately from the definition of F_* that $F_*(\mathbf{x}_u) = \overline{\mathbf{x}}_u$ and $F_*(\mathbf{x}_v) = \overline{\mathbf{x}}_v$. Thus the equations above follow from the hypotheses $E = \overline{E}$, $F = \overline{F}$, $G = \overline{G}$. Hence F is a local isometry. Reversing the argument, we deduce the converse assertion. ◆

This result can sometimes be used to *construct* local isometries. In the simplest case, suppose that M is the image of a single patch $\mathbf{x}: D \to M$.

Then if $\mathbf{y}: D \to N$ is a patch in another surface, a mapping $F: M \to N$ is defined by

$$F(\mathbf{x}(u,v)) = \mathbf{y}(u,v) \quad \text{for } (u,v) \text{ in } D.$$

If $E = \overline{E}, F = \overline{F}, G = \overline{G}$, then by the above criterion, F is a local isometry.

4.6 Example (1) *Local isometry of a plane onto a cylinder.* The plane $\mathbf{R}^2$ may be considered as a surface, with natural frame field U_1, U_2. If

$$\mathbf{x}: \mathbf{R}^2 \to M$$

is a parametrization of some surface, then Exercise 1 of this section shows that $\mathbf{x}$ is a local isometry if

$$\mathbf{x}_*(U_i) \cdot \mathbf{x}_*(U_j) = U_i \cdot U_j \quad \text{for } 1 \le i, j \le 2.$$

Since $\mathbf{x}_*(U_1) = \mathbf{x}_u$, $\mathbf{x}_*(U_2) = \mathbf{x}_v$, and $U_i \cdot U_j = \delta_{ij}$, this is the same as requiring $E = 1, F = 0, G = 1$.

To take a concrete case, the parametrization

$$\mathbf{x}(u, v) = \left(r \cos \frac{u}{r}, \, r \sin \frac{u}{r}, \, v \right)$$

of the cylinder $M: x^2 + y^2 = r^2$ has $E = 1, F = 0, G = 1$. Thus $\mathbf{x}$ is a local isometry that wraps the plane $\mathbf{R}^2$ around the cylinder, with horizontal lines going to cross-sectional circles and vertical lines to rulings of the cylinder.

(2) *Local isometry of a helicoid onto a catenoid.* Let H be the helicoid that is the image of the patch

$$\mathbf{x}(u,v) = (u\cos v, \ u\sin v, v).$$

Furnish the catenoid C with the canonical parametrization $\mathbf{y} \colon \mathbf{R}^2 \to C$ discussed in Example 7.4 of Chapter 5. Thus

$$\mathbf{y}(u,v) = (g(u), h(u)\cos v, h(u)\sin v),$$

$$g(u) = \sinh^{-1}u, \quad h(u) = \sqrt{1+u^2}.$$

Let $F \colon H \to C$ be the mapping such that

$$F(\mathbf{x}(u,v)) = \mathbf{y}(u,v).$$

To prove that F is a local isometry, it suffices to check that

$$E = 1 = \overline{E}, \quad F = 0 = \overline{F}, \quad G = 1 + u^2 = h^2 = \overline{G}.$$

F carries the *rulings* (v constant) of H onto *meridians* of the surface of revolution C, and wraps the *helices* (u constant) of H around the *parallels* of C. In particular, the central axis of H (z axis) is wrapped around the minimal circle $x = 0$ of C.

Figure 6.10 shows how a sample strip of H is carried over to C.

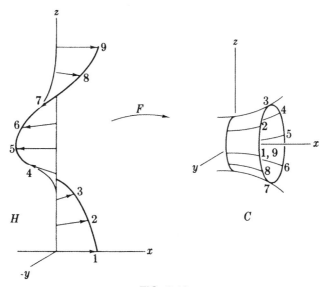

FIG. 6.10

Suppose that the helicoid H (or at least a finite region of it) has been stamped like an automobile fender out of a flexible sheet of steel—the patch **x** does this. Then H may be wrapped into the shape of a catenoid with no further distortion of the metal (Exercise 5 of Section 5).

A similar experiment may be performed by cutting a hole in a ping-pong ball representing a sphere in $\mathbf{R}^3$. Mild pressure will then deform the ball into various nonround shapes, all of which are isometric. For arbitrary isometric surfaces M and $\overline{M}$ in $\mathbf{R}^3$, however, it is generally not possible to bend M (through a whole family of isometric surfaces) so as to produce $\overline{M}$.

There are special types of mappings other than (local) isometries that are of interest in geometry.

4.7 Definition   A mapping of surfaces $F: M \to N$ is *conformal* provided there exists a real-valued function $\lambda > 0$ on M such that

$$\|F_*(\mathbf{v}_p)\| = \lambda(\mathbf{p})\|\mathbf{v}_p\|$$

for all tangent vectors to M. The function λ is called the *scale factor* of F.

Note that if F is a conformal mapping for which λ has constant value 1, F is a local isometry. Thus a conformal mapping is a generalized isometry for which lengths of tangent vectors need not be preserved—but at each point **p** of M the tangent vectors at **p** all have their lengths stretched by the same factor.

The criteria in Lemma 4.5 and in Exercise 1 below may easily be adapted from isometries to conformal mappings by introducing the scale factor (or its square). In Lemma 4.5, for example, replace $E = \overline{E}$ by $E \lambda^2(\mathbf{x}) = \overline{E}$, and similarly for the other two equations.

An essential property of conformal mappings is discussed in Exercise 8.

Exercises

1.   If $F: M \to N$ is a mapping, show that the following conditions on its tangent map at one point **p** are logically equivalent:
  (a) F_* preserves inner products.
  (b) F_* preserves lengths of tangent vectors, that is, $\| F_*(\mathbf{v}) \| = \| \mathbf{v} \|$ for all **v** at **p**.
  (c) F_* preserves frames: If $\mathbf{e}_1$, $\mathbf{e}_2$ is a tangent frame at **p**, then

$$F_*(\mathbf{e}_1), F_*(\mathbf{e}_2)$$

  is a tangent frame at $F(\mathbf{p})$.

(d) For some one pair of linearly independent tangent vectors $\mathbf{v}$ and $\mathbf{w}$ at $\mathbf{p}$,

$$\|F_*(\mathbf{v})\| = \|\mathbf{v}\|, \quad \|F_*(\mathbf{w})\| = \|\mathbf{w}\| \quad \text{and} \quad F_*(\mathbf{v}) \cdot F_*(\mathbf{w}) = \mathbf{v} \cdot \mathbf{w}.$$

[*Hint:* It suffices, for example, to prove $(a) \Rightarrow (c) \Rightarrow (d) \Rightarrow (b) \Rightarrow (a)$.]
These are general facts from linear algebra; in this context they provide useful criteria for F to be a local isometry.

2. Show that each of the following conditions is necessary and sufficient for $F: M \to N$ to be a local isometry.
(a) F preserves the speeds of curves: $\| F(\alpha)' \| = \| \alpha' \|$ for all curves α in M.
(b) F preserves lengths of curves: $L(F(\alpha)) = L(\alpha)$ for all curve segments α in M.

3. Prove that intrinsic distance ρ for a surface M has the same general properties as Euclidean distance d (Ex. 2 of Sec. 2.1), namely,
(a) Positive definiteness: $\rho(\mathbf{p}, \mathbf{q}) \geq 0$; $\rho(\mathbf{p}, \mathbf{q}) = 0$ if and only if $\mathbf{p} = \mathbf{q}$,
(b) Symmetry: $\rho(\mathbf{p}, \mathbf{q}) = \rho(\mathbf{q}, \mathbf{p})$,
(c) Triangle inequality: $\rho(\mathbf{p}, \mathbf{r}) \leq \rho(\mathbf{p}, \mathbf{q}) + \rho(\mathbf{q}, \mathbf{r})$, for all points of M.
The only difficult case is $\rho(\mathbf{p}, \mathbf{q}) = 0 \Rightarrow \mathbf{p} = \mathbf{q}$, which we postpone to Exercise 1 of Section 8.1. (*Hint for* (c): Piecewise differentiable curves are allowed in the definition of ρ.)

4. Give an example and a proof to show: Local isometries can shrink but not increase intrinsic distance.

5. Let $\alpha, \beta: I \to \mathbf{R}^3$ be unit-speed curves with the same curvature function $\kappa > 0$. For simplicity, assume that the parametrization

$$\mathbf{x}(u, v) = \alpha(u) + vT(u)$$

of the $v > 0$ tangent surface of α is actually a patch. Find a local isometry from this surface to:
(a) The $v > 0$ tangent surface of β.
(b) A region D in the plane.

6. Show that the preceding exercise applies to the $v > 0$ tangent surface of a helix, and find the image region D in the plane.

7. Let M be the Euclidean plane $\mathbf{R}^2$ with the origin removed. Show that the intrinsic distance from $(-1, 0)$ to $(1, 0)$ is 2, but that every curve joining these points has length strictly greater than 2. (*Hint:* Ex. 11 of Sec. 2.2.)

8. (a) Modify the conditions in Exercise 1 so that they provide criteria for F to be a conformal mapping.

(b) Show that a parametrization $\mathbf{x}: D \to M$ is a conformal mapping if and only if $E = G$ and $F = 0$.

(c) Prove that a conformal mapping *preserves angles* in this sense: If ϑ is an angle between $\mathbf{v}$ and $\mathbf{w}$ at $\mathbf{p}$, then ϑ is also an angle between $F_*(\mathbf{v})$ and $F_*(\mathbf{w})$ at $F(\mathbf{p})$.

9. If $F: M \to \overline{M}$ is an isometry, prove that the inverse mapping $F^{-1}: \overline{M} \to M$ is also an isometry. If $F: M \to N$ and $G: N \to P$ are (local) isometries, prove that the composite mapping $GF: M \to P$ is a (local) isometry.

10. Let $\mathbf{x}$ be a parametrization of all of M, $\overline{\mathbf{x}}$ a parametrization in N. If $F: M \to N$ is a mapping such that $F(\mathbf{x}(u, v)) = \overline{\mathbf{x}}(f(u), g(v))$, then

(a) Describe the effect of F on the parameter curves of $\mathbf{x}$.

(b) Show that F is a local isometry if and only if

$$E = \overline{E}(f, g)\left(\frac{df}{du}\right)^2, \quad F = \overline{F}(f, g)\frac{df}{du}\frac{dg}{dv}, \quad G = \overline{G}(f, g)\left(\frac{dg}{dv}\right)^2.$$

(In the general case, f and g are functions of both u and v, and this criterion becomes more complicated.)

(c) Find analogous conditions for F to be a conformal mapping.

11. Let M be a surface of revolution, and let $F: H \to M$ be a local isometry of the helicoid that (as in Example 4.6) carries rulings to meridians and helices to parallels. Show that M must be a catenoid. (*Hint:* Use Ex. 10.)

12. Let M be the image of a patch $\mathbf{x}$ with $E = 1$, $F = 0$, and G a function of u only ($G_v = 0$). If the derivative $d(\sqrt{G})/du$ is bounded, show that there is a local isometry of M into a surface of revolution.

Thus any small enough region in M is isometric to a region in a surface of revolution.

13. Let $\mathbf{x}$ be the geographical patch in the sphere Σ of radius r (Example 2.2 of Ch. 4). Stretch $\mathbf{x}$ in the north-south direction to produce a conformal mapping. Explicitly, let

$$\mathbf{y}(u, v) = \mathbf{x}(u, g(v)) \quad \text{with } g(0) = 0,$$

and determine g such that $\mathbf{y}$ is conformal. Find the scale factor of $\mathbf{y}$ and the domain D such that $\mathbf{y}(D)$ omits only a semicircle of Σ. (Mercator's map of the earth derives from $\mathbf{y}$: its inverse is *Mercator's projection.*)

14. Show that stereographic projection $P: \Sigma_0 \to \mathbf{R}^2$ (Example 5.2 of Ch. 4) is conformal, with scale factor

$$\lambda(\mathbf{p}) = 1 + \frac{\|P(\mathbf{p})\|^2}{4}.$$

15. Let M be a surface of revolution whose profile curve is not closed, and hence has a one-to-one parametrization. Find a conformal mapping $F: M \to \mathbf{R}^2$ such that meridians go to lines through the origin and parallels go to circles centered at the origin.

6.5 Intrinsic Geometry of Surfaces in R³

In Chapter 3 we defined Euclidean geometry to consist of those concepts preserved by Euclidean isometries. The same definition applies to surfaces: The *intrinsic geometry* of $M \subset \mathbf{R}^3$ consists of those concepts—called *isometric invariants*—that are preserved by all isometries $F: M \to \overline{M}$. For example, Theorem 4.3 shows that intrinsic distance is an isometric invariant. We can now state Gauss's question (mentioned early in the chapter) more precisely: *Which of the properties of a surface M in $\mathbf{R}^3$ belong to its intrinsic geometry?* The definition of isometry (Definition 4.2) suggests that isometric invariants must depend only on the dot product as applied to *tangent* vectors to M. But the shape operator derives from a *normal* vector field, and the examples in Section 4 show that isometric surfaces in $\mathbf{R}^3$ can have quite different shapes. In fact, these examples provide a formal proof that shape operators, principal directions, principal curvatures, and mean curvature definitely do not belong to the intrinsic geometry of $M \subset \mathbf{R}^3$.

To build a systematic theory of intrinsic geometry, we must look back at Section 1 and see how much of our work there is intrinsic to M. Using the dot product only on tangent vectors to M, we can still define a tangent frame field E_1, E_2 on M. Thus from an adapted frame field we can salvage the two tangent vector fields E_1, E_2—and hence also their dual 1-forms θ_1, θ_2. It is somewhat surprising to find that these completely determine the connection form ω_{12}.

5.1 Lemma The connection form $\omega_{12} = -\omega_{21}$ is the only 1-form that satisfies the first structural equations

$$d\theta_1 = \omega_{12} \wedge \theta_2, \quad d\theta_2 = \omega_{21} \wedge \theta_1.$$

Proof. Apply these equations to the tangent vector fields E_1, E_2. Since $\theta_i(E_j) = \delta_{ij}$, the definition of wedge product gives

$$\omega_{12}(E_1) = d\theta_1(E_1, E_2),$$

$$\omega_{12}(E_2) = -\omega_{21}(E_2) = d\theta_2(E_1, E_2).$$

Thus by Lemma 2.1, $\omega_{12} = -\omega_{21}$ is uniquely determined by θ_1, θ_2. ◆

5.2 Remark In fact, this proof shows how to construct $\omega_{12} = -\omega_{21}$ without the use of Euclidean covariant derivatives (as in Section 1). Given E_1, E_2 and thus θ_1, θ_2, take the equations in the above proof as the *definition* of ω_{12} on E_1 and E_2. Then the usual linearity condition

$$\omega_{12}(V) = \omega_{12}(v_1 E_1 + v_2 E_2) = v_1 \omega_{12}(E_1) + v_2 \omega_{12}(E_2)$$

makes ω_{12} a 1-form on M, and one can easily check (by reversing the argument above) that $\omega_{12} = -\omega_{21}$ satisfies the first structural equations.

If $F: M \to \overline{M}$ is an isometry, we can transfer a tangent frame field E_1, E_2 on M to a tangent frame field $\overline{E}_1, \overline{E}_2$ on $\overline{M}$: For each point $\mathbf{q}$ in $\overline{M}$ there is a unique point $\mathbf{p}$ in M such that $F(\mathbf{p}) = \mathbf{q}$. Then define

$$\overline{E}_1(\mathbf{q}) = F_*(E_1(\mathbf{p})),$$

$$\overline{E}_2(\mathbf{q}) = F_*(E_2(\mathbf{p}))$$

(Fig. 6.11).

In practice, we often abbreviate these formulas somewhat casually to

$$\overline{E}_1 = F_*(E_1), \quad \overline{E}_2 = F_*(E_2).$$

Because F_* preserves dot products, $\overline{E}_1, \overline{E}_2$ is a frame field on $\overline{M}$, since

$$\overline{E}_i \cdot \overline{E}_j = F_*(E_i) \cdot F_*(E_j) = E_i \cdot E_j = \delta_{ij}.$$

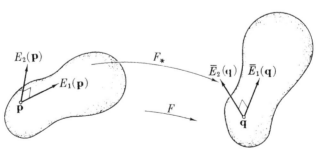

FIG. 6.11

5.3 Lemma Let $F: M \to \overline{M}$ be an isometry, and let E_1, E_2 be a tangent frame field on M. If $\overline{E}_1$, $\overline{E}_2$ is the transferred frame field on $\overline{M}$, then

(1) $\theta_1 = F^*(\overline{\theta}_1)$, $\theta_2 = F^*(\overline{\theta}_2)$;

(2) $\omega_{12} = F^*(\overline{\omega}_{12})$.

Proof. (1) It suffices to prove that θ_i and $F^*(\overline{\theta}_i)$ have the same value on E_1 and E_2. But for $1 \leqq i, j \leqq 2$ we have

$$F^*(\overline{\theta}_i)(E_j) = \overline{\theta}_i(F_*E_j) = \overline{\theta}_i(\overline{E}_j) = \delta_{ij} = \theta_i(E_j).$$

(2) Consider the structural equation $d\overline{\theta}_1 = \overline{\omega}_{12} \wedge \overline{\theta}_2$ on $\overline{M}$. If we apply F^*, then by the results in Chapter 4, Section 5,

$$d(F^*\overline{\theta}_1) = F^*(d\overline{\theta}_1) = F^*(\overline{\omega}_{12}) \wedge F^*(\overline{\theta}_2).$$

Hence, by (1), we have

$$d\theta_1 = F^*(\overline{\omega}_{12}) \wedge \theta_2.$$

The other structural equation,

$$d\overline{\theta}_2 = \overline{\omega}_{21} \wedge \overline{\theta}_1,$$

gives a corresponding equation, so

$$d\theta_1 = F^*(\overline{\omega}_{12}) \wedge \theta_2,$$
$$d\theta_2 = F^*(\overline{\omega}_{21}) \wedge \theta_1.$$

But now (2) is an immediate consequence of the uniqueness property (Lemma 5.1), since

$$F^*(\overline{\omega}_{21}) = F^*(-\overline{\omega}_{12}) = -F^*(\overline{\omega}_{12}). \qquad \blacklozenge$$

From this rather routine lemma we easily derive a proof of the celebrated *theorema egregium* of Gauss.

5.4 Theorem Gaussian curvature is an isometric invariant. Explicitly, if $F: M \to \overline{M}$ is an isometry, then

$$K(\mathbf{p}) = \overline{K}(F(\mathbf{p}))$$

for every point $\mathbf{p}$ in M.

Proof. For an arbitrary point $\mathbf{p}$ of M, pick a tangent frame field E_1, E_2 on some neighborhood of $\mathbf{p}$ and transfer via F_* to $\overline{E}_1$, $\overline{E}_2$ on $\overline{M}$. By the previous lemma, $F_*(\overline{\omega}_{12}) = \omega_{12}$. According to Corollary 2.3, we have

$$d\overline{\omega}_{12} = -\overline{K}\overline{\theta}_1 \wedge \overline{\theta}_2.$$

Apply F_* to this equation. By the results in Chapter 4, Section 5, we get

$$d(F_*\overline{\omega}_{12}) = F_*(d\overline{\omega}_{12}) = -F_*(\overline{K})F_*(\overline{\theta}_1) \wedge F_*(\overline{\theta}_2),$$

where $F^*(\overline{K})$ is simply the composite function $\overline{K}(F)$. Thus by the preceding lemma,

$$d\omega_{12} = -\overline{K}(F)\theta_1 \wedge \theta_2.$$

Comparison with $d\omega_{12} = -K\theta_1 \wedge \theta_2$ yields $K = \overline{K}(F)$; hence, in particular, $K(\mathbf{p}) = \overline{K}(F(\mathbf{p}))$. ◆

Gauss's theorem is one of the great discoveries of nineteenth-century mathematics, and we shall see in the next chapter that its implications are far-reaching. The essential step in the proof is the second structural equation

$$d\omega_{12} = -K\theta_1 \wedge \theta_2.$$

Once we prove Lemma 5.1, all the ingredients of this equation, except K, are known to derive from M alone—thus K must also. This means that the inhabitants of $M \subset \mathbf{R}^3$ can determine the Gaussian curvature of their surface even though they cannot generally find S and have no conception of the shape of M in $\mathbf{R}^3$.

This remarkable situation is perhaps best illustrated by the formula $K = k_1k_2$: An isometry need not preserve the principal curvatures, nor their sum, but *it must preserve their product.* Thus the shapes that isometric surfaces may have—although possibly quite different—are by no means unrelated.

A local isometry is, as we have shown, an isometry on all sufficiently small neighborhoods. Thus it follows from Theorem 5.4 that *local isometries preserve Gaussian curvature.* For example, in Example 4.6 the plane and the cylinder both have $K = 0$. (This is why we did not hesitate to call the curved cylinder "flat." Intrinsically it is as flat as a plane.) In the second part of Example 4.6, at corresponding points

$$\mathbf{x}(u, v) \quad \text{and} \quad F(\mathbf{x}(u, v)) = \mathbf{y}(u, v),$$

both the helicoid and catenoid have Gaussian curvature: $-1/(1+u^2)^2$ (see Examples 4.3 and 7.4 of Chapter 5).

Gauss's *theorema egregium* can obviously be used to show that given surfaces are *not* isometric. For example, there can be no isometry of the sphere

Σ (or even a very small region of it) onto part of the plane, since their Gaussian curvatures are different. This is the dilemma of the mapmaker: The intrinsic geometry of the earth's surface is misrepresented by any flat map.

The next section is computational; thereafter we will find more isometric invariants.

Exercises

1. *Geodesics belong to intrinsic geometry:* In fact, if α is a geodesic in M and $F: M \to N$ is a (local) isometry, then $F(\alpha)$ is a geodesic of N. (*Hint:* Ex. 1 of Sec. 1)

2. Use Exercise 1 to derive the geodesics of the circular cylinder. Generalize to an arbitrary cylinder.

3. For a connected surface, the values of its Gaussian curvature fill an interval. If there exists a local isometry of M onto N (in particular, if M and N are isometric), show that M and N have the same curvature interval. Give an example to show that the converse is false.

4. Prove that no two of the following surfaces are isometric: sphere, torus, helicoid, cylinder, saddle surface.

5. *Bending of the helicoid into the catenoid* (4.6). For each number t in the interval $0 \leqq t \leqq \pi/2$, let $\mathbf{x}_t: \mathbf{R}^2 \to \mathbf{R}^3$ be the mapping such that

$$\mathbf{x}_t(u, v) = \cos t\,(Sc, Ss, v) + \sin t\,(-Cs, Cc, u),$$

where $C = \cosh u$, $S = \sinh u$, $c = \cos v$, and $s = \sin v$.

Now $\mathbf{x}_0$ is a patch covering the helicoid, and $\mathbf{x}_{\pi/2}$ is a parametrization of the catenoid—these are mild variants of our usual parametrizations, and the catenoid now has the z axis as its axis of rotation. If we imagine t to be the time, then $\mathbf{x}_t$ for $0 \leqq t \leqq \pi/2$ describes a *bending* of the helicoid M_0 that carries it onto the catenoid $M_{\pi/2}$ through a whole family of intermediate surfaces $M_t = \mathbf{x}_t(\mathbf{R}^2)$. Prove:

(a) M_t *is a surface.* (Show merely that $\mathbf{x}_t$ is regular.)

(b) M_t *is isometric to the helicoid* M_0 *if* $t < \pi/2$. (Show that $F_t: M_0 \to M_t$ is an isometry, where

$$F_t(\mathbf{x}_0(u, v)) = \mathbf{x}_t(u, v).$$

Also show that for $t < \pi/2$, $F_{\pi/2}$ is local isometry.)

(c) *Each M_t is a minimal surface.* (Compute $\mathbf{x}_{uu} + \mathbf{x}_{vv} = 0$.)

(d) *Unit normals are parallel on orbits.* Along the curve $t \to \mathbf{x}_t(u, v)$ by which the point $\mathbf{x}_0(u, v)$ of M_0 moves to $M_{\pi/2}$, the unit normals U_t of successive surfaces are *parallel*.

(e) *Gaussian curvature is constant on orbits.* Find $K_t(\mathbf{x}_t(u, v))$, where K_t is the Gaussian curvature of M_t.

A brilliant series of illustrations of this bending is given in Struik [S]. They are not diminished by the following computer tour de force.

6. (*Computer continuation.*) (a) Plot the surface M_t for at least six values of t from $t = 0$ (helicoid) to $t < \pi/2$ (catenoid).

(b) Animate the series of plots in (a).

7. Show that *every* local isometry of the helicoid H to the catenoid C must carry the axis of H to the central circle of C, and the rulings of H to the meridians of C, as in Example 4.6.

6.6 Orthogonal Coordinates

We have seen that the intrinsic geometry of a surface $M \subset \mathbf{R}^3$ may be expressed in terms of the dual forms θ_1, θ_2 and connection form ω_{12} derived from a tangent frame field E_1, E_2. These forms satisfy—

the first structural equations:

$$d\theta_1 = \omega_{12} \wedge \theta_2,$$

$$d\theta_2 = \omega_{21} \wedge \theta_1;$$

the second structural equation:

$$d\omega_{12} = -K\theta_1 \wedge \theta_2.$$

In this section we develop a practical way to compute these forms—and hence a new way to find the Gaussian curvature of M.

The starting point is an *orthogonal coordinate* patch $\mathbf{x}: D \to M$, one for which $F = \mathbf{x}_u \cdot \mathbf{x}_v = 0$. Since $\mathbf{x}_u$ and $\mathbf{x}_v$ are orthogonal, dividing by their lengths $\|\mathbf{x}_u\| = \sqrt{E}$ and $\|\mathbf{x}_v\| = \sqrt{G}$ will produce frames.

6.1 Definition   The *associated frame field* E_1, E_2 of an orthogonal patch $\mathbf{x}: D \to M$ consists of the orthogonal unit vector fields E_1 and E_2 whose values at each point $\mathbf{x}(u, v)$ of $\mathbf{x}(D)$ are

$$\frac{\mathbf{x}_u(u,v)}{\sqrt{E(u,v)}} \quad \text{and} \quad \frac{\mathbf{x}_v(u,v)}{\sqrt{G(u,v)}}.$$

In Exercise 7 of Section 4.4, we associated with each patch $\mathbf{x}$ the coordinate functions $\tilde{u}$ and $\tilde{v}$, which assign to each point $\mathbf{x}(u, v)$ the numbers u and v, respectively. For example, for the geographical patch $\mathbf{x}$ of Example 2.2 of Chapter 4, the coordinate functions are the longitude and latitude functions on the sphere Σ. In the extreme case when $\mathbf{x}$ is the identity map of $\mathbf{R}^2$, the coordinate functions are just the natural coordinate functions $(u, v) \to u$, $(u, v) \to v$ on $\mathbf{R}^2$.

For an orthogonal patch $\mathbf{x}$ with associated frame field E_1, E_2, we shall express θ_1, θ_2, and ω_{12} in terms of the coordinate functions $\tilde{u}$, $\tilde{v}$. Since $\mathbf{x}$ is fixed throughout the discussion, we shall run the risk of omitting the inverse mapping $\mathbf{x}^{-1}$ from the notation. With this convention, the coordinate functions $\tilde{u} = u(\mathbf{x}^{-1})$ and $\tilde{v} = v(\mathbf{x}^{-1})$ are written simply u and v, and similarly $\mathbf{x}_u$ and $\mathbf{x}_v$ now become tangent vector fields on M itself. Thus the associated frame field of $\mathbf{x}$ has the concise expression

$$E_1 = \frac{\mathbf{x}_u}{\sqrt{E}}, \quad E_2 = \frac{\mathbf{x}_v}{\sqrt{G}}. \tag{1}$$

The dual forms θ_1, θ_2 are characterized by $\theta_i(E_j) = \delta_{ij}$, and in the exercise referred to above it is shown that

$$du(\mathbf{x}_u) = 1, \quad dv(\mathbf{x}_u) = 0,$$
$$du(\mathbf{x}_v) = 0, \quad dv(\mathbf{x}_v) = 1.$$

Thus we deduce from (1) that

$$\theta_1 = \sqrt{E}\,du, \quad \theta_2 = \sqrt{G}\,dv. \tag{2}$$

By using the structural equations, we find analogous formulas for ω_{12} and K. Recall that for a function f, $df = f_u\,du + f_v\,dv$, where the subscripts indicate partial derivatives. Hence

$$d\theta_1 = d(\sqrt{E}) \wedge du = (\sqrt{E})_v dv\,du = \frac{-(\sqrt{E})_v}{\sqrt{G}} du \wedge \theta_2,$$

$$d\theta_2 = d(\sqrt{G}) \wedge dv = (\sqrt{G})_u du\,dv = \frac{-(\sqrt{G})_u}{\sqrt{E}} dv \wedge \theta_1,$$

where we have used the alternation rule for wedge products and substituted $dv = \theta_2/\sqrt{G}$ and $du = \theta_1/\sqrt{E}$ from (2). Comparison with the first structural equations $d\theta_1 = \omega_{12} \wedge \theta_2$ and $d\theta_2 = -\omega_{12} \wedge \theta_1$ shows that

$$\omega_{12} = \frac{-(\sqrt{E})_v}{\sqrt{G}} \, du + \frac{(\sqrt{G})_u}{\sqrt{E}} \, dv. \tag{3}$$

The logic is simple: By the computations above, this form satisfies the first structural equations; hence by uniqueness (Lemma 5.1), *it must be ω_{12}*.

6.2 Example   *Geographical coordinates on the sphere.* For the geographical patch $\mathbf{x}$ in the sphere Σ (Example 2.2 of Chapter 4), we found $E = r^2 \cos^2 v$, $F = 0$, $G = r^2$. Thus by formula (2) above,

$$\theta_1 = r \cos v \, du, \quad \theta_2 = r \, dv.$$

Now $(\sqrt{E})_v = -r \sin v$ and $(\sqrt{G})_u = 0$; hence by (3),

$$\omega_{12} = \sin v \, du.$$

The associated frame field of this patch is the same one obtained in Example 1.6 from the spherical frame field in $\mathbf{R}^3$. With the notational shift $u \to \vartheta$, $v \to \varphi$, the forms above are (necessarily) also the same. But now we have a simple way to compute them *directly in terms of the surface* with no appeal to the geometry of $\mathbf{R}^3$.

Finally, we derive a new expression for the Gaussian curvature. In this context, exterior differentiation of ω_{12} as given in (3) yields

$$d\omega_{12} = -\left(\frac{(\sqrt{E})_v}{\sqrt{G}} \right)_v dv\,du + \left(\frac{(\sqrt{G})_u}{\sqrt{E}} \right)_u du\,dv$$

From (2) we get

$$\theta_1 \wedge \theta_2 = \sqrt{EG} \, du \, dv;$$

hence

$$-dv \, du = du \, dv = \frac{1}{\sqrt{EG}} \theta_1 \wedge \theta_2.$$

Thus the formula above becomes

$$d\omega_{12} = \frac{1}{\sqrt{EG}} \left\{ \left(\frac{(\sqrt{G})_u}{\sqrt{E}} \right)_u + \left(\frac{(\sqrt{E})_v}{\sqrt{G}} \right)_v \right\} \theta_1 \wedge \theta_2.$$

Now compare this with the second structural equation, $d\omega_{12} = -K\theta_1 \wedge \theta_2$. ◆

6.3 Proposition   If $\mathbf{x}: D \to M$ is an orthogonal patch, the Gaussian curvature K is given in terms of $\mathbf{x}$ by

$$K = \frac{-1}{\sqrt{EG}} \left\{ \left(\frac{(\sqrt{G})_u}{\sqrt{E}} \right)_u + \left(\frac{(\sqrt{E})_v}{\sqrt{G}} \right)_v \right\}.$$

This is perhaps the most elegant of the dozens of formulas that have been found for Gaussian curvature. By contrast with the formula in Corollary 4.1 of Chapter 5, the functions L, M, N (which describe the shape operator) no longer appear. Indeed, since K is now expressed solely in terms of E, F, G, by using Lemma 4.5 we get another proof of the isometric invariance of Gaussian curvature.

Exercises

1. Compute the dual 1-forms, connection form ω_{12}, and Gaussian curvature for the associated frame field of the following orthogonal patches:

(a) $\mathbf{x}(u, v) = (u\cos v, u\sin v, bv)$, helicoid.

(b) $\mathbf{x}(u, v) = (u\cos v, u\sin v, u^2/2)$, paraboloid of revolution.

(c) $\mathbf{x}(u, v) = (u\cos v, u\sin v, au)$, cone.

2. A parametrization $\mathbf{x}: D \to M$ is *isothermal* provided $E = G$ and $F = 0$. (Thus a fine network of parameter curves cuts M into small regions that are almost square.) By Exercise 8 of Section 4, $\mathbf{x}$ is a conformal mapping with scale function λ such that $E = G = \lambda^2$. Prove:

(a) $K = -\Delta\log\lambda/\lambda^2 = -\Delta\log E/(2E)$, where Δ is the *Laplacian*: $\Delta f = f_{uu} + f_{vv}$.

(b) The mean curvature H is zero if and only if $\mathbf{x}_{uu} + \mathbf{x}_{vv} = 0$.

3. If $\mathbf{x}$ is a principal patch (Ex. 8 of Sec. 5.4), prove

(a) $\omega_{13} = \dfrac{L}{\sqrt{E}} du$, $\omega_{23} = \dfrac{N}{\sqrt{G}} dv$.

(b) $L_v = HE_v$, $N_u = HG_u$.

(*Hints:* For (a), compare Cor. 1.5 with Ex. 8 of Sec. 5.4. For (b), use (a) and the Codazzi equations from Thm. 1.7.)

6.7 Integration and Orientation

A primary goal of this section is to define the integral of a 2-form over a compact oriented surface. This notion does not involve geometry at all; it

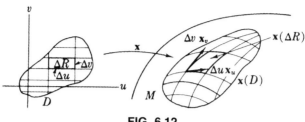

FIG. 6.12

belongs to the integral calculus on surfaces (Chapter 4, Section 6). However, we motivate the definition by considering some geometric applications. Orientation will be involved, for its connection with integration is already shown by the elementary calculus formula $\int_b^a f(x)\,dx = -\int_a^b f(x)\,dx$.

Perhaps the simplest use of double integration in geometry is in finding the area of a surface. To discover a proper definition of area, we start with a patch $\mathbf{x}: D \to M$ and ask what the area of its image should be. Let ΔR be a small coordinate rectangle in D with sides Δu and Δv. Now, $\mathbf{x}$ distorts ΔR into a small curved region $\mathbf{x}(\Delta R)$ in M, marked off by four segments of parameter curves, as shown in Fig. 6.12.

We have seen that the segment from $\mathbf{x}(u, v)$ to $\mathbf{x}(u + \Delta u, v)$ is linearly approximated by the vector $\Delta u\, \mathbf{x}_u$ evaluated at (u, v), and the segment from $\mathbf{x}(u, v)$ to $\mathbf{x}(u, v + \Delta v)$ is approximated by $\Delta v\, \mathbf{x}_v$. Thus the region $\mathbf{x}(\Delta R)$ is approximated by the parallelogram in the tangent plane at $\mathbf{x}(u, v)$ that has these vectors as its sides. From Chapter 2, Section 1 we know that this parallelogram has area

$$\|\Delta u\, \mathbf{x}_u \times \Delta v\, \mathbf{x}_v\| = \|\mathbf{x}_u \times \mathbf{x}_v\|\Delta u\, \Delta v = \sqrt{EG - F^2}\,\Delta u\, \Delta v.$$

We conclude that the area of $\mathbf{x}(\Delta R)$ should be approximately $\sqrt{EG - F^2}$ times the area $\Delta u \Delta v$ of ΔR. So at each point of D, the familiar expression $\sqrt{EG - F^2}$ *gives the rate at which* $\mathbf{x}$ *is expanding area*. Thus it would be natural to define the *area* of the whole region $\mathbf{x}(D)$ to be

$$\iint_D \sqrt{ED - F^2}\; du\, dv.$$

But since D is an open set, such integrals may well be improper. To avoid this we must modify the definition of coordinate patch.

7.1 Definition The *interior* $R°$ of a rectangle $R: a \leq u \leq b, c \leq v \leq d$ is the open set $a < u < b, c < v < d$. A 2-segment $\mathbf{x}: R \to M$ is *patchlike* provided the restricted mapping $\mathbf{x}: R° \to M$ is a patch in M.

The continuous function $\sqrt{EG - F^2}$ is bounded on the closed rectangle R, so the area of a patchlike 2-segment is well defined and finite, since in the double integral above the open set, D will be replaced by R.

A patchlike 2-segment $\mathbf{x}: R \to M$ is not required to be one-to-one on the boundary of R, so its image may not be very rectangular.

7.2 Example Areas of surfaces. We begin with two familiar cases.

(1) *The sphere of radius r*. If the formula defining the geographical patch is applied to the closed rectangle R: $-\pi \leqq u \leqq \pi, \; -\pi/2 \leqq v \leqq \pi/2$, the result is a 2-segment covering the whole sphere. Since

$$E = r^2 \cos^2 v, \quad F = 0, \quad G = r^2,$$

we get $\sqrt{EG - F^2} = r^2 \cos v$. Thus the area of the sphere is

$$A = \int_{-\pi/2}^{\pi/2} \int_{-\pi}^{\pi} r^2 \cos v \, du \, dv = 4\pi r^2,$$

a result well known to Euclid.

(2) *Torus of radii R > r > 0*. From Example 2.5 of Chapter 4 we derive a patchlike 2-segment covering the torus. Here

$$\sqrt{EG - F^2} = r(R + r \cos u),$$

with $-\pi \leqq u, v \leqq \pi$, so the area of the torus is

$$A = \int_{-\pi}^{\pi} \int_{-\pi}^{\pi} r(R + r \cos u) \, du \, dv = 4\pi^2 Rr.$$

(3) *The bugle surface* (Example 7.6 of Chapter 5). Every surface of revolution M has a canonical parametrization with $E = 1$, $F = 0$, $G = h^2$. On the rectangle R: $a \leqq u \leqq b, 0 \leqq v \leqq \pi$, $\mathbf{x}$ is a patchlike 2-segment whose image is the closed region Z_{ab} of M between the parallels $u = a$ and $u = b$ (Fig. 6.13). Thus the area of the *zone* Z_{ab} is

$$A_{ab} = \int_a^b \int_0^{2\pi} h \, du \, dv = 2\pi \int_a^b h \, du.$$

In the case of the bugle surface, $h(u) = ce^{-u/c}$. Hence

$$A_{ab} = 2\pi c \int_a^b e^{-u/c} du = 2\pi c^2 (e^{-a/c} - e^{-b/c}).$$

To find the area of the entire bugle—a noncompact surface—let Z_{ab} expand, with $a \to 0$ and $b \to \infty$. The corresponding areas are positive and increasing, and hence approach a limit. The limit of A_{ab} as $a \to 0$ and $b \to \infty$ is $2\pi c^2$. Hence, in particular, the bugle has *finite* area. ◆

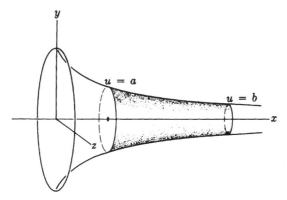

FIG. 6.13

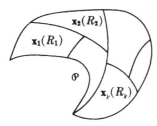

FIG. 6.14

To define the area of a complicated region, we follow the usual scheme from elementary calculus: Break the region into simple pieces and add their areas.

7.3 Definition A *paving* of a region $\mathscr{P}$ in a surface M consists of a finite number of patchlike 2-segments $\mathbf{x}_1, \ldots, \mathbf{x}_k$ whose images fill M in such a way that each point of M is in at most one set $\mathbf{x}_i(R_i^o)$.

In short, the images of the segments cover all of $\mathscr{P}$, overlapping only on their boundaries (see Fig. 6.14).

Not every region is pavable. Since pavings are finite, compactness is certainly necessary (Lemma 7.2 in Chapter 4). In fact, a compact region is pavable if its boundary consists of a finite number of regular curve segments. In particular, *an entire compact surface is always pavable.* The area of a pavable region is defined to be the sum of the areas of $\mathbf{x}_1(R_1), \ldots, \mathbf{x}_k(R_k)$ for any paving of $\mathcal{P}$.

The preceding exposition shows that a treatment of area does not require differential forms, but integration of 2-forms will give area and much more besides (see Remark 7.8). The first question is this: Which 2-form should be integrated over a patchlike 2-segment to get the area of its image? By Definition 6.3 of Chapter 4,

$$\iint_x \mu = \iint_R \mu(\mathbf{x}_u, \mathbf{x}_v)\, du\, dv.$$

Thus we need a 2-form whose value on $\mathbf{x}_u$, $\mathbf{x}_v$ is

$$\|\mathbf{x}_u \times \mathbf{x}_v\| = \sqrt{EG - F^2}.$$

This suggests

7.4 Definition An *area form* on a surface M is a differentiable 2-form μ whose value on any pair of tangent vectors is

$$\mu(\mathbf{v}, \mathbf{w}) = \pm\left((\mathbf{v} \cdot \mathbf{v})(\mathbf{w} \cdot \mathbf{w}) - (\mathbf{v} \cdot \mathbf{w})^2\right)^{1/2} = \pm\|\mathbf{v} \times \mathbf{w}\|.$$

(The sign ambiguity cannot be avoided.)

Because forms are bilinear, it would be equivalent to require that $\mu(E_1, E_2) = \pm 1$ for every frame E_1, E_2 on M.

7.5 Lemma A surface M has an area form if and only if it is orientable. On a connected orientable surface there are exactly two area forms, which are negatives of each other. (These are denoted by $\pm dM$.)

In fact, by Definition 7.4 of Chapter 4, M is orientable if and only if there is a nonvanishing 2-form on it. So if M has an area form, it is certainly orientable. The remainder of the proof follows the same pattern as for Proposition 7.5 in Chapter 4. Indeed, for $M \subset \mathbf{R}^3$ there is a natural one-to-one correspondence between unit normals U and area forms dM given by

$$dM(\mathbf{v}, \mathbf{w}) = \pm U \cdot \mathbf{v} \times \mathbf{w}.$$

To *orient* a connected orientable surface M is to choose one of its two area forms—or equivalently, one of its two unit normals.

Finding area is not a typical integration problem, because area is always positive. Thus to find area by integrating an area form dM we have to be careful about signs. Let $\mathbf{x}$ be a patchlike 2-segment in a surface oriented by area form dM. By definition,

$$\iint_{\mathbf{x}} dM = \iint_R dM(\mathbf{x}_u, \mathbf{x}_v)\, du\, dv.$$

Now there are two cases:

(1) If $dM(\mathbf{x}_u, \mathbf{x}_v) > 0$, we say that $\mathbf{x}$ is *positively oriented*. Then by the definition of area form,

$$dM(\mathbf{x}_u, \mathbf{x}_v) = \sqrt{EG - F^2}\,;$$

hence $\iint_{\mathbf{x}} dM$ is the area of $\mathbf{x}(R)$.

(2) If $dM(\mathbf{x}_u, \mathbf{x}_v) < 0$, we say that $\mathbf{x}$ is *negatively oriented*. Then

$$dM(\mathbf{x}_u, \mathbf{x}_v) = -\sqrt{EG - F^2}\,;$$

hence $\iint_{\mathbf{x}} dM$ is *minus* the area of $\mathbf{x}(R)$.

Thus, to find the area of a pavable region $\mathscr{P}$ by integrating the area form, we cannot use an arbitrary paving. The paving must be *positively oriented*, that is, consist only of positively oriented patchlike 2-segments. Then

$$\text{area}(\mathscr{P}) = \sum_i \text{area}(\mathbf{x}_i(R_i)) = \sum_i \iint_{\mathbf{x}i} dM.$$

Now we replace the area form by an arbitrary 2-form to get the definition we are looking for.

7.6 Definition   Let v be a 2-form on a pavable oriented region $\mathscr{P}$ in a surface. The *integral of v over $\mathscr{P}$* is

$$\iint_{\mathscr{P}} v = \sum_i \iint_{\mathbf{x}i} v,$$

where $\mathbf{x}_1, \ldots, \mathbf{x}_k$ is a positively oriented paving of $\mathscr{P}$.

There is a consistency problem with this definition that appears already in elementary calculus. Any two choices of paving for $\mathscr{P}$ must produce the same value for the integral. A formal proof of this is not elementary and belongs properly to analysis rather than to geometry.

Area forms make it easy to describe integration of *functions*. If f is a continuous function on a pavable region $\mathscr{P}$, then its integral over $\mathscr{P}$ is defined to be $\iint_{\mathscr{P}} f\, dM$. Evidently, this is an analogue of the usual integral $\iint f\, dx\, dy$ of elementary calculus, since $dx \wedge dy$ is the area form of the Euclidean plane.

7.7 Remark Improper integrals. We have defined integration only over compact regions; however, a *positive* function $f > 0$ can be integrated over an arbitrary surface M (or open region) by defining its integral to be the least upper bound of the integrals of f over all pavable regions $\mathscr{P}$ in M:

$$\iint_M f \, dM = \mathrm{lub} \iint_{\mathscr{P}} dM.$$

This becomes $+\infty$ if no upper bound exists. Setting $f = 1$ gives the area of M, as in Example 7.2(3). (For $f < 0$, switch to greatest lower bounds and $-\infty$.)

7.8 Remark Geometry uses two related notions of integration, distinguished by the effect of change of variables. Arc length, area, and even the integral of a function belong to *absolute* integration (or *measure theory*), which is independent of parametrization and involves the absolute value of the relevant Jacobian. By contrast, integration of differential forms depends on orientation and involves the signed Jacobian. This version has a crucial advantage: Stokes theorem (Theorem 6.5 of Chapter 4) and its many consequences.

Exercises

1. For a Monge patch $\mathbf{x}(u, v) = (u, v, f(u, v))$, show that the area of $\mathbf{x}(D)$ is given by the usual formula from elementary calculus. Deduce that $A(\mathbf{x}(D)) \geq A(D)$.

2. (*Theorem of Pappus.*) Let M be a surface of revolution whose profile curve has finite length L. Find a formula for the area of M and interpret it as $A = 2\pi \bar{h} L$, where $\bar{h}$ is the average distance of M from the axis of revolution.

3. Let $\mathbf{x}$ be the usual parametrization of the torus of revolution T (Example 2.5 of Ch. 4), with T oriented by the outward unit normal U. Compute the integral $\iint_T v$ where in each case, v is the 2-form on T such that:

(a) $v(\mathbf{x}_u, \mathbf{x}_v)$ is the square of the distance from $\mathbf{x}(u, v)$ to the origin $\mathbf{0}$ in $\mathbf{R}^3$.
(b) $v(\mathbf{x}_u, \mathbf{x}_v) = U \cdot \mathbf{x}_u \times \mathbf{x}_v$.

4. Let M be a compact surface oriented by dM, and let $-M$ be the same surface oriented by $-dM$. Prove

(a) $\iint_M (c_1 v_1 + c_2 v_2) = c_1 \iint_M v_1 + c_2 \iint_M v_2$ (c_1, c_2 constant).
(b) $\iint_{-M} v = -\iint_M v$. (*Hint:* Use Ex. 5.)

(c) $\iint_M f \, dM = \iint_{-M} f(-dM).$

(d) If $f \leqq g$, then $\iint_M f \, dM \leqq \iint_M g \, dM.$

(Note the effect of $f = 0$ or $g = 0$.)

5.   If $F: M \rightarrow N$ is an orientation-preserving diffeomorphism of compact oriented surfaces, show that

$$\iint_M F^*(v) = \iint_N v$$

for any 2-form v on N. (*Hint:* Ex. 7 of Sec. 4.6.)

6.   A diffeomorphism $F: M \rightarrow N$ is *area-preserving* provided the area of any pavable region $\mathscr{P}$ in M is the same as the area of its image $F(\mathscr{P})$ in N. Prove:
(a) A diffeomorphism $F: M \rightarrow N$ is area-preserving if and only if

$$EG - F^2 = \overline{E}\overline{G} - \overline{F}^2$$

for all patchlike 2-segments $\mathbf{x}$ in M with $\overline{\mathbf{x}} = F(\mathbf{x})$ in N. (Here "all" can be replaced by "sufficiently many to cover M.")
(b) Isometries are area-preserving, isometric surfaces have the same area (include the noncompact case).
(c) The cylindrical projection in Example 5.2(1) of Chapter 4 is area-preserving but not an isometry. (Deduce the standard formula for the area of a zone in the sphere.)
(d) A mapping $F: M \rightarrow N$ is an isometry if and only if it is conformal and area-preserving.

6.8   Total Curvature

One of the most important geometric invariants of a surface is gotten by integration of curvature.

8.1 Definition   Let K be the Gaussian curvature of a compact surface M oriented by area form dM. Then

$$\iint_M K \, dM$$

is the *total Gaussian curvature* of M.

The same definition applies to any pavable region in M.

To compute total curvature of M we add the total curvatures of each patchlike 2-segment $\mathbf{x}$ of a paving of M. With the usual notation for the domain R of $\mathbf{x}$,

$$\iint_{\mathbf{x}} K\, dM = \iint_R \mathbf{x}^*(K\, dM) = \iint_R K(\mathbf{x})\mathbf{x}^*(dM)$$

$$= \int_a^b \int_c^d K(\mathbf{x})\sqrt{EG - F^2}\; du\; dv.$$

As before, $K(\mathbf{x})$ can be computed explicitly using Corollary 4.1 of Chapter 5 or by Proposition 6.3.

8.2 Example   Total curvature of some surfaces.

(1) *Constant curvature.* If the Gaussian curvature of M is constant, then the total curvature of M is

$$\iint_M K\, dM = K\iint_M dM = KA(M).$$

Thus a sphere of radius r has total curvature $4\pi = (1/r^2)(4\pi r^2)$, and the bugle surface has total curvature $-2\pi = (-1/c^2)(2\pi c^2)$.

(2) *Torus.* Let $\mathbf{x}$ be the 2-segment in Example 7.2 that covers the torus T. Then the area form dT has coordinate expression

$$\mathbf{x}^*(dT) = \sqrt{EG - F^2}\; du\; dv = r(R + r\cos u)\, du\; dv.$$

In Example 7.1 of Chapter 5 we computed

$$K(\mathbf{x}) = \frac{\cos u}{r(R + r\cos u)}.$$

Hence the torus has total Gaussian curvature

$$\iint_T K\, dT = \int_{-\pi}^{\pi} \int_{-\pi}^{\pi} \cos u\; du\; dv = 0.$$

Thus the negative curvature of the inner half of the torus has exactly balanced the positive curvature of its outer half.

(3) *Catenoid.* This surface is not compact, and its area is infinite; nevertheless, its total curvature is finite. When the parametrization in Example 7.1 of Chapter 5 is restricted to the rectangle R: $-a \leq u \leq a$, $0 \leq v \leq 2\pi$, it becomes a patch-like 2-segment covering the zone Z_a between the parallels $u = -a$ and $u = +a$ (Fig. 6.15). From this example we get

$$K(\mathbf{x}) = \frac{-1}{c^2 \cosh^4(u/c)}$$

and $\sqrt{EG} = c\cosh^2(u/c)$. Hence the zone has total curvature

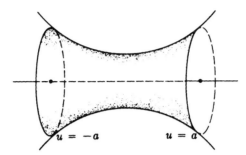

FIG. 6.15

$$\iint_X K\, dM = -\int_{-a}^{a}\int_0^{2\pi} \frac{1}{\cosh^2(u/c)}\, du\, dv = -4\pi \tanh\left(\frac{a}{c}\right).$$

As $a \to \infty$, Z_a expands to fill the whole surface. So the total curvature of the catenoid is

$$\iint_M K\, dM = -4\pi \lim_{a\to\infty} \tanh\left(\frac{a}{c}\right) = -4\pi.$$

The total curvatures computed above are all multiples of 2π, and none depends on the particular parameters (radius r, constant c, . . .) of its surface. This is no accident; a major reason appears in Section 6 of the next chapter, but see also Exercise 6 of this section. ◆

8.3 Definition Let M and N be surfaces oriented by area forms dM and dN. Then the *Jacobian* of a mapping $F: M \to N$ is the real-valued function J_F on M such that

$$F^*(dN) = J_F\, dM.$$

The Jacobian has considerable geometric significance. Let $\mathbf{v}$, $\mathbf{w}$ be tangent vectors to M at a point $\mathbf{p}$. Then

$$J_F(\mathbf{p})dM(\mathbf{v},\mathbf{w}) = F^*(dN)(\mathbf{v},\mathbf{w}) = dN(F^*(\mathbf{v}), F^*(\mathbf{w})). \qquad (*)$$

Suppose F is regular at $\mathbf{p}$. Then if vectors $\mathbf{v}$ and $\mathbf{w}$ at $\mathbf{p}$ are independent, so are their images $F_*\mathbf{v}$ and $F_*\mathbf{w}$ at $F(\mathbf{p})$. Thus $dM(\mathbf{v}, \mathbf{w})$ and $dN(F_*(\mathbf{v}), F_*(\mathbf{w}))$ are both nonzero; hence $J_F(\mathbf{p}) \neq 0$. Evidently the converse is also true, so F is regular at $\mathbf{p}$ *if and only if* $J_F(\mathbf{p}) \neq 0$.

The sign of $J_F(\mathbf{p}) \neq 0$ is also informative, because if $dM(\mathbf{v}, \mathbf{w}) > 0$, then $J_F(\mathbf{p})$ and $dN(F^*(\mathbf{v}), F^*(\mathbf{w}))$ have the same sign. Thus F is

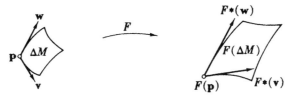

FIG. 6.16

orientation-preserving at **p** if $J_F(\mathbf{p}) > 0$, and
orientation-reversing at **p** if $J_F(\mathbf{p}) < 0$.

Furthermore, taking absolute values in (∗) gives

$$|J_F(\mathbf{p})|\,|dM(\mathbf{v},\mathbf{w})| = |dN(F_*\mathbf{v},\,F_*\mathbf{w})|. \qquad (**)$$

The discussion of area in Section 7 shows that the area of a small region ΔM in M marked off by "short" vectors **v** and **w** is approximately $|dM(\mathbf{v},\mathbf{w})|$. Since $|dN(F^*(\mathbf{v}),\,F^*(\mathbf{w})|$ approximates the area of the image region $F(\Delta M)$ (Fig. 6.16), we interpret (∗∗) as

$$|J_F(\mathbf{p})|(\text{area of }\Delta M) \approx \text{area of } F(\Delta M).$$

Thus $|J_F(\mathbf{p})|$ *is the rate at which F is expanding area at* **p**.

In view of the pointwise definitions above, a mapping $F\colon M \to N$ of oriented surfaces is said to be *orientation-preserving* if $J_F > 0$, *orientation-reversing* if $J_F < 0$.

For preservation of area, see Exercise 6 of Section 7. A signed version of area contains more information. We call

$$\iint_M J_F\,dM = \iint_M F^*(dN)$$

the *algebraic area* of $F(M)$. The discussion above shows that roughly speaking, each small region ΔM in M contributes to this total the *signed* area of its image $F(\Delta M)$ in N, the sign being

(1) Positive, if the orientation of $F(\Delta M)$ agrees with that of N;
(2) Negative, if these orientations disagree (so F has turned ΔM over);
(3) Zero, if F collapses ΔM to a curve or point.

Let us consider what this means in the case of the Gauss map G of an oriented surface M in $\mathbf{R}^3$. As defined in Exercise 4 of Section 5, G maps M to the unit sphere Σ by moving the selected unit normal $U(\mathbf{p})$ at **p** in M, by distant parallelism, to the origin of $\mathbf{R}^3$—there it points to $G(\mathbf{p})$ in Σ (see Fig. 6.17).

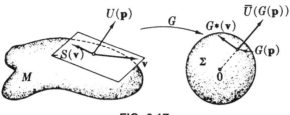

FIG. 6.17

8.4 Theorem   The Gaussian curvature K of an oriented surface $M \subset \mathbf{R}^3$ is the Jacobian of its Gauss map.

(Here the unit sphere is oriented, as usual, by its outward normal $\overline{U}$ or the corresponding area form $d\Sigma$.)

Proof.   If $U = \Sigma g_i U_i$ is the unit normal orienting M, then the Gauss map is $G = (g_1, g_2, g_3)$. Note that if S is the shape operator determined by U, then

$$-S(\mathbf{v}) = \nabla_{\mathbf{v}} U = \sum \mathbf{v}[g_i] U_i(\mathbf{p}),$$

and by Proposition 7.5 of Chapter 1,

$$G_*(\mathbf{v}) = \sum \mathbf{v}[g_i] U_i(G(\mathbf{p})).$$

Hence $G_*(\mathbf{v})$ and $-S(\mathbf{v})$ are parallel for any tangent vector $\mathbf{v}$ to M, as suggested in Fig. 6.17. (Recall that *parallel* here means having the same natural Euclidean coordinates.)

To prove the theorem we must show that

$$K \ dM = G_*(d\Sigma),$$

so we evaluate these 2-forms on an arbitrary pair of tangent vectors to M. Lemma 3.4 of Chapter 5 gives

$$(K \ dM)(\mathbf{v}, \mathbf{w}) = K(\mathbf{p}) \ dM(\mathbf{v}, \mathbf{w}) = K(\mathbf{p})U(\mathbf{p}) \cdot \mathbf{v} \times \mathbf{w}$$

$$= U(\mathbf{p}) \cdot S(\mathbf{v}) \times S(\mathbf{w}).$$

On the other hand,

$$G_*(d\Sigma)(\mathbf{v}, \mathbf{w}) = d\Sigma(G_*(\mathbf{v}), G_*(\mathbf{w})) = \overline{U}(G(\mathbf{p})) \cdot G_*(\mathbf{v}) \times G_*(\mathbf{w}).$$

A triple scalar product depends only on the Euclidean coordinates of its vectors, so we can replace $G_*(\mathbf{v})$, $G_*(\mathbf{w})$ by the vectors $-S(\mathbf{v})$, $-S(\mathbf{w})$ parallel to them. Furthermore, by the definition of G and the special character of the unit sphere Σ, the vectors $U(\mathbf{p})$ and $\overline{U}(G(\mathbf{p}))$ are also parallel (Fig. 6.17).

Thus the two triple scalar products above are the same—and the proof is complete.   ◆

8.5 Corollary The total Gaussian curvature of an oriented surface $M \subset \mathbf{R}^3$ equals the algebraic area of the image of its Gauss map $G: M \to \Sigma$.

To prove this it suffices to integrate the form $K dM = G^*(d\Sigma)$ over M.
The following readily proved special case is easier to use since it involves only ordinary area.

8.6 Corollary Let $\mathscr{R}$ be an oriented region in $M \subset \mathbf{R}^3$ on which
(1) the Gauss map G is one-to-one (the values of U at different points are never parallel), and
(2) either $K \geq 0$ or $K \leq 0$ (K does not change sign).
Then the total curvature of $\mathscr{R}$ is $\pm$area of $G(\mathscr{R})$, where the sign is that of K. (Evidently, this area does not exceed 4π.)

For example, consider the torus T as in Fig. 5.21. Its Gauss map G sends the outer half $\mathscr{O}$ of T (where $K \geq 0$) in one-to-one fashion onto the whole sphere Σ—and does the same for the inner half $\mathscr{I}$ (where $K \leq 0$). Thus the torus has total curvature $A(\Sigma) - A(\Sigma) = 0$, as found in Example 8.2(2) by an explicit integration.

Next, consider the catenoid C, with rotation axis the x axis as in Fig. 6.15. Considering the normal vectors to the profile curve $y = c \cosh(x/c)$ shows that the Gauss map G carries this curve in one-to-one fashion onto an open semicircle of Σ. Then, considering U on the parallels of C shows that G is a one-to-one map onto Σ—omitting only the two points $\pm(1, 0, 0)$. Thus the total curvature of the catenoid is -4π, as computed analytically in Example 8.2(3).

An unexpected application of total Gaussian curvature is to curves, where it provides a proof of the main assertion of Fenchel's Theorem (Exercise 18 in Section 2.4), namely,

A simple closed curve α in $\mathbf{R}^3$ has total curvature $\displaystyle\int_\alpha \kappa \, ds \geq 2\pi$.

Proof. Let M be a tube around α, with parametrization $\mathbf{x}$ as in Exercise 17 of Section 5.4. This exercise shows that the region B in M on which Gauss curvature is nonnegative is $\mathbf{x}(D)$, where D is the rectangle $0 \leq u \leq L$, $\pi/2 \leq v \leq 3\pi/2$ and L is the length of α.

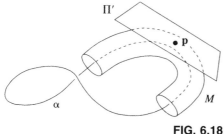 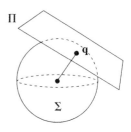

FIG. 6.18

Using the formulas for K and W in that exercise, we find the remarkable result

$$\iint_B K \, dT = -\int_0^L \kappa \, ds \int_{\pi/2}^{3\pi/2} \cos v \, dv = 2 \int_0^L \kappa \, ds. \qquad (*)$$

Next, we show geometrically that the total curvature of B is at least 4π. Let II be the Euclidean tangent plane to the unit sphere $\Sigma \subset \mathbf{R}^3$ at a point $\mathbf{q}$, so II is the plane through $\mathbf{q}$ orthogonal to the vector from $\mathbf{0}$ to $\mathbf{q}$. Take a distant plane parallel to II and move it parallel to itself until it reaches tangency to M—at, say, the point $\mathbf{p}$ (Fig. 6.18). This plane II′ is tangent to T, and M lies entirely on one side of it. Thus, of the curvature choices in Remark 3.3 of Chapter 5, only $K(\mathbf{p}) < 0$ is ruled out—hence $K(\mathbf{p}) \geqq 0$.

Evidently, the Gauss map G of M sends $\mathbf{p}$ to $\mathbf{q}$ in Σ. Thus G maps B *onto* Σ. Then $K(\mathbf{p}) \geqq 0$ implies

$$\iint_B K \, dM \geq 4\pi.$$

In combination with $(*)$, this proves the result.   ◆

On an oriented surface the ambiguity in the measurement of angles mentioned in Section 1 of Chapter 2 can be decisively reduced. Let M be a surface oriented by the unit normal U. If $\mathbf{v}$ is a vector tangent to M, then properties of the cross product show that the tangent vector

$$J(\mathbf{v}) = U \times \mathbf{v}$$

is orthogonal to $\mathbf{v}$ and has the same length. So J is a linear operator on each tangent space to M that rotates each vector through $+90°$. We call J the *rotation operator of M*.

8.7 Definition   Let $\mathbf{v}$ and $\mathbf{w}$ be unit tangent vectors at a point of an oriented surface M. A number φ is an *oriented angle from* $\mathbf{v}$ *to* $\mathbf{w}$ provided

$$\mathbf{w} = \cos \varphi \; \mathbf{v} + \sin \varphi \; J(\mathbf{v}).$$

There is a unique choice of φ in the interval $0 \le \varphi < 2\pi$. Then every oriented angle from $\mathbf{v}$ to $\mathbf{w}$ is given by $\varphi + 2\pi k$ for some integer k. The same scheme gives angles for any pair of nonzero tangent vectors: Simply divide each by its length to produce unit vectors.

8.8 Lemma Let $\alpha\colon I \to M$ be a curve in an oriented surface M. If V and W are nonvanishing tangent vector fields on α, there is a differentiable function φ on I such that for each t in I, $\varphi(t)$ is an oriented angle from $V(t)$ to $W(t)$.

Proof. As just mentioned, angles are unchanged if V and W are reduced to unit vector fields. Then V, $J(V)$ is a frame field on α. Orthonormal expansion gives $W = fV + gJ(V)$, where

$$f = W \cdot V \quad \text{and} \quad g = W \cdot J(V).$$

Since $1 = W \cdot W = f^2 + g^2$, we can apply Exercise 12 of Section 2.1 to get a differentiable function φ on I such that

$$f = \cos \varphi, \quad g = \sin \varphi.$$

Evidently, φ is the required function. ◆

We call φ an *angle function* from V to W, and sometimes write $\varphi = \angle(V, W)$. The exercise mentioned in the proof also implies that φ is uniquely determined by its value at any one $t_0 \in I$. Thus any two choices of φ differ by a constant integer multiple of 2π. Many applications of angle functions will appear later on.

Consistency is the essence of orientability. In studying a surface M oriented by an area form dM, we prefer using *positively oriented* patches, $dM(\mathbf{x}_u, \mathbf{x}_v) > 0$ and *positively oriented* frame fields, $dM(E_1, E_2) = +1$. (Recall that the only possible values of an area form on a frame are ± 1.)

When a frame field E_1, E_2 on M is positively oriented, then the wedge product of its dual 1-forms is precisely the selected area form:

$$dM = \theta_1 \wedge \theta_2.$$

To prove this, it suffices to note that both sides have the value $+1$ on E_1, E_2. A minus sign appears in this equation if the frame field is negatively oriented.

The properties of J show that if $\mathbf{u}$ is a unit tangent vector, then $\mathbf{u}$, $J(\mathbf{u})$ is a frame—in fact, a positively oriented frame. Thus any nonvanishing vector field V on M determines a positively oriented frame field, namely,

$$E_1 = \frac{V}{\|V\|}, \quad E_2 = J(E_1) = \frac{J(V)}{\|V\|}.$$

We call this the *associated frame field* of V.

Exercises

1.   Using the natural orientation of $\mathbf{R}^2$ given by the area form $du \wedge dv$, prove:
(a) For a mapping $F = (f, g)$: $\mathbf{R}^2 \to \mathbf{R}^2$, the definition of Jacobian in the text gives the usual formula $J = f_u g_v - f_v g_u$.
(b) The Jacobian of a patch $\mathbf{x}$: $D \to M$ in an oriented surface M is $\pm\sqrt{EG - F^2}$, where the sign depends on whether $\mathbf{x}$ is positively or negatively oriented.

2.   Let D be a small disk in M centered at $\mathbf{p}$, and let α parametrize its boundary ∂D. By drawing a few unit normals to M along α, sketch the image of α under the Gauss map of M if:
(a) $\mathbf{p}$ is any point of an ellipsoid M.
(b) $\mathbf{p}$ is any point of a circular cylinder.
(c) $\mathbf{p}$ is the origin in M: $z = xy$.

3.   (a) Prove that the Gauss map of a surface $M \subset \mathbf{R}^3$ is conformal if and only if its principal curvature functions satisfy $k_1^2 = k_2^2 > 0$.
(b) Deduce that the Gauss map of a surface M is conformal if and only if M is either (i) a minimal surface without planar points or (ii) part of a sphere.

4.   (a) Show that total curvature is an isometric invariant for compact orientable surfaces. (*Hint:* Use Ex. 5 of Sec. 7.)
(b) Extend (a) to the noncompact case, assuming $K \geqq 0$ or $K \leqq 0$.

5.   (*Total curvature of surfaces of revolution.*) On a surface of revolution with profile curve α, let Z_{ab}, as usual, be the zone bounded by the parallels through $\alpha(a)$ and $\alpha(b)$, $a < b$. Show that the total curvature of Z_{ab} is $2\pi(\sin \varphi_a - \sin \varphi_b)$, where φ_a and φ_b are the slope angles of α at a and b—measured relative to the axis of revolution. (See Fig. 6.19.)

For α on an open interval, $-\infty \leqq A < u < B \leqq \infty$, the total curvature can be treated as an improper integral, giving the (possibly infinite) result

$$2\pi\left(\lim_{a \to A} \sin \varphi_a - \lim_{b \to B} \sin \varphi_b\right)$$

provided both limits exist.

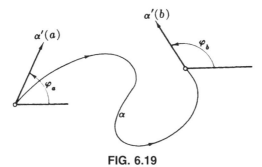

FIG. 6.19

6. (*Continuation.*)
(a) Show that every surface of revolution with closed profile curve has total curvature zero, and every augmented surface of revolution with two intercepts has total curvature 4π.
(b) Find the total curvatures of (i) paraboloid of revolution, (ii) catenoid, (iii) bugle surface.

7. Let $F: M \to N$ be a mapping of oriented surfaces. Prove:
(a) F is area-preserving (Ex. 6 of Sec. 7) if and only if F is a diffeomorphism with Jacobian $J_F = \pm 1$.
(b) If F is an isometry, then it is area-preserving, but not conversely.

8. Let M be a noncylindrical ruled surface whose rulings are entire straight lines; assume $K < 0$.
(a) Show that the total curvature of M is $-2L(\delta)$, where δ is a director curve with $\|\delta\| = 1$. (*Hint:* See Ex. 11 of Sec. 5.6.)
(b) Find the total curvature of the saddle surface $z = xy$ by this method (see Ex. 12).

9. (*Gauss maps of some minimal surfaces.*) In each case show that the Gauss map covers the sphere omitting exactly n points.
(a) Catenoid: $n = 2$.
(b) Helicoid: $n = 2$.
(c) Scherk's surface (Ex. 5 of Sec. 5.5): $n = 4$.
What are the total curvatures of these surfaces?

10. Let $\mathscr{E} = \mathbf{x}(\mathbf{R}^2)$ be Enneper's minimal surface (Ex. 15 of Sec. 5.6). (The formula for $\mathbf{x}$ is that of $\mathbf{x}_1$ in the next exercise.)
(a) Compute the unit normal

$$U = \frac{1}{1 + u^2 + v^2}(-2u,\ 2v,\ 1 - u^2 - v^2).$$

(b) Express U in terms of polar coordinates on $\mathbf{R}^2$, and show that (i) the Gauss map $G \approx U$ is one-to-one and (ii) G maps $\mathscr{E}$ onto the unit sphere Σ minus the south pole $(0, 0, -1)$.
(c) Deduce that $\mathscr{E}$ has total curvature -4π.
It is known that every complete nonflat minimal surface has total curvature $-4\pi m$ for some $1 \leqq m \leqq \infty$, and that only the catenoid and Enneper's surface have the extreme value -4π. (See R. Osserman, *A Survey of Minimal Surfaces*, Dover, New York, 1986.)

11.   (*Computer continuation.*)
(a) On the domain $-2.5 \leqq u, v \leqq 2.5$, plot the surface given by

$$\mathbf{x}_t(u, v) = (u, v, u^2 - v^2) + t\left(-\frac{u^3}{3} + uv^2, -\frac{v^3}{3} + vu^2, 0\right).$$

for at least six values of t from $t = 0$ (saddle surface) to $t = 1$ (Enneper's surface $\mathscr{E}$). Animate this series of plots, observing the formation of the two curves at which $\mathscr{E}$ cuts across itself.
(b) Plot $\mathscr{E}$, on the domain given in (a), as viewed successively from along the x axis, y axis, and z axis.

12.   Parametrize the saddle surface S using polar coordinates r, ϑ (instead of rectangular u, v). Then
(a) Find the total curvature of S by an explicit integration.
(b) Express the unit normal U to S in terms of polar coordinates. Show that the resulting Gauss map is one-to-one. Determine its image and use Corollary 8.6 to check the result in (a).

13.   (*Computer optional.*) (a) Find a parametrization of the monkey saddle M by polar coordinates (see Ex. 19 of Sec. 5.4).
(b) Plot the region $r \leqq 1$.
(c) Find the total curvature of M.

6.9   Congruence of Surfaces

Two surfaces M and $\overline{M}$ in $\mathbf{R}^3$ are *congruent* provided there is an isometry $\mathbf{F}$ of $\mathbf{R}^3$ that carries M exactly onto $\overline{M}$. Thus congruent surfaces differ only in their positions in $\mathbf{R}^3$. For example, the surfaces

$$M: z = xy \quad \text{and} \quad \overline{M}: z = \frac{x^2 - y^2}{2}$$

are congruent under a $45°$ rotation about the z axis.

To simplify the exposition, we assume that the surfaces in this section are orientable as well as connected.

9.1 Theorem   If $\mathbf{F}$ is a Euclidean isometry such that $\mathbf{F}(M) = \overline{M}$, then the restriction of $\mathbf{F}$ to M is an isometry of surfaces,

$$F = \mathbf{F}|M\colon M \to \overline{M}.$$

Furthermore, if M and $\overline{M}$ are suitably oriented, then F preserves shape operators, that is,

$$F_*(S(\mathbf{v})) = \overline{S}(F_*(\mathbf{v}))$$

for all tangent vectors $\mathbf{v}$ to M.

In short, congruent surfaces are isometric and have essentially the same shape operators.

Proof.   We know from Chapter 4, Section 5, that the restriction

$$F = \mathbf{F}|M\colon M \to \overline{M}$$

is a smooth mapping. Also, the differential maps of F and $\mathbf{F}$ agree on tangent vectors $\mathbf{v}$ to M. In fact, such a $\mathbf{v}$ is the initial velocity of a curve α that lies in M; hence $F(\alpha) = \mathbf{F}(\alpha)$. Thus

$$F_*(\mathbf{v}) = F(\alpha)'(0) = \mathbf{F}(\alpha)'(0) = \mathbf{F}_*(\mathbf{v}).$$

It follows that F_* preserves dot products of tangent vectors to M, since $\mathbf{F}_*$ has this property for all pairs of tangent vectors. Furthermore, $F\colon M \to \overline{M}$ is one-to-one (since $\mathbf{F}$ is) and onto (by hypothesis). Hence F is an isometry of surfaces.

To show that F preserves shape operators, we arrange to have corresponding unit normals on M and $\overline{M}$. Let U be a unit normal to M. Since $\mathbf{F}_*$ preserves dot products and $\mathbf{F}$ carries M to $\overline{M}$, it follows that $\mathbf{F}_*(U)$ is a unit vector field everywhere normal to $\overline{M}$. Thus $\mathbf{F}_*(U)$ is one of the two unit normals on $\overline{M}$, say,

$$\mathbf{F}_*(U(\mathbf{p})) = \overline{U}(\mathbf{F}(\mathbf{p})).$$

(See Fig. 6.20.)

If S and $\overline{S}$ are the shape operators of M and $\overline{M}$ derived from U and $\overline{U}$, we will show first that

$$\mathbf{F}_*(S(\mathbf{v})) = \overline{S}(\mathbf{F}_*(\mathbf{v}))$$

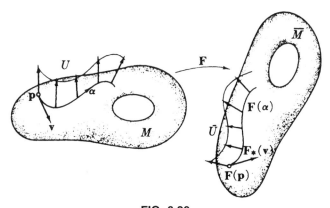

FIG. 6.20

for all tangent vectors to M.

Again, let α be a curve in M with initial velocity $\mathbf{v}$. Then $\mathbf{F}(\alpha)$ is a curve in $\overline{M}$ with initial velocity $\mathbf{F}_*(\mathbf{v})$. For U restricted to α and $\overline{U}$ restricted to $\mathbf{F}(\alpha)$, we have $\mathbf{F}(U) = \overline{U}$ (Fig. 6.20). Since $\mathbf{F}_*$ preserves derivatives of vector fields,

$$\mathbf{F}(S(\mathbf{v})) = -\mathbf{F}(U'(0)) = -\overline{U}'(0) = \overline{S}(\mathbf{F}_*(\mathbf{v})).$$

But $\mathbf{v}$ and $S(\mathbf{v})$ are tangent to M, so $\mathbf{F}_*$ can be replaced here by F_*.   ◆

Although *congruent surfaces are isometric*, it is not usually true that *isometric surfaces are congruent*. For example, the surfaces in Fig. 6.9 are all isometric to a plane rectangle, but evidently no two of them are congruent.

Our goal now is to prove the converse of the preceding theorem, namely, if M and $\overline{M}$ are isometric *and* have essentially the same shape operators, then they are congruent. This is the analogue of the basic congruence theorem for curves (Theorem 5.3 of Chapter 3). The condition M isometric to $\overline{M}$ corresponds to the hypothesis that α and β are unit-speed curves defined on the same interval, and "same shape operators" corresponds to

$$\kappa = \overline{\kappa}, \quad \tau = \overline{\tau}.$$

9.2 Theorem Let M and $\overline{M}$ be oriented surfaces in $\mathbf{R}^3$. Let $F: M \to \overline{M}$ be an isometry of oriented surfaces in $\mathbf{R}^3$ that preserves shape operators, so

$$F_*(S(\mathbf{v})) = \overline{S}(F_*(\mathbf{v}))$$

for all tangent vectors to M. Then M and $\overline{M}$ are congruent; in fact, there is a Euclidean isometry $\mathbf{F}$ such that $\mathbf{F}|M = F$.

(If it should happen that $F_*(S(\mathbf{v})) = -\overline{S}(F_*(\mathbf{v}))$ for all tangent vectors, then reversing the orientation of either M or $\overline{M}$ will give the hypothesis as stated.)

Proof. There is only one candidate for $\mathbf{F}$, which we describe as follows. Fix a point $\mathbf{p}_0$ in M and a tangent frame $\mathbf{e}_1$, $\mathbf{e}_2$ at $\mathbf{p}_0$. Let E_3 and $\overline{E}_3$ be the unit normals to M and $\overline{M}$ that give S and $\overline{S}$. By Theorem 2.3 of Chapter 3, there is a unique Euclidean isometry $\mathbf{F}$ such that

$$\mathbf{F}_*(\mathbf{e}_1) = F_*(\mathbf{e}_1), \quad \mathbf{F}_*(\mathbf{e}_2) = F_*(\mathbf{e}_2), \quad \mathbf{F}_*(E_3(\mathbf{p}_0)) = \overline{E}_3(F(\mathbf{p}_0)).$$

These conditions imply that $(\mathbf{F}|M)_* = F_*$ at $\mathbf{p}_0$—a necessary condition for $\mathbf{F}|M = F$.

Since M is connected, if $\mathbf{p}$ is an arbitrary point of M, there is a curve α in M from $\alpha(0) = \mathbf{p}_0$ to $\mathbf{p}$. *It will suffice to show that* $\mathbf{F}(\alpha) = F(\alpha)$, for then certainly $\mathbf{F}(M) = \overline{M}$ and $\mathbf{F}|M = F$.

There is no loss of generality in assuming that α lies in the domain of a tangent frame field E_1, E_2 on M. If not, we could break up α into segments for which this is true and in sequence repeat essentially the following proof.

We adjust this frame field, using the same constant rotation (and perhaps reflection) at each point, so that its value at $\mathbf{p}_0$ is the selected frame $\mathbf{e}_1$, $\mathbf{e}_2$. This precaution is necessary in order that all choices of $\mathbf{p}$ in M will relate properly to the Euclidean isometry $\mathbf{F}$ defined above.

Next, this frame field is transferred, by the isometry F, to a frame field $\overline{E}_1$, $\overline{E}_2$ on $\overline{M}$. Appending the selected unit normals gives an adapted frame field E_1, E_2, E_3 on a domain in M, and correspondingly $\overline{E}_1$, $\overline{E}_2$, $\overline{E}_3$ on $\overline{M}$. Preparations are completed by restricting these frame fields to the curves α and $\overline{\alpha} = F(\alpha)$, respectively, as shown in Fig. 6.21.

The proof consists in applying Theorem 5.7 of Chapter 3 to the curves α and $\overline{\alpha}$. So we must verify the conditions (1) and (2) in that theorem.

Since the isometry F preserves velocities of curves and covariant derivatives of vector fields,

$$\alpha' \cdot E_k = F_*(\alpha') \cdot F_*(E_k) = F(\alpha)' \cdot \overline{E}_k = \overline{\alpha}' \cdot \overline{E}_k \quad (k = 1, 2).$$

Being tangent to M, α' is orthogonal to E_3; similarly for $\overline{\alpha}'$ and $\overline{E}_3$. Thus $\overline{\alpha}' \cdot \mathbf{E}_3 = 0 = \overline{\alpha}' \cdot \overline{E}_3$. So we now have

$$\alpha' \cdot E_i = \overline{\alpha}' \cdot \overline{E}_i \quad (1 \leq i \leq 3). \tag{1}$$

Let ω_{ij} and $\overline{\omega}_{ij}$ be the connection forms of the adapted frame fields $\{E_i\}$ and $\{\overline{E}_i\}$. By Lemma 5.3, $F^*(\overline{\omega}_{12}) = \omega_{12}$, and we compute

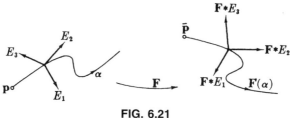

FIG. 6.21

$$E_1' \bullet E_2 = \nabla_{\alpha'} E_1 \bullet E_2 = \omega_{12}(\alpha') = F^*(\overline{\omega}_{12})(\alpha')$$
$$= \overline{\omega}_{12}(F_*(\alpha')) = \overline{\omega}_{12}(\overline{\alpha}') = \nabla_{\overline{\alpha}'} \overline{E}_1 \bullet \overline{E}_2 = \overline{E}_1' \bullet \overline{E}_2.$$

Since F preserves shape operators,

$$E_1' \bullet E_3 = \nabla_{\alpha'} E_1 \bullet E_3 = -\nabla_{\alpha'} E_3 \bullet E_1 = S(\alpha') \bullet E_1$$
$$= F_*(S(\alpha')) \bullet F_* E_1 = \overline{S}(\overline{\alpha}') \bullet \overline{E}_1 = \nabla_{\overline{\alpha}'} \overline{E}_1 \bullet \overline{E}_3 = \overline{E}_1' \bullet \overline{E}_3.$$

The same result holds with 1 replaced by 2, so $E_2' \bullet E_3 = \overline{E}_2' \bullet \overline{E}_3$. By skew-symmetry in i and j, we conclude that

$$E_i' \bullet E_j = \overline{E}_i' \bullet \overline{E}_j \quad (1 \le i, j \le 3). \tag{2}$$

Equations (1) and (2) above are those of Theorem 5.7 in Chapter 3 (with β replaced by $\overline{\alpha}$). Furthermore, by construction, the Euclidean isometry produced by the theorem is the isometry $\mathbf{F}$ defined above. ◆

This theorem provides a formal proof that the shape operators of a surface in $\mathbf{R}^3$ do, in fact, completely describe its shape.

Exercises

1. A surface $M \subset \mathbf{R}^3$ is *rigid* provided every surface in $\mathbf{R}^3$ isometric to M is congruent to M (thus M has only one possible shape in $\mathbf{R}^3$).

Deduce from Liebmann's theorem that *spheres are rigid*, that is, if $M \subset \mathbf{R}^3$ is isometric to a "round" sphere Σ: $\| \mathbf{p} - \mathbf{p}_0 \| = r$, then M *is* a round sphere.

2. If $\alpha, \beta: I \to \mathbf{R}^3$ are unit-speed curves with the same curvature $\kappa > 0$ and torsion τ, show that their tangent surfaces are congruent. (Compare Ex. 5 in Sec. 4, where only κ is the same.)

3. Let M and N be congruent surfaces in $\mathbf{R}^3$, with $\mathbf{F}$ a Euclidean isometry such that $\mathbf{F}(M) = N$. Prove that the isometry $\mathbf{F}|M$ preserves Gaussian and mean curvature, principal curvatures, principal directions, umbilics, asymptotic and principal curves, and geodesics. Which of these are preserved by arbitrary isometries $F: M \to N$?

4. For the sphere $\Sigma \subset \mathbf{R}^3$, show:
    (a) If $F: \Sigma \to \Sigma$ is an isometry, there is a Euclidean isometry $\mathbf{F}$ such that $F = \mathbf{F}|\Sigma$.
    (b) Same as (a), with $F: \Sigma \to \Sigma$ replaced by $F: \Sigma \to M \subset \mathbf{R}^3$.

5. In each of the following cases, show that the surfaces are congruent by finding an explicit Euclidean isometry $\mathbf{F} = TC$ that carries one surface to the other.
    (a) $z = xy$ and $y = xz$.
    (b) $x^2 - y^2 - z^2 = 1$ and $2yz - x^2 = 1$.

6. If M is a surface in $\mathbf{R}^3$, a Euclidean isometry $\mathbf{F}$ such that $\mathbf{F}(M) = M$ is called a *Euclidean symmetry* of M. Show that
    (a) The set of all Euclidean symmetries of M forms a subgroup $S(M)$ of the group $\mathscr{E}(3)$ of all isometries of $\mathbf{R}^3$ (Ex. 7 of Sec. 3.1). $S(M)$ is called the *Euclidean symmetry group* of M.
    (b) The Euclidean symmetric groups of congruent surfaces are isomorphic.

7. (a) Show that every Euclidean symmetry of the saddle surface

$$M: z = xy$$

is an orthogonal transformation C.
    (b) By considering the effect of C on the asymptotic unit vectors $\pm U_x$ and $\pm U_y$ of M at $\mathbf{0}$, show that M has exactly eight Euclidean symmetries and find their matrices.

8. Find all the Euclidean symmetries of the ellipsoid

$$\frac{x^2}{a^2} + \frac{y^2}{b^2} + \frac{z^2}{c^2} = 1, \quad \text{where } a > b > c.$$

(*Hint:* For the Gaussian curvature of the ellipsoid, see Example 5.2 of Ch. 5.)

6.10 Summary

The geometrical study of a surface M in $\mathbf{R}^3$ separates into three distinct categories:

(1) The intrinsic geometry of M.
(2) The shape of M in $\mathbf{R}^3$.
(3) The Euclidean geometry of $\mathbf{R}^3$.

We saw in Chapters 2 and 3 that the geometry of $\mathbf{R}^3$ is based on the dot product and consists of those concepts preserved by the isometries of $\mathbf{R}^3$. We have now found that the geometry of M is also based on the dot product—applied only to vectors tangent to M—and that it consists of those features preserved by the isometries of M.

The shape of M in $\mathbf{R}^3$ is the link between these two geometries. For example, Gaussian curvature K is a crucial invariant of the intrinsic geometry of M and (as Theorem 9.2 shows) the shape operator S dominates category (2). Thus the equation

$$K = \det S$$

shows that the geometries (1) and (3) can be harmonized only by means of restrictions on (2). Stated bluntly: *Only certain shapes in $\mathbf{R}^3$ are possible for a surface M with prescribed Gaussian curvature.* A strong result of this type is Liebmann's theorem, which asserts that a compact surface in $\mathbf{R}^3$ with K constant has only *one* possible shape—spherical.

Chapter 7
Riemannian Geometry

In studying the geometry of a surface in $\mathbf{R}^3$ we found that some of its most important geometric properties belong to the surface itself and not to the surrounding Euclidean space. Gaussian curvature is a prime example; although defined in terms of shape operators, it beongs to this intrinsic geometry since it passes the test of isometric invariance. In the 1850s Riemann drew the correct conclusion: There must exist a geometrical theory of surfaces *completely independent of* $\mathbf{R}^3$, a geometry built from the start solely of isometric invariants. In this chapter, we describe the fundamentals of the resulting theory, concentrating on its dominant features: Gaussian curvature and geodesics. Our constant guides will be the two special cases that led to its discovery: the intrinsic geometry of surfaces in $\mathbf{R}^3$ and Euclidean geometry—particularly that of the plane $\mathbf{R}^2$.

7.1 Geometric Surfaces

Evidence from earlier work on the intrinsic geometry of surfaces in $\mathbf{R}^3$ suggests that what is needed to do geometry on a surface is a dot product on tangent vectors. But to escape from confinement in $\mathbf{R}^3$, we must begin with an abstract surface M (Section 8 of Chapter 4). Since M need not be in $\mathbf{R}^3$ there is no dot product, and hence no geometry. However, the dot product is just one instance of the general notion of *inner product*, and Riemann's idea was to *replace the dot product by an arbitrary inner product on each tangent plane of M.*

1.1 Definition An *inner product* on a real vector space V is a function that assigns to each pair of vectors $\mathbf{v}$, $\mathbf{w}$ in V a number $\langle \mathbf{v}, \mathbf{w} \rangle$ with the following properties.

(1) Bilinearity:

$$\langle a_1\mathbf{v}_1 + a_2\mathbf{v}_2, \mathbf{w} \rangle = a_1\langle \mathbf{v}_1, \mathbf{w} \rangle + a_2\langle \mathbf{v}_2, \mathbf{w} \rangle,$$

$$\langle \mathbf{v}, b_1\mathbf{w}_1 + b_2\mathbf{w}_2 \rangle = b_1\langle \mathbf{v}, \mathbf{w}_1 \rangle + b_2\langle \mathbf{v}, \mathbf{w}_2 \rangle.$$

(2) Symmetry: $\langle \mathbf{v}, \mathbf{w} \rangle = \langle \mathbf{w}, \mathbf{v} \rangle$.
(3) Positive Definiteness:

$$\langle \mathbf{v}, \mathbf{v} \rangle \geqq 0; \quad \text{and} \quad \langle \mathbf{v}, \mathbf{v} \rangle = 0 \text{ if and only if } \mathbf{v} = 0.$$

On the vector space $\mathbf{R}^2$ the dot product $\mathbf{v} \cdot \mathbf{w} = v_1w_1 + v_2w_2$ is, of course, an inner product, but there are infinitely many others; for instance,

$$\langle \mathbf{v}, \mathbf{w} \rangle = 3v_1w_1 + 2v_2w_2.$$

(See Exercise 4.)

Basic features of the dot product remain valid for arbitrary inner products. The *length* of a vector $\mathbf{v}$ is $\|\mathbf{v}\| = \sqrt{\langle \mathbf{v}, \mathbf{v} \rangle}$, and vectors are orthogonal if $\langle \mathbf{v}, \mathbf{w} \rangle = 0$. The Schwarz inequality

$$\langle \mathbf{v}, \mathbf{w} \rangle \leq \|\mathbf{v}\| \, \|\mathbf{w}\|$$

allows the angle $0 \leq \vartheta \leq \pi$ between vectors to be defined by

$$\cos \vartheta = \frac{\langle \mathbf{v}, \mathbf{w} \rangle}{\|\mathbf{v}\| \, \|\mathbf{w}\|}.$$

Replacing surfaces in $\mathbf{R}^3$ by abstract surfaces and dot products by arbitrary inner products yields the following fundamental result.

1.2 Definition A *geometric surface* is an abstract surface M furnished with an inner product $\langle \, , \rangle$ on each of its tangent planes. These inner products are required to vary smoothly in the sense that if V and W are differentiable vector fields on M, then $\langle V, W \rangle$ is a differentiable real-valued function on M.

We emphasize that each tangent plane $T_p(M)$ of M has its own inner product, and these are unrelated except for the differentiability requirement—an obvious necessity for a theory founded on calculus. Here $\langle V, W \rangle$ has its usual pointwise meaning as the function assigning to each point $\mathbf{p}$ the number $\langle V(\mathbf{p}), W(\mathbf{p}) \rangle$.

The geometric structure provided by this collection of inner products can be described as a *metric tensor* g on M, that is, a function on all ordered pairs of tangent vectors $\mathbf{v}$, $\mathbf{w}$ at points $\mathbf{p}$ of M such that

$$g_p(\mathbf{v}, \mathbf{w}) = \langle \mathbf{v}, \mathbf{w} \rangle_p.$$

Thus g is analogous to a differential 2-form on M (Definition 4.1 of Chapter 4). But forms are *skew-symmetric*, $\omega(\mathbf{w}, \mathbf{v}) = -\omega(\mathbf{v}, \mathbf{w})$, while the metric tensor is *symmetric*, $g(\mathbf{w}, \mathbf{v}) = g(\mathbf{v}, \mathbf{w})$. (Informally, *metric tensor* is often shortened to just *metric*.)

Definition 1.2 can be summarized as

$$\text{surface} + \text{metric tensor} = \text{geometric surface}.$$

We emphasize that the same surface furnished with two different metric tensors constitutes two different geometric surfaces.

The Euclidean plane $\mathbf{R}^2$, with its usual dot product, is the best-known geometric surface. Simple though it may be, $\mathbf{R}^2$ is the basic testing ground for the geometry of surfaces. Of course, a surface M in $\mathbf{R}^3$ is a geometric surface, with the dot product of $\mathbf{R}^3$ applied, as usual, to tangent vectors on M. This gives the so-called *induced metric* on M, and unless some other geometric structure is explicitly mentioned, it is always assumed that a surface in $\mathbf{R}^3$ has the induced metric.

1.3 Remark *Construction methods.*

(1) *Conformal Change.* A simple way to get new geometric structures is to distort old ones. For example, if $h > 0$ is a differentiable function on a region in the Euclidean plane $\mathbf{R}^2$, define an inner product at each point $\mathbf{p}$ by

$$\langle \mathbf{v}, \mathbf{w} \rangle = \frac{\mathbf{v} \bullet \mathbf{w}}{h^2(\mathbf{p})}.$$

The resulting geometric surface M is said to be *conformal* with *ruler function* h. In fact, the inner product of M gives the same angle measurements as the dot product on the Euclidean plane.

We call $h > 0$ a ruler function since larger values of h give smaller values for the length of vectors (see Section 2). Examples will soon show that unless h is quite special, the surface M has properties quite different from the Euclidean plane. In fact, *locally* every geometric surface can be so expressed.

(2) *Pullback.* A metric tensor g can be pulled back by a suitable mapping in the same way as differential forms are. Here is an important application. Let $F\colon M \to N$ be a regular mapping from an abstract surface M to a geometric surface N with metric tensor $g = \langle \, , \, \rangle$. The pullback $F^*(g)$ of g to M is given in terms of inner products as

$$\langle \mathbf{v}, \mathbf{w} \rangle_M = \langle F_*(\mathbf{v}), F_*(\mathbf{w}) \rangle_N.$$

By definition, each tangent map F_* is a linear isomorphism (linear, one-to-one, and onto), so it is easy to check that this definition gives an inner product on each tangent space $T_p(M)$. (The differentiability condition follows from the fact that F is a diffeomorphism on small enough neighborhoods.) Thus $F^*(g)$ is a metric tensor on M—in fact, it is the unique one that makes F a local isometry.

(3) *Coordinate description.* Let $\mathbf{x}$ be a coordinate patch in an abstract surface M (without geometry). If M were geometric, the functions

$$E = \langle \mathbf{x}_u, \mathbf{x}_u \rangle, \quad F = \langle \mathbf{x}_u, \mathbf{x}_v \rangle, \quad G = \langle \mathbf{x}_v, \mathbf{x}_v \rangle$$

would be defined as before. But reversing this logic, if suitable functions E, F, G are *given*, then these equations determine a unique metric tensor $\langle\,,\,\rangle$ on the image of $\mathbf{x}$ (see Exercise 4).

Many examples of these methods appear later on.

From the modest beginning in Definition 1.2, a geometric theory can be built that vastly enlarges that of Chapters 5 and 6. Definitions and theorems from before that are clearly intrinsic in character will be used without further discussion. In particular, an isometry $F: M \to N$ of geometric surfaces is still defined by Definition 4.2 of Chapter 6, and the geometry of a geometric surface M consists by definition of its isometric invariants.

Now that the calculus of $\mathbf{R}^3$ is gone, frame field computations on a surface itself become more important. A *frame field* on an arbitrary geometric surface M consists, as usual, of two orthogonal unit vector fields E_1, E_2 defined on some open set in M. The orthonormality conditions

$$\langle E_1, E_1 \rangle = 1, \quad \langle E_1, E_2 \rangle = 0, \quad \langle E_2, E_2 \rangle = 1$$

are now expressed, of course, in terms of the metric tensor of M. Two ways to construct frame fields are given in Exercises 7 and 9 of Section 2.

As before, dual 1-forms θ_1, θ_2 are uniquely determined by $\theta_i(E_j) = \delta_{ij}$, and by Lemma 5.1 of Chapter 6, the connection form $\omega_{12} = -\omega_{21}$ is uniquely characterized by the first structural equations

$$d\theta_1 = \omega_{12} \wedge \theta_2, \quad d\theta_2 = \omega_{21} \wedge \theta_1.$$

We emphasize that these forms θ_1, θ_2, ω_{12} are not invariants of the surface M; a different choice of frame field $\overline{E}_1$, $\overline{E}_2$ will produce different forms $\overline{\theta}_1$, $\overline{\theta}_2$, $\overline{\omega}_{12}$. To obtain invariant results we will need to know how two such sets of forms are related.

On a small enough neighborhood of a point $\mathbf{p} \in M$, careful use of the inverse functions $\cos^{-1}$ and $\sin^{-1}$ will yield a differential angle function φ such that

$$\overline{E}_1 = \cos \varphi \, E_1 + \sin \varphi \, E_2.$$

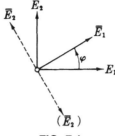

FIG. 7.1

As Fig. 7.1 indicates, there are now two choices for E_2. Either

$$\overline{E}_2 = -\sin \varphi \, E_1 + \cos \varphi \, E_2,$$

so E_1, E_2 and $\overline{E}_1, \overline{E}_2$ have the *same orientation*, or

$$\overline{E}_2 = \sin \varphi \, E_1 - \cos \varphi \, E_2,$$

and they have *opposite orientation*.

In working in an oriented region, it is natural to use positively oriented frame fields. Evidently any two such have the same orientation.

1.4 Lemma Let E_1, E_2 and $\overline{E}_1, \overline{E}_2$ be frame fields on the same region in M. If these frame fields have

(1) the same orientation, then

$$\overline{\omega}_{12} = \omega_{12} + d\varphi, \quad \text{and} \quad \overline{\theta}_1 \wedge \overline{\theta}_2 = \theta_1 \wedge \theta_2;$$

(2) opposite orientation, then

$$\overline{\omega}_{12} = -(\omega_{12} + d\varphi), \quad \text{and} \quad \overline{\theta}_1 \wedge \overline{\theta}_2 = -\theta_1 \wedge \theta_2.$$

Proof. We prove only (1), since (2) follows by changes of sign. By the basis formulas in Lemma 2.1 of Chapter 6, the equations

$$\overline{E}_1 = \cos \varphi \, E_1 + \sin \varphi \, E_2,$$
$$\overline{E}_2 = -\sin \varphi \, E_1 + \cos \varphi \, E_2$$

yield

$$\theta_1 = \cos \varphi \, \overline{\theta}_1 - \sin \varphi \, \overline{\theta}_2, \quad \theta_2 = \sin \varphi \, \overline{\theta}_1 + \cos \varphi \, \overline{\theta}_2. \tag{$*$}$$

Taking the exterior derivative of the first of these gives

$$d\theta_1 = -\sin\varphi\, d\varphi \wedge \overline{\theta}_1 + \cos\varphi\, d\overline{\theta}_1 - \cos\varphi\, d\varphi \wedge \overline{\theta}_2 - \sin\varphi\, d\overline{\theta}_2.$$

Now substitute the first structural equations for $d\overline{\theta}_1$, $d\overline{\theta}_2$ to obtain

$$d\theta_1 = (\overline{\omega}_{12} - d\varphi) \wedge (\sin\varphi\,\overline{\theta}_1 + \cos\varphi\,\overline{\theta}_2)$$
$$= (\overline{\omega}_{12} - d\varphi) \wedge \theta_2.$$

In the same way, we get

$$d\theta_2 = -(\overline{\omega}_{12} - d\varphi) \wedge \theta_1.$$

Because the form $\omega_{12} = -\omega_{21}$ *uniquely* satisfies the first structural equations, we conclude from the last two equations that $\overline{\omega}_{12} = \omega_{12} + d\varphi$, as required. Then direct computation of $\theta_1 \wedge \theta_2$ using (∗) shows that this 2-form equals $\overline{\theta}_1 \wedge \overline{\theta}_2$.   ◆

In this chapter, as earlier, the restriction to low dimensions is not essential. A surface is the 2-dimensional case of the general notion of *manifold* (Chapter 4, Section 8). A manifold of arbitrary dimension furnished with a metric tensor is called a *Riemannian manifold*. Euclidean geometry, as discussed in Chapter 3, is the special case of Riemannian geometry produced on $\mathbf{R}^n$ by the usual dot product.

Thus a geometric surface is the same thing as a 2-*dimensional Riemannian manifold*, and the subject of this chapter is 2-*dimensional Riemannian geometry*.†

Exercises

1. In a conformal geometric surface with ruler function h (Remark 1.3(1)), show that:

(a) The speed of a curve $\alpha = (\alpha_1, \alpha_2)$ is $\sqrt{\alpha_1'^2 + \alpha_2'^2} / h(\alpha)$.

(b) hU_1, hU_2 is a frame field with dual 1-forms du/h, dv/h.

(c) The area forms are $\pm\, du \wedge dv/h^2$

(d) The identity map from the Euclidean plane to this surface is a conformal map with scale factor $\lambda = 1/h$ (Def. 4.7 of Ch. 6). (Using h rather than λ as the descriptive function leads to simpler formulas.)

2. The *Poincaré half-plane* is the upper half-plane $v > 0$ in $\mathbf{R}^2$ with metric tensor $\langle \mathbf{v}, \mathbf{w} \rangle = \mathbf{v} \bullet \mathbf{w}/v^2(\mathbf{p})$. If α is the curve $\alpha(t) = (r\cos t, r\sin t)$, $0 < t < \pi$, with constant $r > 0$,

---

† We would prefer to call a geometric surface a *Riemann(ian) surface*, but this term has a firmly established and distinctly different meaning.

(a) Show that the speed of α is csc t.

(b) Deduce that although the Euclidean length of α is πr, its Poincaré length is infinite.

(c) Find the area of the region between α and the u axis.

For a surface M in $\mathbf{R}^3$ oriented by a unit normal U, the rotation operator J was defined by $J(\mathbf{v}) = U \times \mathbf{v}$. Thus if M is an arbitrary geometric surface, we need a new definition for J so that the results at the end of Section 8, Chapter 6, will remain valid on M.

3. (*Rotation operator.*) Let M be a geometric surface oriented by an area form dM. Prove:

(a) On each tangent space to M there exists a unique "rotation by $+90°$," that is, a linear operator $J: T_p(M) \to T_p(M)$ such that

$$\|J(\mathbf{v})\| = \|\mathbf{v}\|, \quad \langle J(\mathbf{v}), \mathbf{v}\rangle = 0, \quad dM(\mathbf{v}, J(\mathbf{v})) > 0 \ (\text{if } \mathbf{v} \neq 0).$$

(*Hint:* If J is the linear operator such that $J(\mathbf{e}_1) = \mathbf{e}_2$, $J(\mathbf{e}_2) = -\mathbf{e}_1$ holds for one positively oriented frame, show that these relations hold for *every* such frame.)

(b) J is differentiable (that is, V differentiable implies $J(V)$ differentiable), skew-symmetric (that is, $\langle J(\mathbf{v}), \mathbf{w}\rangle = -\langle \mathbf{v}, J(\mathbf{w})\rangle$), and $J^2 = -I$, that is, $J(J(\mathbf{v})) = -\mathbf{v}$ for all $\mathbf{v}$.

(c) If M is oriented instead by $-dM$, then its rotation operator is $-J$.

(d) If M is a surface in $\mathbf{R}^3$ oriented by a unit normal U and dM is the area form such that $dM(\mathbf{v}, \mathbf{w}) = U \bullet \mathbf{v} \times \mathbf{w}$, then for J as in (a), $J(\mathbf{z}) = U \times \mathbf{z}$ for all $\mathbf{z}$.

4. (*Coordinate definition of a metric.*)

(a) If a, b, and c are numbers such that $a > 0$, $c > 0$, $ac - b^2 > 0$, then the formula

$$\langle \mathbf{x}_u, \mathbf{x}_v\rangle = av_1w_1 + b(v_1w_2 + v_2w_1) + cv_2w_2$$

defines an inner product on $\mathbf{R}^2$. (*Hint:* $\left(\sqrt{a}v_1 \pm \sqrt{c}v_2\right)^2 \geq 0$.)

(b) Let $\mathbf{x}: D \to M$ be a coordinate patch in an abstract surface M (without geometry). Given differentiable functions E, F, and G on D such that

$$E > 0, \quad G > 0, \quad EG > F^2$$

prove that there is a unique metric tensor $\langle,\rangle$ on the image of $\mathbf{x}$ such that

$$E = \langle \mathbf{x}_u, \mathbf{x}_u\rangle, \quad F = \langle \mathbf{x}_u, \mathbf{x}_v\rangle, \quad G = \langle \mathbf{x}_v, \mathbf{x}_v\rangle.$$

5. (*Line-element.*)
(a) For an arbitrary inner product, prove the *polarization identity*

$$\langle \mathbf{v}, \mathbf{w} \rangle = \frac{1}{4} \left(\|\mathbf{v} + \mathbf{w}\|^2 - \|\mathbf{v} - \mathbf{w}\|^2 \right).$$

(b) Deduce that the metric tensor of a geometric surface M is completely determined by the function q whose value on each tangent vector $\mathbf{v}$ is $\|\mathbf{v}\|^2$.

Classically, q is called the *line-element*, or *first fundamental form*, of M and is expressed in coordinates as

$$q = ds^2 = E \, du^2 + 2F \, du \, dv + G \, dv^2.$$

(Here $du \, dv$, for example, is ordinary multiplication, *not* wedge product.)
(c) Account for the unusual notation ds^2 by finding the relevant formula for the speed ds/dt of a curve $\alpha(t) = \mathbf{x}(u(t), v(t))$ (compare Ex. 5 of Sec. 5.4).

6. If $F: M \to N$ is a regular mapping of oriented geometric surfaces, show that the following are equivalent:
(a) F is orientation-preserving and conformal (Def. 4.7 of Ch. 6).
(b) F preserves rotation operators; that is, for all tangent vectors $\mathbf{v}$,

$$F_* (J_M(\mathbf{v})) = J_N (F_* (\mathbf{v})).$$

(c) F preserves oriented angles; that is, if φ is an oriented angle from $\mathbf{v}$ to $\mathbf{w}$, then it is also an oriented angle from $F_*(\mathbf{v})$ to $F_*(\mathbf{w})$.

7. Prove that a regular mapping $F = (f, g)$ from a region $\mathcal{U} \subset \mathbf{R}^2$ into $\mathbf{R}^2$ is conformal and orientation-preserving if and only if the *Cauchy-Riemann equations* $f_u = g_v$, $f_v = -g_u$ hold.
If $\mathbf{R}^2$ is considered as the complex plane, with $z = (u, v) = u + iv$, these equations are necessary and sufficient for $z \to F(z)$ to be a complex analytic function. Show also that the scale factor of the resulting conformal map is $|dF/dz|$.

8. (*Continuation.*) If the origin is deleted from $\mathbf{R}^2$, show that the mapping F in (2) of Example 7.3 of Chapter 1 is conformal and orientation-preserving. What is the complex function in this case?

9. Let M be a region $\mathcal{U} \subset \mathbf{R}^2$ furnished with a conformal metric given by ruler function h. If F is a Euclidean isometry $\mathcal{U} \to \mathcal{U}$ that preserves h (that is, $h(F) = h$), show that F is an isometry $M \to M$.

7.2 Gaussian Curvature

For geometric surfaces we need a new definition of Gaussian curvature. The definition $K = \det S$ for surfaces in $\mathbf{R}^3$ is meaningless now, since it uses shape operators. But this original definition made K an isometric invariant, so it is reasonable to look at the proof of the *theorema egregium*—specifically to Corollary 2.3 of Chapter 6—to find a satisfactory generalization.

2.1 Theorem   On a geometric surface M there is a unique real-valued function K such that for every frame field on M the *second structural equation*

$$d\omega_{12} = -K\theta_1 \wedge \theta_2$$

holds. K is called the *Gaussian curvature* of M.

Proof.   For a frame field E_1, E_2, it follows from the basis formulas in Lemma 2.1 of Chapter 6 that there is a unique function K such that

$$d\omega_{12} = -K\theta_1 \wedge \theta_2.$$

But *a priori* this function depends on the particular frame field: Another frame $\overline{E}_1$, $\overline{E}_2$ might have a different function $\overline{K}$ for which

$$d\overline{\omega}_{12} = -\overline{K}\,\overline{\theta}_1 \wedge \overline{\theta}_2.$$

Evidently we must show that where the domains of the frame fields overlap, $K = \overline{K}$. Such domains cover all of M, so we will then have a single function K with the required property.

Suppose first that the two frame fields have the same orientation. By Lemma 1.4, $\overline{\omega}_{12} = \omega_{12} + d\varphi$. Hence $d\overline{\omega}_{12} = d\omega_{12}$, since $d^2 = 0$. But then

$$\overline{K}\,\overline{\theta}_1 \wedge \overline{\theta}_2 = K\theta_1 \wedge \theta_2.$$

Since

$$\overline{\theta}_1 \wedge \overline{\theta}_2 = \theta_1 \wedge \theta_2 \neq 0,$$

we conclude that $\overline{K} = K$.

When the orientations are opposite, we get $d\overline{\omega}_{12} = -d\omega_{12}$, but still find $\overline{K} = K$, since

$$\overline{\theta}_1 \wedge \overline{\theta}_2 = -\theta_1 \wedge \theta_2. \qquad \blacklozenge$$

As mentioned above, Corollary 2.3 of Chapter 6 shows that this general definition of Gaussian curvature agrees with the definition $K = \det S$ when M is a surface in $\mathbf{R}^3$. The proof of isometric invariance presented there is entirely intrinsic in character and thus holds for arbitrary geometric surfaces. Furthermore, the orthogonal curvature formula in Proposition 6.3 of Chapter 6 was derived from $d\omega_{12} = -K\theta_1 \wedge \theta_2$, without reference to the shape operator; *hence it is valid for any geometric surface.*

To summarize: Geometrical investigations in terms of a frame field E_1, E_2 are expressed in terms of its structural equations

$$d\theta_1 = \omega_{12} \wedge \theta_2, \quad d\theta_2 = \omega_{21} \wedge \theta_1,$$

$$d\omega_{12} = -K\theta_1 \wedge \theta_2.$$

The first structural equations determine the connection form $\omega_{12} = -\omega_{21}$ of the frame field, but the second structural equation defines the Gaussian curvature *of the geometric surface*—independent of choice of frame field. Chapter 6, Section 6 showed how ω_{12} and K can be explicitly computed from these implicit definitions.

Gaussian curvature is the central property of a geometric surface M. We will see that it influences—often decisively—virtually every aspect of the geometry of M.

Let us make certain that the new definition of curvature gives the correct result for the Euclidean plane $\mathbf{R}^2$. Its natural frame field $U_1 = (1, 0)$, $U_2 = (0, 1)$ has dual 1-forms $\theta_1 = du$, $\theta_2 = dv$. Then $d\theta_1 = d\theta_2 = 0$, so the identically zero 1-form $\omega_{12} = 0$ satisfies the first structural equations and hence is the connection form of U_1, U_2. Obviously, $d\omega_{12} = 0$, so by the preceding theorem, $K = 0$. It can be no surprise that the Euclidean plane is flat, for $\mathbf{R}^2$ is isometric to, say, the xy plane in $\mathbf{R}^3$, which has $K = 0$ since its shape operators all vanish.

2.2 Example *A flat torus.* Let T be a torus of revolution considered as an abstract surface, without geometry. If $\mathbf{x}$ is its usual parametrization (Example 2.5 of Chapter 4), then the definitions

$$\langle \mathbf{x}_u, \mathbf{x}_u \rangle = 1, \quad \langle \mathbf{x}_u, \mathbf{x}_v \rangle = 0, \quad \langle \mathbf{x}_v, \mathbf{x}_v \rangle = 1$$

determine a unique inner product on each tangent plane of T. The resulting geometric surface T_0 is certainly different from the torus of revolution: it is *flat*. To see this, recall that

$$\mathbf{x}_*(U_1) = \mathbf{x}_u \quad \text{and} \quad \mathbf{x}_*(U_2) = \mathbf{x}_v.$$

Thus $\mathbf{x}_*$ carries frames to frames, and consequently $\mathbf{x}$ is a local isometry of the Euclidean plane $\mathbf{R}^2$ onto T_0—and local isometries preserve curvature. ◆

In effect, T_0 is constructed by gluing opposite sides of a Euclidean rectangle, as specified by the map $\mathbf{x}$. This metric on the flat torus is not the induced metric derived, as usual, from the dot product of $\mathbf{R}^3$. If it were, then since T_0 is compact, by Theorem 3.5 of Chapter 6 it would have to have positive curvature somewhere. Thus T_0 *can never be found in* $\mathbf{R}^3$.

This shows that the class of all geometric surfaces is larger than that of surfaces in Euclidean 3-space—in fact, it is immensely larger. In the course of this chapter and the next, we hope to persuade the reader that geometric surfaces are the natural objects to study and that surfaces in $\mathbf{R}^3$—however visually appealing—are only an interesting special case.

We now find a formula for the Gaussian curvature of the conformal geometric surfaces described in Remark 1.3.

2.3 Corollary For the plane $\mathbf{R}^2$ with metric tensor $\langle \mathbf{v}, \mathbf{w} \rangle = \mathbf{v} \cdot \mathbf{w}/h^2(\mathbf{p})$, the Gaussian curvature is

$$K = h(h_{uu} + h_{vv}) - (h_u^2 + h_v^2).$$

Proof. Let M denote the plane with its new metric. Then the identity map $\mathbf{R}^2 \to M$ is a patch with $E = G = 1/h^2$ and $F = 0$. Thus the result follows directly from the orthogonal curvature formula in Proposition 6.3 of Chapter 6. (Since the patch is conformal, we could also get K as $h^2 \Delta(\log h)$ from Exercise 2 of Section 6.6) ◆

2.4 Example (1) *The stereographic sphere*. It was shown in Example 5.5 of Chapter 4 that stereographic projection P is a diffeomorphism of the punctured sphere Σ_0 onto the Euclidean plane $\mathbf{R}^2$. Now consider Σ_0 as merely an abstract surface, and assign it the pullback metric tensor of Remark 1.3(2) that makes P an isometry. If Σ_0 appears round in Fig. 7.2, it is only because we look at it with Euclidean eyes—intrinsically, this Σ_0 is as flat as the plane.

(2) *The stereographic plane*. Now reverse the process in (1). Consider Σ_0 with its usual geometric structure as a surface in $\mathbf{R}^3$ and ignore the geometry of $\mathbf{R}^2$.

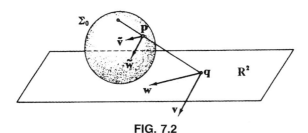

FIG. 7.2

The inverse $P^{-1}: \mathbf{R}^2 \to \Sigma_0$ of stereographic projection is a diffeomorphism since P is. Hence the pullback operation makes $\mathbf{R}^2$ a geometric surface—the *stereographic plane*—that is isometric to $\Sigma_0 \subset \mathbf{R}^3$ and hence has curvature $K = 1$.

We examine this stereographic plane more closely. If $\mathbf{v}$ and $\mathbf{w}$ are tangent vectors to $\mathbf{R}^2$ at $\mathbf{q} = P(\mathbf{p})$, let $\tilde{\mathbf{v}}$ and $\tilde{\mathbf{w}}$ be the unique tangent vectors to Σ_0 that P_* carries to $\mathbf{v}$ and $\mathbf{w}$, respectively. By Exercise 4.14 of Chapter 6, we know that in terms of Euclidean dot products,

$$\mathbf{v} \bullet \mathbf{w} = P_*(\tilde{\mathbf{v}}) \bullet P_*(\tilde{\mathbf{w}}) = \left(1 + \frac{\|\mathbf{q}\|^2}{4}\right)^2 \tilde{\mathbf{v}} \bullet \tilde{\mathbf{w}}.$$

But $(P^{-1})_* = (P_*)^{-1}$ carries $\mathbf{v}$ and $\mathbf{w}$ back to $\tilde{\mathbf{v}}$ and $\tilde{\mathbf{w}}$, so for the pulled back inner product on $\mathbf{R}^2$ we find

$$\langle \mathbf{v}, \mathbf{w} \rangle = (P^{-1})_*(\mathbf{v}) \bullet (P^{-1})_*(\mathbf{w}) = \tilde{\mathbf{v}} \bullet \tilde{\mathbf{w}} = \left(1 + \frac{\|\mathbf{q}\|^2}{4}\right)^{-2} \mathbf{v} \bullet \mathbf{w}.$$

It follows immediately that the new metric on the stereographic plane is conformal as in Remark 1.3, with ruler function

$$h(u, v) = 1 + \frac{u^2 + v^2}{4}.$$

To visualize this unusual plane, it helps to imagine that *rulers get longer as they move farther from the origin*. Since P is now an isometry, the intrinsic distance from $\mathbf{p}$ to $\mathbf{q}$ in Fig. 7.3 is exactly the same as the distance from $\mathbf{p}_*$ to $\mathbf{q}_*$. Also, circles $u^2 + v^2 = r^2$ with r very large actually have very small stereographic radii, since they correspond under P to small circles about the (missing) north pole of Σ_0.

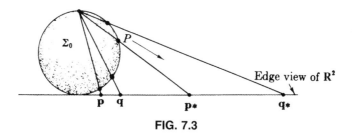

FIG. 7.3

2.5 Example   *The hyperbolic plane.* Let us experiment with a change of sign in the stereographic inner product above, setting

$$h(u, v) = 1 - \frac{u^2 + v^2}{4}.$$

Since $h > 0$ is necessary, this *hyperbolic inner product* $\langle \mathbf{v}, \mathbf{w} \rangle = \mathbf{v} \cdot \mathbf{w}/h^2$ is used only on the open disk $u^2 + v^2 < 4$ of radius 2 in $\mathbf{R}^2$. The resulting geometric surface is the Poincaré disk model of the *hyperbolic plane H*. It follows easily from Corollary 2.3 that *the hyperbolic plane has constant Gaussian curvature* $K = -1$. ◆

As a point (u, v) approaches the *rim* of H, that is, the circle $u^2 + v^2 = 4$ (not part of H!), $h(u, v)$ approaches zero. In the language used above, rulers *shrink* as they approach the rim, so H is a good deal bigger than Euclidean intuition may suggest.

To verify this, let us find the arc length function $s(t)$ of the Euclidean ray

$$\gamma(t) = (t\cos \vartheta,\ t\sin \vartheta) \quad (0 \leqq t < 2),$$

which runs, at angle ϑ, from the center of H out to the rim. On γ, the ruler function becomes $1 - t^2/4$. Since $\gamma' = (\cos \vartheta, \sin \vartheta)$,

$$\frac{ds}{dt} = \|\gamma'(t)\| = \frac{1}{h(\gamma)} = \frac{1}{1 - (t/2)^2}.$$

Then

$$s(t) = \int_0^t \frac{du}{1 - (u/2)^2} = 2\ \tanh^{-1} \frac{t}{2} = \log \frac{2 + t}{2 - t}.$$

Thus, as t approaches 2, $s(t)$ approaches ∞. This "short" segment has infinite hyperbolic length. (Although distance in H is quite non-Euclidean, recall that by conformality, angle measurement is the same.)

Further properties of the hyperbolic plane will be developed as we go along. We will see that it—and not the bugle surface (Example 7.6 in Chapter 5)—is the true analogue of the sphere for constant negative curvature.

Any regular mapping $F: M \to N$ can be used to pull a metric tensor on N back to M, but unless F is a diffeomorphism, it is usually not possible to *push forward* a metric from M to N. The trouble is that when F carries points $\mathbf{p}_1 \neq \mathbf{p}_2$ of M to the same point of N, the inner products at these points may transfer differently to $F(\mathbf{p}_1) = F(\mathbf{p}_2)$. This difficulty can be eliminated by imposing a consistency condition, as follows.

2.6 Proposition Let F be a regular mapping of a geometric surface M onto a surface N without geometry. Suppose that whenever $F(\mathbf{p}_1) = F(\mathbf{p}_2)$ there is an isometry G_{12} from a neighborhood of $\mathbf{p}_1$ to a neighborhood of $\mathbf{p}_2$ such that

$$FG_{12} = F, \quad G_{12}(\mathbf{p}_1) = \mathbf{p}_2.$$

Then there is a unique metric tensor on N that makes F a local isometry.

Proof. If F is to be a local isometry, there is no choice in defining the inner product of tangent vectors $\mathbf{v}$, $\mathbf{w}$ at $\mathbf{q}$ in N. F is regular, so at any point $\mathbf{p}_1$ in M such that $F(\mathbf{p}_1) = \mathbf{q}$ there are unique vectors $\mathbf{v}_1$, $\mathbf{w}_1$ such that

$$F_*(\mathbf{v}_1) = \mathbf{v}, \quad F_*(\mathbf{w}_1) = \mathbf{w}.$$

Hence we must define $\langle \mathbf{v}, \mathbf{w} \rangle_N$ to be $\langle \mathbf{v}_1, \mathbf{w}_1 \rangle_M$.

For this to be a valid definition, it must not depend on the choice of $\mathbf{p}_1$. Explicitly, if $\mathbf{p}_2$ is another point such that $F(\mathbf{p}_2) = \mathbf{q}$ and if $\mathbf{v}_2$, $\mathbf{w}_2$ are the unique vectors there such that

$$F_*(\mathbf{v}_2) = \mathbf{v}, \quad F_*(\mathbf{w}_2) = \mathbf{w},$$

then we require

$$\langle \mathbf{v}_2, \mathbf{w}_2 \rangle_M = \langle \mathbf{v}_1, \mathbf{w}_1 \rangle_M. \tag{$*$}$$

To prove this, let $G = G_{12}$ be an isometry as in the statement of the proposition. Since $FG = F$, the chain rule gives $F_*G_* = F_*$. Consequently, G_* carries the vectors $\mathbf{v}_1$, $\mathbf{w}_1$ to the vectors at $\mathbf{p}_2$ that F_* carries to $\mathbf{v}$, $\mathbf{w}$, namely, to $\mathbf{v}_2$, $\mathbf{w}_2$. Since G is an isometry, $(*)$ holds.   ◆

2.7 Example   *The projective plane.* Example 8.2 of Chapter 4 defined the projective plane P as an abstract surface by identifying antipodal points of a unit sphere $\Sigma \subset \mathbf{R}^3$. Now we give P a geometric structure. Recall that the projection $F: \Sigma \to P$ is related to the antipodal map $A(\mathbf{p}) = -\mathbf{p}$ by $FA = F$. Since A is an isometry, the preceding result applies, with all maps G_{12} being A. The resulting local isometry $\Sigma \to P$ shows that like the sphere, P has constant positive curvature $K = 1$ and all its geodesics are closed.

The same scheme, applied when Σ has radius r, gives the *projective plane of radius r*, with curvature $K = -1/r^2$.   ◆

Like the flat torus, P cannot be found in $\mathbf{R}^3$. For the flat torus this failure could be blamed on its *geometry*, since tori of revolution abound in $\mathbf{R}^3$. But for P the obstruction is *topological*, because (by Theorem 7.10 of Chapter 4) compact surfaces in $\mathbf{R}^3$ must be orientable—and P is not.

Now that we have a geometry for surfaces that may not occur in $\mathbf{R}^3$, a natural place to look for them is in higher-dimensional Euclidean spaces. In the simplest case, let $\mathbf{x}: D \to \mathbf{R}^n$, $n \geq 3$, be a regular map, so $\mathbf{x}(D)$ is at least an immersed surface. Then pullback of the dot product metric of $\mathbf{R}^n$ makes D a geometric surface, with

$$\langle \mathbf{v}, \mathbf{w} \rangle = \mathbf{x}_* (\mathbf{v}) \bullet \mathbf{x}_* (\mathbf{v}).$$

We think of D as a geometric description of $\mathbf{x}(D)$, as in the $n = 3$ case.

2.8 Example *Tangent surfaces.* For any $n \geq 3$, let β be a unit-speed curve in $\mathbf{R}^n$ with $\kappa = \|T'\| > 0$. The tangent surface is given as before by the formula $\mathbf{x}(u, v) = \beta(u) + vT(u)$, with $v \neq 0$. Then $\mathbf{x}_u = T + vT'$ and $\mathbf{x}_v = T$, so

$$E = 1 + v^2\kappa^2(u), \quad F = 1, \quad G = 1.$$

Since $EG - F^2 = v^2\kappa^2(u) > 0$, $\mathbf{x}$ is regular. The resulting immersed surface is flat, just as in the $\mathbf{R}^3$ case. This follows from the general curvature formula in Exercise 9, but it is clear without computation since we know that K derives solely from E, F, and G, and these are given by the same expressions as in the 3-dimensional case (Exercise 13 in Section 5.4). ◆

Exercises

1. Show that the Poincaré half-plane (Ex. 2 of Sec. 1) has constant negative Gaussian curvature $K = -1$.

2. For the conformal structure (Rmk. 1.3) on the entire plane with ruler function $h = \text{sech}(uv)$, find the dual forms and connection forms of the frame field hU_1, hU_2 and derive the Gaussian curvature K. Check by finding K from Corollary 2.3.

3. Find the area of the disk $u^2 + v^2 \leq a^2$ in the hyperbolic plane. (*Hint:* Use polar coordinates $\mathbf{x}(r, v) = (r\cos v, r\sin v)$.) What is the area of the entire hyperbolic plane?

4. The *hyperbolic plane $H(r)$ of pseudo-radius r* is the disk $u^2 + v^2 < 4r^2$ with conformal metric given by the ruler function

$$h(u, v) = 1 - \frac{u^2 + v^2}{4r^2}.$$

Thus the hyperbolic plane of Example 2.5 is $H(1)$. Show that $H(r)$ has constant Gaussian curvature $K = -1/r^2$.

5. In Example 2.2 suppose the flat torus T_0 is constructed from a torus of revolution T with radii $R > r > 0$. Find the area of T_0. How does this compare with the area of T with its usual geometry as a surface in $\mathbf{R}^3$?

6. (a) Show that the geometric surface in Exercise 2 is isometric to a helicoid. (*Hint:* Use $\mathbf{x}(u, v) = (f(u)\cos v, f(u)\sin v, v)$ for a suitable function $f(u)$.)
(b) Check that the two surfaces have the same Gaussian curvature at corresponding points.

7. A *scale change* of M stretches every dimension of M by a constant factor $c > 0$. Formally, if M is a geometric surface with inner product $\langle,\rangle$, let $\overline{M}$ be the same abstract surface with inner product

$$\langle \mathbf{v}, \mathbf{w} \rangle^- = c^2 \langle \mathbf{v}, \mathbf{w} \rangle.$$

We say that $\overline{M}$ is M *scaled by* $c > 0$. Show that:

(a) $\|\mathbf{v}\|^- = c\|\mathbf{v}\|$ for all $\mathbf{v}$, but angles between vectors are the same.
(b) For any curve segment α, $\overline{L}(\alpha) = cL(\alpha)$.

(c) Frame fields E_1, E_2 on M and E_1/c, E_2/c on $\overline{M}$ have dual 1-forms $\overline{\theta}_i = c\theta_i$, but their connection forms are equal.
(d) A region $\mathscr{R}$ has M area A if and only if it has $\overline{M}$ area c^2A.
(e) $\overline{K} = K/c^2$.

Taking $c > 1$ for definiteness, the scale change expands every dimension of M by the factor c—which produces *smaller* curvature (compare a sphere of radius 2 with a unit sphere). It will follow from Lemma 4.5 that M and $\overline{M}$ have the same geodesics—though speeds differ.

8. Prove:
(a) The sphere $\Sigma(r) \subset \mathbf{R}^3$ of radius r is isometric to the unit sphere Σ scaled by r.
(b) The hyperbolic plane $H(r)$ of Exercise 4 preceding is isometric to $H = H(1)$ scaled by r.

Because the geometric effects of scale change are so simple, it usually suffices to work with standard models such as Σ and H.

9. (*Classical tensor formula for Gaussian curvature.*) For an arbitrary patch $\mathbf{x}$, the *associated frame field* is

$$E_1 = \frac{\mathbf{x}_u}{\sqrt{E}}, \quad E_2 = \frac{1}{W\sqrt{E}}(E\mathbf{x}_v - F\mathbf{x}_u),$$

where as usual, $W = \sqrt{EG - F^2}$.

(E_2 is found by subtracting from $\mathbf{x}_v$ its E_1 component and reducing the result to unit length—the so-called Gram-Schmidt process.)

(a) Check that E_1, E_2 is orthonormal.

(b) Express the dual 1-forms θ_1, θ_2 in terms of du and dv.

(c) Find the functions P and Q such that $\omega_{12} = P\,du + Q\,dv$.

(d) Deduce the curvature formula

$$K = \frac{1}{2W}\left[\frac{\partial}{\partial u}\left(\frac{FE_v - EG_u}{EW}\right) + \frac{\partial}{\partial v}\left(\frac{2EF_u - FE_u - EE_v}{EW}\right)\right].$$

This formula is usually stated in terms of *Christoffel symbols* Γ_{ij}^k, a set of functions derived from E, F, and G.

(e) For an orthogonal patch, show that (d) reduces to Proposition 6.3 of Chapter 6.

(f) Test (d) for the saddle surface in Example 4.3(2) of Chapter 5.

10. (*Continuation.*) If the parameter curves of a patch all have unit speed, show that Gaussian curvature is given by $K = -\vartheta_{uv}/\sin\vartheta$, where $0 < \vartheta < \pi$ is the coordinate angle.

11. (a) Show that the mapping $\mathbf{x}: \mathbf{R}^2 \to \mathbf{R}^4$ given by

$$\mathbf{x}(u, v) = (\cosh u \cos v, \cosh u \sin v, \sinh u \cos v, \sinh u \sin v)$$

is conformal, hence regular. (In fact, $\mathbf{x}$ parametrizes a surface in $\mathbf{R}^4$.)

(b) Find its Gaussian curvature. (*Hint:* Ex. 2 of Sec. 6.6.)

12. (*Double saddle.*) The image of the patch $\mathbf{x}(u, v) = (u, v, uv, (u^2 - v^2)/2)$ is a surface in $\mathbf{R}^4$. Find:

(a) its Gaussian curvature;

(b) the area of the region for which $0 \leqq u, v \leqq 1$.

13. (*Computer.*)

(a) Write the computer command based on Exercise 9 that, given the *functions E, F, G*, returns K.

(b) Show how (a) can be used to find the curvature of $\mathbf{x}: D \to \mathbf{R}^n$ for arbitrary n.

(c) Test (a) on the three preceding exercises and on Example 2.8.

7.3 Covariant Derivative

The covariant derivative ∇ of $\mathbf{R}^3$ (Chapter 2, Section 5) is an essential part of Euclidean geometry. It is used, for example, to define the shape operator of a surface in $\mathbf{R}^3$ and, in modified form, to define the acceleration of a curve in $\mathbf{R}^3$ (Chapter 2, Section 2). In this section we show that *each geometric surface has its own notion of covariant derivative.*

As in Euclidean space, a covariant derivative ∇ on a geometric surface M assigns to each pair of vector fields V, W on M a new vector field $\nabla_V W$. Intuitively, the value of $\nabla_V W$ at a point $\mathbf{p}$ will be the rate of change of W in the $V(\mathbf{p})$ direction. We must certainly require that ∇ have the familiar linear and Leibnizian properties (1)–(4) in Corollary 5.4 of Chapter 2—of course, with the Euclidean dot product replaced by the inner product of M.

Furthermore, since the connection form ω_{12} of a frame field E_1, E_2 measures the rate at which E_1 is turning toward E_2, we must also require

$$\omega_{12}(V) = \langle \nabla_V E_1, E_2 \rangle. \tag{$*$}$$

These conditions are enough to completely determine ∇.

3.1 Lemma   Assume that there is a covariant derivative ∇ on M with the usual linear and Leibnizian properties, and such that $(*)$ holds for frame fields E_1, E_2. Then ∇ obeys the *connection equations*

$$\nabla_V E_1 = \omega_{12}(V) E_2, \quad \nabla_V E_2 = \omega_{21}(V) E_1.$$

Furthermore, for a vector field $W = f_1 E_1 + f_2 E_2$,

$$\nabla_V W = (V[f_1] + f_2 \omega_{21}(V)) E_1 + (V[f_2] + f_1 \omega_{12}(V)) E_2.$$

We call this last equation the *covariant derivative formula.* Note that in this formula, $V[f_1]$ and $V[f_2]$ only tell how W is changing *relative to the frame field.* The other two terms tell how the frame field itself is rotating, so the combined formula makes $\nabla_V W$ an *absolute* rate of change.

Proof.   By orthonormal expansion,

$$\nabla_V E_1 = \langle \nabla_V E_1, E_1 \rangle E_1 + \langle \nabla_V E_1, E_2 \rangle E_2.$$

The first summand vanishes since $\langle E_1, E_1 \rangle$ is constant. In fact, by a familiar Leibnizian argument,

$$0 = V[\langle E_1, E_1 \rangle] = 2 \langle \nabla_V E_1, E_1 \rangle.$$

Using $(*)$ in the second summand gives $\nabla_V E_1 = \omega_{12}(V) E_2$, as required. The other connection equation is derived similarly.

Finally, the properties of ∇ yield

$$\nabla_V W = \nabla_V (f_1 E_1 + f_2 E_2) = \nabla_V (f_1 E_1) + \nabla_V (f_2 E_2)$$
$$= V[f_1]E_1 + f_1 \nabla_V E_1 + V[f_2]E_2 + f_2 \nabla_V E_2.$$

Substituting the connection equations then gives the covariant derivative formula. ◆

This lemma shows how to define the covariant derivative of M. The following result (here stated for dimension 2) has been called the fundamental theorem of Riemannian geometry.

3.2 Theorem On each geometric surface M there exists a unique covariant derivative ∇ with the linear and Leibnizian properties (1)–(4) in Corollary 5.4 of Chapter 2 and satisfying equation (*) for every frame field.

Proof. The preceding lemma shows that there is *at most* one such covariant derivative since it gives an explicit formula for $\nabla_V W$. The proof that there is *at least* one splits into two parts.

Local Definition. For a frame field E_1, E_2 on a region $\mathcal{U}$ in M, use the covariant derivative formula in Lemma 3.1 as the definition of $\nabla_V W$. It is a routine exercise in calculus on a surface to verify that ∇ has the required properties. The linearity conditions (1) and (2) are simple, so we will prove the Leibnizian property (3), namely,

$$\nabla_V (fY) = V[f]Y + f\nabla_V Y.$$

Write $Y = g_1 E_1 + g_2 E_2$. Then $fY = fg_1 E_1 + fg_2 E_2$, and by the covariant derivative formula,

$$\nabla_V (fY) = (V[fg_1] + fg_2 \omega_{21}(V))E_1 + (V[fg_2] + fg_1 \omega_{12}(V))E_2.$$

The Leibnizian product rule $V[fg_i] = fV[g_i] + g_i V[f]$ yields

$$\nabla_V (fY) = f(V[g_1] + g_2 \omega_{21}(V))E_1 + f(V[g_2] + g_1 \omega_{12}(V))E_2$$
$$+ g_1 V[f]E_1 + g_2 V[f]E_2.$$

But the right side here is $f\nabla_V Y + V[f]Y$, as required.

To prove (*), use the covariant derivative formula with $W = E_1$. Then $f_1 = 1$ and $f_2 = 0$, so the formula collapses immediately to (*).

Consistency. For two different frame fields, do the local definitions agree? If so, we have a single covariant derivative well-defined on all of M. So let $\overline{\nabla}$ derive from a frame field $\overline{E}_1, \overline{E}_2$ on a domain $\overline{\mathcal{U}}$. We must show that $\nabla_v W = \overline{\nabla}_V W$ on the intersection of $\mathcal{U}$ and $\overline{\mathcal{U}}$. Because of the properties of ∇, it will suffice to show

$$\nabla_V \overline{E}_1 = \overline{\nabla}_V \overline{E}_1, \quad \nabla_V \overline{E}_2 = \overline{\nabla}_V \overline{E}_2.$$

We use Lemma 1.4, assuming for simplicity that the two frame fields have the same orientation. When ∇_V is applied to the equation

$$\overline{E}_1 = \cos \varphi \; E_1 + \sin \varphi \; E_2,$$

the covariant derivative formula gives

$$\nabla_V \overline{E}_1 = \sin \varphi \; (-V[\varphi] + \omega_{21}(V)) E_1$$
$$+ \cos \varphi \; (V[\varphi] + \omega_{12}(V)) E_2.$$

By Lemma 1.4, $\overline{\omega}_{12} = \omega_{12} + d\varphi$. Since $d\varphi(V) = V[\varphi]$, substituting $\omega_{12} = \overline{\omega}_{12} - d\varphi$ into the preceding equation reduces it to

$$\nabla_V \overline{E}_1 = \overline{\omega}_{12}(V)(-\sin \varphi \; E_1 + \cos \varphi \; E_2)$$
$$= \overline{\omega}_{12}(V) \overline{E}_2 = \overline{\nabla}_V \overline{E}_1.$$

In the same way, $\nabla_V \overline{E}_2 = \overline{\nabla}_V \overline{E}_2$ follows from $\overline{E}_2 = -\sin \varphi E_1 + \cos \varphi E_2$. ◆

3.3 Example *The covariant derivative of* $\mathbf{R}^2$. The natural frame field U_1, U_2 has $\omega_{12} = 0$. Thus, for a vector field $W = f_1 U_1 + f_2 U_2$, the covariant derivative formula (Lemma 3.1) reduces to

$$\nabla_V W = V[f_1] U_1 + V[f_2] U_2.$$

This is just Lemma 5.2 of Chapter 2, applied on $\mathbf{R}^2$ instead of $\mathbf{R}^3$, so our abstract definition of covariant derivative produces correct Euclidean results.

Note that the covariant derivative formula shows that (as in the Euclidean case) the value of the vector field $\nabla_V W$ at a point $\mathbf{p}$ depends only on W and the tangent vector $V(\mathbf{p})$. Thus $\nabla_V W$ is meaningful for an individual tangent vector.

The covariant derivative on a geometric surface M can be adapted to a vector field Y along a curve $\alpha: I \to M$. (Recall that Y assigns a vector $Y(t)$ in $T_{\alpha(t)}(M)$ for each t in I.) The covariant derivative Y' of Y ought to be $\nabla_{\alpha'} Y$, but neither α' nor Y is defined on an open set of M as required by the definition of ∇. The simplest solution is to define Y' by a frame field formula modeled on the covariant derivative formula in Lemma 3.1.

So for a frame field E_1, E_2, write $Y = f_1 E_1 + f_2 E_2$, and then define

$$Y' = (f_1' + f_2\omega_{21}(\alpha'))E_1 + (f_2' + f_1\omega_{12}(\alpha'))E_2. \tag{δ}$$

Standard proofs show that this definition is independent of the choice of frame field and has the same linear and Leibnizian properties as in the Euclidean case.

The velocity α' of a curve in M is a vector field on M, so we can take its covariant derivative to get the *acceleration* α'' of α.

In the frequently occurring case of vector fields of constant length, there is a more intuitive description of the covariant derivative along a curve—one that relates more directly to other geometric invariants.

3.4 Lemma   Let E_1, $E_2 = J(E_1)$ be a positively oriented frame field on M, where J is the rotation operator from Exercise 3 of Section 7.1. Let Y be a vector field of constant length $c > 0$ along a curve α in M. If φ is an angle function from E_1 to Y, then

$$Y' = (\varphi' + \omega_{12}(\alpha'))J(Y).$$

Again, φ' tells the rate at which Y is rotating relative to the frame field, and the connection term tells how the frame field is rotating along α, so their sum gives an absolute rate for Y.

Proof.   We write $Y/c = \cos\varphi E_1 + \sin\varphi E_2$. Then applying the covariant derivative formula gives

$$Y'/c = \sin\varphi\,(-\varphi' + \omega_{21}(\alpha'))E_1 + \cos\varphi\,(\varphi' + \omega_{12}(\alpha'))E_2$$
$$= (\varphi' + \omega_{12}(\alpha'))(\cos\varphi\,E_2 - \sin\varphi\,E_1) = (\varphi' + \omega_{12}(\alpha'))\,J(Y/c).$$

Multiplication by c completes the proof.   ◆

A distinctive feature of Euclidean geometry is the ability to move a tangent vector **v** at one point to a *parallel* vector at any other point by simply keeping the same natural coordinates for **v**. As we shall see, this phenomenon of "distant parallelism" rarely occurs on an arbitrary geometric surface. However, it is always possible to define parallelism of a vector field *along a curve*.

3.5 Definition   A vector field V on a curve α in a geometric surface is *parallel* provided its covariant derivative vanishes: $V' = 0$.

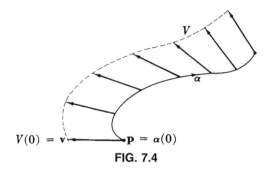

FIG. 7.4

As in the Euclidean case, a parallel vector field has constant length, since $\|V\|^2 = \langle V, V \rangle$ and $\langle V, V \rangle' = 2\langle V', V \rangle = 0$.

3.6 Lemma Let α be a curve in a geometric surface M, and let $\mathbf{v}$ be a tangent vector at $\mathbf{p} = \alpha(t_0)$. Then there is a unique parallel vector field V on α such that $V(t_0) = \mathbf{v}$ (see Fig. 7.4).

Proof. The vector field V must satisfy the conditions

$$V'(t) = 0 \quad \text{for all } t, \quad \text{and} \quad V(t_0) = \mathbf{v}.$$

The existence and uniqueness of V is assured by differential equations theory, but the following explicit proof may be more illuminating.

We can suppose that α lies entirely in the domain of a positively oriented frame field E_1, $E_2 = J(E_1)$ on M, for otherwise α could be broken into subsegments for which this is true. Since V must have constant length $c = \|\mathbf{v}\|$, we can write

$$V = c \cos \varphi \, E_1 + c \sin \varphi \, E_2,$$

where the angle function $\varphi = \varphi(t)$ is to be determined.

By the preceding lemma, V will be parallel, that is, $V' = 0$, if and only if $\varphi' + \omega_{12}(\alpha') = 0$. Furthermore, $V(t_0) = \mathbf{v}$ will hold if $\varphi(t_0)$ is an angle from $E_1(\mathbf{p})$ to $\mathbf{v}$. There is exactly one function φ satisfying these conditions, namely

$$\varphi(t) = \varphi(t_0) - \int_{t_0}^{t} \omega_{12}(\alpha') \, du. \qquad \blacklozenge$$

For a parallel vector field V on α, we say that the vector $V(t)$ at each point $\alpha(t)$ is gotten from $\mathbf{v}$ at $\mathbf{p} = \alpha(t_0)$ by *parallel translation along* α.

In Euclidean space, parallelism means keeping the same natural coordinates, so parallel translation is path-independent. But in a geometric surface, parallel translation from $\mathbf{p}$ to $\mathbf{q}$ along different paths usually gives different results. Equivalently, if a vector **v** at **p** is parallel-translated around a closed curve, the result **v*** need not be **v**—a phenomenon called *holonomy*.

If $\alpha: [a,b] \to M$ is a closed curve in the domain of a frame field, the formula $\varphi' + \omega_{12}(\alpha') = 0$ in Lemma 3.6 shows that all parallel vector fields along α are rotated through the angle

$$\varphi(b) - \varphi(a) = -\int_\alpha \omega_{12}.$$

We call this the *holonomy angle* ψ_α of α. (Additive multiples of 2π can be ignored in ψ_α since they do not affect the determination of **v***.)

3.7 Example *Holonomy on a sphere Σ of radius r.* Suppose the closed curve α parametrizes a circle on Σ. There is no loss of generality in assuming that α is a circle of latitude, say the u-parameter curve

$$\alpha(u) = \mathbf{x}(u, v_0) \quad \text{for} \quad 0 \leq u \leq 2\pi,$$

where **x** is the geographical parametrization of Σ. For the associated frame field $E_1 = \alpha'/\sqrt{E}$, $E_2 = J(E_1)$ of **x**, Example 6.2 of Chapter 6 shows that $\omega_{12} = \sin v \, du$. Since v has constant value v_0 on α, the holonomy angle ψ_α of α is

$$\varphi(2\pi) - \varphi(0) = -\int_\alpha \sin v \, du = -\int_0^{2\pi} \sin v_0 \, du = -2\pi \sin v_0.$$

Thus only on the equator $v = 0$ does a vector **v** return to itself after parallel translation around α. If α is in the northern hemisphere ($v_0 > 0$), then $\varphi' < 0$, so an initially north-pointing parallel vector rotates eastward as it traverses α (Fig. 7.5).

Up to this point, everything in this section belongs to the geometry of an arbitrary surface, so it may be well to look back at the case of a surface M in $\mathbf{R}^3$. There we now have two ways to take covariant derivatives: one from the intrinsic geometry of M as a geometric surface, the other the Euclidean covariant derivative of $\mathbf{R}^3$. The two are usually different, but there is a simple relationship between them. In the following lemma, ∇ denotes the covariant derivative of M as a geometric surface and $\tilde{\nabla}$ is the Euclidean covariant derivative.

3.8 Lemma If V and W are tangent vector fields on a surface M in $\mathbf{R}^3$, then, as in Fig. 7.6,

(1) $\nabla_V W$ is the component of $\tilde{\nabla}_V W$ tangent to M.
(2) If S is the shape operator of M derived from a unit normal U, then

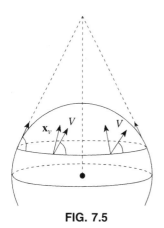

FIG. 7.5

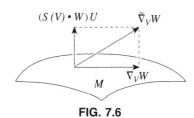

FIG. 7.6

$$\tilde{\nabla}_V W = \nabla_V W + (S(V) \cdot W)U.$$

Proof.   Since (2) implies (1), we need only prove (2). Let E_1, E_2, $E_3 = U$ be an adapted frame on M. Suppose first that W is the vector field E_1. By the *Euclidean* connection equations (Theorem 7.2 of Chapter 2),

$$\tilde{\nabla}_V E_1 = \sum_{j=1}^{3} \omega_{1j}(V)E_j = \omega_{12}(V)E_2 + \omega_{13}(V)E_3.$$

The connection equations for M (Lemma 3.1) give

$$\nabla_V E_1 = \omega_{12}(V)E_2.$$

Since $E_3 = U$, the definition of ω_{ij} in Section 7 of Chapter 2 gives

$$\omega_{13}(V)E_3 = (\tilde{\nabla}_V E_1 \cdot E_3)E_3 = -(\tilde{\nabla}_V E_3 \cdot E_1)E_3 = (S(V) \cdot E_1)U.$$

Substitution then gives (2) for $W = E_1$. The same result holds for E_2.

In the general case, write $W = f_1 E_1 + f_2 E_2$. Then (2) follows, by a routine computation, from the special cases above.   ◆

Evidently we have been using the intrinsic covariant derivative of $M \subset \mathbf{R}^3$ all along, without giving it formal recognition.

For a patch $\mathbf{x}$ in an arbitrary geometric surface M, we shall inevitably use the notation $\mathbf{x}_{uu}$ for the covariant derivative of the vector field $\mathbf{x}_u$ along u-parameter curves—with corresponding meanings for $\mathbf{x}_{uv} = \mathbf{x}_{vu}$ and $\mathbf{x}_{vv}$. Thus when M is a surface in $\mathbf{R}^3$ we will need a new notation, say $\tilde{\mathbf{x}}_{uu}, \ldots$ for the corresponding Euclidean covariant derivatives.

Exercises

1. In the Poincaré half-plane,
(a) show that the connection form of the frame field $E_1 = vU_1$, $E_2 = vU_2$ is $\omega_{12} = du/v = \theta_1$.
Then for each of the following curves, express velocity and acceleration in terms of this frame field:
(b) The curve α in Exercise 2 of Section 1. (*Hint:* α' and α'' are collinear.)
(c) The Euclidean straight line

$$\beta(t) = (ct, st) \quad \text{for all } t \text{ such that } st > 0,$$

where c and s are constants such that $c^2 + s^2 = 1$. Check results using the formula $\langle \beta', \beta' \rangle' = 2\langle \beta', \beta'' \rangle$.

2. Let V be a parallel vector field on a curve α in M. Show that a vector field W on α is parallel if and only if it has constant length and the angle between V and W is constant.

3. Let α be a curve in $M \subset \mathbf{R}^3$. If Y is a vector field on α tangent to M, prove this analogue of Lemma 3.8:

$$\dot{Y} = Y' + (S(\alpha') \cdot Y)U,$$

where $\dot{Y}$ denotes the Euclidean derivative of Y. Hence the Euclidean acceleration of α is $\ddot{\alpha} = \alpha'' + (S(\alpha') \cdot \alpha')U$.

4. Let $\mathbf{x}$ be the geographical parametrization of a sphere Σ of radius r, and let α be the circle of latitude

$$\alpha(u) = \mathbf{x}(u, v_0), \quad \text{for } 0 \leq u \leq 2\pi.$$

(In view of the symmetry of Σ this is a typical circle on Σ.)
(a) Show that α has intrinsic acceleration

$$\alpha'' = r \cos v_0 \sin v_0 \, E_2,$$

where $E_2 = \mathbf{x}_v/\sqrt{G}$.

(b) Compute separately the Euclidean acceleration of α and $S(\alpha')$. Then check the last formula in Exercise 3.

5. (*Curvature and Holonomy.*) Let α be a closed curve in a geometric surface M.
(a) If α is the boundary curve of a smooth oriented disk $\mathscr{D}$ in M, show that the holonomy angle ψ_α of α is $\iint_{\mathscr{D}} K dM$. (*Hint:* There is always a frame field on $\mathscr{D}$.)

(b) With notation as in (a), deduce the following characterization of the Gaussian curvature of M at a point $\mathbf{p}$:

$$K(\mathbf{p}) = \lim_{\mathcal{D} \to p} \frac{\psi_\alpha}{\text{area } \mathcal{D}}.$$

6. If there is a nonvanishing vector field W on M such that $\nabla_V W = 0$ for all V, show that M is flat. (*Hint:* There is a frame with $E_1 = W/c$.)

7. (*Isometries preserve covariant derivatives.*) For an isometry $F: M \to N$, prove the following two cases:

(a) If V and W are vector fields on M and $\overline{V}$ and $\overline{W}$ are their transferred vector fields on $\overline{M}$, then

$$\overline{\nabla_V W} = \overline{\nabla}_{\overline{V}} \overline{W}.$$

(*Hint:* By the linearity of ∇ it suffices to assume $W = fE_1$ for a frame field E_1, E_2. Strictly speaking, the transferred vector field $\overline{V}$ is $(F*(V))(F^{-1})$, but to avoid clutter, write simply $\overline{V} = F*(V)$.)

(b) If Y is a vector field on a curve α in M, then

$$F * (Y') = \overline{Y}',$$

where $\overline{Y}$ is the vector field $F*(Y)$ on the curve $\overline{\alpha} = F(\alpha)$ in $\overline{M}$. Hence in particular, acceleration is preserved: $F*(\alpha'') = F(\alpha)''$.

This is the analogue of the Euclidean result, Corollary 4.1 of Chapter 3.

8. Let $\mathbf{x}$ be an orthogonal patch in a geometric surface M. Prove that $\mathbf{x}_{uv} = \mathbf{x}_{vu}$ by first showing that for the associated frame field $E_1 = \mathbf{x}_u/\sqrt{E}$, $E_2 = \mathbf{x}_v/\sqrt{G}$ we have:

(a) $\omega_{12}\mathbf{x}_v = \dfrac{\langle \mathbf{x}_{uv}, \mathbf{x}_v \rangle}{\sqrt{EG}}$. (*Hint:* Use the (*) property of ∇.)

(b) $\omega_{12}\mathbf{x}_v = \dfrac{\langle \mathbf{x}_{vu}, \mathbf{x}_v \rangle}{\sqrt{EG}}$. (*Hint:* Use the formula for ω_{12} in Sec. 6 of Ch. 6.)

9. (*Continuation.*) To show that $\mathbf{y}_{uv} = \mathbf{y}_{vu}$ for an arbitrary patch, write $\mathbf{y}(u, v) = \mathbf{x}(\overline{u}(u, v), \overline{v}(u, v))$ with $\mathbf{x}$ orthogonal, and compute $\mathbf{y}_{uv} - \mathbf{y}_{vu}$.

7.4 Geodesics

Geodesics in an arbitrary geometric surface generalize straight lines in Euclidean geometry. We have seen that a Euclidean straight line $\gamma(t) = \mathbf{p} + t\mathbf{q}$ is characterized infinitesimally by vanishing of acceleration; thus

4.1 Definition A curve γ in a geometric surface is a *geodesic* provided its acceleration is zero, $\gamma'' = 0$.

In other words, the velocity γ' of a geodesic is parallel: Geodesics never turn. In particular, since γ parallel implies $\|\gamma'\|$ constant, geodesics have constant speed.

Acceleration is preserved by isometries (Exercise 7 of Section 3), so it follows that geodesics are isometric invariants. In fact, if $F: M \to N$ is merely a local isometry, then F carries each geodesic γ of M to a geodesic $F(\gamma)$ of N. (This follows since F is an isometry on small enough neighborhoods.)

For a surface M in $\mathbf{R}^3$, a curve in M was defined to be a geodesic if its Euclidean acceleration is always normal to M. This is consistent with the general definition above, for by Exercise 3 of Section 3 the intrinsic acceleration of a curve in $M \subset \mathbf{R}^3$ is the component tangent to M of its Euclidean acceleration. Thus the former is zero if and only if the latter is normal to M.

Now we find coordinate conditions for a curve α in a geometric surface M to be a geodesic. If E_1, E_2 is a frame field on M, then throughout this section we write the velocity and acceleration of α as

$$\alpha' = v_1 E_1 + v_2 E_2 \quad \text{and} \quad \alpha'' = A_1 E_1 + A_2 E_2.$$

This means that α *is a geodesic if and only if* $A_1 = A_2 = 0$. (Note that v_1, v_2, A_1, and A_2 are real-valued functions on the domain I of α.)

From Section 3 these components of acceleration can be expressed in terms of the connection form of E_1, E_2 as

$$A_1 = v_1' + v_2 \omega_{21}(\alpha'),$$
$$A_2 = v_2' + v_1 \omega_{12}(\alpha').$$

Now we describe A_1 and A_2 in coordinate terms.

4.2 Theorem Let $\mathbf{x}$ be an orthogonal coordinate patch in a geometric surface M. A curve $\alpha(t) = \mathbf{x}(a_1(t), a_2(t))$ is a geodesic of M if and only if

$$A_1 = a_1'' + \frac{1}{2E}\left(E_u a_1'^2 + 2E_v a_1' a_2' - G_u a_2'^2\right) = 0,$$

$$A_2 = a_2'' + \frac{1}{2G}\left(-E_v a_1'^2 + 2G_u a_1' a_2' + G_v a_2'^2\right) = 0.$$

Note the symmetry of these two equations under the reversals $1 \leftrightarrow 2$, $u \leftrightarrow v$, $E \leftrightarrow G$. In this context we always understand that the functions E and

G and their partial derivatives E_u, E_v, ... are evaluated on (a_1, a_2) and hence become functions on the domain I of α.

Proof.   The velocity of α is $\alpha' = \alpha'_1 \mathbf{x}_u + \alpha'_2 \mathbf{x}_v$, so in terms of the associated frame field of $\mathbf{x}$ (Chapter 6, Section 6),

$$\alpha' = (a'_1 \sqrt{E})E_1 + (a'_2 \sqrt{G})E_2.$$

Thus the acceleration components A_1, A_2 become

$$A_1 = (a'_1 \sqrt{E})' + (a'_2 \sqrt{G})\omega_{21}(\alpha'),$$

$$A_2 = (a'_2 \sqrt{G})' + (a'_1 \sqrt{E})\omega_{12}(\alpha'). \tag{1}$$

The formula for ω_{12} in Section 6 of the Chapter 6 gives

$$\omega_{12}(\alpha') = \omega_{12}(a'_1 \mathbf{x}_u + a'_2 \mathbf{x}_v) = -\frac{(\sqrt{E})_v}{\sqrt{G}} a'_1 + \frac{(\sqrt{G})_u}{\sqrt{E}} a'_2. \tag{2}$$

When this is substituted into (1), we get

$$A_1 = (a'_1 \sqrt{E})' + (\sqrt{E})_v a'_1 a'_2 - \frac{\sqrt{G}(\sqrt{G})_u}{\sqrt{E}} a'^2_2,$$

$$A_2 = (a'_2 \sqrt{G})' + (\sqrt{G})_u a'_1 a'_2 - \frac{\sqrt{E}(\sqrt{E})_v}{\sqrt{G}} a'^2_1. \tag{3}$$

Standard calculus computations transform (3) into the form stated in the proposition. We remind the reader that in a Leibnizian expansion such as

$$(a'_1 \sqrt{E})' = a''_1 \sqrt{E} + a'_1 \frac{E'}{2\sqrt{E}},$$

the notation E is short for $E(a_1, a_2)$, so

$$E' = E_u a'_1 + E_v a'_2. \qquad \blacklozenge$$

4.3 Theorem   Given a tangent vector $\mathbf{v}$ to M at a point $\mathbf{p}$, there is a unique geodesic γ, defined on an interval I around 0, such that

$$\gamma(0) = \mathbf{p}, \quad \text{and} \quad \gamma'(0) = \mathbf{v}.$$

Thus there are plenty of geodesics in any geometric surface—each determined by its initial position and velocity. In $\mathbf{R}^2$, for example, the geodesic determined by $\mathbf{v}$ at $\mathbf{p}$ is $\gamma(t) = \mathbf{p} + t\mathbf{v}$.

Proof. Let $\mathbf{x}$ be an orthogonal patch around $\mathbf{p} = \mathbf{x}(u_0, v_0)$, and write $\mathbf{v} = c\mathbf{x}_u + d\mathbf{x}_v$. The geodesic equations in Theorem 4.2 have the form

$$a_1'' = f_1(a_1, a_2, a_1', a_2'),$$
$$a_2'' = f_2(a_1, a_2, a_1', a_2') \tag{1}$$

for differentiable functions f_2 and f_2. Furthermore, the initial conditions in the present theorem are equivalent to

$$a_1(0) = u_0, \quad a_1'(0) = c,$$
$$a_2(0) = v_0, \quad a_2'(0) = d. \tag{2}$$

The fundamental existence and uniqueness theorem of differential equations theory asserts that there is an interval I around 0 on which unique functions a_1, a_2 are defined that satisfy (1) and (2). Thus $\gamma = \mathbf{x}(a_1, a_2)$ is the unique geodesic defined on I such that $\gamma(0) = \mathbf{p}$ and $\gamma'(0) = \mathbf{v}$. ◆

This result is not entirely satisfactory since the interval I may be unnecessarily small. However, it is easy to make it as large as possible. Suppose $\gamma_1: I_1 \to M$ and $\gamma_2: I_2 \to M$ are geodesics that both satisfy the initial conditions in the theorem. The uniqueness theorem for differential equations implies that $\gamma_1 = \gamma_2$ on the intersection of I_1 and I_2. Applying this consistency result to all such geodesics gives a single *maximal* geodesic $\gamma: I \to M$ satisfying the initial conditions. (The interval I is the largest possible.) Intuitively, this simply means we let the geodesic run as far as it can.

We ordinarily assume that geodesics are maximal, and denote by γ_v the one with initial velocity $\mathbf{v}$.

4.4 Definition A geometric surface M is *complete* provided every maximal geodesic in M is defined on the whole real line $\mathbf{R}$.

Briefly: geodesics run forever. A constant curve is trivially a geodesic, but excluding this case, every geodesic has constant nonzero speed. Thus completeness is equivalent to the requirement that all nonconstant geodesics have infinite length—in both directions. For example, $\mathbf{R}^2$ is certainly complete, and the explicit computations in Example 6.9 of Chapter 5 show that spheres in $\mathbf{R}^3$ are complete as well. More generally, all compact geometric surfaces are complete (Ch. 8), as are all surfaces in $\mathbf{R}^3$ of the form $M: f = c$. Removal of even a single point from a complete surface destroys the property, since geodesics that formerly passed through the point will be obliged to stop.

It is rarely possible to find explicit solutions to the geodesic differential equations $A_1 = 0$, $A_2 = 0$ in Theorem 4.2, but the situation can often be improved by the following result, which shows that either one of these equations can be replaced by a simpler equation. For definiteness, we will replace $A_2 = 0$.

4.5 Lemma   Let E_1, E_2 be a frame field, and let α be a constant speed curve such that α' and E_2 are never orthogonal. If $A_1 = 0$, then $A_2 = 0$; hence α is a geodesic.

Proof.   Taking the derivative of $\langle \alpha', \alpha' \rangle = $ const gives $\langle \alpha'', \alpha' \rangle = 0$. Hence

$$0 = \langle A_1 E_1 + A_2 E_2, \alpha' \rangle = A_1 \langle E_1, \alpha' \rangle + A_2 \langle E_2, \alpha' \rangle.$$

By hypothesis, $A_1 = 0$ and $\langle E_2, \alpha' \rangle \neq 0$, so $A_2 = 0$.   ◆

In coordinate terms this means that the rather complicated second-order differential equation $A_2 = 0$ in Theorem 4.2 has been replaced by the first-order differential equation.

$$E a_1'^2 + G a_2'^2 = \text{const.}$$

Note that by continuity the lemma remains valid if α' and E_1 are orthogonal at isolated points.

There is a Frenet theory of curves in a geometric surface that generalizes that for curves in the Euclidean plane (Exercise 8 in Section 2.3). Since M has only two dimensions, torsion cannot be defined. But when M is oriented, curvature can be given a geometrically significant sign as follows.

If $\beta: I \rightarrow M$ is a unit-speed curve in an oriented surface, then as usual, $T = \beta'$ is the unit tangent vector field of β. But to get the *principal normal vector field* of β, we "rotate T through $+90°$," defining $N = J(T)$, where J is the rotation operator in Exercise 3 of Section 1 (see Fig. 7.7). Then the *geodesic curvature* κ_g of β is the unique real-valued function on I for which the *Frenet formula*

$$T' = \kappa_g N$$

holds. Explicitly, $\kappa_g = \langle T', N \rangle$. Thus κ_g is not restricted to nonnegative values as in the case of curves in $\mathbf{R}^3$: $\kappa_g > 0$ means that T—hence β—is turning in

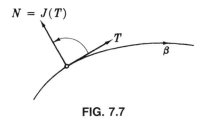

FIG. 7.7

the *positive* direction (that is, toward $N = J(T)$), while $\kappa_g < 0$ means *negative* turning (toward $-N$).

4.6 Corollary Let β be a unit-speed curve in a region oriented by a frame field E_1, E_2. If φ is an angle function from E_1 to β' along β, then

$$\kappa_g = \frac{d\varphi}{ds} + \omega_{12}(\beta').$$

Proof. Setting $Y = T$ and $\alpha = \beta$ in Lemma 3.4 gives

$$T' = (\varphi' + \omega_{12}(\beta'))\,J(T).$$

Since $J(T) = N$, the result follows by comparison with $T' = \kappa_g N$. ◆

In $\mathbf{R}^2$ the natural frame field has $\omega_{12} = 0$, so φ is the usual slope angle of the curve β. Then the result above—called *Liouville's formula*—reduces to $\kappa_g = d\varphi/ds$. In $\mathbf{R}^2$ this is often taken as the definition of curvature (see Exercise 8 of Section 2.3).

For an arbitrary-speed regular curve α in M, the present modest Frenet apparatus is defined—just as in Section 4 of Chapter 2—by reparametrization. The same proof as before shows

$$\alpha' = vT, \quad \alpha'' = \frac{dv}{dt}T + \kappa_g v^2 N, \qquad (*)$$

where $v = \|\alpha'\|$ is the speed function of α.

4.7 Lemma A regular curve α in M is a geodesic if and only if α has constant speed and geodesic curvature $\kappa_g = 0$.

Proof. In $(*)$, since $v > 0$, we have $\alpha'' = 0$ if and only if $\dfrac{dv}{dt} = \kappa_g = 0$. ◆

The equations (∗) also show that a regular curve α *has geodesic curvature zero if and only if its velocity* α′ *and acceleration* α″ *are always collinear.* Such curves α are sometimes called geodesics; to get a geodesic in the strict sense of Definition 4.1, it suffices to reparametrize α to give it constant speed. In contexts where parametrization may be important, we call a curve with $\kappa_g = 0$ a *pregeodesic*.

4.8 Remark   *Abbreviations.* In computations involving curves, where there are fixed coordinates, say u and v, it is often convenient to use abbreviations such as $u(t)$ for $a_1(t) = u(\alpha(t))$, and hence $u'(t)$ for $a'_1(t)$. This is helpful in computing with differential equations—in particular, with the geodesic differential equations.

Exercises

1.   Show that a reparametrization $t \rightarrow \alpha(f(t))$ of a nonconstant geodesic α is again a geodesic if and only if f has the form $f(t) = at + b$.

2.   If γ_v is the unique geodesic in M with initial velocity **v**, show that for any number c, $\gamma_{cv}(t) = \gamma_v(ct)$ for all t.

3.   Find the routes of the geodesics in the stereographic sphere in (1) of Example 2.4. (*Hint:* No computation is needed.)

4.   Let $\mathbf{p}_1$ and $\mathbf{p}_2$ be points on the equator of a sphere Σ. Let β be the closed curve that starts at the north pole **n**, follows a meridian to $\mathbf{p}_1$, the equator to $\mathbf{p}_2$, then a meridian back to **n**. Prove that the holonomy angle of β is the angle at **n** between the two meridians.

5.   (*Closed geodesics.*)
   (a) If the boundary curve β: $[a, b] \rightarrow M$ of a smooth disk $\mathscr{D}$ is a geodesic, prove $\iint_{\mathscr{D}} K dM$ is an integer multiple of 2π. (Sec. 7 shows that it is exactly 2π.)
   (b) Deduce that there are no smoothly closed geodesics (equivalently, periodic geodesics) on the following surfaces. Use this version of the Jordan curve theorem: Any piecewise-smooth closed curve in a simply connected surface is the boundary of a simple region.
   (i) A simply connected surface with $K \leq 0$.
   (ii) A paraboloid of revolution.

6.   In the projective plane $P(r)$ of radius r (Example 2.7), prove:
   (a) The geodesics are simple closed curves of length πr.

(b) There is a unique geodesic route through any two distinct points.
(c) Two distinct geodesic routes meet in exactly one point.
(*Hint:* The projection $F: \Sigma(r) \to P(r)$ is a local isometry.)

7. At each point **p** of a geometric surface M there is an $\varepsilon > 0$ such that every geodesic starting at **p** runs for at least length ε. This follows from differential equations theory (see Sec. 1 of Ch. 8).

(a) Deduce that in a connected surface M, any two points can be joined by a broken geodesic segment (that is, a piecewise differentiable curve whose smooth subsegments are geodesic). (*Hint:* Use the open set criterion for connectedness, Ex. 9 of Sec. 4.7.)

(b) Give an example of a connected surface M for which broken geodesics with at least three geodesic subsegments will be needed in order to connect all pairs of points. Find M such that arbitrarily many breaks will be necessary to connect all pairs.

8. Let α be a curve with speed function $v > 0$ in an oriented surface M.

(a) Show that the geodesic curvature κ_g of α is $\dfrac{\langle \alpha'', J(\alpha') \rangle}{v^3}$

(b) Deduce that if M is an oriented surface in $\mathbf{R}^3$, then

$$\kappa_g = \frac{U \cdot \alpha' \times \alpha''}{v^3} = \kappa \cos \vartheta,$$

where $\vartheta = \angle(B, U)$. Here κ is the curvature and B the binormal vector of α as a curve in $\mathbf{R}^3$.

7.5 Clairaut Parametrizations

We consider a special situation where extensive information about geodesics can be obtained with a minimum of computation.

5.1 Definition A *Clairaut parametrization* $\mathbf{x}: D \to M$ is an orthogonal parametrization for which both E and G depend only on u, that is, $F = 0$ and $E_v = G_v = 0$.

For example, the usual parametrization of a surface of revolution is of this type.

5.2 Lemma If $\mathbf{x}$ is a Clairaut parametrization, then

(1) All the u-parameter curves of $\mathbf{x}$ are *pregeodesics*, and
(2) A v-parameter curve $u = u_0$ is a geodesic if and only if $G_u(u_0) = 0$.

Proof. For (1) it suffices by a remark in Section 4 to show that $\mathbf{x}_u$ and $\mathbf{x}_{uu}$ are collinear. Since $\mathbf{x}_u$ and $\mathbf{x}_v$ are orthogonal, this is equivalent to $\langle \mathbf{x}_v, \mathbf{x}_{uu} \rangle = 0$. The following equations imply this result.

$$0 = E_v = \langle \mathbf{x}_u, \mathbf{x}_u \rangle_v = 2\langle \mathbf{x}_u, \mathbf{x}_{uv} \rangle,$$

$$0 = F_u = \langle \mathbf{x}_u, \mathbf{x}_v \rangle_u = \langle \mathbf{x}_{uu}, \mathbf{x}_v \rangle + \langle \mathbf{x}_u, \mathbf{x}_{vu} \rangle.$$

Similarly, for (2) the v-parameter curve $u = u_0$ is pregeodesic if and only if

$$\langle \mathbf{x}_{vv}(u_0, v), \mathbf{x}_u(u_0, v) \rangle = 0.$$

The following equations show that this is true if and only if $G_u(u_0) = 0$:

$$0 = F_v = \langle \mathbf{x}_{uv}, x_v \rangle + \langle \mathbf{x}_u, \mathbf{x}_{vv} \rangle,$$

$$G_u(u_0) = G_u(u_0, v) = 2\langle \mathbf{x}_{vu}(u_0, v), \mathbf{x}_v(u_0, v) \rangle \quad \text{for all } v.$$

(Recall that $\mathbf{x}_{uv} = \mathbf{x}_{vu}$.) So far, we have not used the condition $G_v = 0$; its effect is to show that the v-parameter pregeodesics are in fact geodesics, since it means that they have constant speed. $\blacklozenge$

In the case of a surface of revolution, for example, this lemma provides another proof that the meridians are geodesics and that a parallel $u = u_0$ is geodesic if and only if $h'(u_0) = 0$. (See Exercise 3 of Section 5.6.)

A surface M that has a Clairaut parametrization need not be a surface of revolution, but we continue to call its u-parameter geodesics *meridians*. Ignoring the parametrization of these meridians, we can think of them as fibers of which M is composed—and measure the behavior of arbitrary geodesics relative to them.

5.3 Theorem Let $\alpha = \mathbf{x}(a_1, a_2)$ be a unit-speed geodesic with $\mathbf{x}$ a Clairaut parametrization. If φ is the angle from $\mathbf{x}_u$ to α', then the function

$$c = G(a_1)a_2' = \sqrt{G}(a_1) \sin \varphi$$

is *constant* along α. Hence α cannot leave the region where $G \geqq c^2$.

We call the constant $c = c(\alpha)$ thus associated with each geodesic α the *slant* of α since—in combination with G—it determines the angle φ at which α cuts across the meridians of $\mathbf{x}$ (see Fig. 7.8).

Proof. Because $E_v = G_v = 0$ for a Clairaut parametrization, the geodesic equation $A_2 = 0$ of Theorem 4.2 reduces to

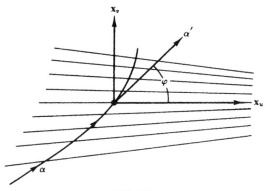

FIG. 7.8

$$a_2'' + \frac{G_u}{G} a_1' a_2' = 0.$$

This is equivalent to the constancy of $c = Ga_2'$, since

$$c' = (Ga_2')' = G'a_2' + Ga_2'' = G_u a_1' a_2' + Ga_2''.$$

To show that $c = \sqrt{G} \sin \varphi$, compare the two equations

$$\langle \alpha', \mathbf{x}_v \rangle = \langle a_1' \mathbf{x}_u + a_2' \mathbf{x}_v, \mathbf{x}_v \rangle = c,$$

$$\langle \alpha', \mathbf{x}_v \rangle = \|\alpha'\| \|\mathbf{x}_v\| \cos\left(\frac{\pi}{2} - \varphi\right) = \sqrt{G} \sin \varphi.$$

It follows immediately from $|\sin \varphi| \leq 1$ that $G \geq c^2$. ◆

A geodesic with $c = 0$ has $\varphi = 0$ or π, and hence it is one of the geodesic meridians, so we may as well assume $c \neq 0$. Moving along α in a direction in which G is increasing, the meridians are spreading apart and the constancy of $c^2 = G(a_1) \sin^2 \varphi$ shows that $|\sin \varphi|$ is decreasing, thus forcing α to turn more toward the direction of the meridians. On the other hand, if G is decreasing along α, then α cuts across the meridians at ever-increasing angles. Note that $\sin \varphi$ cannot change sign, for if it is zero even at a single point, then the geodesic α is tangent to the meridian; hence by the uniqueness of geodesics it parametrizes that meridian.

Such considerations give remarkable information about the general global behavior of geodesics in a Clairaut parametrization.

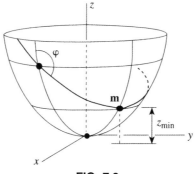

FIG. 7.9

5.4 Example (*Global trajectories.*) What happens to the geodesics passing through an arbitrary point p_0 on the paraboloid of revolution M: $z = x^2 + y^2$? The parametrization $x(u, v) = (u\cos v, u\sin v, u^2)$—for which u is distance to the z axis—is Clairaut with $E = 1 + 4u^2$, $F = 0$, $G = u^2$. Thus the slant of a unit-speed geodesic $\alpha(t) = x(a_1(t), a_2(t))$ is

$$c = \sqrt{G(a_1(t))} \sin \varphi(t) = a_1(t) \sin \varphi(t),$$

where $\varphi(t)$ is the oriented angle from x_u to $\alpha'(t)$ (see Fig. 7.9).

Suppose α starts *upward* from p_0, that is, with initial angle $0 < \varphi_0 < \pi/2$ (by symmetry, we may as well take $\varphi \geq 0$). Then a_1 increases; hence $|\sin \varphi|$ decreases, so α' turns steadily toward x_u, and α escapes to $z = \infty$.

If $\varphi_0 = \pi$, then α parametrizes a meridian of this surface of revolution, so α runs directly down through the vertex 0 of M and back up to $z = \infty$.

The interesting case is $\pi/2 < \varphi_0 < \pi$. Now α starts *downward*, but it never reaches 0—only meridians can do that. It is easy to predict *at the start* how low α gets. Initially, since $u = a_1$ is decreasing, $|\sin \varphi|$ is increasing, so α' is turning away from the meridians. These changes continue until φ reaches $\pi/2$. This will occur precisely when $c = u_0 \sin \varphi_0 = u_1$, hence at height

$$z_{\min} = u_1^2 = u_0^2 \sin \varphi_0^2 > 0.$$

The geodesic α cannot asymptotically approach the nongeodesic v-parameter curve at this height, any more than a straight line in the plane can asymptotically approach a circle. So α meets and bounces off the v-parameter curve (Fig. 7.9) and turning steadily back toward x_u, escapes to $z = \infty$, as does every geodesic on M.

Not only does the notion of slant give remarkable qualitative information about where geodesics can go in a Clairaut surface, but more than that, it leads to a single integral formula for pregeodesics. Thus the *routes* followed by the geodesics are specified—though not their geodesic parametrizations.

5.5 Proposition If **x** is a Clairaut parametrization, then every geodesic α such that α' is never orthogonal to meridians can be parametrized as $\beta(u) = \mathbf{x}(u, v(u))$, where

$$\frac{dv}{du} = \pm \frac{c\sqrt{E}}{\sqrt{G}\sqrt{G - c^2}},$$

with c the slant of α. Hence, by the fundamental theorem of calculus,

$$v(u) = v(u_0) \pm \int_{u_0}^{u} \frac{c\sqrt{E}\,dt}{\sqrt{G}\sqrt{G - c^2}}$$

Proof. Since α has unit speed,

$$1 = E(a_1)a_1'^2 + G(a_1)a_2'^2.$$

By Theorem 5.3,

$$a_2' = \frac{c}{G(a_1)}. \tag{C-2}$$

Now substitute (C-2) into the preceding equation, and solve for

$$a_1' = \pm \frac{\sqrt{G - c^2}}{\sqrt{EG}}. \tag{C-1}$$

The nonorthogonality condition means that a_1' is never zero on the domain I of α. Thus, as in elementary calculus, the functions $u = a_1(s)$, $v = a_2(s)$ can be reparametrized by $s = (a_1^{-1})(u)$ to give $u = u$, $v = v(u)$. Then (C-1) and (C-2) give

$$\frac{dv}{du} = \frac{a_2'(s)}{a_1'(s)} = \pm \frac{c\sqrt{E}}{\sqrt{G}\sqrt{G - c^2}}.$$

as required.   ◆

In the Clairaut case, one can check that equations (C-1) and (C-2) above are not only necessary but also sufficient for $\alpha = \mathbf{x}(a_1(s), a_2(s))$ to be a *unit-speed* geodesic. Note that dropping the plus-or-minus sign from (C-1) loses only a reverse parametrization, since c can have either sign. These equations are convenient for numerical computation (see Exercises 9–12), although they fail at turning points since there $G = c^2$. They are, of course, much simpler than the second-order differential equations in Theorem 4.2.

5.6 Example   *Routes of geodesics.*

(1) *The Euclidean plane.* To illustrate the preceding proposition, we find the routes of the well-known geodesics of $\mathbf{R}^2$ in terms of a polar parametrization

$$\mathbf{x}(u, v) = (u \cos v, u \sin v).$$

Since $E = 1$, $F = 0$, $G = u^2$, this is a Clairaut parametrization. The u-parameter geodesics are just straight lines through the origin. By Proposition 5.5, all the others can be parametrized as $\beta(u) = \mathbf{x}(u, v(u))$, where

$$\frac{dv}{du} = \frac{\pm c}{u\sqrt{u^2 - c^2}} = \pm \frac{d}{du}\left(\cos^{-1}\frac{c}{u}\right).$$

Hence $v - v_0 = \pm\cos^{-1}(c/u)$, or equivalently, $u\cos(v - v_0) = c$, which is the polar equation of a straight line. Here the slant has geometric significance as the shortest distance from the origin to the line, and the point on the line nearest the origin is a turning point.

(2) *The hyperbolic plane.* Again we try the polar parametrization. The ruler function h that describes the conformal geometric structure of H depends only on Euclidean distance u to the origin $\mathbf{0}$; in fact, $h(u) = h(\mathbf{x}(u, v)) = 1 + u^2/4$. Thus

$$E = \langle \mathbf{x}_u, \mathbf{x}_u \rangle = \frac{1}{h^2}, \quad F = 0, \quad G = \langle \mathbf{x}_v, \mathbf{x}_v \rangle = \frac{u^2}{h^2}.$$

As in (1), $\mathbf{x}$ is a Clairaut parametrization, so the u-parameter curves— Euclidean lines through the origin—are the routes of geodesics. By Proposition 5.5, the routes of all other geodesics are parametrized by $\beta(u) = \mathbf{x}(u, v(u))$, where

$$\frac{dv}{du} = \frac{\pm\, ch/u^2}{\sqrt{1 - (ch/u)}}. \tag{1}$$

This equation can be integrated explicitly by setting

$$w = \frac{a}{u}\left(1 + \frac{u^2}{4}\right), \quad \text{where } a = \frac{c}{\sqrt{1 + c^2}}.$$

Then a straightforward computation gives

$$\frac{dv}{du} = \frac{\pm dw/du}{\sqrt{1-\omega^2}}. \tag{2}$$

Hence $v - v_0 = \pm\cos^{-1} w$, that is,

$$\cos(v - v_0) = w = \frac{a}{u}\left(1 + \frac{u^2}{4}\right),$$

which we rearrange as

$$u^2 + 4 - \frac{4u}{a}\cos(v - v_0) = 0. \tag{3}$$

The Euclidean law of cosines, $A^2 + B^2 - 2AB\cos v = C^2$, applied to a diagram like Fig. 7.10, shows that the polar equation of a circle of radius r with center at $\mathbf{x}(u_0, v_0)$ is

$$u^2 + u_0^2 - 2u_0 u\cos(v - v_0) = r^2.$$

Comparison with equation (3) shows that the route of β is a Euclidean circle C with $u_0 = r^2 + 4$. Since $u_0 > 2$, the center of the circle lies outside the hyperbolic plane H: $x^2 + y^2 < 4$. As can be seen from Fig. 7.10, the relation $u_0 = r^2 + 4$ implies that C is orthogonal to the rim $x^2 + y^2 = 4$ of H. Of course, β is restricted to the open arc of C inside the rim.

Conclusion: *The routes of the geodesics of the hyperbolic plane are the portions in H of* (1) *all Euclidean straight lines through the origin* $\mathbf{0}$ *and* (2) *all Euclidean circles orthogonal to the rim of H* (Fig. 7.11). ◆

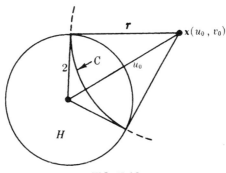

FIG. 7.10

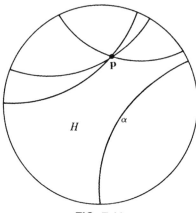

FIG. 7.11

As these examples show, the geodesics of the hyperbolic plane bear comparison with those of the Euclidean plane. Around 300 B.C., Euclid established a remarkable set of axioms for the straight lines of his plane. The goal was to derive its geometry from axioms so overwhelmingly reasonable as to be "self-evident." The most famous of these is equivalent to the *parallel postulate*: If **p** is a point not on a line α, then there is a unique line β through **p** that does not meet α.

From the beginning this postulate was regarded as somewhat less certain than the others. For example, the axiom that two points determine a unique straight line might be checked by laying down a (perhaps long, but still finite) straightedge touching both points. But for the parallel postulate, one would have to travel the whole infinite length of β to be sure it never touches α. Thus, over the centuries, tremendous efforts were expended in trying to deduce the parallel postulate from the other axioms. The hyperbolic plane H offers the most convincing proof that this cannot be done. For if "straight line" is replaced by "route of a geodesic," then every Euclidean axiom holds in H *except* the parallel postulate. For example, given any two points it is easy to see that one and only one geodesic route runs through them. But it is clear from Fig. 7.11 that in H there are an infinite number of geodesic routes through **p** that do not meet α.

When the implications of this discovery were worked out, what was destroyed was not the modest hope of proving the parallel postulate, but the whole idea that the Euclidean plane is, in some philosophical sense, an Absolute, whose properties are self-evident. It had become just one geometric surface among the infinitely many discovered by Riemann.

FIG. 7.12

Exercises

1. In the Poincaré half-plane, show that the routes of geodesics are all vertical lines and all semicircles with centers on the u-axis (see Fig. 7.12). (*Hint:* $x(u, v) = (u, v)$ is a Clairaut patch "relative to v," so in the text equations, reverse u and v, and E and G.)

2. (*Barrier curves.*) Let $\mathbf{x}$ be a Clairaut parametrization, and let $\alpha = \mathbf{x}(a_1, a_2)$ be a unit-speed geodesic with slant c. Suppose that α starts at the point $\mathbf{p}_0 = \mathbf{x}(u_0, v_0)$ and, for definiteness, that $a_1'(0) > 0$. If there is a number $u > u_0$ such that $G(u) = c^2$, let u_1 be the smallest such number. Then the v-parameter curve $\beta(v) = \mathbf{x}(u_1, v)$ is called a *barrier curve for* α. Prove:
(a) α comes arbitrarily close to β.
(b) If β is a geodesic (that is, if $G_u(u_0) = 0$), then α does not meet β—hence asymptotically approaches it. (See Ex. 11.)

3. (*Continuation.*) If the barrier curve is not a geodesic, it can be shown that α does meet β, say at the point $\alpha(t^*)$. Prove that $a'(t^*) = 0$ and that a' changes sign at t^*. Thus α has a turning point at $\alpha(t^*)$, bouncing off β as in Fig. 7.13. (*Hint:* Show that $\alpha''(t^*) \neq 0$.)
In Example 5.6, what are the barrier curves for the geodesics of $\mathbf{R}^2$? the geodesics of H?

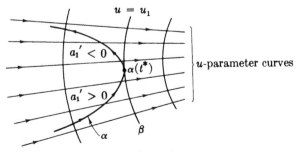

FIG. 7.13

4. Let α be a geodesic on a surface of revolution.
(a) Show that the slant of α is $c = h \sin \varphi$, where $h(t)$ is the distance to the axis of revolution, and φ gives the angles at which α cuts the meridians.
(b) Deduce that α cannot cross a parallel of radius $|c|$.

5. On a torus of revolution T with the usual parametrization, prove:
(a) If a geodesic α is, at some point, tangent to the top circle ($u = \pi/2$), then α remains always on the outer half of $T (-\pi/2 \leq u \leq \pi/2)$ and travels around T, oscillating between the top circle and bottom circle.
(b) Every geodesic of T that crosses the inner equator ($u = -\pi$) also crosses the outer equator ($u = 0$), and furthermore, unless it is a meridian, it will spiral around the torus, crossing both equators infinitely many times.
(c) Every geodesic of T—except a parametrization of the inner or outer equator—*crosses* the outer equator.

6. Let M be the catenoid in Example 7.1 of Chapter 5, with $c = 1$. Let C be its central circle $x = 0$. If α is a unit-speed geodesic starting at a typical point, say $\mathbf{p}_0 = (u_0, \cosh u_0, 0)$, $u_0 > 0$, let $\varphi(t)$ be the angle from the positive direction on meridians to $\alpha'(t)$. In each of the following cases, find initial values $0 \leq \varphi_0 \leq \pi$ of φ (if any) such that α:
(a) stays in the region $u > u_0$.
(b) starts toward the central circle C, but turns back before reaching it.
(c) asymptotically approaches C.
(d) bounces off C and returns to approach $x = +\infty$.
(e) crosses C at least twice.
(f) crosses C once and continues to $x = -\infty$.
(*Hint:* Compare with Ex. 4.)

7. Prove that no geodesic on the bugle surface (Example 7.6 of Ch. 5) can be defined on the whole real line.

It is often convenient to regard a point of $\mathbf{R}^2$ as a complex number

$$z = u + iv = (u, v).$$

Thus the hyperbolic plane can be described as the disk $|z| < 2$ with conformal structure given by the ruler function $h(z) = 1 - |z|^2/4$. Here $|z|$ is the magnitude of z, characterized by

$$|z|^2 = z\bar{z} = u^2 + v^2.$$

8. (*The Poincaré half-plane P is isometric to the hyperbolic plane H.*) In terms of complex numbers, P is the half-plane $\mathscr{I}z > 0$, with conformal structure given by $h(z) = \mathscr{I}z$. ($\mathscr{I}z$ is the imaginary part v of $z = u + iv$.) Let $F: H \to P$ be the mapping

$$F(z) = \frac{z + 2i}{iz + 2} \quad (|z| < 2).$$

Show that:

(a) $\mathscr{F}(z) = \left(4 - |z|^2\right)/|iz + 2|^2$.

(b) F is a diffeomorphism of H onto P. (Compute F^{-1} explicitly.)

(c) Relative to Euclidean geometry, F is conformal, with scale factor $\lambda(z) = 4/|iz + 2|^2$. (See Ex. 7 of Sec. 1.)

(d) $F \colon H \to D$ is an isometry.

Make a sketch of H and P indicating the images in P of each of the four quadrants of H. (*Hint:* the u and v axes in H are the routes of geodesics.)

9. (*Numerical integration, computer graphics.*) (a) Let $\mathbf{x}(u, v)$ be a Clairaut parametrization with metric components $E(u)$, $G(u)$. Write the computer commands that produce:

(a) Numerical solution of the first-order geodesic equations (C-1) and (C-2) in the proof of Proposition 5.5 (changing a_1 to $u(t)$, etc.). Use initial conditions $u(0) = u_0$, $v(0) = v_0$, with slant c, on the interval $t_{min} \leqq t \leqq t_{max}$.

(b) A plot of the geodesic given in (a).

(Dropping the $\pm$ in (C-1) costs only a reverse parametrization. The differential equations behave badly when the geodesic meets a barrier.)

10. (*Continuation.*) Let $\mathbf{x}(u, v) = (u \cos v, u \sin v)$ be the Clairaut parametrization of the hyperbolic plane $H(1)$ as in Example 5.6.

(a) Plot the geodesic that starts at $u(0) = 1$, $v(0) = 0$ with slant $c = 1.330$ (avoiding the turning point problem of $c = 4/3$) and runs to near the rim of H.

(b) Show the following on a single plot: the geodesic of (a), its other branch with $c = -1.330$, the relevant arc of its barrier curve $u = 1$, and the rim $u = 2$. (*Hint:* See Fig. 7.11.)

11. (*Continuation.*) Let M be the surface of revolution with parametrization $\mathbf{x}(u, v) = (u, f(u) \cos v, f(u) \sin v)$, where $f(u) = (3 + u^2)/(4 + u^2)$.

(a) Plot the zone of M between $u = -5$ and $u = 2$.

(b) Using Exercise 9, plot the geodesic γ that starts at $\mathbf{x}(-4, 0)$ and asymptotically approaches the neck $u = 0$ of M on an interval $0 \leqq t \leqq b$, where b is 20 or more.

12. (*Continuation.*) On the paraboloid given by $\mathbf{x}(u, v) = (u \cos v, u \sin v, u^2)$, let γ be the geodesic that has a turning point (meets its barrier curve) at $u(0) = 1$, $v(0) = 0$. Plot:

(a) The xy projection ($u \cos v, u \sin v$) of one branch of γ, defined on an interval $0 \leqq t \leqq b$, with b large.

(b) On a single figure, the branch in (a) and its symmetrical branch.

13. A *Liouville parametrization* $\mathbf{x}: D \to M$ is an orthogonal parametrization for which $E = G = U(u) + V(v)$. (Thus Clairaut parametrization is the
special case where U or V is zero.) If $\alpha = \mathbf{x}(a_1, a_2)$ is a unit speed geodesic,
with $\mathbf{x}$ Liouville, prove that

$$U(a_1)\sin^2 \varphi - V(a_2)\cos^2 \varphi$$

is constant along α, where φ is the angle from $\mathbf{x}_u$ to α'.
    (*Hint:* First show that $\sin^2 \varphi = (U(a_1) + V(a_2))a_1'^2$ and $\cos^2 \varphi =$
$(U(a_1) + V(a_2))a_2'^2$.)

14. (*Continuation of Exercise 8.*) (a) Show that the restriction to H of any
orthogonal transformation of $\mathbf{R}^2$ (e.g., a Euclidean rotation about $\mathbf{0}$) is an
isometry of H, and that for any real number a, the Euclidean translation
$T_a(z) = z + a$ is an isometry of P. (*Hint:* See Ex. 9 of Sec. 1.)
    (b) If $F: H \to P$ is the isometry in Exercise 8, show by explicit computtion
that the isometry $F^{-1}T_a F: H \to H$ carries $\mathbf{0}$ to a point w_a in H with
$|w_a| = 4a^2/(4 + a^2)$.
    (c) Deduce that given any point w of H there is an isometry of H that
sends $\mathbf{0}$ to w.

Thus, all *points* of the hyperbolic plane are geometrically equivalent. A mild
extension of (c) shows that all *frames* are geometrically equivalent.

7.6   The Gauss-Bonnet Theorem

We have seen that the Gaussian curvature K of a geometric surface M has a
strong influence on other properties of M, notably the shape of M when it
is a surface in $\mathbf{R}^3$. Now we will show that the influence of Gaussian curvature penetrates to the *topological* conformation of M—to properties independent of the geometry of M.
    To show this, the main step is a theorem that relates the total curvature of
a 2-segment to the total amount its boundary curve turns. The geodesic curvature of a curve α in an oriented surface M tells its rate of turning relative
to arc length s. So to find the total turning, we integrate with respect to arc
length, adjusting suitably when α is merely a regular curve.

6.1 Definition   Let $\alpha: [a, b] \to M$ be a regular curve segment in an
oriented geometric surface M. The *total geodesic curvature* of α is

$$\int_\alpha \kappa_g ds = \int_{s(a)}^{s(b)} \kappa_g(s(t))\frac{ds}{dt} dt.$$

For example, if α makes one counterclockwise trip around a circle C of radius r in the plane $\mathbf{R}^2$, then α has constant geodesic curvature $\kappa_g = 1/r$. Thus, regardless of the size of r,

$$\int_\alpha \kappa_g ds = \frac{1}{r} 2\pi r = 2\pi.$$

The usual orientation of $\mathbf{R}^2$ makes $\kappa_g > 0$ for a (left-turning) counterclockwise trip, but $\kappa_g < 0$ for a (right-turning) clockwise trip. So a clockwise trip around C would have total curvature -2π.

Integrating the formula in Corollary 4.6 gives

6.2 Lemma   Let $\alpha: [a, b] \to M$ be a regular curve segment in a region of M oriented by a frame field E_1, E_2. Then

$$\int_\alpha \kappa_g ds = \varphi(b) - \varphi(a) + \int_\alpha \omega_{12},$$

where φ is an angle function from E_1 to α' along α, and ω_{12} is the connection form of E_1, E_2.

For integration we have used 2-segments $\mathbf{x}: R \to M$ that are one-to-one and regular only on the interior R° of R, since 2-dimensional integration can afford to neglect 1-dimensional sets. But now the boundary of $\mathbf{x}$ becomes important, so we require $\mathbf{x}$ to be one-to-one and regular on the entire rectangle R. Equivalently, $\mathbf{x}: R \to M$ is the restriction to R of a patch defined on some open set containing R.

Then the edge curves α, β, γ, δ of $\mathbf{x}$ (Definition 6.4 of Chapter 4) are one-to-one regular curve segments, and the boundary of $\mathbf{x}$,

$$\partial\mathbf{x} = \alpha + \beta - \gamma - \delta,$$

is a single piecewise regular curve enclosing the rectangular region $\mathbf{x}(R) \subset M$.

Now we want to define the total geodesic curvature of $\partial\mathbf{x}$. If (contrary to fact) $\partial\mathbf{x}$ were a single smoothly closed curve γ, this total would be just $\int_\gamma \kappa_g ds$. But since $\partial\mathbf{x}$ has corners, it is not enough to get the total turning on the edge curves, namely,

$$\int_{\partial\mathbf{x}} \kappa_g ds = \int_\alpha \kappa_g ds + \int_\beta \kappa_g ds + \int_{-\gamma} \kappa_g ds + \int_{-\delta} \kappa_g ds$$

$$= \int_\alpha \kappa_g ds + \int_\beta \kappa_g ds - \int_\gamma \kappa_g ds - \int_\delta \kappa_g ds.$$

We must also add the angles through which a unit tangent on $\partial\mathbf{x}$ *would have to turn* at the four corners of the rectangular region $\mathbf{x}(R)$. These angles replace

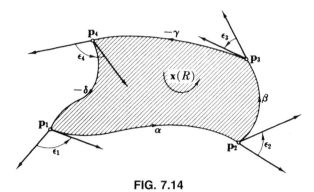

FIG. 7.14

the total curvature of small curved segments needed to round off the corners and make $\partial\mathbf{x}$ a single smooth curve.

For R: $a \leqq u \leqq b$, $c \leqq v \leqq d$ these corners,

$$\mathbf{p}_1 = \mathbf{x}(a, c), \quad \mathbf{p}_2 = \mathbf{x}(b, c), \quad \mathbf{p}_3 = \mathbf{x}(b, d), \quad \mathbf{p}_4 = \mathbf{x}(a, d),$$

are called the *vertices* of $\mathbf{x}(R)$.

In general, if a regular curve segment α ends at the starting point of another segment β, say $\alpha(1) = \beta(0)$, then the *turning angle* ε from α to β is defined to be the oriented angle from $\alpha'(1)$ to $\beta'(0)$ that is smallest in absolute value. For a 2-segment—if no other orientation has been specified—we use the orientation determined by $\mathbf{x}$, represented by the area form dM such that $dM(\mathbf{x}_u, \mathbf{x}_v) > 0$. The following terminology is familiar in the case of a polygon in the Euclidean plane.

6.3 Definition   Let $\mathbf{x}: R \to M$ be a one-to-one regular 2-segment with vertices $\mathbf{p}_1$, $\mathbf{p}_2$, $\mathbf{p}_3$, $\mathbf{p}_4$. The *exterior angle* ε_j of $\mathbf{x}$ at $\mathbf{p}_j$ $(1 \leqq j \leqq 4)$ is the turning angle at $\mathbf{p}_j$ derived from the edge curves $\alpha, \beta, -\gamma, -\delta, \alpha, \ldots$ in order of occurrence in $\mathbf{x}$ (Fig. 7.14). The *interior angle* ι_j at $\mathbf{p}_j$ is $\pi - \varepsilon_j$.

This general definition will be needed later, but in the case at hand, exterior angles can be expressed directly in terms of the usual coordinate angle $0 < \vartheta < \pi$ from $\mathbf{x}_u$ to $\mathbf{x}_v$ as

$$\varepsilon_1 = \pi - \vartheta_1, \quad \varepsilon_2 = \vartheta_2, \quad \varepsilon_3 = \pi - \vartheta_3, \quad \varepsilon_4 = \vartheta_4,$$

where ϑ_j is the coordinate angle at $\mathbf{p}_j$. For example, in Fig. 7.15 consider the situation at $\mathbf{p}_3$. By the definition of edge curves, β' is $\mathbf{x}_v$, but $(-\gamma)'$ is $-\mathbf{x}_u$, since $-\gamma$ is an orientation-reversing reparametrization of γ. Thus $\varepsilon_3 + \vartheta_3 = \pi$.

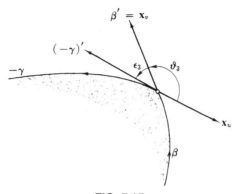

FIG. 7.15

We can now prove one of the fundamental theorems of differential geometry.

6.4 Theorem   Let $\mathbf{x}: R \to M$ be a one-to-one regular 2-segment in a geometric surface M. If dM is the area form determined by $\mathbf{x}$, then

$$\iint_{\mathbf{x}} K dM + \int_{\partial \mathbf{x}} \kappa_g ds + \varepsilon_1 + \varepsilon_2 + \varepsilon_3 + \varepsilon_4 = 2\pi,$$

where ε_j is the exterior angle at the vertex $\mathbf{p}_j$ of $\mathbf{x}$ $(1 \le j \le 4)$.

We emphasize that the 2-segment $\mathbf{x}$ supplies the necessary orientation; M itself need not be oriented—or even orientable.

This result is called the *Gauss-Bonnet formula with exterior angles*. Since $\varepsilon_j = \pi - \iota_j$ for $1 \le j \le 4$, the formula can be rewritten in terms of the interior angles of $\mathbf{x}(R)$ as

$$\iint_{\mathbf{x}} K dM + \int_{\partial \mathbf{x}} \kappa_g ds = \iota_1 + \iota_2 + \iota_3 + \iota_4 - 2\pi.$$

Proof.   The associated frame field $E_1 = \mathbf{x}_u / \sqrt{E}$, $E_2 = J(E_1)$ of $\mathbf{x}$ has $dM(E_1, E_2) = +1$. Then the second structural equation becomes

$$d\omega_{12} = -K\theta_1 \wedge \theta_2 = -K\, dM.$$

The power for this proof is supplied by Stokes' theorem (6.5 of Ch. 4), which gives

$$\iint_{\mathbf{x}} K\, dM + \int_{\partial \mathbf{x}} \omega_{12} = 0. \tag{1}$$

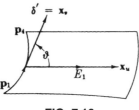

FIG. 7.16

Now we use Lemma 6.2 to evaluate

$$\int_{\partial x} \omega_{12} = \int_{\alpha} \omega_{12} + \int_{\beta} \omega_{12} - \int_{\gamma} \omega_{12} - \int_{\delta} \omega_{12}. \tag{2}$$

Beginning with α, we have $\alpha' = x_u = \sqrt{E}\, E_1$, so the angle from E_1 to α' is identically zero. Thus by Lemma 6.2,

$$\int_{\alpha} \omega_{12} = \int_{\alpha} \kappa_g ds. \tag{3}$$

Now we try a harder case, say $\int_{\delta} \omega_{12}$. Here the angle from $E_1 = x_u/\sqrt{E}$ to $\delta' = x_v$ is just the coordinate angle ϑ from x_u to x_v. (See Fig. 7.16.) Hence by Lemma 6.2,

$$\int_{\delta} \omega_{12} = \vartheta_1 - \vartheta_4 + \int_{\delta} \kappa_g ds,$$

where ϑ_j is the coordinate angle at the vertex p_j ($1 \leqq j \leqq 4$). But since $\vartheta_1 = \pi - \varepsilon_1$ and $\vartheta_4 = \varepsilon_4$, this becomes

$$\int_{\delta} \omega_{12} = \pi - \varepsilon_1 - \varepsilon_4 + \int_{\delta} \kappa_g ds. \tag{4}$$

In an entirely similar way we find

$$\int_{\beta} \omega_{12} = -\pi + \varepsilon_2 + \varepsilon_3 + \int_{\beta} \kappa_g ds. \tag{5}$$

and

$$\int_{\gamma} \omega_{12} = \int_{\gamma} \kappa_g ds. \tag{6}$$

Equations (3)–(6) turn (2) into

$$\int_{\partial x} \omega_{12} = \int_{\alpha} \kappa_g ds + \int_{\beta} \kappa_g ds - \int_{\gamma} \kappa_g ds - \int_{\delta} \kappa_g ds$$
$$-2\pi + (\varepsilon_1 + \varepsilon_2 + \varepsilon_3 + \varepsilon_4)$$
$$= \int_{\partial x} \kappa_g ds - 2\pi + (\varepsilon_1 + \varepsilon_2 + \varepsilon_3 + \varepsilon_4).$$

Substitution in (1) then yields the required formula.   ◆

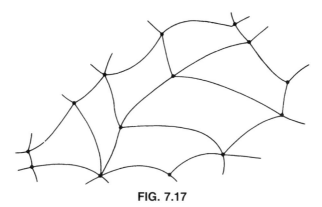

FIG. 7.17

Our goal now is to extend the reach of the Gauss-Bonnet formula to an entire surface.

A *rectangular decomposition* $\mathcal{D}$ of a surface M is a finite collection of one-to-one regular 2-segments $\mathbf{x}_1, \ldots, \mathbf{x}_f$ whose images cover M in such a way that if any two intersect, they do so in either a single common vertex or a single common edge.

Evidently, a rectangular decomposition is a special kind of paving (Definition 7.3 of Chapter 6), but the regions $\mathbf{x}_i(R_i)$ now have well-defined regular edges. Furthermore, they are required to fit together very neatly, as in Fig. 7.17. (Compare the more casual paving in Fig. 6.14.)

6.5 Theorem   Every compact surface M has a rectangular decomposition.

This result is certainly plausible, for if M were made of paper, we could just take a pair of scissors and cut out rectangular pieces until all of M was gone. In fact, any compact region bounded by a finite number of piecewise regular curve segments has a rectangular decomposition. A rigorous proof would be tedious and not very instructive.

The topological analogue of a diffeomorphism is a *homeomorphism*, a continuous mapping $F: M \to N$ that has a continuous inverse map $F^{-1}: N \to M$. When such an F exists, M and N are said to be *homeomorphic*. A *topological invariant* is a property that is preserved by homeomorphisms, hence shared by homeomorphic surfaces. These are the properties that can be defined solely in terms of open sets.

Every diffeomorphism is a homeomorphism, because differentiable functions are continuous. But the converse is false; indeed, it is already clear in elementary calculus that a continuous function $f: \mathbf{R} \to \mathbf{R}$ can have a jagged graph, while differentiability requires smoothness. Nevertheless, in dimension 2 we have the remarkable result that *two surfaces are diffeomorphic if and only if they are homeomorphic.* This is a considerable simplification, since it allows many topological invariants to be discussed in differentiable terms. Here is a famous instance.

We shall understand that a rectangular decomposition $\mathscr{D}$ carries with it not only its rectangular regions $\mathbf{x}_i(\mathbf{R}_i)$—called *faces*—but also the *vertices* and *edges* of these regions.

6.6 Theorem If $\mathscr{D}$ is a rectangular decomposition of a compact surface M, let v, e, and f be the numbers of vertices, edges, and faces in $\mathscr{D}$. Then the integer $v - e + f$ is the same for every rectangular decomposition of M. This integer $\chi(M)$ is called the *Euler characteristic* of M.

An elementary topological proof of this famous theorem is outlined in Chapter 1, Section 8 of [Ma].

The fact that the decomposition $\mathscr{D}$ is based on *rectangles* is merely a convenience for integration. We could just as well cut M into arbitrary polygons (see Exercise 5 of Section 6). In the resulting *polygonal decomposition*, the different polygons of course are still required to fit neatly, but (as in Fig. 7.18) they need not have the same number of sides. When only triangles are used, the decomposition is called a *triangulation* of M.

6.7 Example *Euler characteristic.*

(1) Any sphere Σ has $\chi(\Sigma) = 2$. The surface of a tetrahedron provides a triangulation of Σ, and we count $v = 4$, $e = 6$, $f = 4$; hence $\chi(\Sigma) = 2$. By inflating a cube, as in Fig. 7.18, we get a rectangular decomposition of Σ with $v = 8$, $e = 12$, $f = 6$, so again $\chi(\Sigma) = 2$. Inflating a prism gives a polygonal decomposition with $v = 6$, $e = 9$, $f = 5$, but still $\chi(\Sigma) = 2$.

(2) A torus T has $\chi(T) = 0$. Picture T as a torus of revolution, and cut it along any three meridians and three parallels. This gives a rectangular decomposition with $v = 9$, $e = 18$, $f = 9$; hence $\chi = 0$.

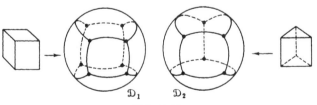

$\mathscr{D}_1$ $\mathscr{D}_2$

FIG. 7.18

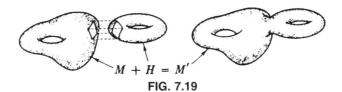

$M + H = M'$

FIG. 7.19

(3) Adding a handle to a compact surface reduces its Euler characteristic by 2.

This last assertion requires explanation. A *handle* is a torus with the interior of one face removed. To add a handle to a surface *M*—also in some rectangular decomposition—remove the interior of a face of *M*, and to the resulting rim, smoothly attach the rim of the handle, with the vertices and edges of the two rims coinciding (Fig. 7.19).

This operation produces a new surface *M'* already furnished with a rectangular decomposition. Its Euler characteristic is

$$\chi(M') = \chi(M) - 2.$$

In fact, the decomposition has exactly two faces fewer than *M* and the torus combined. Coalescing the two rims has also eliminated four vertices and four edges, but this has no effect on χ.

It is easy to see that *diffeomorphic surfaces have the same Euler characteristic*, for if $x_1, \ldots, x_f$ is a decomposition of *M* and $F: M \to N$, then $F(x_1)$, $\ldots, F(x_f)$ is a decomposition of *N* with the same *v*, *e*, and *f*.

For example, no matter how wildly we distort the "round" sphere

$$\Sigma: \ x^2 + y^2 + z^2 = 1,$$

the resulting surface will still have Euler characteristic 2.

Suppose we start from a sphere Σ and successively add *h* handles ($h \geq 0$) to obtain a surface $\Sigma[h]$. What the handles look like and where they are attached is irrelevant in view of the following remarkable result.

6.8 Theorem   If *M* is a compact, connected, orientable surface, there is a unique integer $h \geq 0$ such that *M* is diffeomorphic to $\Sigma[h]$.

In this case, we say that *M* itself is a sphere with *h* handles. The theorem is proved in Chapter 1 of [Ma], where by a remark above, *homeomorphism* can everywhere be replaced by *diffeomorphism*.

In Fig 7.20 the surfaces all have just one handle, so they are all diffeomorphic. If *M* has *h* handles, then by (1) and (3) of the preceding example, $\chi(M) = 2 - 2h$, since

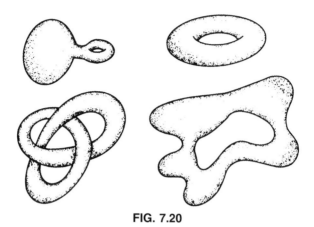

FIG. 7.20

$$\chi(M) = \chi(\Sigma[h]) = \chi(\Sigma) - 2h = 2 - 2h.$$

6.9 Corollary   Compact orientable surfaces M and N have the same Euler characteristic if and only if they are diffeomorphic.

Proof.   If $\chi(M) = \chi(N)$, then M and N have the same number of handles; hence by the theorem they are diffeomorphic. The converse was noted above.   ◆

This is the sense in which $\chi(M)$ is characteristic of M. Though the number of handles (or *genus*) of M is visually appealing, $\chi(M)$ is usually easier to compute and in the long run turns out to be more fundamental.

Returning to geometric surfaces, we can now prove this spectacular consequence of the Gauss-Bonnet formula.

6.10 Theorem (Gauss-Bonnet)   *The total Gaussian curvature M of a compact orientable geometric surface M is 2π times its Euler characteristic:*

$$\iint_M K \, dM = 2\pi\chi(M).$$

Proof.   Orient M by an area form dM, and let $\mathscr{D}$ be a rectangular decomposition of M whose 2-segments are all positively oriented. (A wrong orientation of $\mathbf{x}$ can be corrected by reversing u and v.) Then $\mathscr{D}$ is an oriented paving of M as defined in Chapter 6, Section 7. By definition, the total curvature of M is

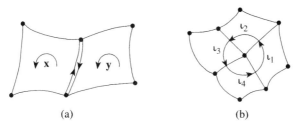

FIG. 7.21

$$\iint_M K \, dM = \sum_{i=1}^{f} \iint_{\mathbf{x}_i} K \, dM. \tag{1}$$

Apply the Gauss-Bonnet formula to each summand. In terms of interior angles, the result is

$$\iint_{\mathbf{x}_i} K \, dM = -\int_{\partial \mathbf{x}_i} \kappa_g ds - 2\pi + \iota_1 + \iota_2 + \iota_3 + \iota_4. \tag{2}$$

Now consider what happens when (2) is substituted into (1).

Because M is a surface—locally like $\mathbf{R}^2$—each edge of the decomposition $\mathscr{D}$ will occur in exactly *two* faces, say $\mathbf{x}_i(R_i)$ and $\mathbf{x}_j(R_j)$. Let α_i and α_j be the parametrizations of this edge occurring in the oriented boundaries $\partial \mathbf{x}_i$ and $\partial \mathbf{x}_j$, respectively.

By construction, the regions $\mathbf{x}_i(R_i)$ and $\mathbf{x}_j(R_j)$ have the same orientation as M, so α_i and α_j are orientation-reversing reparametrizations of each other, as shown in Fig. 7.21(a). Thus

$$\int_{\alpha_i} \kappa_g ds + \int_{\alpha_j} \kappa_g ds = 0.$$

It follows that summing over all faces yields

$$\sum_{i=1}^{f} \int_{\partial \mathbf{x}_i} \kappa_g ds = 0. \tag{3}$$

In fact, we have just seen that the integrals over edge curves cancel in pairs.

As usual, v, e, and f are the numbers of vertices, edges, and faces in the decomposition. Substituting (2) into (1) gives

$$\iint_M K \, dM = -2\pi f + \mathscr{I}, \tag{4}$$

where $\mathscr{I}$ is the sum of all interior angles of all the 2-segments in the decomposition. But the sum of the interior angles at each vertex is just 2π (Fig. 7.21(b)), so $\mathscr{I} = 2\pi v$. Thus

$$\iint_M K \, dM = -2\pi f + 2\pi v. \tag{5}$$

A simple combinatorial observation will complete the proof. The faces of the decomposition $\mathscr{D}$ are rectangular: Each face has four edges. But each edge belongs to two faces. Thus $4f$ *counts e twice*; that is, $4f = 2e$. Equivalently, $-f = f - e$, so (5) becomes

$$\iint_M K \, dM = 2\pi(v - e + f) = 2\pi\chi(M). \qquad \blacklozenge$$

The most striking aspect of the Gauss-Bonnet theorem is that it directly links topology and geometry. Because the Euler characteristic is a topological invariant, the theorem shows that *total Gaussian curvature is a topological invariant.*

For example, the flat torus in Example 2.2 has $K = 0$, hence total curvature zero. Earlier, an explicit calculation showed that although the (diffeomorphic) torus of revolution in $\mathbf{R}^3$ has variable curvature, its total curvature is also zero. The Gauss-Bonnet theorem makes this evident without calculation—and if the various tori in Fig. 7.20 were realized formally, they too would have total curvature zero.

The Gauss-Bonnet theorem provides a way to attack some seemingly formidable problems. For instance, Example 2.4(1) shows that if a single point is removed from a sphere Σ, there exists a metric on the punctured sphere with $K = 0$. But

There can be no metric on an entire sphere for which $K \leqq 0$.

Indeed, $K \leqq 0$ would imply $\iint_\Sigma K d\Sigma \leqq 0$, but we know that $2\pi\chi(\Sigma) > 0$. Reversing this argument,

If a compact orientable geometric surface M has $K > 0$, then M is diffeomorphic to a sphere.

Since $K > 0$, M has positive total curvature, hence positive Euler characteristic. But $\chi(M) = 2 - 2h$, so $h = 0$, and hence M is diffeomorphic to $\Sigma[0] = \Sigma$.

Further applications of the Gauss-Bonnet theorem are given in the next section.

Exercises

1. Find the total Gaussian curvature of:
   (a) An ellipsoid.   (b) The surface in Fig. 4.8.
   (c) $M: x^2 + y^4 + z^6 = 1$.

2. Prove that for a compact orientable geometric surface M:

$K > 0 \Rightarrow M$ is diffeomorphic to a sphere,

$K = 0 \Rightarrow M$ is diffeomorphic to a torus,

$K < 0 \Rightarrow M$ is diffeomorphic to a sphere with $h \geq 2$ handles.

3. Let M be a compact orientable geometric surface with h handles. Prove that there exists a point $\mathbf{p}$ in M at which

$$K(\mathbf{p}) > 0 \quad \text{if } h = 0,$$

$$K(\mathbf{p}) = 0 \quad \text{if } h = 1,$$

$$K(\mathbf{p}) < 0 \quad \text{if } h \geq 2.$$

4. If M is a compact orientable geometric surface in $\mathbf{R}^3$ that is not diffeomorphic to a sphere, show that there is a point $\mathbf{p}$ of M at which $K(\mathbf{p}) < 0$. (Compare Thm. 3.5 of Ch. 6.)

5. In each case, show that the change in polygonal decomposition of a compact surface M does not change the Euler characteristic $v - e + f$ of M:

(a) Given a rectangular decomposition of M, cut each rectangle along a diagonal to form two triangles, thus producing a triangulation of M.

(b) Given a triangulation of M, cut each triangle into three rectangles by lines from a central point to a midpoint of each side, thus producing a rectangular decomposition of M.

(c) Given an arbitrary polygonal decomposition of a compact surface M, cut each polygon into triangles, thus producing a triangulation of M.

6. (a) For a regular curve segment $\alpha\colon [a, b] \to M$, show that the total geodesic curvature $\int_\alpha \kappa_g ds$ is

$$\int_\alpha \frac{\langle \alpha'', J(\alpha') \rangle}{\langle \alpha', \alpha' \rangle}\, dt.$$

(*Hint:* Ex. 8 of Sec. 4.)

(b) Let $\mathbf{x}$ be a positively oriented orthogonal patch in M. Deduce the following formulas for the total geodesic curvature of the parameter curves:

$$-\frac{1}{2} \int_{u_1}^{u_2} \frac{E_v}{\sqrt{EG}} (u, v_0)\, du, \quad \frac{1}{2} \int_{u_1}^{v_2} \frac{G_u}{\sqrt{EG}} (u_0, v)\, dv.$$

Assume for simplicity that M is in $\mathbf{R}^3$. Thus

$$\langle \mathbf{x}_{uu}, \mathbf{x}_v \rangle = -\frac{1}{2} E_v \quad \text{and} \quad \langle \mathbf{x}_{vv}, \mathbf{x}_u \rangle = -\frac{1}{2} G_u,$$

where either intrinsic or Euclidean derivatives give the same result.

7. Let **x**: $R \to \Sigma(r)$ be the restriction of the geographical patch (Ex. 2.2 of Ch. 4) to the rectangle R: $0 \leqq u, v \leqq \pi/4$. Check the Gauss-Bonnet formula by computing separately each of its terms.

8. If F: $M \to N$ is a mapping of compact oriented surfaces, the *degree* d_F of F is the algebraic area of $F(M)$ divided by the area of N. Thus d_F gives the total algebraic number of times F wraps M around N. (It can be shown that d_F is always an integer.)

Prove the *Hopf theorem*: If M is a compact oriented surface in $\mathbf{R}^3$, the degree of its Gauss map is $\chi(M)/2$.

9. Consider a polygonal decomposition whose faces are hexagons, with exactly three edges meeting at each vertex. Is there such a decomposition on the sphere? the torus?

7.7 Applications of Gauss-Bonnet

The Gauss-Bonnet theorem (6.10) was proved by cutting an entire surface M into rectangular regions and applying the Gauss-Bonnet formula (6.4) to each. The scheme works because these rectangles are all consistently oriented by an orientation of M, and thus the integrals $\int \kappa_g ds$ on their boundaries cancel in pairs. Here in essence is the fundamental idea of algebraic topology—specifically, *homology theory*. (Indeed, considerations of this kind led Poincaré to its invention.) By applying this scheme to suitable regions in M we can extend the range of the Gauss-Bonnet theorem.

7.1 Definition An *oriented polygonal region* $\mathscr{P}$ in a surface M is a (necessarily compact) oriented region furnished with a positively oriented rectangular decomposition $\mathbf{x}_1, \ldots, \mathbf{x}_f$.

A *boundary segment of* $\mathscr{P}$ is a curve segment β that is an edge curve of exactly one of the rectangles $\mathbf{x}_i(R_i)$. For simplicity, we add the requirement that a vertex of the decomposition cannot belong to more than two boundary segments.

We will define the boundary of the region $\mathscr{P}$ as a generalization of the boundary of a single 2-segment. Each boundary segment σ of $\mathscr{P}$ is an edge curve of exactly one rectangle, say $\mathbf{x}(R)$, and has an orientation given by the definition of the boundary $\partial\mathbf{x}$ (Definition 6.4 in Chapter 4). Recall that in

$$\partial \mathbf{x} = \alpha + \beta - \gamma - \delta,$$

the parametrizations that γ and δ inherit from $\mathbf{x}$ determine the "wrong" orientation (see Fig. 4.37). The minus signs correct this—as would an orientation-reversing reparametrization.

With the proper orientations, the boundary segments of $\mathscr{P}$ comprise a finite number of simply closed polygonal curves $\beta_1, \ldots, \beta_k$. As in the case of a single rectangle, these orientations obey the informal rule, "Travel around the boundary keeping the region always on the left."

7.2 Definition   The *oriented boundary* $\partial\mathscr{P}$ of an oriented polygonal region $\mathscr{P}$ is the formal sum of the simple closed, oriented polygonal curves β_i described above:

$$\partial \mathscr{P} = \beta_1 + \cdots + \beta_k.$$

This care with orientation is needed to reach the simplicity of the following generalized Stokes' theorem.

7.3 Theorem   If ϕ is a 1-form on an oriented polygonal region $\mathscr{P}$, then

$$\iint_{\mathscr{P}} d\phi = \int_{\partial\mathscr{P}} \phi.$$

In particular, if $\mathscr{P}$ is an entire compact oriented surface M, then $\iint_M d\phi = 0$.

Proof.   By definition,

$$\iint_{\mathscr{P}} d\phi = \sum_{i=1}^{f} \iint_{\mathbf{x}_i} d\phi.$$

Then Stokes' theorem for rectangles (Theorem 6.5 of Chapter 4) gives

$$\iint_{\mathscr{P}} d\phi = \sum_{i=1}^{f} \int_{\partial\mathbf{x}_i} \phi. \tag{$*$}$$

As in the proof of the Gauss-Bonnet theorem, those edges that belong to *two* rectangles acquire opposite orientations from them, so the resulting two integrals cancel. These double edges can thus be ignored, leaving exactly the edges in the oriented boundary $\partial\mathscr{P}$. Hence

$$\sum_{i=1}^{f} \int_{\partial\mathbf{x}_i} \phi = \sum_{j=1}^{k} \int_{\beta_j} \phi = \int_{\partial\mathscr{P}} \phi.$$

Substituting this into $(*)$ gives the primary result. If $\mathscr{P} = M$, there are no boundary edges, so all edge contributions cancel.   ◆

The following result suggests a new meaning for the Euler characteristic.

7.4 Corollary The following properties of a compact orientable surface are equivalent.

(1) There is a nonvanishing tangent vector field on M.
(2) $\chi(M) = 0$.
(3) M is diffeomorphic to a torus.

Proof. (1)→(2) Let V be a nonvanishing vector field on M. Then for any geometric structure on M, the associated frame field

$$E_1 = \frac{V}{\|V\|}, \quad E_2 = J(E_1)$$

is defined on the entire surface. If ω_{12} is its connection form, we know that $d\omega_{12} = -KdM$. Using both the Gauss-Bonnet theorem and Stokes' theorem, we find

$$2\pi\chi(M) = \iint_M K\,dM = -\iint_M d\omega_{12} = 0.$$

(2)→(3) From the preceding section, $\chi(M) = 0$ implies that M has exactly one handle, and hence is a torus.

(3)→(1) Use $\mathbf{x}_u$ or $\mathbf{x}_v$ from the usual parametrization of a torus of revolution. ◆

The following result generalizes the Gauss-Bonnet formulas and theorem.

7.5 Theorem If $\mathscr{P}$ is an oriented polygonal region in a geometric surface, then

$$\iint_{\mathscr{P}} K\,dM + \int_{\partial\mathscr{P}} \kappa_g\,ds + \sum \varepsilon_j = 2\pi\chi(\mathscr{P}),$$

where $\sum \varepsilon_j$ is the sum of the exterior angles (Definition 6.3) of all the closed boundary curves comprising $\partial\mathscr{P}$.

The proof is virtually the same as for the Gauss-Bonnet theorem (6.10), but now the boundary curves survive and give the additional terms. (See Fig. 7.22.)

The simplest case of this theorem occurs when the polygonal region is a single triangle (hence has Euler characteristic 1).

7.6 Corollary If Δ is a triangle in an oriented geometric surface M, then

$$\iint_{\Delta} K\,dM + \int_{\partial\Delta} \kappa_g\,ds = 2\pi - (\varepsilon_1 + \varepsilon_2 + \varepsilon_3) = (\iota_1 + \iota_2 + \iota_3) - \pi.$$

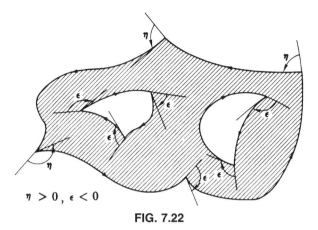

$\eta > 0, \epsilon < 0$

FIG. 7.22

If the edges of the triangle are geodesics, then of course the geodesic curvature terms vanish. If, further, Gaussian curvature K is constant, this result reduces to

$$KA = \iota_1 + \iota_2 + \iota_3 - \pi,$$

where A is the area of the triangle. Thus the celebrated theorem of plane geometry that the sum of the interior angles of a triangle is π depends on the fact that $\mathbf{R}^2$ is flat. Indeed, for constant $K \neq 0$, the sum of the angles is determined by the area of the triangle—and vice versa. Fig. 7.23 shows how a geodesic triangle manages to have $\iota_1 + \iota_2 + \iota_3$ greater than π on a sphere ($K > 0$) and less than π on a hyperbolic plane ($K < 0$).

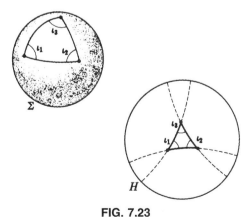

FIG. 7.23

Corollary 7.4 linked Euler characteristic zero to the existence of nonvanishing vector fields. Now we look for a generalization. On an arbitrary compact surface M it is always possible to define a vector field that is differentiable and nonvanishing except at a finite number of points. If we think of the vectors of the field as the velocity vectors of the smooth steady-state flow of a fluid, then these points are *singularities*, where the flow stops or is turbulent. We want to assign to each such point a number, called its *index*, that measures how bad the singularity is.

A point $\mathbf{p}$ is an *isolated singular point* of a vector field V if V is nonvanishing and differentiable on some neighborhood $\mathcal{N}$ of $\mathbf{p}$—except at the point $\mathbf{p}$ itself.

For example, on a sphere $\Sigma \subset \mathbf{R}^3$, let N be the unit tangent vector field that everywhere points due north. Evidently, there is no way to define V differentiably at the poles $(0, 0, \pm 1)$, so they are isolated singular points. We view the south pole $(0, 0, -1)$ as the *source* of a fluid flowing toward a *sink* at the north pole $(0, 0, 1)$.

To define the index of an isolated singular point $\mathbf{p}$ of V, let $\mathcal{D}$ be an oriented disk centered at $\mathbf{p}$ in a coordinate patch of M and small enough so that it contains no other singular points of V. Then the behavior of V *on the boundary C of $\mathcal{D}$* is already enough to tell how bad the singularity is. The idea is to compare V on C with the restriction to C of a smooth vector field X that has no singularities anywhere in $\mathcal{D}$ (for example, one of the partial velocities of the patch).

Let $\alpha \colon [a, b] \to C$ be a parametrization of C as the oriented boundary $\partial \mathcal{D}$ of $\mathcal{D}$. Let $\varphi = \angle_\alpha(X, V)$ be an angle function from X_α to V_α (these vector fields restricted to α).

As V_α tranverses C, changes in the angle φ measure the rotation of V_α relative to X_α. After one circuit these vectors return to their initial values; hence the *total rotation* $\varphi(b) - \varphi(a)$ is an integer multiple of 2π. For example, in Fig. 7.24, where V_α and X_α are initially equal, the total rotation is 2π.

7.7 Definition   With notation as above, the *index of V at $\mathbf{p}$* is the integer

$$\operatorname{ind}(V, \mathbf{p}) = \frac{\varphi(b) - \varphi(a)}{2\pi}.$$

We will soon see that this definition depends only on V and $\mathbf{p}$ (and the orientation of $\mathcal{D}$), but not on the other choices involved. This can be observed experimentally in Fig. 7.25. There, in each case, the general character of the vector field V near the singularity is described by drawing some of its integral curves (Exercise 13 of Section 4.8) since V supplies all their velocity vectors.

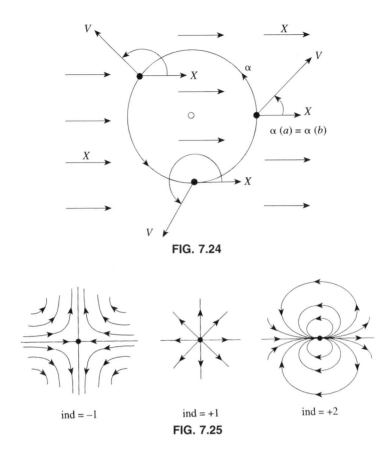

FIG. 7.24

ind = −1   ind = +1   ind = +2

FIG. 7.25

Picture X as a unit vector field in the positive x-direction, and count the number of counterclockwise rotations of V relative to X on any circle around the central singularity. For example, Fig. 7.25 indicates that *sources* have index +1. Reversing the direction of the vectors in this figure shows that *sinks* have the same index, +1.

Indices are just what is needed to generalize Corollary 7.4.

7.8 Theorem (Poincaré-Hopf)   Let V be a vector field on a compact oriented surface M. If V is differentiable and nonvanishing except at isolated singular points $\mathbf{p}_1, \ldots, \mathbf{p}_k$, then the Euler characteristic of M is the sum of their indices.

$$\chi(M) = \sum_{i=1}^{k} \mathrm{ind}(V, \mathbf{p}_i).$$

Proof. The theorem is topological, but to simplify the proof we can assume, without loss of generality, that M is a geometric surface.

The case $k = 0$ is just Corollary 7.4, so suppose $k > 0$. Since the points $\mathbf{p}_i$ are isolated, by the Hausdorff axiom there is a coordinate disk $\mathcal{D}_i$ around each such that no two disks meet. Let $\mathcal{M}$ be the polygonal region that remains when the interiors of all the disks are removed from M. Thus

$$\iint_M K\, dM = \iint_{\mathcal{M}} K\, dM + \sum_{i=1}^{k} \iint_{\mathcal{D}_i} K\, dM. \tag{1}$$

Since V has no singularities on $\mathcal{M}$, it has an associated frame field $E_1 = V/\|V\|$, $E_2 = J(E_1)$ on $\mathcal{M}$. If $\overline{\omega}_{12}$ is the connection form of this frame field, then by Stokes' theorem,

$$\iint_{\mathcal{M}} K\, dM = -\int_{\partial \mathcal{M}} \overline{\omega}_{12} = \sum_{i=1}^{k} \int_{C_i} \overline{\omega}_{12}. \tag{2}$$

The initial minus sign disappears because by the definition of index, each C_i derives its orientation from $\mathcal{D}_i \subset M$, so $\mathcal{M}$ gives it the reverse orientation.

For each $i = 1, \ldots, k$, let X_i be a differentiable unit vector field defined on the disk $\mathcal{D}_i$. Again, for each i, we have a frame field $X_i, J(X_i)$ on the disk $\mathcal{D}_i$. Denote their connection forms collectively by ω_{12}.

Let $\gamma_i: [a, b] \to C_i$ be a positively oriented parametrization of $C_i = \partial \mathcal{D}_i$. By Stokes' theorem,

$$\iint_{\mathcal{D}_i} K\, dM = -\int_{\gamma_i} \omega_{12}. \tag{3}$$

Substituting (2) and (3) into (1) gives

$$\iint_M K\, dM = \sum_{i=1}^{k} \int_{\gamma_i} (\overline{\omega}_{12} - \omega_{12}). \tag{4}$$

Now let P_i be a parallel unit vector field along γ_i. Oriented angles are determined only up to addition of a multiple of 2π; at the initial point a we choose them so that

$$\angle(X_i, P_i) + \angle(P_i, V) = \angle(X_i, V). \tag{5}$$

By continuity, this relation persists along γ_i. Write $\varphi = \angle(X_i, P_i)$ and $\overline{\varphi} = \angle(V, P_i) = -\angle(P_i, V)$. Then integrating (5) shows that

$$\int_a^b (\varphi' - \overline{\varphi}')\, dt \tag{6}$$

is the total rotation of V relative to X_i along C_i. We saw in Section 3 that since P_i is parallel, the integrand here equals $\overline{\omega}_{12}(\gamma') - \omega_{12}(\gamma')$. The rotation term is just $2\pi\, \mathrm{ind}(V, \mathbf{p}_i)$, so (6) becomes

$$\int_{\gamma_i} (\overline{\omega}_{12} - \omega_{12}) = 2\pi\, \mathrm{ind}(V, \mathbf{p}_i).$$

Substituting this into (4) gives

$$\iint_M K \, dM = 2\pi \sum_{i=1}^{k} \text{ind}(V, \mathbf{p}_i).$$

The Gauss-Bonnet theorem then completes the proof.   ◆

Thus for any choice of vector field on M with only finitely many singularities (necessarily isolated), the sum of their indices is the same—and depends only on the *topology* of M.

To see that the index is independent of the various choices involved in its definition, suppose that in the theorem we use different ingredients $\mathscr{D}$, X, α at one singular point, say $\mathbf{p}_1$. In the Poincaré-Hopf equation, $x(M) = \sum_i \text{ind}(V, \mathbf{p}_i)$, every term except $\text{ind}(V, \mathbf{p}_1)$ is unchanged—hence $\text{ind}(V, \mathbf{p}_1)$ is unchanged.

7.9 Example The Poincaré-Hopf theorem provides an efficient way to compute Euler characteristics. For instance, we saw earlier that the due north vector field N on the sphere Σ has two singularities: a source and a sink. Each has index +1, so we find again that $\chi(\Sigma) = 2$.

Fig. 7.26 shows a top view of a double torus, that is, a sphere with two handles. A few integral curves are drawn for a vector field V on M with one visible singularity. There is a symmetrical pattern of curves on the bottom half of M, with a corresponding singularity. Both singularities are of the meeting of two streams type, which has index −1. Hence $\chi(M) = -2$, as found in Section 6 by quite different means.

The definition of index makes sense for a point $\mathbf{p}$ where the vector field V is nonsingular, that is, differentiable and nonzero. But then the index is zero, since we can use V itself as the base vector field X in the definition of index; so the angle φ is identically zero.

The converse is not true, that is, $\text{ind}(V, \mathbf{p}) = 0$ does not imply that V is nonsingular at $\mathbf{p}$. But it is almost true, for it can be shown that V can be modi-

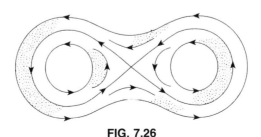

FIG. 7.26

fied near **p** so as to eliminate all singularities there and introduce no new ones elsewhere. Thus isolated singular points with index zero are said to be *removable*. From the viewpoint of Theorem 7.8, a vector field whose singularities are all removable is as good as one with no singularities at all.

7.10 Remark The vector fields in this section have been assumed to be differentiable, but only continuity is actually required. In general, continuous consequences can often be derived from differentiable theorems since (roughly speaking) continuous functions can always be approximated by differentiable functions. In Corollary 7.4, for example, we need only assume that the nonvanishing vector field V in assertion (1) is continuous, since a sufficiently close differentiable approximation will also be nonvanishing.

Exercises

1. If P is an oriented geodesic n-polygon (that is, the sides of P are geodesics), show that

$$\iint_P K \, dM = 2\pi - \sum_{j=1}^{n} \varepsilon_j = (2-n)\pi + \sum_{j=1}^{n} \iota_j,$$

where ε_j and ι_j are the exterior and interior angles of P, respectively.

2. On a surface M with $K \leq 0$, prove that there exist no geodesic n-polygons with $n = 0$, 1, or 2. (In other words, in M, a smoothly closed geodesic or a broken geodesic with either one or two corners cannot bound a simple region.)

3. On the standard sphere $\Sigma \subset \mathbf{R}^3$, since geodesics follow great circles they are simply closed, and any two must meet.
(a) If M is a compact surface with $K > 0$, prove that any two simply closed geodesics must meet.
(b) Describe an example where M is diffeomorphic to Σ and $K \geq 0$ but the conclusion of (a) fails.

4. (a) If P is a geodesic n-polygon in the Euclidean plane, show that $n \geq 3$ and the sum of the exterior angles is 2π, independent of n.
(b) On a sphere Σ, for which values of $n \geq 0$, do there exist closed broken geodesics with n corners?

5. As in the definition of index, let α parametrize a curve surrounding a common isolated singularity **p** of vector fields V and W. Prove that V and W have the same index at **p** if either:

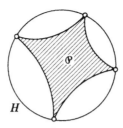

FIG. 7.27

(a) There is a continuous family of vector fields V_s, for $0 \leqq s \leqq 1$, nonvanishing on α, such that $V_0 = V$ and $V_1 = W$, or
(b) $0 < \angle(V, W) < 2\pi$. (In particular, $\pm V$ and $\pm J(V)$ all have the same index.)

6. In each case, sketch and describe a vector field on the sphere $\Sigma \subset \mathbf{R}^3$ that has (a) exactly two singular points, but no sources or sinks; (b) only one singular point (*Hint:* stereographic projection); (c) six singular points (index zero not allowed).

7. In the hyperbolic plane with $K = -1/r^2$, let P_n be a geodesic n-polygon whose $n \geqq 3$ vertices are on the rim of H—hence not actually in H (Fig. 7.27). Find the area of P_n. (b) Deduce that the area of $H(r)$ is infinite.

8. Stand the double torus M of Fig. 7.26 on end, and at each point, let V be the component of $U_z = (0, 0, 1)$ that is tangent to M.
(a) Determine the indices of the singularities of V, and check that their sum is $\chi(M)$.
(b) Do the same for a vector field whose integral curves are the level curves $z = \text{const}$ in M.

9. Use notation as in the definition of index of a vector field at an isolated singularity $\mathbf{p}$ (Def. 7.7), so $\varphi = \angle(X_\alpha, V_\alpha)$ on $[a, b]$.
(a) If $\|X_\alpha\| = 1$, show that

$$\varphi(b) - \varphi(a) = \int_a^b \frac{fg' - gf'}{f^2 + g^2} \, dt,$$

where $f = \langle V_\alpha, X_\alpha \rangle$, $g = \langle V_\alpha, J(X_\alpha) \rangle$.
(b) Deduce that the index of V at $\mathbf{p}$ is the winding number of the plane curve $(f, g): [a, b] \to \mathbf{R}^2 - 0$ (Ex. 5 of Sec. 4.6).

10. Referring to Fig. 7.25: (a) Sketch some integral curves of the meeting of *three* streams and find the index of the central singularity. What is the index for the meeting of $n > 3$ streams?

(b) Do the same for the singularity with *three* leaves. What is the index for $n > 3$ leaves?

(c) Deduce that there exist isolated singularities of every integer index.

11. For the vector field $V = -uU_1 + vU_2$ on $\mathbf{R}^2$,

(a) Solve the differential equations $u' = -u$, $v' = v$ explicitly to find the integral curve of V starting at an arbitrary point (a, b) in $\mathbf{R}^2$. (For integral curves, see Ex. 13 of Sec. 4.8.)

(b) Sketch sufficiently many of these integral curves to decide what the index of V at $(0, 0)$ is.

(c) Verify the index in (b) by using the integral in Exercise 9 above. (*Hint:* Take $X = (1, 0)$.)

12. (*Continuation by computer.*) For the vector field $V = (2u^2 - v^2)U_1 + 3uvU_2$ on $\mathbf{R}^2$,

(a) Numerically solve the differential equations for the integral curves of V, and plot sufficiently many to decide what the index of V is at $(0, 0)$. (*Hint:* Try the integral curve starting at the point $(0, 1)$. See the Appendix for numerical solution of differential equations.)

(b) Verify the index in part (a) by a numerical integration derived from Exercise 9.

13. (*Computer.*) Same as the preceding exercise but with one sign changed: $V = (2u^2 - v^2)U_1 - 3uvU_2$.

7.8 Summary

A geometric surface—that is, a 2-dimensional Riemannian manifold—consists of an abstract surface furnished with an inner product on each tangent space. The simplest case is a surface in $\mathbf{R}^3$, using the dot product of $\mathbf{R}^3$. Many geometric features of surfaces in $\mathbf{R}^3$ survive unchanged in the more general setting (e.g., length of a curve, area of a surface). Others must have their earlier definitions modified (e.g., geodesics, Gaussian curvature K). Still others, that involve $\mathbf{R}^3$ in an essential way, cannot be generalized at all (e.g., mean curvature H).

The new definitions are not revolutionary. They appear naturally when vectors normal to $M \subset \mathbf{R}^3$ are ignored. Previously, a geodesic had acceleration normal to M, so now it has acceleration zero. When the normal vector E_3 is dropped from an adapted frame E_1, E_2, E_3, the connection forms reduce to a single 1-form ω_{12}, and the elegant formula $d\omega_{12} = -K\,dM$ serves to define Gaussian curvature K.

The Gaussian curvature of a surface is its dominant geometric property, and as we have seen, curvature enters into almost every geometrical investigation. But its deepest consequence is the link between geometry and topology established by the Gauss-Bonnet theorem: The curvature of a compact surface completely determines its topological structure.

Chapter 8
Global Structure of Surfaces

In this chapter we investigate the global structure of geometric surfaces, that is, 2-dimensional Riemannian manifolds. We want to know what the possible surfaces are and what they are like. In maximum generality this goal is unrealistic, but under reasonable hypotheses, good results can be obtained.

The central theme of this chapter is *the influence of Gaussian curvature on geodesics.* A significant preliminary result is that in any surface, the geometry of a neighborhood of a point is completely described by curvature and the geodesics radiating out from that point.

This local result can be extended to show that geodesics can grip an entire surface. The first step is to show that geodesics starting at any point **p** of a complete surface eventually reach *every* point of that surface (Section 2). Gaussian curvature controls the spreading and contracting of these geodesics as they radiate out to cover the surface, but their global pattern can be quite complicated.

Nevertheless, by using topological methods (Section 4), considerable global information can be derived from this pattern. In particular, we give detailed results in two broad cases: surfaces with constant curvature and surfaces whose curvature obeys either $K \leqq 0$ or $K \geqq k > 0$.

8.1   Length-Minimizing Properties of Geodesics

The preceding chapters viewed geodesics as *straightest* curves (no turning); now we examine their character as *shortest* curves. For the Euclidean plane, the general problem of shortest routes is simple: Given any two points **p** and **q** in $\mathbf{R}^2$, there is a unique straight-line segment (geodesic) from **p** to **q**, and this is shorter than any other curve from **p** to **q**.

For an arbitrary geometric surface the situation is more interesting. In the first place, there may be no shortest curve from **p** to **q** (Exercise 7 in Section 6.4). And even if there is one, it may not be unique. For example, we will soon prove the expected result that on the sphere, every semicircle from, say, the north pole to the south pole is shortest.

1.1 Definition Let α be a curve segment from **p** to **q** in M. Then

(1) α is *a shortest curve segment* from **p** to **q** provided that if β is any other curve segment from **p** to **q**, then

$$L(\beta) \geqq L(\alpha).$$

(2) α is *the* shortest curve segment from **p** to **q** provided it is a shortest curve and any other shortest curve segment from **p** to **q** is a reparametrization of α.

In case (1) we also say that α *minimizes arc length* from **p** to **q**. In terms of intrinsic distance ρ in M (Definition 4.1 in Chapter 6), an equivalent definition is $L(\alpha) = \rho(\mathbf{p}, \mathbf{q})$.

In case (2) we say that α uniquely *minimizes arc length* from **p** to **q**. Uniqueness must be interpreted liberally here, since monotone reparametrization does not change arc length (Exercise 7 in Section 2.2).

All such shortest curves will turn out to be geodesics, and our goal in this section is to show that "short" geodesic segments in an arbitrary geometric surface M behave as well as geodesics in $\mathbf{R}^2$. To do so we use a remarkable mapping that compares the region around any point **p** in M with the Euclidean plane. The tangent plane $T_p(M)$ at **p** provides the plane, and the mapping is defined as follows, with γ_v, as usual, denoting the geodesic of M whose initial velocity is **v**.

1.2 Definition For a point **p** of a geometric surface M, the *exponential map* $\exp_p$ is given by

$$\exp_p(\mathbf{v}) = \gamma_v(1)$$

for all **v** in $T_p(M)$ such that γ_v is defined on the interval $[0, 1]$.

The exponential map is differentiable and is defined at least on a neighborhood of **0** in $T_p(M)$—a consequence of the fact that solutions of ordinary differential equations depend differentiably on initial conditions as well as on the parameter.

The geometrical meaning of $\exp_p$ is clarified by

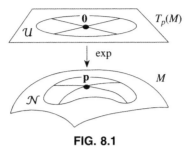

FIG. 8.1

1.3 Lemma   The exponential map $\exp_p: T_p(M) \rightarrow M$ carries radial lines from $\mathbf{0}$ in $T_p(M)$ to geodesics starting at $\mathbf{p}$ in M. Explicitly,

$$\exp_p(t\mathbf{v}) = \gamma_v(t)$$

for all t such that $\exp_p(t\mathbf{v})$ is well-defined (Fig. 8.1).

Proof.   For fixed t in $\mathbf{R}$ and $\mathbf{v}$ in T_p, the geodesic $s \rightarrow \gamma_v(ts)$ has initial velocity $t\gamma'_v(0)$. So does the geodesic γ_{tv}. Hence by the uniqueness of geodesics (Theorem 4.3 of Chapter 7), these geodesics are equal, that is, $\gamma_v(ts) = \gamma_{tv}(s)$ whenever both sides make sense. In particular, setting $s = 1$ gives

$$\gamma_v(t) = \gamma_{tv}(1) = \exp_p(t\mathbf{v}). \qquad \blacklozenge$$

Thus the exponential map at $\mathbf{p}$ collects all the geodesics starting at $\mathbf{p}$ into a single mapping. The best case is when the exponential map is defined on the whole tangent space, as discussed in the next section. But in any case, the exponential map $\exp_p$ is well behaved near $\mathbf{p}$.

1.4 Lemma   For each $\mathbf{p}$ in M the exponential map at $\mathbf{p}$ carries some neighborhood $\mathcal{U}$ of $\mathbf{0}$ in $T_p(M)$ diffeomorphically onto a neighborhood $\mathcal{N}$ of $\mathbf{p}$ in M (Fig. 8.1).

Here $\mathcal{N}$ is called a *normal neighborhood* of $\mathbf{p}$. When $\mathcal{U}$ is an open disk $\|\mathbf{v}\| < \varepsilon$, the normal neighborhood is written as $\mathcal{N}_\varepsilon$ and is said to have *radius ε.*

Proof.   (For the first time we treat the tangent plane as a surface in its own right—and work with vectors tangent to it.) In view of the inverse function theorem, it will suffice to show that the tangent map of $\exp_p$ at $\mathbf{0}$ is a linear isomorphism, for then $\exp_p$ will be diffeomorphism on some neighborhood of $\mathbf{0}$.

A tangent vector $\mathbf{v}_0$ to $T_p(M)$ at $\mathbf{0}$ is the initial velocity of the ray $\rho(t) = t\mathbf{v}$, and we saw above that

$$\exp_p(\rho(t)) = \gamma_v(t).$$

Since tangent maps preserve velocities,

$$\exp_{p*}(\mathbf{v}_0) = \exp_{p*}(\rho'(0)) = \gamma_v'(0) = \mathbf{v}.$$

Thus the tangent map of $\exp_p$ at $\mathbf{0}$ is just the natural isomorphism $\mathbf{v}_0 \to \mathbf{v}$. ◆

Note that for a given point $\mathbf{p}$ there is always a largest—possibly infinite—normal radius ε.

Now we apply the exponential map to the problem at hand. In the Euclidean plane, if one is interested in distance to the origin, it is natural to use polar coordinates, since the distance from $(0, 0)$ to $(u\cos v, u\sin v)$ is simply u. Polar coordinates can be installed in an arbitrary geometric surface M with any point $\mathbf{p}$ as the pole. The first step is to put these coordinates into the tangent plane at $\mathbf{p}$ by choosing a frame $\mathbf{e}_1, \mathbf{e}_2$ at $\mathbf{p}$ and defining

$$\tilde{\mathbf{x}}(u, v) = u \cos v \, \mathbf{e}_1 + u \sin v \, \mathbf{e}_2 \quad (0 \leqq u \leqq \varepsilon).$$

Then the exponential map at $\mathbf{p}$ moves this parametrization into M.

1.5 Definition (1) Let $\mathbf{e}_1, \mathbf{e}_2$ be a frame at a point $\mathbf{p}$ of M. The mapping

$$\mathbf{x}(u, v) = \exp_p(\tilde{\mathbf{x}}(u, v)) = \exp_p(u \cos v \, \mathbf{e}_1 + u \sin v \, \mathbf{e}_2)$$

is called a *geodesic polar mapping*.

(2) If $\mathcal{N}_\varepsilon$ is a normal ε-neighborhood of $\mathbf{p}$ in M, and the mapping $\mathbf{x}(u, v)$ is defined only for $0 \leqq u < \varepsilon$, then $\mathbf{x}$ is called a *geodesic polar parametrization of $\mathcal{N}_\varepsilon$ with pole $\mathbf{p}$* (Fig. 8.2).

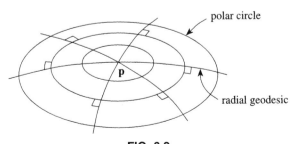

FIG. 8.2

Even though the full mapping $\mathbf{x}$ in (1) is usually neither one-to-one nor regular, we will be able to use it along particular radial geodesics. Restricted in (2) to a normal ε-neighborhood, $\mathbf{x}$ is strictly speaking only a parametrization of the punctured neighborhood $\mathcal{N}_\varepsilon - \mathbf{p}$. However, the ambiguities when $u = 0$ are familiar from Euclidean polar coordinates—and we need $u = 0$.

The u-parameter curve $v = v_0$ of $\mathbf{x}$ is the radial geodesic with initial velocity $\mathbf{v} = \cos v_0 \mathbf{e}_1 + \sin v_0 \mathbf{e}_2$ since, using Lemma 1.3,

$$\mathbf{x}(u, v_0) = \exp_p(u \cos v_0 \; \mathbf{e}_1 + u \sin v_0 \; \mathbf{e}_2) = \gamma_v(u).$$

Since $\|\mathbf{v}\| = 1$, this geodesic has unit speed, so its length from $\mathbf{p} = \mathbf{x}(u, v)$ is just u.

The v-parameter curve $u = u_0 > 0$,

$$\mathbf{x}(u_0, v) = \exp_p(u_0 \cos v \; \mathbf{e}_1 + u_0 \sin v \; \mathbf{e}_2),$$

parametrizes a closed curve called the *polar circle of radius* u_0 and *pole* $\mathbf{p}$ (see Fig. 8.2).

Note that if $\mathbf{q} = \mathbf{x}(u_0, v_0)$ is any point of $\mathcal{N}_\varepsilon$ except $\mathbf{p}$, then (but for reparametrization) there is only one unit-speed geodesic from $\mathbf{p}$ to $\mathbf{q}$ *that lies entirely in* $\mathcal{N}_\varepsilon$, namely, the radial geodesic

$$\gamma(u) = \mathbf{x}(u, v_0) \quad (0 \leq u \leq u_0).$$

1.6 Lemma   For a geodesic polar parametrization,

$$E = 1, \quad F = 0, \quad \text{and} \quad (\text{except at the pole}) \; G > 0.$$

Proof.   Since the u-parameter curves are unit-speed geodesics,

$$E = \langle \mathbf{x}_u, \mathbf{x}_u \rangle = 1, \quad \mathbf{x}_{uu} = 0.$$

Thus

$$F_u = \langle \mathbf{x}_u, \mathbf{x}_v \rangle_u = \langle \mathbf{x}_u, \mathbf{x}_{vu} \rangle = \langle \mathbf{x}_u, \mathbf{x}_{uv} \rangle = \frac{1}{2} E_v = 0.$$

Thus F is constant on each u-parameter curve.

The v-parameter curve $v \rightarrow \mathbf{x}(0, v)$ is the constant curve at $\mathbf{p}$, so $\mathbf{x}_v(0, v) = 0$ for all v. This means that $F(0, v) = 0$ for all v. Since $F_u = 0$, we conclude that F is identically zero.

The results so far—that $E = 1$, $F = 0$—are valid if $\mathbf{x}$ is merely a geodesic polar mapping. But now we assume that $\mathbf{x}$ is a parametrization of an ε-neighborhood. Thus, restricted to $0 < u < \varepsilon$, $\mathbf{x}$ is a regular mapping, so

$$G = EG - F^2 > 0. \qquad \blacklozenge$$

Here $E = 1$ means that the exponential map preserves radial distances, and $F = 0$ means that polar circles are always orthogonal to radial geodesics (Fig. 8.2). Only G can be non-Euclidean, that is, $\neq 1$. So a normal neighborhood in any surface M can be manufactured from a (flat) neighborhood of $\mathbf{0}$ in $\mathbf{R}^2 \approx T_p(M)$ merely by stretching the polar circles $u = $ const.

1.7 Example We explicitly work out geodesical polar parametrizations in two classic cases.

(1) *The unit sphere Σ in $\mathbf{R}^3$.* For simplicity, we take the pole $\mathbf{p}$ to be the north pole $(0, 0, 1)$. (By a scale change and a Euclidean rotation of Σ, essentially the same results will hold for any point in a sphere of any radius.) To get geodesics radiating out from $\mathbf{p}$, we change the geographical parametrization to

$$\mathbf{x}(u, v) = (\sin u \cos v, \sin u \sin v, \cos u).$$

Each u-parameter curve is a unit-speed parametrization of a circle of longitude through $\mathbf{p}$, and hence is geodesic. The initial velocity of these radial geodesics is

$$\mathbf{x}_u(0, v) = (\cos v, \sin v, 0) = \cos v\, U_1 + \sin v\, U_2.$$

So for the frame $\mathbf{e}_1 = U_1$, $\mathbf{e}_2 = U_2$ (Fig. 8.3), the uniqueness of geodesics implies that

$$\mathbf{x}(u, v) = \gamma_{\cos v\, \mathbf{e}_1 + \sin v\, \mathbf{e}_2}(u) = \exp_p(u \cos v\, \mathbf{e}_1 + u \sin v\, \mathbf{e}_2).$$

Hence $\mathbf{x}$ is indeed a polar geodesic parametrization. It is clear that the largest normal neighborhood of $\mathbf{p}$ has radius π and fills all of Σ except the south pole $(0, 0, -1)$. The radial geodesics are longitudinal semicircles, and the polar circles are the circles of latitude.

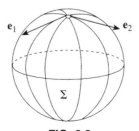

FIG. 8.3

(2) *The hyperbolic plane H* (Example 2.5 of Chapter 7). We take the pole to be the origin $\mathbf{0}$ and let $\mathbf{e}_1$, $\mathbf{e}_2$ be the natural frame U_1, U_2. (Since $h(0, 0) = 1$, this Euclidean frame is also a hyperbolic frame.) According to Example 5.6 of Chapter 7, the geodesics of H through the origin follow Euclidean straight lines. Thus for any number v, the curve

$$\alpha(t) = (t \cos v, t \sin v)$$

is at least a pregeodesic, and we found the arc length function of α to be

$$s(t) = 2 \tanh^{-1}\left(\frac{t}{2}\right).$$

Thus, shifting notation from s to u, all the radial unit-speed geodesics can be collected in the mapping

$$\mathbf{x}(u, v) = 2 \tanh\frac{u}{2}(\cos v, \sin v).$$

Since

$$\mathbf{x}_u(0, v) = \cos v \, \mathbf{e}_1 + \sin v \, \mathbf{e}_2,$$

it follows, as in (1), that $\mathbf{x}$ is a geodesic polar parametrization. The largest normal neighborhood in this case is the entire surface H. The radial geodesics are Euclidean lines, and polar circles are the Euclidean circles with center $\mathbf{0}$ (both with non-Euclidean metric properties). ◆

1.8 Theorem   For each point $\mathbf{q}$ of a normal neighborhood $\mathcal{N}_\varepsilon$ of $\mathbf{p}$, the radial geodesic segment of $\mathcal{N}_\varepsilon$ from $\mathbf{p}$ to $\mathbf{q}$ uniquely minimizes arc length.

Proof.   Let $\mathbf{x}$ be a polar parametrization of $\mathcal{N}_\varepsilon$. If $\mathbf{q} = \mathbf{x}(u_0, v_0)$, then—admitting the value 0 for u—the radial segment out to $\mathbf{q}$ is

$$\gamma(u) = \mathbf{x}(u, v_0) \quad (0 \le u \le u_0).$$

Now let α be an arbitrary curve segment from $\mathbf{p}$ to $\mathbf{q}$ in M. We parametrize α on the same interval as γ without changing its arc length. For γ to be a shortest curve from $\mathbf{p}$ to $\mathbf{q}$, we must prove

$$L(\gamma) \le L(\alpha).$$

Consider first the case where α stays in the neighborhood $\mathcal{N}_\varepsilon$ (Fig. 8.4).

We can assume that once having left $\mathbf{p}$, α never returns—for if it did, throwing away the resulting loop would shorten α. Thus it is possible to write

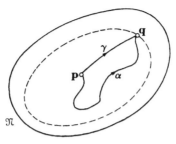

FIG. 8.4

$$\alpha(u) = \mathbf{x}(a_1(u),\, a_2(u)) \quad (0 \le u \le u_0).$$

Since $\alpha(0) = \mathbf{p}$ and $\alpha(u_0) = \mathbf{q}$, we have

$$a_1(0) = 0, \quad a_1(u_0) = u_0,$$
$$a_2(0) = v_0, \quad a_2(u_0) = v_0. \tag{1}$$

Because $\mathbf{x}$ has $E = 1$ and $F = 0$, the speed of α is $\|\alpha'\| = \sqrt{a_1'^2 + Ga_2'^2}$. Now,

$$\sqrt{a_1'^2 + Ga_2'^2} \ge \sqrt{a_1'^2} = |a_1'| \ge a_1'. \tag{2}$$

Hence

$$L(\alpha) = \int_0^{u_0} \sqrt{(a_1')^2 + G(a_1')^2}\, du \ge \int_0^{u_0} a_1'\, du$$
$$= a_1(u_0) - a_1(0) = u_0, \tag{3}$$

where the last step uses (1). Since the radial geodesic has unit speed,

$$L(\gamma) = \int_0^{u_0} du = u_0,$$

and so we conclude that $L(\gamma) \le L(\alpha)$.

If α does not stay in $\mathcal{N}_{\varepsilon}$, then strict inequality, $L(\gamma) < L(\alpha)$, will hold. In fact, to leave $\mathcal{N}_{\varepsilon}$, α must cross the polar circle $u = u_0$; so it already has length $u_0 = L(\gamma)$—and it has further to go.

Now we prove the uniqueness assertion:

> If $L(\alpha) = L(\gamma)$, then α is a reparametrization of γ.

The argument above shows that if $L(\alpha) = L(\gamma)$, then α stays inside $\mathcal{N}_{\varepsilon}$ and the inequality in (3) becomes an equality. The latter implies

$$\sqrt{a_1'^2 + G a_2'^2} = a_1'.$$

Since $G > 0$, we conclude from (2) that

$$a_1' \geqq 0 \quad \text{and} \quad a_2' = 0.$$

Thus a_2 has constant value v_0, so

$$\alpha(u) = \mathbf{x}(a_1(u),\ v_0) = \gamma(a_1(u)).$$

This expresses α as a monotone reparametrization of γ.   ◆

This fundamental result shows, as promised earlier, that if points $\mathbf{p}$, $\mathbf{q}$ in a surface M are close enough together, then—as in Euclidean space for arbitrary points—there is a unique geodesic segment from $\mathbf{p}$ to $\mathbf{q}$ that is shorter than any other curve in M from $\mathbf{p}$ to $\mathbf{q}$.

1.9 Example   *Length-minimizing geodesics on the sphere.* Let $\Sigma \subset \mathbf{R}^3$ be a sphere of radius r. It follows from Example 1.7(1) that for each point $\mathbf{p}$ of Σ the largest normal neighborhood has radius πr and covers the entire sphere except for the point $-\mathbf{p}$ antipodal to the pole $\mathbf{p}$. Hence the preceding theorem implies:

(a) If distinct points $\mathbf{p}$ and $\mathbf{q}$ are not antipodal, $\mathbf{q} \neq \pm\mathbf{p}$, then the radial geodesic γ from $\mathbf{p}$ to $\mathbf{q}$ is the unique shortest curve joining $\mathbf{p}$ and $\mathbf{q}$. But we know all the geodesics of Σ, so γ can only be the one that follows the shorter arc of the great circle through $\mathbf{p}$ and $\mathbf{q}$.

(b) Intrinsic distance on Σ is given by the formula

$$\rho(\mathbf{p}, \mathbf{q}) = r\vartheta \quad (0 \leqq \vartheta \leqq \pi),$$

where ϑ is the angle from $\mathbf{p}$ to $\mathbf{q}$ in $\mathbf{R}^3$ (Fig. 8.5). If $\mathbf{q} \neq \pm\mathbf{p}$, this follows from (a), since

$$\rho(\mathbf{p}, \mathbf{q}) = L(\gamma) = r\vartheta.$$

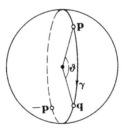

FIG. 8.5

Otherwise, $\rho(\mathbf{p}, \mathbf{p}) = 0$, and as $\mathbf{q}$ moves toward the antipodal point $-\mathbf{p}$ of $\mathbf{p}$, continuity implies $\rho(\mathbf{p}, -\mathbf{p}) = \pi r$. Consequently,

(c) There are infinitely many minimizing geodesics from any point $\mathbf{p}$ to its antipodal point $-\mathbf{p}$, namely, constant speed parametrizations of the semicircles running from $\mathbf{p}$ to $-\mathbf{p}$. (*Proof.* These all have length $\pi r = \rho(\mathbf{p}, -\mathbf{p})$.)

(d) No geodesic γ of length $L(\gamma) > \pi r$ can minimize arc length between its end points. This follows immediately from the fact that intrinsic distance on Σ never exceeds πr. It is clear geometrically, since if γ starts at $\mathbf{p}$, its length exceeds πr as soon as it passes the antipodal point $-\mathbf{p}$. But then the other arc of the same great circle is shorter than γ. ◆

Although the geodesic structure of the hyperbolic plane is simpler than that of the sphere, Theorem 1.8 is still informative. Since the entire hyperbolic plane is a normal neighborhood of the point $\mathbf{0}$, it tells us that the intrinsic distance from $\mathbf{0}$ to any point $\mathbf{p}$ is the length of the radial geodesics from $\mathbf{0}$ to $\mathbf{p}$. The preceding example gives this explicitly as

$$\rho(\mathbf{0}, \mathbf{p}) = 2 \tanh^{-1} \frac{\|\mathbf{p}\|}{2},$$

where $\|\mathbf{p}\|$ is the Euclidean distance from $\mathbf{0}$ to $\mathbf{p}$. It follows that every geodesic γ of H has infinite length—hence H *is complete*. In fact, using the triangle inequality,

$$L(\gamma\|[t_0, t]) \geqq \rho(\gamma(t_0), \gamma(t)) \geqq \rho(\mathbf{0}, \gamma(t)) - \rho(\mathbf{0}, \gamma(t_0)).$$

Then as $t \to \infty$, $\rho(\mathbf{0}, \gamma(t)) \to \infty$; hence $L(\gamma\|[t_0, t]) \to \infty$.

The differential equations result used to establish the domain and differentiability of the exponential map will, in fact, yield this local uniformity property of normal radii: For each point $\mathbf{p}$ of a surface M, there is an $\varepsilon > 0$ such that every point $\mathbf{q}$ with $\rho(\mathbf{p}, \mathbf{q}) < \varepsilon$ has a normal neighborhood of radius ε. (For a proof, see [doC], for example.)

Using this property, we can confirm the maxim that a shortest road has no turning.

1.10 Corollary If α is a shortest piecewise regular curve in M from $\mathbf{p}$ to $\mathbf{q}$, then α is an (unbroken) geodesic.

Proof. First we show that the regular segments of α are geodesics; then that α has no corners. We can suppose α has unit speed.

Assume that a regular segment $\alpha_i\|[b_i, b_{i+1}]$ of α is *not* a geodesic. Then $\alpha_i''(t_0)$ is nonzero for some t_0 in $[b_i, b_{i+1}]$, and by continuity we can suppose $t_0 < b_{i+1}$. Then a subsegment $\alpha_i\|[t_0, t_0 + \varepsilon]$ is contained in a normal ε-

FIG. 8.6

neighborhood of $\alpha_i(t_0)$. Note that $\alpha_i|[t_0, t_0 + \varepsilon]$ cannot be reparametrized to be geodesic. Thus by the uniqueness feature of Theorem 1.8, the radial geodesic segment σ from $\alpha_i(t_0)$ to $\alpha_i(t_0 + \varepsilon)$ is strictly shorter (Fig. 8.6). Replacing $\alpha_i|[t_0, t_0 + \varepsilon]$ by σ changes α to a strictly shorter curve from **p** to **q**, contradicting the assumption that α is shortest. Thus α is a possibly broken geodesic.

Now we assume that α actually has a corner, say at $\alpha_{i-1}(b_i) = \alpha_i(b_i)$, and again deduce a contradiction. By the remark preceding this lemma, there is an $\varepsilon > 0$ such that $\alpha_{i-1}(b_i)$ is contained in in a normal neighborhood $\mathcal{N}$ of $\alpha_{i-1}(b_i - \varepsilon)$. By continuity, some initial subsegment $\alpha_i|_{[b_i, t_1]}$ of α_i is still in $\mathcal{N}$.

Thus the combined curve from $\alpha_{i-1}(b_i - \varepsilon)$ to $\alpha_i(t_1)$ has a corner and lies in $\mathcal{N}$. So it is strictly longer than the radial geodesic τ of $\mathcal{N}$ joining these points (Fig. 8.6). Then as before, replacing the combined curve by τ shortens α, giving the required contradiction. $\blacklozenge$

Exercises

1.   Prove:
(a) A normal ε-neighborhood of **p** in M consists of all points **q** in M such that $\rho(\mathbf{p}, \mathbf{q}) < \varepsilon$.
(b) If $\mathbf{p} \neq \mathbf{q}$ in M, then $\rho(\mathbf{p}, \mathbf{q}) > 0$. (This completes the proof that intrinsic distance ρ is a metric on M; see Ex. 3 of Sec. 6.4.)

2.   (*Normal coordinates.*) Let $\mathcal{N} = \exp_p(\mathcal{U})$ be a normal neighborhood of a point **p** in M, and let $\mathbf{e}_1, \mathbf{e}_2$ be a frame at **p**. Prove:
(a) The mapping

$$\mathbf{n}(x, y) = \exp_p(x\mathbf{e}_1 + y\mathbf{e}_2)$$

is a coordinate patch on $\mathcal{N}$.

(b) At **p** (but generally not elsewhere) $E = 1$, $F = 0$, $G = 1$. Thus normal coordinates are Euclidean at **p**, hence at least approximately Euclidean near **p**.
(c) Coordinate straight lines through **p** are geodesics of M.

(d) With suitable choices, $\mathbf{n}$ for $\mathbf{R}^2$ is the identity map $\mathbf{n}(x, y) = (x, y)$. So for arbitrary M, normal coordinates generalize the natural (rectangular) coordinates of $\mathbf{R}^2$.

3. At the point $\mathbf{p} = (r, 0, 0)$ of the cylinder M: $x^2 + y^2 = r^2$, let $\mathbf{e}_1 = (0, 1, 0)$ and $\mathbf{e}_2 = (0, 0, 1)$. Find an explicit formula for the normal parametrization in Exercise 2. What is the largest normal neighborhood of the point $\mathbf{p}$?

4. (*Continuation.*) Prove:
(a) A geodesic starting at an arbitrary point $\mathbf{p} = (a, b, c)$ in the cylinder M does not minimize arc length after it passes through the antipodal line $t \to (-a, -b, t)$. (Only vertical geodesics through $\mathbf{p}$ fail to meet this line.)
(b) If $\mathbf{q}$ is not on the antipodal line of $\mathbf{p}$, there is a unique shortest geodesic from $\mathbf{p}$ to $\mathbf{q}$.
(c) Derive a formula for intrinsic distance on the cylinder.

5. Let M be an augmented surface of revolution (Ex. 12 of Sec. 4.1). Prove, without computation:
(a) If M has only one intercept $\mathbf{p}$ on the axis of revolution, then every geodesic segment γ starting at $\mathbf{p}$ uniquely minimizes arc length.
(b) If M has a second intercept $\mathbf{q}$, then the conclusion in (a) holds if and only if γ does not reach $\mathbf{q}$.

6. In M let α be a curve segment in M from $\mathbf{p}$ to $\mathbf{q}$, and β a curve segment from $\mathbf{q}$ to $\mathbf{r}$. Joining α and β does not usually produce a differentiable curve from $\mathbf{p}$ to $\mathbf{r}$ since $\alpha + \beta$ will usually have a corner at $\mathbf{q}$.

In this case, prove that there is a piecewise smooth curve γ from $\mathbf{p}$ to $\mathbf{r}$ that is arbitrarily close to $\alpha + \beta$ but strictly shorter: $L(\gamma) < L(\alpha) + L(\beta)$. (*Hint:* See proof of Cor. 1.10.)

Techniques from advanced calculus show that the corner can actually be smoothed away, leaving γ differentiable throughout.

7. (*Intrinsic distance is continuous.*)
(a) For $\mathbf{p}_0$ in M, show that the real-valued function $\mathbf{p} \to \rho(\mathbf{p}_0, \mathbf{p})$ is continuous; in fact, if $\rho(\mathbf{p}, \mathbf{q}) < \varepsilon$, then $|\rho(\mathbf{p}_0, \mathbf{p}) - \rho(\mathbf{p}_0, \mathbf{q})| < \varepsilon$.
(b) State precisely and prove that the function $(\mathbf{p}, \mathbf{q}) \to \rho(\mathbf{p}, \mathbf{q})$ is continuous.

8. The radial geodesics from the point $(0, 1)$ in the Poincaré half-plane P are given by $(x - a)^2 + y^2 = a^2 + 1$ for all a (see Ex. 1 of Sec. 7.5).
(a) Show that the curves given by $x^2 + (y - b)^2 = b^2 - 1$ for all $b > 1$ are everywhere orthogonal to these geodesics.
(b) Deduce that the curves in (a) are the polar circles about the pole $(0, 1)$.
(c) (*Computer graphics.*) Plot a few curves from each family on the same figure.

8.2 Complete Surfaces

A geometric surface M is complete provided all its geodesics can be defined on the entire real line (Definition 4.4 of Chapter 7). Thus the plane $\mathbf{R}^2$ is complete, but any smaller open set in $\mathbf{R}^2$ is not. Completeness is a very natural condition. Physically, we can picture a point in a frictionless surface M as a penny $\mathbf{p}$, constrained by a normal force to remain in M. Once $\mathbf{p}$ is given an initial velocity, its motion is completely determined, and since there is no acceleration, it traces out a geodesic of M. It is reasonable to think that this motion can be stopped only by a flaw in the surface.

In this section we consider some consequences of completeness. The most remarkable is that the geodesics of a complete surface, assumed only to run forever, actually *go everywhere*—and in the shortest possible arc length.

2.1 Theorem (Hopf-Rinow)   Any two points $\mathbf{p}$ and $\mathbf{q}$ in a complete connected geometric surface M can be joined by a shortest geodesic segment.

Proof. The scheme is an ingenious one, which begins by picking a promising candidate for the shortest curve (see Theorem 10.9 of [Mi]). Let

$$\beta(v) = \mathbf{x}(a, v) \quad (0 \le v \le 2\pi)$$

parametrize a polar circle C_a of radius a in a normal neighborhood of $\mathbf{p}$. By Exercise 7 of the preceding section, the function $v \to \rho(\beta(v), \mathbf{q})$ is continuous on the closed interval $[0, 2\pi]$, so it takes on a minimum value at some v_0. Thus $\beta(v_0)$ *is the point of C_a nearest to* $\mathbf{q}$. This makes it reasonable to hope that the radial geodesic through $\beta(v_0)$, namely

$$\gamma(u) = \mathbf{x}(u, v_0),$$

will hit target point $\mathbf{q}$.

Since M is complete, γ is defined for all $u \ge 0$. We will, in fact, show

$$\gamma(r) = \mathbf{q}, \quad \text{where } r = \rho(\mathbf{p}, \mathbf{q}). \tag{1}$$

(See Fig. 8.7.) Furthermore, since γ has unit speed, (1) implies

$$L(\gamma) = r = \rho(\mathbf{p}, \mathbf{q}),$$

so γ minimizes arc length as required.

To prove (1) we will show that

$$\rho(\gamma(u), \mathbf{q}) = r - u \quad \text{for all } a \le u \le r. \tag{2}$$

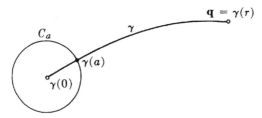

FIG. 8.7

This is a kind of efficiency condition on γ: Each inch that γ advances brings it one inch closer to **q**. Note that (2) does imply (1), since for $u = r$ it gives $\rho(\gamma(r), \mathbf{q}) = 0$, hence $\gamma(r) = \mathbf{q}$.

We separate the proof of (2) into three parts.

Start. Equation (2) holds for $u = a$ (the radius of C_a); that is,

$$\rho(\gamma(a), \mathbf{q}) = r - a. \tag{3}$$

To show this, note that by Theorem 1.8, $\rho(\mathbf{p}, \gamma(a)) = a$. Hence by the triangle inequality,

$$r = \rho(\mathbf{p}, \mathbf{q}) \leq a + \rho(\gamma(a), \mathbf{q}).$$

To get (3) we must reverse this inequality. By the definition of intrinsic distance, for any $\varepsilon > 0$ there is a curve segment α from **p** to **q** such that

$$L(\alpha) \leq \rho(\mathbf{p}, \mathbf{q}) + \varepsilon.$$

Now α must hit the polar circle C_a, say at $\alpha(t_0)$, and we observe that the part of α from **p** to $\alpha(t_0)$ has length $L_1 \geq a$, while the remainder has length

$$L_2 \geq \rho(\alpha(t_0), \mathbf{q}) \geq \rho(\gamma(a), \mathbf{q}).$$

(The latter inequality holds since $\gamma(a)$ is a nearest point to **q** on C.) Thus

$$a + \rho(\gamma(a), \mathbf{q}) \leq L_1 + L_2 \leq \rho(\mathbf{p}, \mathbf{q}) + \varepsilon.$$

Since ε was arbitrary, we get the inequality,

$$a + \rho(\gamma(a), \mathbf{q}) \leq \rho(\mathbf{p}, \mathbf{q}),$$

required to prove (3).

Plan. The set of numbers u in $[a, r]$ for which (2) holds has a least upper bound $b \leq r$. Since the functions involved in (2) are continuous, it follows that (2) holds for $u = b$, that is, $\rho(\gamma(b), \mathbf{q}) = r - b$. Evidently it suffices to show that $b = r$.

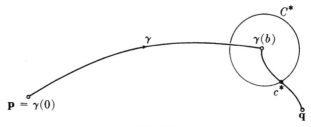

FIG. 8.8

To do this, our plan is to *assume $b < r$ and deduce a contradiction* by proving

$$\rho(\gamma(b + a^*), \mathbf{q}) = r - (b + a^*) \quad \text{for some number } a^* > 0.$$

Indeed, this contradicts the least upper bound definition of b.

Finish. Let C^* be a polar circle of radius $0 < a^* < r - b$ in a normal neighborhood of $\gamma(b)$. By reproducing the argument for the circle C_a, we get a point c^* such that

$$\rho(\mathbf{c}^*, \mathbf{q}) = \rho(\gamma(b), \mathbf{q}) - a^*.$$

(See Fig 8.8.) Since $\rho(\gamma(b), \mathbf{q}) = r - b$, this becomes

$$\rho(\mathbf{c}^*, \mathbf{q}) = r - (b + a^*). \tag{4}$$

It remains only to show that

$$\mathbf{c}^* = \gamma(b + a^*). \tag{5}$$

By the triangle inequality,

$$\rho(\mathbf{p}, \mathbf{c}^*) + \rho(\mathbf{c}^*, \mathbf{q}) \geq \rho(\mathbf{p}, \mathbf{q}) = r.$$

Then (4) implies

$$\rho(\mathbf{p}, \mathbf{c}^*) \geq b + a^*.$$

But there is a broken geodesic from $\mathbf{p}$ to $\mathbf{c}^*$ whose length is $b + a^*$. In fact, as Fig. 8.8 shows, we can travel from $\mathbf{p}$ to $\gamma(b)$ with arc length b, and then from $\gamma(b)$ to c^* on a radial geodesic of length a^*. Thus by Corollary 1.10, this curve is not broken. Hence it is γ all the way, which means that $\gamma(b + a^*)$ is exactly c^*.

Finally, we substitute (5) into (4) to get the required contradiction

$$\rho(\gamma(b + a^*), \mathbf{q}) = r - (b + a^*). \qquad \blacklozenge$$

Here are two fundamental consequences of completeness.

2.2 Corollary   At every point **p** of a complete connected surface M the exponential map $\exp_p$ is defined on the entire tangent space $T_p(M)$ and maps it onto M.

Proof.   By completeness, if **v** is tangent to M at **p**, the geodesic γ_v is defined on the whole real line. But $\gamma_v(1)$ is the definition of $\exp_p(\mathbf{v})$, so $\exp_p$ is defined on all of $T_p(M)$.

By the Hopf-Rinow theorem, for any other point **q** in M there is a geodesic segment γ from $\gamma(0) = \mathbf{p}$ to $\gamma(r) = \mathbf{q}$. Thus $\exp_p(r\gamma'(0)) = \gamma(r) = \mathbf{q}$.   ◆

A connected geometric surface M is said to be *extendible* if it is isometric to an open subset of a strictly larger connected surface $\tilde{M}$. Thus, for example, $\mathbf{R}^2 - 0$ is extendible to $\mathbf{R}^2$, but the following result shows that $\mathbf{R}^2$ itself is inextendible.

2.3 Theorem   A complete connected surface M is inextendible.

Proof.   We assume M is extendible and deduce a contradiction. For simplicity, let M actually be a subset of $\tilde{M}$. Then it is an open set of $\tilde{M}$ (Exercise 15 of Section 4.3). It will suffice to find a geodesic γ of $\tilde{M}$ that meets M. For then the portion of γ in M is a geodesic of M that cannot be extended—in M—over the whole real line. But this contradicts the completeness of M.

Since $\tilde{M}$ is connected, the open set criterion (Exercise 9 of Section 4.7) implies that $\tilde{M} - M$ is not open. Hence there is a point **p** in $\tilde{M} - M$ such that every neighborhood $\mathcal{N}$ of **p** meets M. When $\mathcal{N}$ is a normal neighborhood, some radial geodesic from **p** will meet M and is thus the required geodesic γ.   ◆

Completeness, a geometric condition, is implied by the topological property compactness.

2.4 Corollary   A compact geometric surface is complete.

Proof.   If every unit-speed geodesic in M can be extended by some fixed amount $\eta > 0$, then M is complete, because these small extensions can be repeated indefinitely.

By the remark preceding Corollary 1.10, for each point **p** in M there is an $\varepsilon > 0$ such that every point **q** within distance ε of **p** has a normal neighborhood of radius ε. Since M is compact, a finite number of such neighborhoods, say, $\mathcal{N}_1, \ldots, \mathcal{N}_k$, cover all of M. Let 2η be the smallest of the corresponding radii $\varepsilon_1, \ldots, \varepsilon_k$.

Now we can extend any geodesic by η. Consider $\gamma: [0, b) \to M$. From $\gamma(b - \eta)$ there will be radial geodesics of length 2η in all directions. The one that has initial velocity $\gamma'(b - \eta)$ will then run for arc length 2η and hence extend γ by η.   ◆

In fact, a more elaborate proof shows that *every closed surface M in* $\mathbf{R}^3$ *is complete*. Although it applies only to surfaces in $\mathbf{R}^3$, this result is stronger there than the corollary above. For example, it tells us that any quadric surface in $\mathbf{R}^3$ is complete.

The properties above show that completeness is a crucial prerequisite for the global study of surfaces.

Exercises

1.   Show that the converse of the Hopf-Rinow theorem is false: Give an example of a geometric surface M such that any two points can be joined by a minimizing geodesic segment but M is not complete.

2.   Test the scheme used to prove the Hopf-Rinow theorem (2.1) as follows. Given points $\mathbf{p} = (p_1, p_2)$ and $\mathbf{q} = (q_1, q_2)$ in $M = \mathbf{R}^2$, use only that scheme to find a formula for a geodesic γ starting at $\mathbf{p}$ and aimed at $\mathbf{q}$.

3.   Prove that every complete generalized cylinder C is isometric to the plane $\mathbf{R}^2$ or to a circular cylinder with radius uniquely determined by C.

4.   Let $M \subset \mathbf{R}^3$ be flat. Exercise 2 of Section 6.3 correctly suggests that every point of M has a neighborhood that is ruled. Give an example to show that M itself need not be ruled. (*Hint:* Begin by cutting a small square from each corner of a plane square.)

However, it is known that every *complete* flat surface in $\mathbf{R}^3$ is a generalized cylinder (so Ex. 3 applies). See W. S. Massey, "Surfaces of Gaussian Curvature Zero in Euclidean Space," *Tohoku Math. J.* 14, 1962.

5.   Many surfaces of revolution with constant curvature $K \neq 0$ were found in Section 7 of Chapter 5 and its exercises. Show that except for the sphere, none are complete.

6.   Give an example of a surface in which every pair of points can be joined by a geodesic segment, but not always by a minimizing one.

7.   Let C be the $z > 0$ part of the cone $z^2 = x^2 + y^2$ in $\mathbf{R}^3$. Show that C is not complete, but that any two points of C can be joined by a minimizing geodesic. (*Hint:* Cut C and bend it isometrically into $\mathbf{R}^2$.)

8.3 Curvature and Conjugate Points

The preceding section dealt with curve segments γ from **p** to **q** whose arc length L is a minimum compared to all curve segments joining these points. Now we consider γ on which L has only a *local* minimum. The definition has the same pattern as for a local minimum of a real valued function on **R**, but with points of **R** replaced by curve segments from **p** to **q**.

3.1 Definition A geodesic segment γ from **p** to **q** in M *locally minimizes* arc length from **p** to **q** provided that $L(\alpha) \geqq L(\gamma)$ holds for every curve segment α from **p** to **q** that is sufficiently close to γ.

To clarify the term "sufficiently close," we define α to be ε-*close* to γ provided there is a reparametrization $\tilde{\alpha}$ of α defined on the same interval I as γ and such that

$$\rho(\tilde{\alpha}(t), \gamma(t)) < \varepsilon \quad \text{for all } t \text{ in } I.$$

Then the ending of the preceding definition becomes:

provided there exists an $\varepsilon > 0$ such that $L(\alpha) \geqq L(\gamma)$ holds for any curve segment from **p** to **q** that is ε-close to γ.

The local minimization is *strict* (or *unique*) provided strict inequality $L(\alpha) > L(\gamma)$ holds except when α is a reparametrization of γ.

To get an intuitive picture of this definition, we can imagine that γ is an elastic string—or rubber band—that (1) is constrained to lie in M, (2) is under tension, and (3) has its end points pinned down at **p** and **q**.

Because γ is a geodesic, it is in equilibrium. If it were not a geodesic, its tension would pull it to a new shorter position. But is the equilibrium stable? That is, if γ is pulled aside slightly to a new curve α and released, will it return to its original position? (See Fig. 8.9.) Evidently γ is strictly stable if and only if γ is a strict local minimum in the sense above, for if α is always longer than γ, its tension will pull it back to γ.

The key to local minimization is the notion of *conjugate point*. If γ is a unit-speed geodesic starting at **p**, then γ is a u-parameter curve $v = v_0$ of a geodesic polar mapping **x** with pole **p**. We know that along γ the function $G = \langle \mathbf{x}_v, \mathbf{x}_v \rangle$ is zero at $u = 0$, but is nonzero immediately thereafter (Lemma 1.6). A point $\gamma(s) = \mathbf{x}(s, v_0)$ with $s > 0$ is called a *conjugate point of* $\gamma(0) = \mathbf{p}$ *on* γ provided $G(s, v_0) = 0$. (Such points may or may not exist.)

The geometric meaning of conjugacy rests on viewing $\sqrt{G} = \|\mathbf{x}_v\|$ as the distance between nearby radial geodesics and hence of its derivative $(\sqrt{G})_u$ as

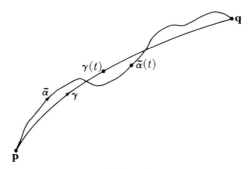

FIG. 8.9

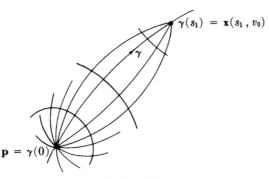

FIG. 8.10

the rate at which the radial geodesics are spreading apart. In fact, the distance from $\mathbf{x}(u, v)$ to $\mathbf{x}(u, v + \varepsilon)$ is approximately $\varepsilon \|\mathbf{x}_v(u, v)\|$, and if $(\sqrt{G})_u > 0$ this distance is increasing, so the radial geodesics are spreading apart, while if $(\sqrt{G})_u < 0$ the distance is decreasing, so the radial geodesics are pulling closer together.

Since G vanishes at a conjugate point $\gamma(s_1) = \mathbf{x}(s_1, v_0)$, we can expect that, for v near v_0, the radial geodesics γ_v will all reach this point at distance $s = s_1$ (Fig. 8.10). Unfortunately, this meeting *may not actually occur*, since G controls only first derivative terms of $\mathbf{x}$, and higher order terms may still be nonzero when G vanishes.

The Euclidean plane $\mathbf{R}^2$ sets the standard rate at which radial geodesics spread apart, and for $\mathbf{x}(u, v) = (u \cos v, u \sin v)$ we have

$$\sqrt{G} = u, \quad \text{hence} \quad (\sqrt{G})_u = 1.$$

In particular, there are no conjugate points. Let us compare this with the cases in Example 1.7, the unit sphere Σ and the hyperbolic plane H.

For Σ, since $\mathbf{x}_v = (-\sin u \sin v, \sin u \cos v, 0)$, we find

$$\sqrt{G} = \sin u, \quad \text{hence} \quad (\sqrt{G})_u = \cos u.$$

Thus radial geodesics from the north pole (meridians of longitude) begin in Euclidean fashion but as $\cos u$ drops below 1, they spread apart ever less rapidly than in $\mathbf{R}^2$. After they pass the equator (at $u = \pi/2$) $\cos u$ turns negative and the geodesics begin crowding closer together. All have their first conjugate point after traveling distance π, since $\sqrt{G}(\pi, v) = \sin \pi = 0$. In this case, of course, the meeting actually takes place—at the south pole of Σ (Fig. 8.3).

For the hyperbolic plane, we know that radial geodesics from the origin follow Euclidean straight lines. From the formula for $\mathbf{x}(u, v)$ in Example 1.7(2), we can compute†

$$\sqrt{G} = \sinh u, \quad \text{hence} \quad (\sqrt{G})_u = \cosh u.$$

Thus the radial geodesics, after beginning in Euclidean fashion, spread apart ever more rapidly than in $\mathbf{R}^2$, as might be guessed from the slogan "rulers shrink as they approach the rim." In particular, there are no conjugate points.

3.2 Theorem (Jacobi) If γ is a geodesic segment from $\mathbf{p}$ to $\mathbf{q}$ such that there are no conjugate points of $\mathbf{p} = \gamma(0)$ on γ, then γ locally minimizes arc length (strictly) from $\mathbf{p}$ to $\mathbf{q}$.

Proof.   For a geodesic polar mapping $\mathbf{x}$ with pole $\mathbf{p}$, we can write

$$\gamma(u) = \mathbf{x}(u, v_0) \quad \text{for} \quad 0 \leqq u \leqq b = L(\gamma).$$

Since there are no conjugate points of $\mathbf{p}$ on γ, we have

$$G(u, v_0) > 0 \quad \text{for} \quad 0 < u < b.$$

As noted earlier, it follows from Lemma 1.6 that $EG - F^2$ reduces to G for a geodesic polar mapping, so $\mathbf{x}$ is regular for all (u, v_0) with $0 < u \leqq b$. Thus the image of γ for $0 < u < b$ is covered by neighborhoods $\mathscr{N}$ that are diffeomorphic images, under $\mathbf{x}$, of open sets in the uv plane.

Then γ can be divided into segments by numbers

$$0 < u_1 < \ldots < u_k < u_{k+1} = b,$$

so that the first segment $\gamma|[0, u_1]$ lies in a normal neighborhood $\mathscr{N}_0$ of $\mathbf{p}$ and each later segment $\gamma|[u_i, u_{i+1}]$, $1 \leqq i \leqq k$, is in one of the neighborhoods $\mathscr{N}$, say $\mathscr{N}_i$.

† We will soon find a method much quicker than direct computation.

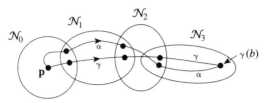

FIG. 8.11

Now we can choose an $\varepsilon > 0$ so small that if a curve α is ε-close to γ, then each subsegment $\alpha|[u_i, u_{i+1}]$ lies in the same open set $\mathcal{N}_i$ as $\gamma|[u_i, u_{i+1}]$ and ε is less than the radius of the normal neighborhood $\mathcal{N}_0$ (Fig. 8.11).

The first segment of α can evidently be written in the polar form

$$\alpha(t) = \mathbf{x}(a_1(t), a_2(t)),$$

with $a_1(0) = 0$, $a_2(0) = v_0$. The local inverses of $\mathbf{x}$ on each $\mathcal{N}_i$ carry the later segments, in succession, back to the uv plane. Thus α can be written as

$$\alpha(t) = \mathbf{x}(a_1(t), a_2(t)),$$

with

$$a_1(0) = 0, \; a_2(0) = v_0; \quad a_1(b) = b, \; a_2(b) = v_0.$$

We must show that $L(\alpha) \geqq L(\gamma)$, with equality only if α is a monotone reparametrization of γ. As in the proof of Theorem 1.8, we find

$$L(\alpha) = \int_0^b \sqrt{a_1'^2 + Ga_2'^2} \geqq \int_0^b a_1' \, dt$$

$$= a_1(b) - a_1(0) = b = L(\gamma).$$

And if $L(\alpha) = L(\gamma)$, then as before, α is a monotone reparametrization of γ. ◆

The study of conjugate points can be radically simplified by freeing it from dependence on geodesic polar mappings. To achieve this, we examine the "spreading coefficient" $(\sqrt{G})_u$ more closely.

3.3 Theorem Let $\mathbf{x}$ be a geodesic polar mapping on whose domain in the uv plane, $G > 0$ if $u > 0$. Then $\sqrt{G} = \|\mathbf{x}_v\|$ satisfies the Jacobi differential equation

$$(\sqrt{G})_{uu} + K\sqrt{G} = 0,$$

subject to the initial conditions

$$\sqrt{G}(0, v) = 0, \quad \lim_{u \to 0}(\sqrt{G})_u(u, v) = 1 \quad \text{for all } v.$$

The restriction $G > 0$ is needed to ensure that $\sqrt{G}$ is differentiable, at least for $u > 0$.

Proof. The Jacobi equation follows immediately from the curvature formula in Proposition 6.3 of Chapter 6. As noted in the proof of Lemma 1.6, $\mathbf{x}(0, v) = \mathbf{p}$ for all v, so $G(0, v) = 0$ for all v. Because $\sqrt{G} = \|\mathbf{x}_v\|$ is not differentiable at $u = 0$, limits are required, and we must show

$$\lim_{u \to 0}(\sqrt{G})_u(u, v) = 1.$$

It is only necessary to consider a single radial geodesic $\gamma(u) = \mathbf{x}(u, v_0)$, setting

$$g(u) = \sqrt{G}(u, v_0) \quad (u > 0).$$

On γ, since $E = 1$, $F = 0$, we get a frame field

$$E_1 = \gamma' = \mathbf{x}_u, \quad E_2 = \mathbf{x}_v/g.$$

Because γ is a geodesic, E_1 is parallel, and by Exercise 2 of Section 7.3, so is E_2. By parallelism, E_2 is well-defined at $u = 0$. Now,

$$E_1(0) = \mathbf{x}_u(0, v_0) = \cos v_0\, \mathbf{e}_1 + \sin v_0\, \mathbf{e}_2.$$

Hence

$$E_2(0) = -\sin v_0\, \mathbf{e}_1 + \cos v_0\, \mathbf{e}_2.$$

Furthermore, since $\mathbf{x}_v = gE_2$ on γ, and E_2 is parallel, we find

$$\mathbf{x}_{uv} = \mathbf{x}_{vu} = g'E_2$$

on γ, for $u > 0$. Taking limits as $u \to 0$ yields

$$\mathbf{x}_{uv}(0, v_0) = \lim_{u \to 0} g'(u)E_2(0). \qquad (*)$$

But $\mathbf{x}_u(0, v) = \cos v\, \mathbf{e}_1 + \sin v\, \mathbf{e}_2$ for all v. Hence

$$\mathbf{x}_{uv}(0, v_0) = -\sin v_0\, \mathbf{e}_1 + \cos v_0\, \mathbf{e}_2 = E_2(0).$$

Comparing this equation with $(*)$ shows that $\lim_{u \to 0} g'(u) = 1$. Since $g = \sqrt{G}$, this is the required limit. ◆

Recall that for the Euclidean plane we found $\sqrt{G} = u$ for $u > 0$, and hence $(\sqrt{G})_u = 1$. Thus the initial conditions in the preceding theorem show that as radial geodesics first leave the pole **p** in any geometric surface, they are spreading at the same rate as in $\mathbf{R}^2$. Thereafter, the Jacobi equation

$$(\sqrt{G})_{uu} = -K\sqrt{G}$$

shows that *the rate of spreading depends on Gaussian curvature.* For $K < 0$, radial geodesics spread *faster* than in $\mathbf{R}^2$, as we saw earlier for the hyperbolic plane. For $K > 0$, they spread slower than in $\mathbf{R}^2$, as on the sphere.

To locate conjugate points it is no longer necessary to construct a geodesic polar mapping, as we have done so far. We can find $\sqrt{G}$ on a geodesic γ simply by solving the Jacobi equation along γ, subject to the Jacobi initial conditions. Explicitly, Theorem 3.3 gives

3.4 Corollary Let γ be a unit-speed geodesic starting at **p** in M. Let g be the unique solution of the Jacobi equation on γ,

$$g'' + K(\gamma)g = 0,$$

that satisfies the initial conditions $g(0) = 0$, $g'(0) = 1$.

Then the conjugate points of $\gamma(0) = \mathbf{p}$ on γ are the points $\gamma(s)$, $s > 0$, at which $g(s) = 0$.

As in the plane $\mathbf{R}^2$, conjugate points may not exist, but if they do, the first one is particularly important because of Theorem 3.2.

3.5 Example *Conjugate Points.*

(1) Let γ be a unit-speed geodesic starting at any point **p** of the sphere Σ of radius r. Since $K = 1/r^2$, the Jacobi equation for γ is $g'' + g/r^2 = 0$, which has the general solution

$$g(s) = A \sin\left(\frac{s}{r}\right) + B \cos\left(\frac{s}{r}\right).$$

The initial conditions $g(0) = 0$, $g'(0) = 1$ then give

$$g(s) = r \sin\left(\frac{s}{r}\right).$$

The first zero $s_1 > 0$ of g occurs at distance $s_1 = \pi r$. Thus the first conjugate point of **p** occurs at the antipodal point $-\mathbf{p}$. This agrees with our earlier computation using geodesic polar parametrization.

(2) On a torus of revolution with radii $R, r > 0$, let γ be a unit-speed parametrization of the outer equator. Now γ is a geodesic, and along it K has constant value $1/(r(R + r))$. Thus by the preceding corollary the first conjugate point of, say, $\gamma(0)$ will occur *at exactly the same distance as if γ were on a sphere with this curvature*, so $s_1 = \pi\sqrt{r(R + r)}$.

3.6 Corollary There are no conjugate points on any geodesic in a surface with curvature $K \leqq 0$. Hence every geodesic segment in such a surface is locally minimizing.

Proof. Apply Corollary 3.4 to a geodesic in M. The initial conditions $g(0) = 0$ and $g'(0) = 1$ show that g starts out as a strictly increasing positive function. Since $K \leqq 0$, the Jacobi equation gives

$$g'' = -Kg \geqq 0.$$

Thus $g'(s) \geq 1$ for all s. Then evidently $g(u) > 0$ for all $u > 0$. ◆

For example, on a (flat) circular cylinder, the helical geodesic γ from **p** to **q** indicated in Fig. 8.12 is indeed stable, as one can verify by experiment. Although locally minimizing, it is certainly not minimizing, since the straight-line segment σ gives a much shorter way to travel from **p** to **q**.

Although geodesics locally minimize arc length before the first conjugate point, they do not locally minimize past it.

3.7 Theorem Let γ be a geodesic segment from **p** to **q**. If there is a conjugate point of **p** along γ before **q**, then γ does not locally minimize arc length between **p** and **q**.

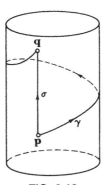

FIG. 8.12

A formal proof requires the calculus of variations (see [Mi] or [dC]), but we can give a persuasive argument. As before, let γ be the u-parameter curve $v = v_0$ of a geodesic polar mapping $\mathbf{x}$. If $\mathbf{q} = \gamma(u_0) = \mathbf{x}(u, v_0)$, then by hypothesis, the function $u \to G(u, v_0)$ is zero at some number s with $0 < s < u_0$.

We have seen that this means that nearby u-parameter geodesic segments *of the same length as* γ tend to meet again at $\gamma(s) = \mathbf{x}(s, v_0)$ (see Fig. 8.10). Suppose that (as on the sphere) this meeting actually occurs for some v_1 arbitrarily near v_0. Then we construct the broken geodesic $\beta : [0, u_0] \to M$ for which

$$\beta(u) = \begin{cases} \mathbf{x}(u, v_1) & \text{if} \quad 0 \le u \le s, \\ \gamma(u) & \text{if} \quad s \le u \le u_0. \end{cases}$$

Thus β has the same length as $\gamma|[0, u_0]$. But it has a corner at $\gamma(s)$, and as the proof of Corollary 1.10 shows, cutting across this corner produces a strictly shorter curve that is arbitrarily near to β and hence to γ. Thus γ does not locally minimize arc length from $\mathbf{p}$ to $\mathbf{q}$.

Note that the proof of Corollary 1.10 remains valid for local minimization, so *a locally minimizing curve from* $\mathbf{p}$ *to* $\mathbf{q}$ *is an (unbroken) geodesic.*

In the critical case where the end point $\gamma(b)$ of a geodesic segment $\gamma : [a, b] \to M$ is exactly the first conjugate point of its initial point $\gamma(a)$, nothing can be said in general: γ may or may not locally minimize arc length.

The Jacobi differential equation can be used to give an intuitive description of Gaussian curvature. First we need

3.8 Lemma If $\mathbf{x}$ is a geodesic polar mapping with pole $\mathbf{p}$, then

$$\sqrt{G}(u, v) = u - K(\mathbf{p})\frac{u^3}{6} + o(u^3) \quad (u \ge 0).$$

Here $o(u^3)$ denotes a function of $u > 0$ and v such that $\lim_{u \to 0} o(u^3)/u^3 = 0$. So in the formula, when u is small enough, $o(u^3)$ is negligible compared to the first two terms.

Proof. Again consider $g(u) = \sqrt{G}(u, v)$ on a radial geodesic $u \to \mathbf{x}(u, v)$. As a solution of the Jacobi equation, g is differentiable at $u = 0$, and hence has a Taylor expansion,

$$g(u) = g(0) + g'(0)u + g''(0)\frac{u^2}{2} + g'''(0)\frac{u^3}{6} + o(u^3).$$

We can evaluate all these coefficients. The Jacobi initial conditions in Corollary 3.4 are $g(0) = 0$, $g'(0) = 1$. Using the first of these in the Jacobi equation gives $g''(0) = 0$. Now differentiate the Jacobi equation to get

$$g''' + K(\gamma)' g + K(\gamma)g' = 0.$$

Thus

$$g'''(0) = -K(\gamma(0)) = -K(\mathbf{p}).$$

Substitution in the Taylor expansion then gives the required result.   ◆

Suppose the inhabitants of a geometric surface M want to determine the Gaussian curvature of M at a point $\mathbf{p}$. By measuring a short distance ε in all directions from $\mathbf{p}$, they obtain the polar circle C_ε of radius ε. If $M = \mathbf{R}^2$, then C_ε is just an ordinary Euclidean circle, with circumference $L(C_\varepsilon) = 2\pi\varepsilon$. But for $K > 0$ the radial geodesics from $\mathbf{p}$ are not spreading as rapidly, so C_ε should be shorter than $2\pi\varepsilon$, and for $K < 0$ they are spreading more rapidly, so C_ε should be longer than $2\pi\varepsilon$.

The dependence of $L(C_\varepsilon)$ on K can be measured with precision. For ε small enough, C_ε is parametrized by $v \to \mathbf{x}(\varepsilon, v)$, where $\mathbf{x}$ is a geodesic polar patch at $\mathbf{p}$. Thus

$$L(C_\varepsilon) = \int_0^{2\pi} \sqrt{G}(\varepsilon, v)dv.$$

Hence by the preceding lemma,

$$L(C_s) = 2\pi\left(\varepsilon - K(\mathbf{p})\frac{\varepsilon^3}{6} + o(\varepsilon^3) \right). \tag{$*$}$$

Thus if surveyors in M measure $L(C_\varepsilon)$ carefully for ε small, they can estimate the Gaussian curvature of M at $\mathbf{p}$ to any desired accuracy. Taking limits yields a precise result.

3.9 Corollary $K(\mathbf{p}) = \lim\limits_{\varepsilon \to 0} \dfrac{3}{\pi\varepsilon^3}(2\pi\varepsilon - L(C_\varepsilon)).$

Let us test the formula $(*)$ on a sphere Σ of radius r in $\mathbf{R}^3$. As Fig. 8.13 shows, the polar circle C_ε with center $\mathbf{p}$ is actually a Euclidean circle of Euclidean radius $r \sin\vartheta$, where $\vartheta = \varepsilon/r$. Then by the Taylor series of the sine function,

$$L(C_s) = 2\pi\left(r \sin\frac{\varepsilon}{r} \right) = 2\pi\left(\varepsilon - \frac{\varepsilon^3}{6r^2} + o(\varepsilon^3) \right).$$

Comparison with $(*)$ gives yet another proof that the sphere has Gaussian curvature $K = 1/r^2$.

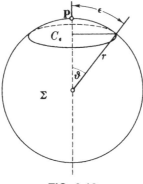

FIG. 8.13

Exercises

1. If $x(u, v)$ is a polar parametrization in a surface of constant curvature k, show that

$$\sqrt{G(u, v)} = \begin{cases} r\sin\dfrac{u}{r} & \text{if} \quad k = \dfrac{1}{r^2}, \\ u & \text{if} \quad k = 0, \\ r\sinh\dfrac{u}{r} & \text{if} \quad k = -\dfrac{1}{r^2}. \end{cases}$$

2. In a normal neighborhood of $p \in M$, call the region D_ε on and within the polar circle C_ε a *polar disk* of radius ε.

(a) Show that the area of a polar disk is

$$A(D_\varepsilon) = \pi\left(\varepsilon^2 - K(p)\frac{\varepsilon^4}{12} + o(\varepsilon^4)\right),$$

and hence

$$K(p) = \frac{12}{\pi}\lim_{\varepsilon \to 0}\frac{\pi\varepsilon^2 - A(D_\varepsilon)}{\varepsilon^4}.$$

(b) Use this formula to find the Gaussian curvature of a sphere of radius r.

3. (a) At the pole 0 in the hyperbolic plane H, find the length of the polar circle C_ε and the area of the polar disk D_ε, where $0 < \varepsilon < \infty$ is the hyperbolic radius.

(b) Deduce from each that $K(p) = -1$.

(As we shall later see, these results hold for any point of H.)

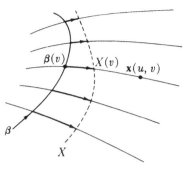

FIG. 8.14

4. Let M be an augmented surface of revolution (Ex. 12 of Sec. 4.1).
(a) If M crosses the axis A at only one point $\mathbf{p}$ (as on a paraboloid of revolution), show that $\mathbf{p}$ has no conjugates on any geodesic.
(b) If M crosses the axis A at only two points $\mathbf{p}$ and $\mathbf{q}$ (as on an ellipsoid of revolution), show that $\mathbf{p}$ and $\mathbf{q}$ are conjugate along every meridian. (*Hint:* Use a canonical parametrization to get the Jacobi equation.)

The following exercises deal with a useful variant of the geodesic polar parametrization in which the pole $\mathbf{p}$ is replaced by an arbitrary regular curve.

5. Let $\beta: I \to M$ be a regular curve in M, and let X be a nonvanishing vector field on β such that $\beta'(v)$ and $X(v)$ are linearly independent for all v. Define

$$\mathbf{x}(u, v) = \gamma_{X(v)}(u) = \exp_{\beta(v)}(uX(v)).$$

Thus the u-parameter curve $v = v_0$ is a geodesic cutting across β with initial velocity $X(v)$ (Fig. 8.14). Prove:
(a) $\mathbf{x}$ is a regular mapping on some open region D in $\mathbf{R}^2$ containing the points $(0, v)$ for all v in I.
(b) By suitable choices of β and X, this parametrization $\mathbf{x}$ becomes (i) the identity map of $\mathbf{R}^2$ (natural coordinates), (ii) the canonical parametrization of a surface of revolution, and (iii) a ruled parametrization of a ruled surface (Def. 2.6 in Ch. 4).

(*Note:* To obtain familiar formulas it may sometimes be necessary to reverse u and v.)

6. (*Continuation.*) If β is a unit-speed curve and X is the unit normal $N = J(\beta')$ in M, show that $E = 1$, $F = 0$, and $\sqrt{G}$ is the solution of the Jacobi equation $(\sqrt{G})_{uu} + K\sqrt{G} = 0$ such that

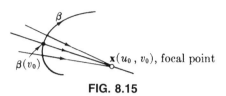

FIG. 8.15

$$\sqrt{G}(0, v) = 1 \quad \text{and} \quad (\sqrt{G})_u(0, v) = -\kappa_g(v).$$

By analogy with conjugate points, if $G(u_0, v_0) = 0$, we say that $\mathbf{x}(u_0, v_0)$ is a *focal point* of β along the normal geodesic $v = v_0$. Here light rays emerging orthogonally from β tend to meet (Fig. 8.15).

7. (a) If β is a circle of latitude on a sphere Σ, show that the north and south poles of Σ are the only focal points of β.
(b) If β is a curve in the Euclidean plane, show that its focal points are exactly its centers of curvature, that is, the points on its evolute. (See Ex. 13 of Sec. 2.4.)

8. (a) Let $\mathbf{x}$ and $\overline{\mathbf{x}}$ be geodesic polar parametrizations of normal ε-neighborhoods $\mathcal{N}_\varepsilon$ and $\overline{\mathcal{N}}_\varepsilon$ (same ε) in two (not necessarily different) geometric surfaces. If $K(\mathbf{x}) = K(\overline{\mathbf{x}})$ on the common domain of $\mathbf{x}$ and $\overline{\mathbf{x}}$, prove that $\mathcal{N}_\varepsilon$ and $\overline{\mathcal{N}}_\varepsilon$ are isometric.
(b) Deduce Minding's theorem that *constant curvature uniquely determines local geometry*: If M and M' have the same constant curvature, then any points $\mathbf{p}$ in M and $\mathbf{p}'$ in M' have isometric neighborhoods.

8.4   Covering Surfaces

There is a close relationship between certain pairs of surfaces that lets information about one—obtained perhaps with difficulty—be transmitted easily to the other. We are interested mostly in surfaces, but we state the basic definition for manifolds of arbitrary dimension so that Curves (1-dimensional manifolds) are included.

4.1 Definition   Let F be a differentiable mapping of a manifold M onto a manifold N of the same dimension. An open set $\mathcal{V}$ in N is *evenly covered* provided the set of points in M that map into $\mathcal{V}$ splits into disjoint open sets, each mapped diffeomorphically onto $\mathcal{V}$ by F. Then F is a *covering map* provided every point of N has an evenly covered neighborhood.

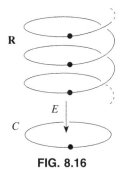

FIG. 8.16

Here "diffeomorphically" means that the restriction of F to any one of these disjoint open sets, say $\mathcal{U}_1$, is a diffeomorphism of $\mathcal{U}_1$ onto $\mathcal{V}$. Thus $F|\mathcal{U}_1$ has a differentiable inverse mapping $\lambda_1 \colon \mathcal{V} \to \mathcal{U}_1$. Note that any open subset of an evenly covered neighborhod is itself evenly covered.

If $F \colon \tilde{M} \to M$ is a covering map, then $\tilde{M}$ is called a *covering manifold* of M, and we say that $\tilde{M}$ *covers* M.

4.2 Example The classic 1-dimensional example of a covering map is the *exponential map*

$$E(\vartheta) = e^{i\vartheta} = (\cos \vartheta, \sin \vartheta),$$

which wraps the line $\mathbf{R}$ around the unit circle $C \colon x^2 + y^2 = 1$ in $\mathbf{R}^2$.† Figure 8.16 presents $\mathbf{R}$ as an infinite coil, with E dropping each point of $\mathbf{R}$ down to C; the evenly covered neighborhoods are readily pictured (see also Exercise 4.2).

The exponential map E resolves the ambiguity between the "geometric angle" formed by two vectors (or two radial lines) and the "numerical angle" that measures it. Here a point (x, y) of C represents the geometric angle between the positive x axis and the line from the origin through the point (x, y), and ϑ is an oriented angle (using the natural orientation of the plane). For any point (x, y) in C, the infinitely many numbers sent to (x, y) by E are the possible measurements of the angle—any two differing by an integer multiple of 2π.

It is clear from the definition that a covering map $\tilde{M} \to M$ is a *local diffeomorphism* onto M, but we will soon see that the converse is not true.

† This mapping, with $\mathbf{R}$ considered as a tangent line to $\mathbf{C}$, was a model for the general notion of the exponential map in Section 1.

4.3 Example   Several covering maps have appeared earlier.

(1) The usual parametrization $\mathbf{x}: \mathbf{R}^2 \to T$ of the torus of revolution.
(2) The map $F(x, y) = (\cos x, \sin x, y)$ that wraps the Euclidean plane $\mathbf{R}^2$ around a cylinder $C \subset \mathbf{R}^3$.
(3) The projection F of the sphere Σ onto the projective plane P (Example 8.2 in Chapter 4).
(4) The natural map of the orientation covering $\hat{M}$ of a surface M onto M (Exercise 6 of Section 4.8).   ◆

A covering map $\tilde{M} \to M$ has the crucial property that curves in M can always be *lifted* to $\tilde{M}$ in the following sense.

4.4 Proposition   Let $F: \tilde{M} \to M$ be a covering map. If $\alpha: I \to M$ is a curve and $\mathbf{p}$ is any point of $\tilde{M}$ such that $F(\mathbf{p}) = \alpha(t_0)$ in M, then there is a unique curve $\tilde{\alpha}: I \to \tilde{M}$ such that

$$\tilde{\alpha}(t_0) = \mathbf{p}, \quad F(\tilde{\alpha}) = \alpha.$$

Proof.   For simplicity, take $I = [0, b]$ and $t_0 = 0$.

Existence of Lifts. Since every point $\alpha(t)$ is in an evenly covered neighborhood, a standard result from advanced calculus asserts that there are numbers

$$0 = t_0 < t_1 < \ldots < t_n = b$$

such that for $i = 1, \ldots, n$, each subinterval $[t_{i-1}, t_i]$ is contained in an evenly covered neighborhood, say $\mathcal{V}_i$. We build $\tilde{\alpha}$ on $[0, b]$ by lifting α on each subinterval in succession.

There is no choice as to the lift of $\alpha|[0, t_1]$. Let $\mathcal{U}_1$ be the covering neighborhood of $\mathcal{V}_1$, that contains $\mathbf{p}$. Since the covering neighborhoods are disjoint, a lift of $\alpha|[0, t_1]$ that starts at $\mathbf{p}$ must lie entirely in $\mathcal{U}_1$. But F carries $\mathcal{U}_1$ diffeomorphically onto $\mathcal{V}_1$, so this lift can only be

$$\tilde{\alpha}_1 = \lambda_1(\alpha_1): [0, t_1] \to \tilde{M},$$

where $\lambda_1: \mathcal{V}_1 \to \mathcal{U}_1$, is the inverse diffeomorphism of $\mathbf{F}|V_1$.

Now we repeat this process, replacing $\mathbf{p}$ by $\tilde{\alpha}(t_1)$ to get a curve $\tilde{\alpha}_{12}$ such that $F(\tilde{\alpha}_{12}) = \alpha|[t_1, t_2]$. This curve starts at the end point of α_1, so the two combine to give a curve $\tilde{\alpha}_2$ such that

$$\tilde{\alpha}_2(0) = \mathbf{p} \quad \text{and} \quad F(\tilde{\alpha}_2) = \alpha|[0, t_2].$$

After a finite number of such repetitions the entire curve is lifted.

Uniqueness of Lifts. Here we only assume that $F: \tilde{M} \to M$ is a local diffeomorphism. Then if $\alpha_1, \alpha_2: [0, b] \to \tilde{M}$ are lifts of the curve α in M such that $\alpha_1(0) = \alpha_2(0)$, we must show that $\alpha_1 = \alpha_2$. The proof is based on the fact that $[0, b]$ is not the union of two disjoint open sets.

Let $A \subset [0, b]$ consist of all t such that $\alpha_1(t) = \alpha_2(t)$. Since curves are continuous, the set $[0, b] - A$ is open. But A is also open, since if t is in A then there is a neighborhood $\mathcal{U}$ of $\alpha_1(t) = \alpha_2(t)$ that is mapped diffeomorphically onto a neighborhood $\mathcal{V}$ of $\alpha(t)$ in M. For t' near t, $\alpha(t')$ is in $\mathcal{V}$, so, arguing as above, both $\alpha_1(t')$ and $\alpha_2(t')$ can only be $\lambda(\alpha(t'))$; hence

$$\alpha_1(t') = \alpha_2(t').$$

Since A actually contains the number 0, we conclude that A is the entire interval $[0, b]$.

Note that the result holds if instead of $\alpha_1(0) = \alpha_2(0)$, we assume that α_1 and α_2 agree at any t_0 in $[0, b]$.   ◆

In short, a curve in M can be uniquely lifted *to any level* in a covering surface of M.

Notation: For any map F, let $F^{-1}(y)$ be the set of all x such that $F(x) = y$.

4.5 Corollary   Let $F: \tilde{M} \to M$ be a covering map with M connected, and let k be a positive integer. If for some one $\mathbf{q}_0$ in M there are exactly k points in $F^{-1}(\mathbf{q}_0)$, then the same is true for every point $\mathbf{q}$ in M. In this case the covering is said to have *multiplicity k*.

Proof.   For any $\mathbf{q}$ in M, let $\alpha: [0, 1] \to M$ be a curve segment from $\mathbf{q}_0$ to $\mathbf{q}$. Then let $\alpha_1, \ldots, \alpha_k$ be the lifts of α starting at the k points of $F^{-1}(\mathbf{q}_0)$. The end points $\tilde{\alpha}_i(1)$ all lie in $F^{-1}(\mathbf{q})$, and they are all different, for by the uniqueness of lifts, if any two end points agreed, the entire curves would be identical, but by construction their initial points differ.

Consequently, $F^{-1}(\mathbf{q})$ contains at least as many points as $F^{-1}(\mathbf{q}_0)$. Clearly the argument holds with $\mathbf{q}_0$ and $\mathbf{q}$ reversed, so the two sets have the same number of points.   ◆

Note that this corollary implies that if $F^{-1}(\mathbf{q})$ is infinite for one point $\mathbf{q}$ in M, it is infinite for every point. Evidently, the exponential map E above has infinite multiplicity, as do the first two maps in Example 4.3.

A covering map of multiplicity 1 is just a diffeomorphism. A covering map of multiplicity 2 is called a *double covering*. In Example 4.3 the last two maps are double coverings. When faced with a choice of two objects at each point of a surface M, it often turns out that the set of all these objects forms a double covering surface of M.

Corollary 4.5 can also be used to prove that a particular map is *not* a covering map. For example, let E_1 be the restriction of the map E above to an interval $J: 0 < t < 3\pi$. Now J is still mapped nicely around the unit circle $C - S^1$; in fact, $E_1: J \to S^1$ is a local diffeomorphism. However E_1 is not a covering map since for some points, $F^{-1}(\mathbf{q})$ contains two points, for others only one. A sketch will show that the even covering condition fails only around the two edge points $E_1(0) = (1, 0)$ and $E_1(3\pi) = (-1, 0)$.

Proposition 4.4 has an analogue that asserts that 2-segments can also be lifted to any level.

4.6 Theorem
Let $F: \tilde{M} \to M$ be a covering map, and let $\mathbf{x}: R \to M$ be a 2-segment, where $D: a \leqq u \leqq b, c \leqq v \leqq d$. If $\mathbf{p}$ is any point of $\tilde{M}$ such that $F(\mathbf{p}) = \mathbf{x}(a, c)$, then there is a unique 2-segment $\tilde{\mathbf{x}}$ in $\tilde{M}$ such that

$$F(\tilde{\mathbf{x}}) = \mathbf{x} \quad \text{and} \quad \tilde{\mathbf{x}}(a, c) = \mathbf{p}.$$

The proof is a straightforward 2-dimensional version of the proof of Proposition 4.4. The rectangle R is chopped into subrectangles, each lying in an evenly covered neighborhood. Then these are uniquely lifted—across one row and back the next—until $\tilde{\mathbf{x}}$ is completed. For details, see the Covering Homotopy Theorem in [ST].

4.7 Remark
Coverings with finite multiplicity. Suppose $F: \tilde{M} \to M$ is a covering with multiplicity k.

(1) If M is compact, then $\tilde{M}$ is compact. *Proof.* M is covered by a finite number of 2-segments. Each of these can be lifted to k 2-segments in $\tilde{M}$. Clearly these finitely many lifts cover all of $\tilde{M}$, so it is compact.

(2) The Euler characteristic of $\tilde{M}$ is k times of that M. *Proof.* For a rectangular decomposition of $\mathcal{D}$ of M, lift (as in (1)) each face in $\mathcal{D}$ to k faces in $\tilde{M}$. Using the uniqueness of curve lifts, we can check that if such faces meet, they do so in a lifted edge or vertex. Each edge has k lifts, and the lifts of a vertex $\mathbf{p}$ are just the k points in $F^{-1}(\mathbf{p})$. Thus the lifted faces constitute a rectangular decomposition of $\tilde{M}$, and $\chi(\tilde{M}) = kv - ke + kf = k(v - e + f) = k\chi(M)$.

A covering surface can be expected to have simpler topology than the surface it covers. Evidently the plane $\mathbf{R}^2$ is simpler than either a torus or a cylinder. Now we show that simply connected covering surfaces are those that cannot, in this sense, be further simplified.

4.8 Theorem If $F: \tilde{M} \to M$ is a covering map with $\tilde{M}$ connected and M simply connected, then F is a diffeomorphism.

Proof. Since F is, in particular, a regular mapping onto M, it will suffice to show that F is one-to-one. So if $\mathbf{p}_1$ and $\mathbf{p}_2$ are points of $\tilde{M}$ such that $F(\mathbf{p}_1) = F(\mathbf{p}_2) = \mathbf{q}$ in M, we must show that $\mathbf{p}_1 = \mathbf{p}_2$.

Since $\tilde{M}$ is connected, there is a curve segment $\tilde{\alpha}: [0, 1] \to \tilde{M}$ running from $\mathbf{p}_1$ to $\mathbf{p}_2$. Now F carries $\tilde{\alpha}$ to a curve α in M, and α is closed since

$$\alpha(0) = F(\tilde{\alpha}(0)) = F(\mathbf{p}_1) = \mathbf{q} = F(\mathbf{p}_2) = F(\tilde{\alpha}(1)) = \alpha(1).$$

Because M is simply connected, α is homotopic to a constant. By Definition 7.6 of Chapter 4, this means that there is a 2-segment $\mathbf{x}$ in M—defined for simplicity on the unit square $0 \leqq u, v \leqq 1$—whose base curve is α, with its other boundary curves constant at $\mathbf{p}$.

By the preceding lemma there is a lift of $\mathbf{x}$ to the $\mathbf{p}_1$ level, that is, a 2-segment $\tilde{\mathbf{x}}$ in $\tilde{M}$ such that

$$F(\tilde{\mathbf{x}}) = \mathbf{x} \quad \text{and} \quad \tilde{\mathbf{x}}(0, 0) = \mathbf{p}_1.$$

To show that $\mathbf{p}_1 = \mathbf{p}_2$, we chase around the rim of the 2-segment $\tilde{\mathbf{x}}$, as in Fig. 8.17.

The edge curve $\tilde{\delta}$ of $\tilde{\mathbf{x}}$ starts at $\mathbf{p}_1$ and is carried by F to the constant curve at $\mathbf{p}$. Hence by the uniqueness of lifts (see proof of Prop. 4.4), $\tilde{\delta}$ can only be the constant curve at $\mathbf{p}_1$.

Similarly the edge curve $\tilde{\gamma}$ starts at $\tilde{\gamma}(0) = \tilde{\delta}(1) = \mathbf{p}_1$ and is carried to a constant curve; hence $\tilde{\gamma}$ is constant at $\mathbf{p}_1$.

Finally, $\tilde{\beta}$ *ends* at $\tilde{\gamma}(1) = \mathbf{p}_1$ and is carried to a constant curve; hence $\tilde{\beta}$ is also constant at $\mathbf{p}_1$. Thus

$$\mathbf{p}_2 = \tilde{\gamma}(1) = \tilde{\beta}(0) = \mathbf{p}_1. \quad \blacklozenge$$

Here are three applications of covering methods that involve orientation covering surfaces (Exercise 6 of Section 4.8).

(1) Proof of Theorem 7.11 of Chapter 4. This asserts that *a simply connected surface M is orientable.* By the preceding theorem, the orientation

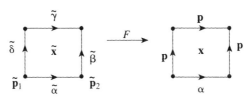

FIG. 8.17

covering surface $\hat{M}$ of M is not connected. For $M \subset \mathbf{R}^3$, it follows from Exercises 6 and 7 of Section 4.8 that M is orientable.

If M is not in $\mathbf{R}^3$ the argument is the same, but an abstract definition of $\hat{M}$ is required. For example, we could take $\hat{M}$ to be the set of all rotation operators J on M—two at each point.

(2) The Poincaré-Hopf theorem for compact nonorientable surfaces N. The proof in Section 7 of Chapter 7 required orientability; to extend to the nonorientable case, suppose that V is a vector field on N with only isolated singularities $\mathbf{p}_1, \ldots, \mathbf{p}_k$.

Let $F: \hat{N} \to N$ be the orientation covering of N. Since F is a local diffeomorphism, there is a unique vector field $\hat{V}$ on N such that $F_*(\hat{V}) = V$.

For each $\mathbf{p}_i$, the two points $\mathbf{q}_i$, $\mathbf{q}'_i$ in $F^{-1}(\mathbf{p}_i)$ are also singular points of $\hat{V}$. Near each of them, $\hat{V}$ is just an isometric copy of V near $\mathbf{p}_i$, so both $\mathbf{q}_i$ and $\mathbf{q}'_i$ have the same index as $\mathbf{p}_i$. In view of the property of Euler characteristic in Remark 4.7,

$$2\chi(N) = \chi(\hat{N}) = \sum_{i=1}^{k} \mathrm{ind}(\hat{V}, \mathbf{q}_i) + \sum_{i=1}^{k} \mathrm{ind}(\hat{V}, \mathbf{q}'_i) = 2\sum_{i=1}^{k} \mathrm{ind}(V, \mathbf{p}_i).$$

Thus $\chi(N) = \Sigma_{i=1}^{k}\mathrm{ind}(V, \mathbf{p}_i)$.

(3) Classification of compact nonorientable surfaces. These surfaces can all be constructed by a scheme similar to that used in the orientable case (Section 7.6).

Instead of handles, *crosscaps* are used. A crosscap is a projective plane P with a hole punched in it. To add a crosscap to a surface M, punch a corresponding hole in it (in both cases, by removing the interior of a face of a rectangular decomposition). Join the boundaries of the holes smoothly to get the new surface M'. It has Euler characteristic

$$\chi(M') = \chi(M) - 1.$$

In fact, its decomposition has two fewer faces than M and P combined, and the changes in vertices and edges cancel. (Recall that $\chi(P) = \chi(\Sigma)/2 = 1$.)

Then, analogous to Theorem 6.8 of Chapter 7, we have

4.9 Theorem   If N is a compact connected nonorientable surface, there is a unique integer $k \geq 0$ such that N is diffeomorphic to a projective plane with k crosscaps.

(See Chapter 1 of [Ma].) In this case we denote N by $P[k]$. The formula above gives

$$\chi(P[k]) = \chi(P) - k = 1 - k \quad (k \geq 0).$$

Thus the nonorientable compact surfaces have Euler characteristics

$$k = 1, 0, -1, -2, -3, \ldots.$$

The preceding theorem and Theorem 6.8 of Chapter 7 constitute the following surface classification theorem.

4.10 Theorem Two compact connected surfaces are diffeomorphic if and only if they have the same Euler characteristic and are both orientable or both nonorientable. Equivalently, every compact connected surface is diffeomorphic to exactly one surface in the following double sequence:

$$\Sigma = \Sigma[0], \quad T = \Sigma[1], \quad \Sigma[2], \ldots \quad \Sigma[k], \ldots$$
$$\downarrow \qquad\quad \downarrow \qquad\quad \downarrow \qquad\quad \downarrow$$
$$P = P[0], \quad K = P[1], \quad P[2], \ldots. \quad P[k], \ldots$$
$$(\chi > 0) \qquad (\chi = 0) \qquad\qquad (\chi < 0)$$

The vertical arrows are orientation covering maps, since

4.11 Corollary The orientation covering surface of $P[k]$ is $\Sigma[k]$, a sphere with k handles.

Proof. If N is a nonorientable connected surface, then (as asserted in Exercise 7 of Section 4.8) its orientation covering surface $\hat{N}$ is connected and orientable. If N is also compact, then by Remark 4.7, $\hat{N}$ is compact and has Euler characteristic $2\chi(N)$. Thus the orientation covering of $P[k]$ is a compact orientable surface with Euler characteristic $2(1 - k) = 2 - 2k$. Hence by Theorem 6.8 of Chapter 7, $\widehat{P[k]}$ is diffeomorphic to $\Sigma[k]$. ◆

A local isometry of geometric surfaces that is also a covering map is called a *Riemannian covering map*. To make use of the power of covering methods, it is important to be able to decide when a given local isometry is a Riemannian covering map. If it is, then curves in N can be lifted to every level in M. We now consider a geometrical converse.

A local isometry $F: M \to N$ is said to have the *geodesic lift property* provided every geodesic segment γ in N can be lifted to every level in M. (Note that such lifts are necessarily geodesics since the local inverses of F are isometries.)

4.12 Theorem If a local isometry $F: M \to N$, with N connected, has the geodesic lift property, then F is a Riemannian covering map.

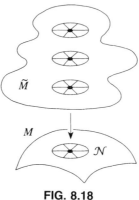

FIG. 8.18

Proof.   To show that F maps M *onto* N, fix a point $\mathbf{p}_0$ in M and let $\mathbf{q}$ be an arbitrary point of N. Since N is connected, there is a broken geodesic segment β: $[0, b] \to N$ from $F(\mathbf{p}_0)$ to $\mathbf{q}$ (Exercise 7 of Section 7.4). Using the geodesic lift property, we lift successively the unbroken subsegments of β to get a broken geodesic $\tilde{\beta}$: $[0, b] \to M$ starting at $\mathbf{p}_0$. Thus

$$F(\tilde{\beta}(b)) = \beta(b) = \mathbf{q}.$$

We must find an evenly covered neighborhood for each point $\mathbf{q}$ in N. It turns out that any normal ε-neighborhood $\mathcal{N}$ of $\mathbf{q}$ will work. To show this, let $\mathbf{p}$ be any point of $F^{-1}(\mathbf{q})$. Each radial geodesic γ starting at $\mathbf{q}$ runs for length ε, hence so does its lift $\tilde{\gamma}$ to M starting at $\mathbf{p}$. Since F is, in particular, a local isometry, every geodesic starting at $\mathbf{p}$ is such a lift. Furthermore, once they leave $\mathbf{p}$, no two such geodesic segments meet again. It follows that these segments fill a normal ε-neighborhood $\mathcal{N}(\mathbf{p})$ of $\mathbf{p}$, and furthermore that F carries $\mathcal{N}(\mathbf{p})$ in one-to-one fashion onto $\mathcal{N}$. As a local isometry, F is a regular mapping; hence $\mathcal{N}(\mathbf{p}) \to \mathcal{N}$ is a diffeomorphism.

Next we show that the neighborhoods $\mathcal{N}(\mathbf{p})$ for all $\mathbf{p}$ in $F^{-1}(\mathbf{q})$ fill the entire set $F^{-1}(\mathcal{N})$ (Fig. 8.18). They are evidently contained in it. For $\mathbf{r}$ in $F^{-1}(\mathcal{N})$, we must show that $\mathbf{r}$ is in some $\mathcal{N}(\mathbf{p})$. Let σ be the radial geodesic from $\mathbf{q}$ to $F(\mathbf{r})$ in $\mathcal{N}$. Then if $\tilde{\sigma}$ is the lift of σ ending at $\mathbf{r}$, we have

$$F(\tilde{\sigma}(0)) = \sigma(0) = \mathbf{q}.$$

Hence $\tilde{\sigma}(0)$ is in $F^{-1}(\mathbf{q})$, and $\mathbf{r}$ is in $\mathcal{N}(\tilde{\sigma}(0))$ as required.

It remains only to note that for points $\mathbf{p} \neq \mathbf{p}'$ of $F^{-1}(\mathbf{q})$, their normal ε-neighborhoods do not meet. In fact, if they did, then radial geodesic segments from $\mathbf{p}$ and $\mathbf{p}'$ meet—contrary to the uniqueness of lifts.   ◆

Exercises

1. Describe a vector field on the projective plane that has exactly one singularity. What is its index?

2. Let $E: \mathbf{R} \to C$ be the exponential map in Example 4.2.
(a) For any integer k, if T_k is the translation $\vartheta \to 2\pi k$ of $\mathbf{R}$, show that $ET_k = E$.
(b) If $\mathcal{N}$ is the open semicircle $y > 0$ in the unit circle C, let $\cos^{-1}$ be the (differentiable) inverse of the restriction of cos to the interval $I: 0 < \vartheta < \pi$ and define $\lambda: \mathcal{N} \to \mathbf{R}$ by $\lambda(x, y) = \cos^{-1}(x)$. Prove that $E(\lambda(x, y)) = (x, y)$ for all (x, y) in $\mathcal{N}$. (*Hint:* Use $y > 0$.)
(c) Same as (b) for the semicircles $y < 0$, $x > 0$, and $x < 0$ in C.
(d) Deduce that E is a covering map.

3. Let $F: M \to N$ be a local diffeomorphism. Show that F is a covering map if $F^{-1}(\mathbf{q})$ contains the same finite number of points for all $\mathbf{q}$ in M.

4. For each integer $n \geq 1$, find a covering map $F_n: C \to C$ of the circle C that has multiplicity n.

5. Prove:
(a) The only compact connected surface that can be a covering surface of a torus T is a torus.
(b) There are coverings $F: T \to T$ of every multiplicity $n \geq 1$.

6. Using sketches as needed, show:
(a) A crosscap contains a Möbius band. (*Hint:* For $F: \Sigma \to P$, remove the arctic and antarctic regions from Σ.)
(b) If a crosscap is added to any surface, the resulting surface is nonorientable.
(c) Cutting the Klein bottle in Fig. 8.20 by its plane of symmetry leaves two Möbius bands.
(d) $P[1]$ is a Klein bottle. (Explain without using the classification theorems.)

8.5 Mappings That Preserve Inner Products

We have seen that a local isometry $F: M \to N$ carries geodesics of M to geodesics of N. The notation γ_v for the geodesic with initial velocity $\mathbf{v}$ allows a more explicit description.

5.1 Lemma   If $F: M \to N$ is a local isometry, and $\mathbf{v}$ is a tangent vector to M, then

$$F(\gamma_v) = \gamma_{F_*(v)}.$$

Proof.   As just noted, $\bar{\gamma} = F(\gamma_v)$ is a geodesic of N. Its initial velocity is the tangent vector

$$\bar{\gamma}'(0) = F_*(\gamma_v'(0)) = F_*(\mathbf{v})$$

at the point $F(\mathbf{p})$ in N. Thus by the uniqueness of geodesics, $\bar{\gamma}$ and $\gamma_{F_*(v)}$ are the same.   ◆

Thinking of an isometry as a rigid motion suggests that if two isometries agree on some neighborhood, then they agree everywhere. In fact, a stronger result is true.

5.2 Theorem   Let F and G be local isometries from a connected surface M to a surface N. If at some one point $\mathbf{p}$ in M their differential maps agree, that is, if

$$F_* = G_*: T_p(M) \to T_q(M), \quad \text{where } \mathbf{q} = F(\mathbf{p}) = G(\mathbf{p}),$$

then $F = G$.

Proof.   If M is complete, the proof is easy. Let $\mathbf{m}$ be an arbitrary point of M. By the Hopf-Rinow theorem (2.1) there is a vector $\mathbf{v}$ at the special point $\mathbf{p}$ such that $\gamma_v(r) = \mathbf{m}$ for some number r. Then the preceding lemma gives

$$F(\gamma_v) = \gamma_{F_*(v)} = \gamma_{G_*(v)} = G(\gamma_v).$$

Hence, in particular,

$$F(\mathbf{m}) = F(\gamma_v(r)) = G(\gamma_v(r)) = G(\mathbf{m}).$$

In the general case, there is at least a *broken* geodesic β from $\mathbf{p}$ to $\mathbf{m}$ (Exercise 7 of Section 7.4). Then $F(\beta) = G(\beta)$ follows by applying the argument above, successively, to each unbroken segment of β. So again $F(\mathbf{m}) = G(\mathbf{m})$.   ◆

An isometry $F: M \to M$ from a surface to itself can be regarded as an intrinsic symmetry of M. Every feature of the geometry of M is the same at $\mathbf{p}$ as at $F(\mathbf{p})$, since this geometry consists of isometric invariants. The results of Exercise 9 of Section 6.4, show at once that the set $I(M)$ of all isometries

$M \to M$ forms a group, just as does the set $\mathscr{E}(n)$ of all isometries of n-dimensional Euclidean space $\mathbf{R}^n$ (Exercise 7 of Section 3.1). $I(M)$ is called the *isometry group* of M.

The group $I(M)$ is intrinsic to M and when M is a surface in $\mathbf{R}^3$ should not be confused with the group $S(M)$ of Euclidean symmetries of M (Exercise 6 of Section 6.9). Intuitively, $S(M)$ gives the symmetries of M seen by Euclidean observers, while $I(M)$ gives those observed by the inhabitants of M. Each Euclidean symmetry restricts to an isometry $F|M : M \to M$, but this does not generally give all the isometries of $M \subset \mathbf{R}^3$, as we now see.

5.3 Example   Let M be a cylinder in $\mathbf{R}^3$ whose cross-sectional curve is a noncircular ellipse of length λ. Parametrize M by

$$\mathbf{x}(u, v) = \alpha(u) + v U_3,$$

where α is a periodic unit-speed parametrization of the ellipse. Now for any number a, the map

$$F_a(\mathbf{x}(u, v)) = \mathbf{x}(u + a, v)$$

is an isometry of M. The inhabitants of M cannot distinguish M from the isometric cylinder C whose cross-sectional curve is a *circle* of circumference λ; so for them, F_a is just a rotation.

But evidently F_a will be the restriction of a Euclidean symmetry $\mathbf{F}$ only in the special case where a is a multiple of $\lambda/2$. Then $\mathbf{F}$ is a 180° rotation of M or the identity map.

For an arbitrary geometric surface M, the isometry group $I(M)$ gives a novel algebraic description of M. Roughly speaking, the more symmetrical M is, the larger $I(M)$ is. For example, the ellipsoid

$$M: \frac{x^2}{a^2} + \frac{y^2}{b^2} + \frac{z^2}{c^2} = 1 \quad (a > b > c)$$

has eight isometries derived from Euclidean isometries: three reflections (one in each coordinate plane), three 180° rotations (one around each coordinate axis), the isometry $\mathbf{p} \to -\mathbf{p}$, and, of course, the identity map of M. In fact, these are the *only* isometries of M (argue from K as in Exercise 8 of Section 6.9).

The smallest possible isometry group $I(M)$ occurs when the identity map is the only isometry of M. We can produce such a surface by putting a bump on the ellipsoid in such a way as to destroy all seven of its nontrivial isometries.

By contrast, a surface M has the maximum possible symmetry if every isometry allowed by Theorem 5.2 actually exists. Explicitly, given any linear isometry ϕ: $T_p(M) \rightarrow T_q(M)$ of tangent spaces, there exists an isometry F: $M \rightarrow M$ such that $F_* = \phi$.

Equivalently, given frames $\mathbf{e}_1$, $\mathbf{e}_2$ and $\bar{\mathbf{e}}_1$, $\bar{\mathbf{e}}_2$ at $\mathbf{p}$ and $\mathbf{q}$, respectively, there exists an isometry F: $M \rightarrow M$ such that

$$F_*(\mathbf{e}_1) = \bar{\mathbf{e}}_1, \quad F_*(\mathbf{e}_2) = \bar{\mathbf{e}}_2.$$

Thus in this case we say that M is *frame-homogeneous*: Any two frames on M are symmetrically positioned. So what was proved in Theorem 2.3 of Chapter 3 is that $\mathbf{R}^3$ is frame-homogeneous, and the same proof is valid for any $\mathbf{R}^n$, in particular, for the Euclidean plane $\mathbf{R}^2$.

5.4 Definition   A geometric surface M is *point-homogeneous* (or merely *homogeneous*) provided that for any points $\mathbf{p}$ and $\mathbf{q}$ of M there is an isometry F: $M \rightarrow M$ such that $F(\mathbf{p}) = \mathbf{q}$.

A frame-homogeneous surface is, of course, homogeneous—but not conversely. A circular cylinder C furnishes an example. Rotations of $\mathbf{R}^3$ about the axis of C and translations of $\mathbf{R}^3$ along this axis produce isometries of C. Evidently we can move any point of C to any other using a rotation and a translation, so C is homogeneous. But C is not frame-homogeneous: All its points are geometrically equivalent, but not all its frames. *Proof.* No isometry F could carry any vector $\mathbf{v}$ tangent to a ruling to a vector $\mathbf{w}$ tangent to a cross-sectional circle, since by Lemma 7.1, F would have to send the one-to-one geodesic γ_v to the closed geodesic γ_w an impossibility since F is one-to-one.

Homogeneity is a strong restriction.

5.5 Theorem   If a geometric surface M is homogeneous, then M is complete and has constant Gaussian curvature.

Proof.   Constancy of curvature follows immediately from the definition of homogeneity and the fact that isometries preserve curvature.

The proof of completeness is more interesting. If M is not complete, there is a unit-speed geodesic α whose largest interval of definition I is not all of $\mathbf{R}$. Suppose I: $t < b$. Let us show that this is impossible.

The radial geodesics in a normal ε-neighborhood of some point $\mathbf{p}$ all run for length $\varepsilon > 0$. Choose t_0 in I so that $b - t_0 < \varepsilon$. Because M is homo-

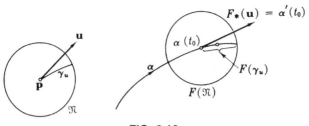

FIG. 8.19

geneous, there is an isometry $F: M \to M$ such that $F(\mathbf{p}) = \alpha(t_0)$. For some unit vector $\mathbf{u}$ at $\mathbf{p}$, $F*(\mathbf{u}) = \alpha'(t_0)$. Thus the geodesic segment $F(\gamma_{\mathbf{u}})$ has initial velocity

$$F* \left(\gamma_u'(0) \right) = F* (\mathbf{u}) = \alpha'(t_0),$$

and this segment runs for distance ε at unit speed (Fig. 8.19). By the uniqueness of geodesics, a reparametrization of γ_u can be attached onto α to produce an unbroken geodesic defined on the interval I': $t < t_0 + \varepsilon$. Since $t + \varepsilon > b$, this contradicts the maximality of the interval I and thus proves that M is complete. ◆

As the title of this section suggests, isometries and local isometries are not the only inner-product preserving mappings of importance in geometry. We consider briefly some other types.

5.6 Definition   Let $F: M \to \mathbf{R}^n$ be a mapping of a geometric surface into Euclidean n-space ($n = 3, 4, \ldots$). If the differential map F_* preserves inner products of tangent vectors, then F is an *isometric immersion*.

If, furthermore, F is one-to-one and its inverse $F^{-1}: F(M) \to M$ is continuous, then F is an *isometric imbedding*.

For example, a proper patch (Section 1 of Chapter 4) is an imbedding of an open set $D \subset \mathbf{R}^2$ into $\mathbf{R}^3$. The continuity requirement for the inverse map F^{-1} is the same as in that special case: It prevents edges of M from being glued onto M, thus making $F(M)$ quite different from M. In the special case where the surface M is compact, it has no "edges," and in fact a theorem of topology asserts that F^{-1} is always continuous.

The following technical result makes it clear that the surfaces $M \subset \mathbf{R}^3$ studied in the early chapters are just those abstract geometric surfaces that can be imbedded in $\mathbf{R}^3$.

5.7 Lemma   If $F: M \to \mathbf{R}^3$ is an isometric imbedding of a geometric surface in $\mathbf{R}^3$, then the image $F(M)$ is a surface in $\mathbf{R}^3$ (Definition 1.2 of Chapter 4), and the function $F: M \to F(M)$ is an isometry.

Proof.   If $\mathbf{x}: D \to M$ is a coordinate patch in M, then the composite mapping $F(\mathbf{x}): D \to \mathbf{R}^3$ is a proper patch. In fact, its inverse function $F(\mathbf{x}(D)) \to D$ is $\mathbf{x}^{-1}F^{-1}$, which is continuous since both $\mathbf{x}^{-1}$ and F^{-1} are. Evidently, $F(\mathbf{x}(D))$ is contained in M, so the definition of surface in $\mathbf{R}^3$ is satisfied.

$F(M)$ uses the dot product of $\mathbf{R}^3$ as its inner product, and by definition the imbedding $F: M \to \mathbf{R}^3$ preserves inner products. Hence, when considered as a mapping of M onto $F(M)$, F also preserves inner products, and since it is one-to-one, it is an isometry.   ◆

We have seen that there are geometric surfaces M that cannot be imbedded in $\mathbf{R}^3$, for example, the flat torus and the projective plane (both defined in Section 2 of Chapter 7). In such cases it is natural to try to imbed M in a higher-dimensional Euclidean space $\mathbf{R}^n$. Finding isometric imbeddings is seldom easy, but the larger n is, the less difficult the task becomes. Roughly speaking, with more dimensions for M to curve through, there is a better chance that a shape can be found for it that is compatible with its intrinsic geometry.

The following example imbeds the flat torus; for the projective plane see Exercise 10.

5.8 Example   Isometric embedding of a flat torus in $\mathbf{R}^4$. The idea here is that since the unit circle C is naturally imbedded in the plane $\mathbf{R}^2$, by taking Cartesian products (Exercise 15 of Section 4.8) we can get an imbedding of the flat torus $C \times C = T_0$ into $\mathbf{R}^2 \times \mathbf{R}^2 = \mathbf{R}^4$.

Start with the mapping $\overline{\mathbf{x}}: \mathbf{R}^2 \to \mathbf{R}^4$ given by

$$\overline{\mathbf{x}}(u, v) = (\cos u, \sin u, \cos v, \sin v).$$

If $\mathbf{x}$ is the parametrization of the flat torus given in Example 2.2 of Chapter 7, then the formula

$$F(\mathbf{x}(u, v)) = \overline{\mathbf{x}}(u, v)$$

is consistent in the sense of Exercise 13 of Section 4.5. Now

$$\mathbf{x}(u, v) = \mathbf{x}(u_1, v_1)$$

$$\Leftrightarrow u_1 = u + 2\pi m, \ v_1 = v + 2\pi n$$

$$\Leftrightarrow \overline{\mathbf{x}}(u, v) = \overline{\mathbf{x}}(u_1, v_1),$$

Thus the exercise shows not only that F is a well-defined mapping $F: M \to \mathbf{R}^4$, but also that it is one-to-one.

To show that F preserves inner products, the method in Lemma 4.5 of Chapter 6 remains valid. We compute

$$\overline{\mathbf{x}}_u = (-\sin u, \cos u, 0, 0),$$

$$\overline{\mathbf{x}}_v = (0, 0, -\sin v, \cos v).$$

Hence, using the dot product of $\mathbf{R}^4$,

$$\overline{E} = 1, \quad \overline{F} = 0, \quad \overline{G} = 1.$$

Since these functions agree with E, F, G for $\mathbf{x}$, inner products are preserved.   ◆

The general situation here is not well understood. It is known that every compact surface can be isometrically imbedded in $\mathbf{R}^{17}$, but it seems likely that 17 can be replaced by a lower dimension.

Exercises

1. Prove that isometric surfaces have isomorphic isometry groups.

2. Prove:
    (a) The torus of revolution T is not homogeneous.
    (b) The flat torus T_0 is homogeneous, but not frame homogeneous. (*Hint:* Show that T_0 has closed geodesics of different lengths.)

3. Suppose that in M any two points can be joined by *at least* one geodesic, and that in N any two points can be joined by *at most* one geodesic. Prove that every local isometry $F: M \to N$ of such surfaces is one-to-one.

4. Show that the set I^+ of all orientation-preserving isometries of a surface M is a normal subgroup of the isometry group $I(M)$. (A subgroup H of a group G is *normal* provided $h \in H$ and $g \in G$ imply $ghg^{-1} \in H$.)

5. (a) If there is a point $\mathbf{p}_0$ in M such that for each $\mathbf{p}$ in M some isometry of M sends $\mathbf{p}_0$ to $\mathbf{p}$ (or the reverse), show that M is homogeneous.
    (b) State and prove the analogous fact for frame-homogeneity.
    (c) Suppose there is a point $\mathbf{p}_0$ in a homogeneous surface M with the property that for any two frames at $\mathbf{p}_0$ there is an isometry F of M such that F_* carries one of the frames to the other. Show that M is frame-homogeneous.

6. For the sphere Σ: $\|\mathbf{p}\| = r$ in $\mathbf{R}^3$, prove:
(a) Every orthogonal transformation C of $\mathbf{R}^3$ restricts to an isometry of Σ.
(b) Σ is frame-homogeneous.
(c) Every isometry of the sphere is the restriction of a Euclidean isometry. (*Hint*: For (b) use Ex. 5(c).)

7. Prove:
(a) If $C: \Sigma \to \Sigma$ is an isometry as in Exercise 6(a), there is a unique isometry $C_p: P \to P$ such that $FC = C_pF$, where F is the projection $\Sigma \to P$.
(b) The projective plane P is frame-homogeneous.

8. If M is a connected surface in $\mathbf{R}^3$ that is not contained in a plane, show that the function $\mathbf{F} \to \mathbf{F}|M$ is a one-to-one homomorphism of the Euclidean symmetry group $S(M)$ onto a subgroup of the isometry group $I(M)$.

9. Let M be the trough-shaped surface $z = \cosh^2 y$ in $\mathbf{R}^3$.
(a) By adjusting the Monge patch $(u, v) \to (u, v, \cosh v)$ find an *isometry* $\mathbf{x}: \mathbf{R}^2 \to M$.
(b) Show that $I(M)$ is isomorphic to the isometry group $\mathscr{E}(2)$ of the Euclidean plane. Which isometries of M derive as in Exercise 8 from Euclidean symmetries of M? Which do not?

10. (*Isometric imbedding of the projective plane.*) Consider the mapping $F: \mathbf{R}^3 \to \mathbf{R}^6$ given by

$$F(x, y, z) = \left(\frac{1}{\sqrt{2}} x^2, \frac{1}{\sqrt{2}} y^2, \frac{1}{\sqrt{2}} z^2, xy, xz, yz \right).$$

(a) If $\mathbf{v}$ is a tangent vector to $\mathbf{R}^3$ at $\mathbf{p} = (x, y, z)$, show that

$$\|F_*(\mathbf{v})\|^2 = \|\mathbf{p}\|^2\|\mathbf{v}\|^2 + (\mathbf{p} \cdot \mathbf{v})^2.$$

Thus the restriction $F_1 = F|\Sigma$ of F to the unit sphere Σ is an isometric immersion of Σ in $\mathbf{R}^6$.
(b) Show that F_1 is one-to-one, hence is an imbedding. (By a theorem of topology, since Σ is compact, the inverse map $(F|\Sigma)^{-1}$ is automatically continuous.)
(c) Derive an isometric imbedding of the projective plane P in $\mathbf{R}^6$. (*Hint:* Prop. 2.6 of Ch. 7.)
(d) Check that each point in the image of F_1 has the same dot product with $(1, 1, 1, 0, 0, 0)$, and improve (c) to an isometric imbedding of P in $\mathbf{R}^5$.

11. The n-dimensional sphere Σ^n: $\|\mathbf{p}\| = r$ in $\mathbf{R}^{n+1}$ becomes a Riemannian manifold, just as in the 2-dimensional case, by using the dot product of $\mathbf{R}^{n+1}$ on its tangent vectors. Show that:

(a) The mapping in Example 5.8 actually gives an isometric imbedding of the flat torus in a 3-dimensional sphere of radius $\sqrt{2}$.
(b) The mapping in the preceding exercise gives an isometric imbedding of the projective plane in a 5-dimensional sphere of radius $1/\sqrt{2}$.

8.6 Surfaces of Constant Curvature

The simplest possibility for the Gaussian curvature of a surface is that it is constant. We have seen several examples of such surfaces, and there are infinitely many more. The goal of this section is to give a reasonable organization for all of them.

A realistic treatment must be limited to complete connected surfaces. Without connectedness, we would have to consider, for example, a surface composed of a random collection of planes, spheres, and tori. Without completeness we would have to deal with enormous collections starting, say, with every connected open subset of the plane.

Given any number k, there is a particularly simple geometric surface $M(k)$ whose Gaussian curvature has the constant value k.

- $k > 0$: $M(k)$ is the sphere $\Sigma \subset \mathbf{R}^3$ of curvature k (hence radius $1/\sqrt{k}$).
- $k = 0$: $M(k)$ is the Euclidean plane $\mathbf{R}^2$.
- $k < 0$: $M(k)$ is the hyperbolic plane H of curvature k (hence pseudo-radius $r = 1/\sqrt{-k}$). (See Exercise 4 of Section 7.2.)

We call $M(k)$ the *standard geometric surface* of constant curvature k. These surfaces are complete and simply connected, and using covering methods we will show that in a sense they dominate all surfaces of constant curvature.

The following preliminary result will simplify matters.

6.1 Lemma Let M be a complete connected surface and let $F: M \rightarrow N$ be a local isometry into a connected surface N. Then F is a Riemannian covering map onto N, and N is complete.

Proof. If the geodesic lift property holds, then Theorem 4.12 will show that F is a Riemannian covering map. So let $\alpha: [0, b] \rightarrow N$ be a geodesic segment and let $\mathbf{p}$ be a point of M such that $F(\mathbf{p}) = \alpha(0)$. There is a unique vector $\mathbf{v}$ at $\mathbf{p}$ such that $F_*(\mathbf{v}) = \alpha'(0)$. Since M is complete, the geodesic γ_v can be defined on the entire real line. Then

$$F(\gamma_v) = \gamma_{F_*(v)} = \alpha,$$

wherever both sides are defined. Thus γ_v provides a lift of α, and furthermore, $F(\gamma_v)$ is an extension of α over the whole real line, so N is complete.   ◆

6.2 Theorem   If N is a complete connected surface with constant curvature k, then the standard surface $M(k)$ is a Riemannian covering surface of N.

Proof.   In view of the preceding lemma we need only prove that there is a local isometry F from $M(k)$ into N. There are three cases.

The Case $k < 0$. It will suffice to work with the $k = -1$ hyperbolic plane H.

As in (2) of Example 1.7, we use the frame U_1, U_2 at $\mathbf{0}$ in H. Let $\mathbf{e}_1$, $\mathbf{e}_2$ be a frame at an arbitrary point of N. Then let $\tilde{\mathbf{x}}$ and $\mathbf{x}$ be the resulting geodesic polar mappings in $M(k)$ and N. For the surface N we assert:

(1) $\mathbf{x}$ is defined on the entire closed half-plane $\mathcal{H}$: $u \geqq 0$ (a consequence of completeness).
(2) The mapping $\mathbf{x}$: $\mathcal{H} \to N$ is regular for $u > 0$. (*Proof:* As noted in the proof of Lemma 1.6, geodesic polar mappings have $E = 1$ and $F = 0$. We saw earlier that the Jacobi equation for $k = -1$ gives $\sqrt{G} = \sinh u$, so $EG - F^2 = \sinh^2 u > 0$ for $u > 0$.)

These general results are valid for $\tilde{\mathbf{x}}$: $\mathcal{H} \to H$ as well, but here we know more. Example 1.7 showed that the whole surface H is a normal neighborhood of the pole $\mathbf{0}$. Thus $\tilde{\mathbf{x}}$ has *only* the usual ambiguities of polar coordinates, that is, the equation $\tilde{\mathbf{x}}(u, v) = \mathbf{q}$ determines u and v uniquely, but for the addition of multiples of 2π. This means that the formula

$$F(\tilde{\mathbf{x}}(u, v)) = \mathbf{x}(u, v)$$

is *consistent* in the sense of Exercise 13 of Section 4.5. Thus it defines a mapping F of H onto N. Condition (1) ensures that F is defined on all of $M(k)$, and the differentiability of F at the pole follows from the differentiability of the exponential map there.

That F is a local isometry follows from the fundamental criterion in Section 4 of Chapter 6. Indeed, using (2) above gives

$$\tilde{E} = 1 = E, \quad \tilde{F} = 0 = F, \quad \tilde{G} = \sinh^2 u = G \quad \text{for } u > 0.$$

At the pole the preservation of inner products is an honest consequence of continuity.

The Case $k = 0$. This proof is a word-for-word copy of the preceding one, except that $M(k) = \mathbf{R}^2$ and $\tilde{G} = G = u^2$.

The Case k > 0. Here a new idea is required, since the largest normal neighborhood $\mathcal{N}$ of a point **p** in the sphere $M(k) = \Sigma$ is not the whole sphere, but the sphere minus the antipodal point −**p** of the pole **p**.

Arguing as in the case $k < 0$, we get a local isometry $F_1: \mathcal{N} \to N$. Now repeat this argument once more at a point **p*** different from both **p** and −**p**. This will produce another local isometry $F_2: \mathcal{N}^* \to N$, where $\mathcal{N}^*$ is all of Σ except −**p***. For the frames that determine F_2 we use an arbitrary frame $\mathbf{e}_1$, $\mathbf{e}_2$ at **p***, but in N the frame $F_{1*}(\mathbf{e}_1)$, $F_{1*}(\mathbf{e}_2)$ at $F_1(\mathbf{p}^*)$.

Thus Theorem 5.2 applies, showing that $F_1 = F_2$ on the (connected) intersection of $\mathcal{N}$ and $\mathcal{N}^*$. But $\mathcal{N}$ and $\mathcal{N}^*$ cover the entire sphere, so taken together F_1 and F_2 constitute a single local isometry F from Σ to N. ◆

6.3 Corollary The standard constant curvature surfaces $M(k)$ are the only geometric surfaces that are complete, simply connected, and have constant curvature.

Proof. We have already seen that the three types of $M(k)$ have the specified properties. Conversely, if M is a complete connected surface with constant curvature k, then by Theorem 6.2 there is a Riemannian covering map $F: M(k) \to M$. But Theorem 4.8 asserts that a simply connected surface has only trivial coverings, so F is an isometry. ◆

Using Theorem 6.2 we give an overview of constant curvature surfaces, omitting some proofs. We assume connectedness for all surfaces, and consider the cases $k > 0$, $k = 0$, $k < 0$ in turn.

Constant Positive Curvature Given any number $k > 0$ there are (up to isometry) exactly two complete geometric surfaces M with curvature k: the sphere and the projective plane. The preceding corollary covers the simply connected case. Since the sphere has Euler characteristic 2, Remark 4.7 implies that it can nontrivially cover only a compact surface of Euler characteristic 1. By Theorem 4.10, the projective plane P is the only such surface.

Flat Surfaces For $k = 0$ we have met, so far, three types of complete surfaces: the plane, cylinder, and flat torus. There are two more types, discovered only in the late 1800s. They can be found by the scheme used to derive the projective plane P from the sphere Σ. Example 8.2 of Chapter 4 used the antipodal map A of the sphere to construct P as an abstract surface; then Example 2.7 of Chapter 7 carried geometry along to make P a *geometric* surface. Starting from a surface M (in the role of Σ), all that is needed is an isometry A of M that has the two essential properties of the antipodal map of Σ:

$A^2 = I$ (the identity map of M), and

A has no fixed points, that is, $A(\mathbf{p})$ never equals $\mathbf{p}$.

Then the abstract surface M/A is defined to be the set of all unordered pairs $\{\mathbf{p}, A(\mathbf{p})\}$, and patches are defined just as for the projective plane $P = \Sigma/A$. Evidently, Proposition 2.6 of Chapter 7 applies, giving M/A a metric tensor that makes the projection $\mathbf{p} \to \{\mathbf{p}, A(\mathbf{p})\}$ a local isometry. Then, by Lemma 6.1, it is a Riemannian covering.

6.4 Example   *Complete flat nonorientable surfaces.*

(1) *A complete flat Möbius band.* Let C be the cylinder $x^2 + y^2 = r^2$ in $\mathbf{R}^3$. For the required isometry A we use the antipodal map $A(x, y, z) = (-x, -y, -z)$. It may help to recognize the resulting surface C/A as a Möbius band by imagining the cylinder as shrunk down to an ordinary band, its vertical lines reduced to intervals $-1 < z < 1$.

However, we need the whole cylinder C, so that it, and hence C/A, will be complete.

(2) *A flat Klein bottle.* The familiar image of the Klein bottle as laboratory glassware (right side of Fig. 8.20) is not very flat. But we can construct a flat one, starting from the flat torus T_0 in Example 2.2 of Chapter 7.

T_0 has the same symmetries in $\mathbf{R}^3$ as the torus of revolution, so once more, the mapping $A(x, y, z) = (-x, -y, -z)$ is an isometry $T_0 \to T_0$. Then T_0/A is a complete flat surface.

The transition in Fig. 8.20 suggests how to see that T_0/A is actually a Klein bottle. On the left we have discarded an open half of T_0 since the deleted points are antipodal to points in the remaining tube (hence nothing is lost in the construction of T_0/A). The identifications imposed by A on the two

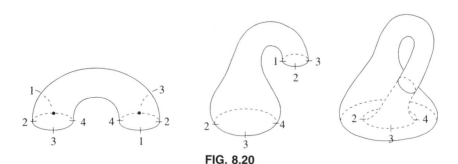

FIG. 8.20

boundary circles give them orientations inconsistent from those that would produce a torus. To match these oriented circles, the tube must be stretched around as suggested in Fig. 8.20 and twisted along the way. The criterion in Exercise 1 of Section 4.8 shows readily that T_0/A is nonorientable.

So now—up to scale change—we have five types of complete flat surfaces.

Euclidean plane, cylinder, flat torus, Möbius band, Klein bottle.

It is known that *there are no more*. Of course, except for the plane, different parameters produce nonisometric examples within a given type, for example, the cylinders over circles of different radii.

All five types appear in the following diagram, where arrows represent Riemannian covering maps.

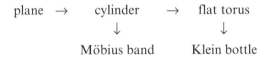

For the horizontal arrows, the multiplicity of the covering is infinite. The vertical arrows are the two double coverings defined above.

Only the flat torus and Klein bottle are compact; only the Möbius band and Klein bottle are nonorientable. In fact, both vertical arrows represent orientation coverings.

Only the plane and cylinder can be isometrically imbedded in $\mathbf{R}^3$. Compact flat surfaces are barred from $\mathbf{R}^3$ by Theorem 3.5 of Chapter 6. For the (non-compact) flat Möbius band, see the reference in Exercise 4 of Section 2.

Constant Negative Curvature Ignoring scale changes, we have found very few surfaces that have constant $k \geqq 0$, but the situation is radically different for $k < 0$.

6.5 Theorem   Every compact surface with a negative Euler characteristic admits a metric of constant negative curvature.

(Of the infinitely many compact surfaces listed in the classification theorem (4.10), all but four have $\chi < 0$.)

We give an informal proof of the theorem in the orientable case. Consider a regular geodesic octagon P in the $k = -1$ hyperbolic plane, as shown in Fig. 8.21. P is centered at $\mathbf{0}$, and *regular* means that its eight angles are equal and also its eight sides. Recall that the flat torus defined in Example 2.2 of Chapter 7 was visualized as a rectangle with opposite sides sewn together.

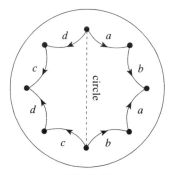

FIG. 8.21

Figure 8.21 uses the same scheme: Like-named sides are sewn together with matching orientations (indicated by the arrows). When this is done, the eight vertices of P are fused into one. All the angles of P meet there; hence if the result is to be a surface, *their sum must be* 2π.

This can always be achieved by adjusting the size of P. To see this, let P_r be the regular octagon that has hyperbolic distance r from $\mathbf{0}$ to every vertex. The Gauss-Bonnet formula proves what can be seen in figures such as Fig. 7.11: for r small, the angle sum is too large, and for r very large, the sum is too small. (Recall that hyperbolic angles are the same as Euclidean.) Since the angle sum depends continuously on r, there is an intermediate value for which the sum is exactly 2π.

Let $\overline{P}$ be the resulting surface. It is compact and since it is cut from the hyperbolic plane has constant negative curvature. We claim that topologically it is a double torus, that is, a sphere with two handles. In Fig. 8.21, note that the line from vertex to vertex becomes a circle in $\overline{P}$. On the right side the four edges a, b, a, b are sewn together by the pattern used for the torus. Thus the right half of the figure produces in $\overline{P}$ a torus with a hole punched in it. The same is true for the left side, so $\overline{P}$ is constructed by joining two handles; it is thus a double torus, that is, a sphere with two handles.

A proof by induction takes care of the general case. To reach the sphere with $h > 2$ handles, start with a regular $4h$-sided polygon and modify Fig. 8.21 by replacing sides c, d, c, d (on the left) by identification pattern of the sphere with $h - 1$ handles. Then, for sides a, b, a, b (on the right) add one more handle just as before.   ◆

It turns out that two compact surfaces with the same curvature $k < 0$ and the same number $h \geqq 2$ of handles need not be isometric. However, they must have the same area $A(k, h) = 4\pi(1 - h)/k$, since by the Gauss-Bonnet theorem,

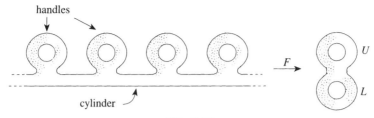

FIG. 8.22

$$kA = \iint K\, dM = 2\pi\chi(M) = 2\pi(2 - 2h) = 4\pi(1 - h).$$

If compactness is weakened to completeness, there are infinitely many non-isometric constant $k < 0$ surfaces, with complicated patterns of tubes and spikes that defy any systematic organization.

As a simple example, let M be the abstract surface built by attaching an infinite number of handles to a cylinder, as in Fig. 8.22. Let F be a covering map of M onto a double torus D that wraps the cylinder repeatedly around the *lower* loop L of D in such a way that each handle of M is carried diffeomorphically onto the *upper* loop U of D. We have just seen that D has a metric with constant negative curvature. The pullback metric on M makes F a local isometry (Remark 1.3 of Section 7) and thus gives M constant negative curvature.

Not one of these many $k < 0$ surfaces can be found in $\mathbf{R}^3$, for a theorem of Hilbert asserts that a complete surface in $\mathbf{R}^3$ cannot have constant negative curvature; this is proved in [dC]. A refinement by Efimov weakens the curvature hypothesis to $K \leqq k < 0$.†

Scale changes effectively reduce the standard surfaces $M(k)$ to just three: the unit sphere Σ, Euclidean plane $\mathbf{R}^2$, and hyperbolic plane H of curvature $k = -1$. The plane, in concept at least, was well known long before Euclid. The sphere was prominent in ancient astronomy and later in long-distance navigation. Only hyperbolic geometry was a self-conscious mathematical invention.

Like the Euclidean plane, the hyperbolic plane was initially studied *synthetically*, that is, in terms of axioms governing relations between abstract points and lines. Only later were these synthetic geometries realized by geometric surfaces whose geodesics are the "lines." The familiar axiom that two points determine a unique line fails for antipodal points on the sphere Σ, but

---

† See T. K. Milnor, "Efimov's Theorem about Complete Immersed Surfaces of Negative Curvature," *Advances in Mathematics* 8 (1972), 472–543.

it is recovered by identifying antipodal pairs, thus forming the projective plane P.

The *classical geometries* are those of the Euclidean plane, sphere, projective plane, and hyperbolic plane. In the sphere any two lines (that is, routes of geodesics) must intersect; hence the same is true for P. Thus the parallel postulate for these four geometries becomes:

Through any point **p** not on a line L, the number of lines through **p** that do not meet L is

hyperbolic: ∞,   Euclidean: 1,   spherical and elliptic: 0.

6.6 Corollary   All four surfaces $\mathbf{R}^2$, Σ, P, H realizing the classical geometries are frame-homogeneous.

Proof.   We already know this for $M = \mathbf{R}^2$, Σ, and H, the latter two from earlier exercises. But Theorem 6.2 gives a simple proof for all three simultaneously. Fix a tangent frame to M—for H locate it at the origin **0**. By the proof of Theorem 6.2 this frame can be carried to an arbitrary frame on M by a Riemannian covering map. The surfaces M are simply connected; hence Theorem 4.8 implies that this covering map is an isometry. This is enough to prove frame-homogeneity (see Exercise 5 of Section 5).

Since the projective plane is not simply connected, it escapes the argument above. But its frame-homogeneity can be derived from that of the sphere, as suggested in Exercise 7 of Section 5.   ◆

Previously, as in this proof, we have given preference to the origin **0** in the hyperbolic plane H, but now we know that all frames on H are geometrically equivalent.

Frame homogeneity is sometimes called the *axiom of free mobility* because the inhabitants of such a surface can then use (as we have) Euclid's intuitive definition of congruence: Two geometric figures are *congruent* provided one can be moved isometrically into coincidence with the other.

The surfaces in the preceding corollary are the *only* frame-homogeneous surfaces. In fact, if $k > 0$, then M is either the sphere or projective plane, both frame-homogeneous. If $k = 0$, then among the five flat surfaces only the plane is frame-homogeneous (see Exercise 2 of Section 5 and Exercise 2 below). More advanced covering methods are needed in the $k < 0$ case; there the hyperbolic plane is the only surface that is even homogeneous.

It is noteworthy that of the multitude of complete, constant curvature surfaces only three types—spheres, cylinders, and the plane—can be found in $\mathbf{R}^3$.

Exercises

1. (a) How do we know, before trying it, that the construction method for Theorem 6.5 cannot produce a $\Sigma[1]$ (sphere with one handle) that has constant $k = -1$?

(b) If we do try this method, exactly where does it fail?

2. Show that the flat Möbius band $\mathscr{M}$ contains a unique shortest closed geodesic. Deduce that $\mathscr{M}$ is not homogeneous (hence not frame-homogeneous).

3. An *incidence geometry* consists of two abstract sets—called the *points* and the *lines*—and a single *incidence relation* described by either "point is on line" or "line is through point." Consider the following axioms:

(i) Through any two distinct points there is a unique line.

(ii) Any two distinct lines pass through a unique point.

(iii) There exists a set of four points, no three of which are on the same line.

For each of the classical geometries $M = \mathbf{R}^2$, Σ, P, H decide which of these axioms are satisfied, with points of M as the *points*, routes of geodesics as *lines*, and the obvious incidence relation. Which (if any) satisfy all three axioms?

4. In the sphere Σ of radius r, let Δ be a triangle whose sides are geodesic segments of lengths a, b, and c (all less than πr). Let ϑ be the interior angle of Δ at the vertex $\mathbf{p}$ opposite side a.

(a) Prove this spherical law of cosines:

$$\cos \frac{a}{r} = \cos \frac{b}{r} \cos \frac{c}{r} + \sin \frac{b}{r} \sin \frac{c}{r} \cos \vartheta.$$

(*Hint:* To determine $\cos \vartheta$, find unit vectors at $\mathbf{p}$ tangent to the sides b and c.)

(b) Show that this formula approximates the usual Euclidean law of cosines when r is large compared to a, b, c.

5. (*Side-angle-side criterion*). Let Δ and Δ' be geodesic triangles in the hyperbolic plane H. If two adjacent sides α, β of Δ have the same lengths as corresponding sides in Δ' and the angles between the sides are equal, prove that the triangles are congruent, that is, show that there exists an isometry F of H such that $F(\Delta) = \Delta'$.

6. Show: (a) Every $\Sigma[n]$, $n \geq 2$, is a covering surface of the double torus $\Sigma[2]$.

(b) $\Sigma[4]$ can be covered by $\Sigma[n]$ if and only if n has the form $3N + 1$. (*Hint:* To show existence, adapt Fig. 8.22.)

8.7   Theorems of Bonnet and Hadamard

The simplest way to weaken the hypothesis that Gaussian curvature is constant is to require that it obey *inequalities*. The natural cases may seem to be $K < 0$ and $K > 0$, but, in fact, they turn out to be $K \leq 0$ and $K \geq k > 0$. The reason stems from the opposite effects produced by the Jacobi equation: As we have seen, $K \leq 0$ implies that there are no conjugate points, and, as we now show, $K \geq k > 0$ implies that there are many.

The following is an instance of the general rule that *conjugate points arrive sooner on surfaces with larger positive curvature.*

7.1 Lemma   Let M be a surface with $K \geq k > 0$ for some constant k. If a geodesic segment σ starting at $\mathbf{p}$ has length $\geq \pi/\sqrt{K}$, there is a conjugate point of $\mathbf{p}$ along σ.

Proof.   The sphere Σ of radius $r = 1/\sqrt{K}$ has curvature $k = 1/r^2$, so M has curvature at least as large as that of Σ. We know that in Σ, conjugate points arrive at arc length $\pi r = \pi/\sqrt{K}$, so the lemma asserts that they arrive no later on M.

To prove this, let σ be a unit-speed geodesic in M defined on the interval $[0, \lambda]$, where $\lambda = \pi/\sqrt{K}$. From Corollary 3.4, the Jacobi equation and initial conditions for M are

$$g'' + K(\sigma)g = 0, \quad \text{with} \quad g(0) = 0, \, g'(0) = 1,$$

and the corollary asserts that there will be a conjugate point of $\sigma(0)$ along σ if and only if the function g is zero at some s_1 with $0 < s_1 \leq \lambda$.

The corresponding data for Σ are

$$f'' + kf = 0, \quad \text{with} \quad f(0) = 0, \quad f'(0) = 1.$$

Here we have the explicit solution

$$f(s) = \frac{\sin \sqrt{k}\,s}{\sqrt{k}}.$$

We will show that *as long as $g(s)$ stays positive, it is no larger than $f(s)$.* Since f is zero at λ, it follows that g has a zero at or before λ, thus proving the lemma.

The assertion is proved by some simple but ingenious calculus: Suppose that $g > 0$ on the open interval $J: 0 < s < b$, where $b \leq \lambda$. Evidently $f > 0$ on J, so the Jacobi equations above give

$$(gf' - fg')' = gf'' - fg'' = fg(K - k) \geq 0 \quad \text{on } J.$$

Thus the function $gf' - fg'$ is nondecreasing (that is, monotonically increasing) on J. Since $gf' - fg'$ is zero at $s = 0$, it follows that $gf' - fg' \geq 0$ on J. Hence

$$\left(\frac{f}{g}\right)' = \frac{gf' - fg'}{g^2} \geq 0$$

So f/g is also nondecreasing on J. This function is equal to 1 at $s = 0$, since by l'Hôpital's rule,

$$\lim_{s \to 0} \frac{f(s)}{g(s)} = \lim_{s \to 0} \frac{f'(s)}{g'(s)} = 1.$$

Thus $f/g \geq 1$ on J, that is, $f \geq g$ there.   ◆

The main theorem on curvature $K \geq k > 0$ continues the comparison with a sphere Σ of curvature k and shows that M also is compact—and has diameter and area no larger than those of Σ. "Diameter" here means *intrinsic* diameter: the least upper bound of all distances $\rho(\mathbf{p}, \mathbf{q})$ between points of M—or ∞ if these distances are unbounded. For a surface in $\mathbf{R}^3$, this should not be confused with the Euclidean diameter, which is generally smaller, since Euclidean distances tend to be shorter (see Fig. 6.8).

For example, consider the sphere Σ of curvature k. It has Euclidean radius $r = 1/\sqrt{k}$, hence intrinsic diameter $\pi r = \pi/\sqrt{k}$, the distance between antipodal points. Its Euclidean diameter, of course, is just $2r = 2/\sqrt{k}$. Note that in terms of curvature its area $4\pi r^2$ becomes $4\pi/k$.

7.2 Theorem (Bonnet)   If M is a complete connected surface with $K \geq k > 0$, then M is compact and has intrinsic diameter $\leq \pi/\sqrt{k}$ and area $\leq 4\pi/k$.

Proof.   If we show that $\rho(\mathbf{p}, \mathbf{q}) \leq \pi/\sqrt{k}$ for all $\mathbf{p}, \mathbf{q}$ in M, then the diameter inequality follows.

By the Hopf-Rinow theorem there is a minimizing geodesic segment σ from $\mathbf{p}$ to $\mathbf{q}$. Then σ is certainly locally minimizing, so Theorem 3.7 asserts that there are no conjugate points of $\mathbf{p}$ on σ before $\mathbf{q}$.

By the preceding lemma, there will be a conjugate point of $\mathbf{p}$ on σ if σ is strictly longer than $\lambda = \pi/\sqrt{k}$. Thus $L(\sigma) \leq \pi/\sqrt{k}$. Since σ is minimizing,

$$\rho(\mathbf{p}, \mathbf{q}) = L(\sigma) \leq \frac{\pi}{\sqrt{k}},$$

as required.

To show that M is compact, consider the exponential map exp at an arbitrary point $\mathbf{p}$. Line segments radiating from $\mathbf{0}$ in the tangent plane $T_p(M)$ are carried by exp to radial geodesics—of the same length—starting at $\mathbf{p}$. The argument above shows that we can travel from $\mathbf{p}$ to any point $\mathbf{q}$ of M by a geodesic of length at most $\lambda = \pi / \sqrt{k}$. Thus exp maps the disk D: $\|\mathbf{v}\| \leq \lambda$ in $T_p(M)$ onto the entire surface M. The disk D is compact, and the continuous image of a compact set is compact (Exercise 2 of Section 4.7); hence M is compact.

For the area assertion, we only need to integrate $\sqrt{EG - F^2}$ over enough of the geodesic polar parametrization at $\mathbf{p}$ to be sure of covering all of M. (No harm will result if some of M is covered more than once.) Let $s_1(v)$ be the distance to the first conjugate point of $\mathbf{p}$ along the radial geodesic $s \to \mathbf{x}(s, v)$, where we write s instead of the usual u. The proof of the preceding lemma showed that on the interval J_v: $0 < s < s_1(v)$,

$$g_v(s) = \sqrt{G}(s, v)$$

is never greater than

$$\frac{\sin(\sqrt{k}\,s)}{\sqrt{k}}.$$

Since $E = 1$ and $F = 0$ for a polar parametrization,

$$\sqrt{EG - F^2}(s, v) = \sqrt{G}(s, v) = g_v(s) \leq \frac{\sin(\sqrt{k}\,s)}{\sqrt{k}}$$

on the interval J_v.

The argument above shows that to cover all of M it suffices to go out to $s_1(v)$ on each radial geodesic. Then, since $\sin(\sqrt{k}\,s)$ remains positive out to $\pi/\sqrt{k} \geq s_1(v)$,

$$\text{area } M \leq \int_0^{2\pi} dv \int_0^{s_1(v)} \frac{\sin(\sqrt{k}\,s)}{\sqrt{k}} \, ds \leq \int_0^{2\pi} dv \int_0^{\pi/\sqrt{k}} \frac{\sin(\sqrt{k}\,s)}{\sqrt{k}} \, ds$$

$$= \frac{2\pi}{\sqrt{k}} \int_0^{\pi/\sqrt{k}} \sin(\sqrt{k}\,s) \, ds = \frac{2\pi}{k} [-\cos(\sqrt{k}\,s)]_0^{\pi/\sqrt{k}}$$

$$= \frac{4\pi}{k}.$$

◆

We saw in Section 2 that compactness implies completeness. Bonnet's theorem provides a converse when $K \geq k > 0$. Just $K > 0$ would not suffice—for example, the paraboloid of revolution is complete and has $K > 0$, but is certainly not compact.

Bonnet's theorem can be strengthened as follows.

7.3 Corollary   Under the hypotheses of Bonnet's theorem, M is diffeomorphic to either a sphere or a projective plane, and in the latter case the bounds on diameter and area can be cut in half.

Proof. M is compact, connected, and has positive curvature; hence, it has positive Euler characteristic. Thus the first assertion is an immediate consequence of the surface classification theorem (4.10). When M is diffeomorphic to a projective plane, its orientation covering surface is diffeomorphic to a sphere. If $F: \tilde{M} \to M$ is the covering map, assign $\tilde{M}$ the pullback metric, thus making F a local isometry. Now apply Bonnet's theorem to $\tilde{M}$. Since this is a double covering, it is clear that the area of M is exactly half that of $\tilde{M}$. We omit the more complicated proof of the corresponding fact for diameters.   ◆

Turning to the study of a surface M with $K \le 0$ we again use comparison—this time with the Euclidean plane in the form of any tangent plane to M. The essential fact is that the exponential maps of M are covering maps. To prove this, two lemmas are needed.

7.4 Lemma   If M is a complete surface with $K \le 0$, then for any point $\mathbf{p}$ in M the exponential map $\exp_p: T_p(M) \to M$ is length nondecreasing, that is,

$$\|(\exp_p)*(\mathbf{s})\| \ge \|\mathbf{s}\|$$

for every tangent vector $\mathbf{s}$ to $T_p(M)$.

Proof.   As in Section 1, let $\tilde{\mathbf{x}}$ be a polar parametrization of $T_p(M)$, with corresponding geodesic polar mapping $\mathbf{x} = \exp_p(\tilde{\mathbf{x}})$. (Henceforth we omit the subscript from exp.)

At any point $\tilde{\mathbf{x}}(u_0, v_0)$ in $T_p(M)$, the tangent vectors $\tilde{\mathbf{x}}_u$ and $\tilde{\mathbf{x}}_v$ are an orthogonal basis for the tangent plane to $T_p(M)$ at $\tilde{\mathbf{x}}(u_0, v_0)$. As we saw in Lemma 1.6, $F = \langle \mathbf{x}_u, \mathbf{x}_v \rangle = 0$, so the vectors

$$\mathbf{x}_u = \exp_*(\tilde{\mathbf{x}}_u) \quad \text{and} \quad \mathbf{x}_v = \exp_*(\tilde{\mathbf{x}}_v)$$

are also orthogonal.

The proof of the lemma separates into three assertions.

(1) It suffices to show that the lengths of $\tilde{\mathbf{x}}_u$ and $\tilde{\mathbf{x}}_v$ are not decreased by the differential map $\exp_*$.

Proof.   To show that this condition is sufficient, suppose that, in fact, $\|\mathbf{x}_u\| \ge \|\tilde{\mathbf{x}}_u\|$ and $\|\mathbf{x}_v\| \ge \|\tilde{\mathbf{x}}_v\|$, and let $\mathbf{s}$ be an arbitrary tangent vector to $T_p(M)$ at $\tilde{\mathbf{x}}(u_0, v_0)$. Now

$$\exp_*(\mathbf{s}) = \exp_*(a\tilde{\mathbf{x}}_u + b\tilde{\mathbf{x}}_v) = a\,\exp_*(\tilde{\mathbf{x}}_u) + b\,\exp_*(\tilde{\mathbf{x}}_v) = a\mathbf{x}_u + b\mathbf{x}_v.$$

Then, since $\mathbf{x}_u$ and $\mathbf{x}_v$ are orthogonal,

$$
\begin{aligned}
\|\exp_*(\mathbf{s})\|^2 &= \|a\mathbf{x}_u\|^2 + \|b\mathbf{x}_v\|^2 \\
&= a^2\|\mathbf{x}_u\|^2 + b^2\|\mathbf{x}_v\|^2 \\
&\geqq a^2\|\tilde{\mathbf{x}}_u\|^2 + b^2\|\tilde{\mathbf{x}}_v\|^2 \\
&= \|\mathbf{s}\|^2.
\end{aligned}
$$

(2) $\|\mathbf{x}_u\| = \|\tilde{\mathbf{x}}_u\|$.

Proof. The definition of polar parametrization in Section 1 shows that evaluated at an arbitrary (u_0, v_0), $\tilde{\mathbf{x}}_u$ is the velocity vector at $u = u_0$ of the *unit-speed* radial line

$$
u \to \tilde{\mathbf{x}}(u, v_0) = u \cos v_0 \, \mathbf{e}_1 + u \sin v_0 \, \mathbf{e}_2
$$

in $T_p(M)$. We know that exp carries this radial line to the *unit-speed* geodesic $u \to \mathbf{x}(u, v_0)$ in M with initial velocity $\cos v_0 \, \mathbf{e}_1 + \sin v_0 \, \mathbf{e}_2$. Hence

$$
\|\tilde{\mathbf{x}}_u(u_0, v_0)\| = 1 = \|\mathbf{x}_u(u_0, v_0)\|.
$$

(3) $\|\mathbf{x}_v\| \geqq \|\tilde{\mathbf{x}}_v\|$.

Proof. This time the comparison between $\tilde{\mathbf{x}}$ in $T_p(M)$ and $\mathbf{x}$ in M involves the Jacobi equation. The definition of polar parametrization in Section 1 shows that

$$
\tilde{\mathbf{x}}_v(u, v) = -u \sin v \, \mathbf{e}_1 + u \cos v \, \mathbf{e}_2.
$$

Thus, for the radial line $v = v_0$,

$$
\|\tilde{\mathbf{x}}_v(u, v_0)\| = u \quad (u \geqq 0). \tag{$*$}
$$

Again exp carries this ray to a geodesic $u \to \mathbf{x}(u, v_0)$, along which

$$
\exp_*(\tilde{\mathbf{x}}_v(u, v_0)) = \mathbf{x}_v(u, v_0) \quad (u \geqq 0).
$$

We know that $g(u) = \|\mathbf{x}_v(u, v_0)\|$ is uniquely determined by

$$
g'' + Kg = 0, \quad g(0) = 0, \quad g'(0) = 1.
$$

Since $K \leqq 0$, this implies

$$
(gg')' = g'^2 + gg'' = g'^2 - Kg^2 \geqq 0.
$$

But then $gg' \geqq 0$ for $u \geqq 0$, so

$$(g'^2)' = 2g'g'' = -2Kgg' \geqq 0 \quad (u \geqq 0).$$

Using the initial conditions on g, we find $(g'^2)(u) \geqq (g'^2)(0) = 1$ for $u \geqq 0$. Since g' is initially positive, $g'(u) \geqq u$ for all $u \geqq 0$. Hence

$$\|\mathbf{x}_v(u, v_0)\| = g(u) \geqq u \quad (u \geqq 0).$$

Combining this with ($*$) completes the proof. ◆

If a map F is length nondecreasing in this sense, then evidently $F_*(\mathbf{v}) = 0$ implies $\mathbf{v} = 0$, so F is a regular mapping. Thus the exponential maps of complete $K \leqq 0$ surfaces are local diffeomorphisms. As in earlier cases, the question now becomes, Are they covering maps? The length nondecreasing property provides the answer.

7.5 Lemma If an exponential map $\exp_p\colon T_p(M) \to M$ of a complete surface is length nondecreasing, then curve segments in M can be lifted to any level in $T_p(M)$.

Proof. Let $\alpha\colon [0\ b] \to M$ be a curve. If $\mathbf{v}$ in $T_p(M)$ is carried to $\alpha(0)$ by $\exp_p$, we must prove that there is a curve segment $\tilde{\alpha}\colon [0, b] \to T_p(M)$ such that $\exp_p(\tilde{\alpha}) = \alpha$.

Assume that such a lift does not exist. Since $\exp_p$ is a local diffeomorphism, there is a partial lift $\tilde{\alpha}_c$ defined on some initial segment $J_c\colon 0 \leqq t < c$ of $[0, b]$. Let $\tilde{\alpha}_d$ be the largest such partial lift. We will show that $\tilde{\alpha}_d$ can, in fact, be extended past d—and this contradiction proves the lemma.

Let $\{t_i\}$ be an increasing sequence in J_d that approaches d. The lifted points $\tilde{\alpha}(t_i)$ in $T_p(M)$ are in the set $\tilde{\alpha}(J_d)$. Since $\exp_p$ does not decrease the lengths of vectors, it cannot decrease the length of curves; hence

$$\|\tilde{\alpha}(t_i) - \mathbf{v}\| \leqq L(\tilde{\alpha}_d)) \leqq L(\exp_p(\tilde{\alpha}_d)) \leqq L(\alpha|J_d) < L(\alpha).$$

Thus these points all lie in a closed disk D of radius $L(\alpha)$ centered at $\mathbf{v}$. A standard theorem of calculus asserts that every infinite sequence in a closed disk in $\mathbf{R}^2$ contains a convergent subsequence. Since $T_p(M)$ is isometric to $\mathbf{R}^2$, we can apply that result to the lifted points $\tilde{\alpha}(t_i)$ to get a subsequence $\{\tilde{\alpha}(t_n)\}$ converging to some point $\mathbf{w}$ of $T_p(M)$. Necessarily, $t_n \to d$.

Again, since $\exp_p$ is a local diffeomorphism, some neighborhood of $\mathbf{w}$ is carried diffeomorphically onto a neighborhood of $\alpha(d)$. But then the inverse λ of this local map lifts a segment of α around $\alpha(d)$ into $T_p(M)$, where (by the uniqueness of lifts) it agrees with $\tilde{\alpha}$ for $t < d$ and hence provides the sought-for extension of $\tilde{\alpha}$ to values of t greater than d.   ◆

Now we can prove the result mentioned earlier.

7.6 Theorem   If M is a complete connected surface with $K \leqq 0$, then for any point **p** in M, the exponential map $\exp_p$: $T_p(M) \to M$ is a covering map.

Proof.   We would like to use Theorem 4.12, which says that a local isometry with the geodesic lift property is a covering map. But the exponential maps in the theorem are generally not local isometries—in terms of the natural Euclidean structure of $T_p(M) \approx \mathbf{R}^2$. But as noted above, these exponential maps are *regular*, so temporarily assigning $T_p(M)$ the pullback metric from M, as in Remark 1.3(2) of Chapter 7, makes them local isometries. The preceding lemma shows, in particular, that geodesics can be lifted; thus Theorem 4.12 applies to complete the proof.   ◆

7.7 Theorem (Hadamard)   If M is a complete, simply connected surface with $K \leqq 0$, then
(1) M is diffeomorphic to the Euclidean plane $\mathbf{R}^2$.
(2) All nonconstant geodesics are one-to-one (so there are no geodesic loops or closed geodesics).
(3) Through any points $\mathbf{p} \neq \mathbf{q}$ in M there is a unique (up to parametrization) geodesic.

Proof.   (1) By the preceding theorem, exp is a covering map, and since M is simply connected, Theorem 4.8 says that exp is a diffeomorphism. So for any point **p**, we have diffeomorphisms

$$\mathbf{R}^2 \approx T_p(M) \approx M.$$

(2) If a nonconstant geodesic has, say, $\gamma(0) = \gamma(1)$, then the exponential map $\exp_{\gamma(0)}$ sends both $\mathbf{0}$ and $\gamma'(0) \neq \mathbf{0}$ to $\gamma(1)$. But exp is one-to-one.
(3) If geodesics α and β each pass through both **p** and **q**, then by reparametrization we can suppose

$$\alpha(0) = \beta(0) = \mathbf{p}, \quad \alpha(1) = \beta(1) = \mathbf{q}.$$

But $\exp_p$ sends $\alpha'(0)$ to $\alpha(1)$ and $\beta'(0)$ to $\beta(1)$, hence

$$\exp_p(\alpha'(0)) = \exp_p(\beta'(0)).$$

Since $\exp_p$ is one-to-one, $\alpha'(0) = \beta'(0)$. Hence by the uniqueness of geodesics, $\alpha = \beta$.   ◆

Thus M has the same differentiable structure as the Euclidean plane $\mathbf{R}^2$ and shares its most famous geometric property: "two points determine a line."

Exercises

1. Prove that the exponential maps of a complete flat connected surface are Riemannian covering maps.

2. Show that under the hypotheses of Bonnet's theorem, if area(M) = $4\pi/k$, then M is isometric to a sphere Σ_k of curvature k. (*Hint:* Compare the final part of the proof of Theorem 1.8, where less-than-or-equal successively becomes equality.)

3. Prove that a compact surface M is intrinsically bounded (has finite intrinsic diameter) and if M is in $\mathbf{R}^3$ has finite Euclidean diameter.

4. If a complete connected surface has finite intrinsic diameter, show that it is compact. (*Hint:* Use the scheme in the proof of Bonnet's theorem.)

8.8 Summary

The global structure of a complete connected surface M can be described in terms of geodesics and Gaussian curvature K. From a point $\mathbf{p}$ in M, run geodesics out radially until (by the Hopf-Rinow theorem) they fill M. The only *geometric* difference from the Euclidean plane is the stretching of the polar circles (orthogonal trajectories of the radial geodesics), and this stretching is controlled by K, via the Jacobi differential equation.

The *topological* difference from $\mathbf{R}^2$ is more serious, but the notion of covering surface and the classification of compact surfaces lead to good results. In particular, we can give a reasonably complete account of all surfaces of constant curvature. When only inequalities are imposed on curvature, we still get the powerful theorems of Bonnet (for $K \geq k > 0$) and Hadamard (for $K \leq 0$).

Generally speaking, the strictly geometric results of the last two chapters hold up well when geometric surfaces are replaced by higher-dimensional Riemannian manifolds. Once the definitions are readjusted, most proofs are essentially the same. For example, the Hopf-Rinow theorem and the Bonnet and Hadamard theorems remain valid since their proofs depend only on what happens on an individual geodesic. Dimension 2 has simplified certain consistency proofs such as Theorems 2.1 and 3.2 in Chapter 7, but these can be avoided entirely by more advanced methods.

The most striking changes come in the links between topology and geometry. In higher dimensions there is no classification of compact manifolds comparable to Theorem 4.10 of this chapter. Where in dimension 2 only the sphere and projective space admitted metrics of positive curvature, now there are many. The Euler characteristic χ generalizes but no longer characterizes, and odd-dimensional compact manifolds all have $\chi = 0$. Nevertheless, the Gauss-Bonnet theorem extends to higher (even) dimensions, and in this and other ways, curvature continues its strong influence over topology.

Appendix
Computer Formulas

The computer commands most useful in this book are given in both the *Mathematica* and *Maple* systems. More specialized commands appear in the answers to several computer exercises. For each system, we assume a familiarity with how to access the system and type into it.

In recent versions of Mathematica, the core commands have generally remained the same. By contrast, *Maple* has made several fundamental changes; however most older versions are still recognized. For both systems, users should be prepared to adjust for minor changes.

Mathematica

1. Fundamentals

Basic features of *Mathematica* are as follows:

(a) There are no prompts or termination symbols—except that a final semicolon suppresses display of the output. Input (new or old) is activated by the command *Shift-return* (or *Shift-enter*), and the input and resulting output are numbered.

(b) Parentheses (. . .) for algebraic grouping, brackets [. . .] for arguments of functions, and braces {. . .} for lists.

(c) Built-in commands typically spelled in full—with initials capitalized—and then compressed into a single word. Thus it is preferable for user-defined commands to avoid initial capitals.

(d) Multiplication indicated by either $*$ or a blank space; exponents indicated by a caret, e.g., x^2. For an integer n only, $nX = n*X$, where X is not an integer.

(e) Single equal sign for assignments, e.g., $x = 2$; colon-equal (:=) for deferred assignments (evaluated only when needed); double equal signs for mathematical equations, e.g., $x + y == 1$.

(f) Previous outputs are called up by either names assigned by the user or **%n** for the nth output.

(g) Exact values distinguished from decimal approximations (floating point numbers). Conversion using **N** (for "numerical"). For example, **E^2*Sin[Pi/3]** returns $e^2\sqrt{3}/2$; then **N[%]** gives a decimal approximation.

(h) Substitution by slash-dot. For example, if **expr** is an expression involving x, then **expr/.x→u^2 + 1** replaces x everywhere in the expression by $u^2 + 1$.

Mathematica has excellent error notification and online help. In particular, for common terms, **?term** will produce a description. Menu items give formats for the built-in commands. The complete general reference book— exposition and examples—is *The Mathematica Book* [W]. For our purposes, the outstanding reference is Alfred Gray's book [G].

> Some basic notation. Functions are given, for example, by

$$f[x\_] := x^3-2x+1 \quad \text{or}$$
$$g[u\_,v\_] := u*Cos[v]-u^2*Sin[v]$$

Here, as always, an underscore "_" following a letter (or string) makes it a variable. Thus the function f defined above can be evaluated at u or 3.14 or $a^2 + b^2$.

> Basic calculus operations.

Derivatives (including partial derivatives) by **D[f[x],x]** or **D[g[u,v],v]**

Definite integrals by **Integrate[f[x],{x,a,b}]**. For numerical integration, prefix an **N** thus: **NIntegrate**.

> Linear algebra. A *vector* is just an n-tuple, that is, a list **v={v1,...,vn}**, whose entries can be numbers or expressions. Addition is given by **v+w** and scalar multiplication by juxtaposition, with **sv=s{v1,...,vn}** yielding **{s*v1,...,s*vn}**. The dot product is given by **v.w** and, for n = 3, the cross product is **Cross[v,w]**.

Mathematica describes a matrix as a list of lists, the latter being its rows. For example, {{a,b},{c,d}} is a matrix and is treated as such in all contexts. To make it look like $\begin{pmatrix} a & b \\ c & d \end{pmatrix}$, apply the command **MatrixForm**. The determinant of a square matrix **m** is given by **Det[m]**.

The full power of the dot operator (.) appears only when matrices are involved. First, if **p** and **q** are properly sized matrices, then **p.q** is their product. Next, if **m** is an m × n matrix and **v** is an n-vector, then **m.v** gives the usual operation of **m** on **v**. Taking m = n = 3 for example, if **m1, m2, m3** are the rows of **m** and **v={v1,v2,v3}**, then *Mathematica* defines

$$\textbf{m.v} \text{ to be } \{\texttt{m1.v,m2.v,m3.v}\}$$

This can be seen to be the result of **m** (in 3 × 3 form) matrix-multiplying the column-vector corresponding to **v**, with the resulting column-vector restated as an n-tuple. In this sense, *Mathematica* obeys the "column-vector convention" from the end of Section 3.1, which identifies n-tuples with n × 1 matrices.

If **A** is any *array*—say, a vector or matrix—then for most commands, **cmd[A]** will apply the command **cmd** to each entry of **A**.

2. Curves

A curve in $\mathbf{R}^3$ can be described by giving its components as expressions in a single variable. Example:

$$\texttt{c[t\_]:=}\{\texttt{Cos[t],Sin[t],2t}\}$$

Then the vector derivative (i.e., velocity) is returned by **D[c[t],t]**.

> Curves with parameters. For example, the curve c above can be generalized to

$$\texttt{helix[a\_,b\_][t\_]:=}\{\texttt{a*Cos[t],a*Sin[t],b*t}\}$$

Then **helix[1,2]=c**.

The following formulas, drawn from Theorem 4.3 of Chapter 2, illustrate aspects of vector calculus in *Mathematica*.

The curvature and torsion functions κ and τ of a curve $c \approx \gamma$ are given by

```
kappa[c_][t_]:=Simplify[Cross[D[c[tt],tt],D[c[tt],
   {tt,2}]].
   Cross[D[c[tt],tt],D[c[tt],{tt,2}]]]^(1/2)/
   Simplify[D[c[tt],tt].D[c[tt],tt]]^(3/2)/.tt->t
```

(Note the description of second derivatives.) The use of the dummy variable **tt** makes **kappa[c]** a real-valued *function* $\mathbf{R} \to \mathbf{R}$. Otherwise, it would be merely an expression in whichever single variable was used.

"Simplify" is the principal *Mathematica* simplification weapon; however, it cannot be expected to give ideal results in every case. ("FullSimplify" is more powerful but slower.) Thus human intervention is often required, either to do hands-on simplification or to use further computer commands such as "Together" or "Factor" or trigonometric simplifications.

```
tau[c_][t_]:=Simplify[
    Det[{D[c[tt],tt],D[c[tt],{tt,2}],D[c[tt],
    {tt,3}]}]]/
    Simplify[Cross[D[c[tt],tt],D[c[tt],{tt,2}]].
    Cross[D[c[tt],tt],D[c[tt],{tt,2}]]]]/.tt->t
```

Here the determinant gives a triple scalar product.

Note: The distinction between functions and mathematical expressions is basic. Thus, with notation as above, **tau** applied to a curve, say, **helix[1,2]**, is a real-valued function **tau[helix[1,2]]** whose value on any variable or number **s** is **tau[helix[1,2]][s]**.

The unit tangent, normal, and binormal vector fields T, N, B of a curve with $\kappa > 0$ are given by

```
tang[c_][t_]:=D[c[tt],tt]/
    Simplify[D[c[tt],tt].D[c[tt],tt]]^(1/2)/.tt->t
```

```
nor[c_][t_]:=Simplify[Cross[binor[c][t],
    tang[c][t]]]
```

```
binor[c_][t_]:=Simplify[Cross[D[c[tt],tt],D[c[tt],
    {tt,2}]]]/
    Simplify[Factor[Cross[D[c[tt],tt],D[c[tt],
    {tt,2}]].
    Cross[D[c[tt],tt],D[c[tt],{tt,2}]]]]^(1/2)/.tt->t
```

Here is how to preserve any such commands for future use: Type (or copy) them into a *Mathematica* notebook, say **frenet**, and use the *Cell* menu to designate the cells containing them as *initialization cells*. When this notebook is saved, a choice will be offered letting you save, not only **frenet**, but also a new file **frenet.m** that contains only the commands. Then these can be read into later work by **<<frenet.m**

3. Surfaces

A coordinate patch, say **x**, is given by listing its components as expressions in two variables. For example,

```
x[u_,v_]:={u*Cos[v],u*Sin[v],2v}
```

> Parameters can be handled as above for curves. For example, the 2 in this formula can be replaced by an arbitrary parameter using

```
helicoid[b_] [u_,v_] :={u*Cos[v] ,u*Sin[v] ,b*v}
```

Then `helicoid[2]` gives the original **x**.

For a patch, the following commands return $E, F, G, W = \sqrt{EG - F^2}$, and L, M, N. We elect to represent our capital letters (E) by double lowercase letters (ee), since many capitals have special meaning for *Mathematica* (for example, $E = 2.7183\ldots$).

```
ee[x_] [u_,v_] :=
  Simplify[D[x[uu,vv] ,uu] .D[x[uu,vv] ,uu] ]/.
  {uu->u,vv->v}
ff[x_] [u_,v_] :=
  Simplify[D[x[uu,vv] ,uu] .D[x[uu,vv] ,vv] ]/.
  {uu->u,vv->v}
gg[x_] [u_,v_] :=
  Simplify[D[x[uu,vv] ,vv] .D[x[uu,vv] ,vv] ]/.
  {uu->u,vv->v}
ww[x_] [u_,v_] :=
  Simplify[Sqrt[ee[x] [u,v] *gg[x] [u,v] -
  ff[x] [u,v] ^2] ]
```

The variant command, say **www**, in which `Sqrt[...]` is replaced by `PowerExpand[Sqrt[...]]` will often give decisively simpler square roots. But one must check that its results are positive, since for example, `PowerExpand[Sqrt[x^2]]` yields **x**.

```
ll[x_] [u_,v_] :=Simplify[Det[{D[x[uu,vv] ,uu,uu] ,
  D[x[uu,vv] ,uu] ,D[x[uu,vv] ,vv] }] /ww[x] [u,v] ]/.
  {uu->u,vv->v}
```

The formulas for **mm** and **nn** are the same except that the double derivative **uu, uu** is replaced by **uu, vv** and **vv, vv**, respectively.

> Gaussian curvature *K*. When the commands for *E, F, G* and L, M, N have been read in, commands for *K* and *H* follow directly from Corollary 4.1 of Chapter 5 (see Exercise 18 of Section 5.4). However, the fastest way to find *K* for a given patch in $\mathbf{R}^3$ is by the following command, based on Exercise 20 of Section 5.4. In it, "Module" creates an enclave in which temporary definitions can be made that let the final formula be expressed more simply.

```
gaussK[x_] [u_,v_] := Module[{xu,xv,xuu,xuv,xvv},
  xu=D[x[uu,vv] ,uu] ;xv= D[x[uu,vv] ,vv] ;
```

```
xuu=D[x[uu,vv],uu,uu];
xuv=D[x[uu,vv],uu,vv];
xvv=D[x[uu,vv],vv,vv];
Simplify[(Det[{xuu,xu,xv}]*Det[{xvv,xu,xv}]-
Det[{xuv,xu,xv}]^2)/
(xu.xu*xv.xv-(xu.xv)^2)^2]]/.{uu->u,vv->v}
```

As with other useful commands, this should be saved for future use.

4. Plots

There are four basic types: **Plot** and **Plot3D** plot the graphs of functions of one and two variables respectively. Examples:

```
Plot[f[x]//Evaluate,{x,a,b}]
Plot3D[g[x,y]//Evaluate,{x,a,b},{y,c,d}]
```

Here **//Evaluate** improves the speed of plotting.

ParametricPlot plots the image of a parametrized curve in the plane $\mathbf{R}^2$.

ParametricPlot3D plots the image of a parametrized curve or patch. For example, a parametrized curve $c(t)$ in $\mathbf{R}^3$ is plotted for $a \leqq t \leqq b$ by

```
ParametricPlot3D[c[t]//Evaluate,{t,a,b}]
```

and if **x** is an explicitly defined patch or parametrization, its image on the rectangle $0 \leqq u \leqq 1, 0 \leqq v \leqq 2\pi$ is plotted by

```
ParametricPlot3D[x[u,v]//Evaluate,{u,0,1},
{v,0,2Pi}]
```

Various refinements are available for plots. For example, if the end of the command above is altered to

```
...{v,0,2Pi},AspectRatio->Automatic]
```

then the same scale is imposed on height and width. Formally, the *option* "AspectRatio" has been reset from its default value. Various adjuncts to a plot can be also be changed. For example, the box surrounding the preceding plot is eliminated by **Boxed->False**. The plot can be made smoother by using **PlotPoints->{m,n}**, where the integers increase the default values governing smoothness in the u and v directions, respectively.

The options available for a command **cmd** are given, along with their default values, by **Options[cmd]**. Then **?opt** will describe a particular option.

Previously drawn plots can be shown on the same page by

```
Show[plot1,plot2,plot3]
```

5. Differential Equations

Explicit solutions in terms of elementary functions are inherently rare, so we describe how to find and plot numerical solutions, which are all that is needed in many contexts. In the command for such a solution, *Mathematica* lumps equations and initial conditions into a single list, then specifies the dependent variables and the interval of the dependent variable.

Example: Solve numerically the differential equations

$$x' = f(x, y, t), \quad y' = g(x, y, t),$$

subject to the initial conditions

$$x(t_0) = x_0, \quad y(t_0) = y_0,$$

on the interval $t_{min} \leqq t \leqq t_{max}$. The format is

```
soln = NDSolve[{x'[t]==f[x[t],y[t],t],
   y'[t]==g[x[t],y[t],t],
   x[t0]==x0,y[t0]==y0},{x,y},{t,tmin,tmax}]
```

Note the double equal signs. Without the N for "numerical," an exact solution would be sought.

NDSolve expresses x and y in terms of *Interpolating Functions*, data sufficient for subsequent plots. If **soln** is an explicit result from the preceding command, the solution is plotted by

```
ParametricPlot[Evaluate[{x[t],y[t]}/.soln],
   {t,tmin,tmax}]
```

Here "/." substitutes **soln** into the coordinates. Note the general equivalence: **Evaluate[X]** is the same as **X//Evaluate**.

Maple

1. Fundamentals

Basic features of *Maple* are as follows:

(a) Input is typed after a prompt and *must* be terminated by a semicolon— or colon, to suppress display of the output. **We do not show these below.** Then press ENTER (or RETURN).
(b) Parentheses used for algebraic grouping and arguments of functions; braces {...} for sets; brackets [...] for lists.

(c) Built-in commands are abbreviated, with multiword commands compressed into a single word; most are written in lower case.

(d) Multiplication *always* indicated by ∗, exponents by a caret, e.g., **x^2**.

(e) Assignments indicated by colon-equal, e.g., **x:=2**; equations by single equal, e.g., **x+y=1**.

(f) Previous outputs are called up by names assigned by the user. (Naming is important since input/outputs are not numbered.) Also, the percent symbol (%) gives the immediately preceding output, and two of these give the one before that.

(g) Exact values distinguished from decimal approximations (floating point numbers). Conversion is accomplished by the "evalf" command. For example, **exp(2)*sin(Pi/3)** returns $e^2\sqrt{3}/2$; then **evalf(%)** gives a decimal approximation.

(h) Substitution by the "subs" command. If *expr* is an expression involving x, then **subs(x=u^2+1,** *expr***)** replaces every x in the expression by $u^2 + 1$.

(i) If A is an *array*—say a matrix or vector—then to apply an operation F to each entry of A, use the command "map" thus: **map(***F,A***)**.

Maple has a distinctive command "unapply" that converts mathematical expressions into functions. For example, if *expr* is an expression involving u and v, then **unapply(***expr***,u,v)** is the corresponding function of u and v.

Many specialized *Maple* commands are collected in *packages*, which are loaded, for example, by **with(plots)**. A list of the commands in the package appears unless output is suppressed. We rarely use packages other than *plots* and *LinearAlgebra* (which is replacing *linalg*).

Maple has reasonable error notification and excellent on-line help. For common terms, **?term** will produce a detailed description (no semicolon required).

The *Maple Learning Guide* is a good introduction to the most recent version of *Maple*; it may be obtained from the website *maplesoft.com*. Of course, there are a variety of more advanced books.

Some basic notations.

Functions can be produced by the arrow notation. Examples:

$$\texttt{f := x->x^3-2*x+1} \text{ or}$$
$$\texttt{g := (u,v)->u*cos(v)-u^2*sin(v)}$$

Derivatives (including partials):

$$\texttt{diff(f(x),x)} \text{ or } \texttt{diff(g(u,v),v)}$$

Definite integral:

> int(f(x),x=a..b) or
> int(g(x,y),x=a..b,y=c..d)

If an explicit integral cannot be found, then **evalf(%)** gives a numerical result. Direct numerical integration is given by **evalf(Int(f(x),x=a..b))**.

Linear algebra. Recent versions of *Maple* have changed considerably (though it still recognizes many old forms). Currently, its commands, whether new or not, are often signalled by new names. Typically, the new command begins with a capital letter and is not abbreviated. These changes are most evident in the package *LinearAlgebra* that is replacing *linalg*.

Maple has always made a fundamental distinction between an n-tuple **[v1,..,vn]**—which is a *list*—and a *vector*, in any notation. The two types cannot directly interact. In the new version, *vector* is replaced by *Vector* (capital V).

Lists are the easiest to deal with. For instance, the usual sum of n-tuples **v=[v1,..,vn]** and **w=[w1,..,wn]** is given by **v+w**, and scalar multiplication of an n-tuple by a number **s** uses an asterisk, with **s*v** giving **[s*v1,..,s*vn]**.

A matrix is produced by applying the command **Matrix** to a list whose entries are lists, the latter being the rows of the matrix. Thus

$$\begin{pmatrix} a & b \\ c & d \end{pmatrix}$$ is described by *Maple* as **Matrix([[a,b],[c,d]])**.

With the package *LinearAlgebra* loaded, the determinant of a square matrix **m** is given by **Determinant(m)**.

When an n × n matrix **C** is considered as a linear transformation on R^n, it cannot directly attack **[v1,..,vn]** to give the image **[w1,..,wn]**. The list **[v1,..,vn]** must first be stood on end as **Vector([v1,..,vn])**, which is, in fact, an n × 1 matrix. Now matrix multiplication is valid, and, with *LinearAlgebra* installed, **Multiply(C,Vector([v1,..,vn]))** is the n × 1 matrix that **convert(%,list)** turns into **[w1,..,wn]**. This identification of an n-tuple with a column vector is just the "column vector convention" at the end of Section 3.1.

Since curves and surfaces are described in terms of lists, we can largely avoid the list/Vector conflict by defining three basic vector operations directly in terms of lists. First, note that the entries of a list **p:=[p1,p2,...,pn]** can be any expressions, and the ith entry is displayed by the command **p[i]**.

An operation applied to a list is automatically applied to each entry. (By contrast, other *arrays* require the command **map**.)

Dot product: `dot:=(p,q)-> simplify(p[1]*q[1]+`
                 `p[2]*q[2]+p[3]*q[3])`

Cross product: `cross:=(p,q)-> simplify`
                 `([p[2]*q[3]-p[3]*q[2],p[3]*q[1]-`
                 `p[1]*q[3],p[1]*q[2]-p[2]*q[1]])`

Triple scalar product: `tsp:=(p,q,r)-> dot(p,cross(q,r))`

The built-in *simplify* above will reduce the number needed in later commands. Note that **tsp(p,q,r)** is just the determinant of the matrix with rows **p,q,r**, so reversal of any two entries gives (only) a sign change.

The three commands can be saved in Maple's concise machine language by:

$$\texttt{save dot,cross,tsp,"dotcrosstsp.m"}$$

(Any name ending in ".m" will do as well.) These commands can then be introduced into later sessions by

$$\texttt{read "dotcrosstsp.m"}$$

(Formerly, *save* and *read* were expressed by **save(cmd1,cmd2,** **'filename.m')** and **read('filename.m')**, using backquotes.)

> Differential forms. The package *difforms* provides the essentials, including the exterior derivative operator **d**. The command **defform** is used to specify the degree of the forms involved. For example, **defforms(x=0,y=0)** tells *Maple* that x and y are 0-forms, that is, real-valued functions. Then the command **d(x^2*sin(y))** yields **2x sin(y)d(x)+x² cos(y)d(y)**.

2. Curves

A curve in $\mathbf{R}^3$ is described by giving its components as expressions in a single variable, for example, **c:= t->[3*cos(t),3*sin(t),2*t]**. Then the vector derivative (i.e., velocity) of **c** is returned by **diff(c(t),t)**, which differentiates each component of the curve by **t**.

> Curves with parameters. For example, using the *unapply* command, the curve **c** can be generalized to

$$\texttt{helix:=(a,b)-> unapply([a*cos(t),a*sin(t),b*t]}$$

Then **helix(3,2)** gives **c** as above.

> Frenet apparatus. We now show how the Frenet formulas in Theorem 4.3 of Chapter 2 can be expressed in terms of *Maple*.

The curvature function κ of a curve $c \sim \gamma$ is given by

```
kappa := c -> unapply(simplify(
  dot(cross(diff(c(t),t),diff(c(t),t,t)),
  cross(diff(c(t),t),diff(c(t),t,t)))^(1/2)/
  dot(diff(c(t),t),diff(c(t),t))^(3/2)),t)
```

Here "unapply" makes **kappa(c)** a real-valued function on the domain of c. Otherwise, it would merely be an expression in t and could not be evaluated on real numbers or other variables.

The command "simplify" is the principal *Maple* simplification weapon, but it not a panacea. It can be augmented by related commands such as "factor" or "expand." Use **?simplify** for information about these.

No set pattern of commands will give good results in every case, and human intervention is often required to get reasonable simplification.

The torsion function **tau** of a curve **c** is given by

```
tau := c -> unapply(simplify(
  tsp(diff(c(t),t),diff(c(t),t,t),
    diff(c(t),t,t,t))/factor(
  dot(cross(diff(c(t),t),diff(c(t),t,t)),
  cross(diff(c(t),t),diff(c(t),t,t)))))),t)
```

The distinction between functions and mathematical expressions is always important. Thus, with notation as above, **tau**, applied to a curve, say **helix(3,2)**, is a real-valued function whose value at a number or variable **s** is given by **tau(helix(3,2))(s)**.

Maple has several varieties of scalar multiplication when *Linear Algebra* is installed, however, since we are working with lists, **s*v** suffices.

The Frenet frame of a curve. The unit tangent, normal, and binormal vector fields T, N, B of a curve c are given by

```
tang := c->unapply(
  dot(diff(c(t),t),diff(c(t),t))^(-1/2)
    *diff(c(t),t),t)

nor := c->unapply(cross(binor(c)(t),tang(c)(t)),t)

binor := c->unapply(simplify(factor(
dot(cross(diff(c(t),t),diff(c(t),t,t)),
  cross(diff(c(t),t),diff(c(t),t,t)))))^(-1/2)*
  cross(diff(c(t),t),diff(c(t),t,t)),t)
```

The presence of square roots in these formulas means that we cannot expect simple results unless the curve itself is quite simple. However, individual values of the vector fields are usually readable.

Once the Frenet commands have been typed, they can be saved in a *Maple* dot-m file by

> **save kappa,tau,tang,nor,binor,"frenet.m"**

and, as usual, these commands can be installed in later work by **read** **"frenet.m"**.

3. Surfaces

A coordinate patch, say **x**, in $\mathbf{R}^3$ is defined as a list-valued function whose entries are expressions in two variables. For example,

> **x:=(u,v)->[3*u*cos(v),3*u*sin(v),2*v]**

Parameters in a patch can be handled as above for curves. For example, the 3 and 2 in this formula can be replaced by an arbitrary parameters **a** and **b** using

> **helicoid:=(a,b)->unapply([a*u*cos(v),a*u*sin(v), b*v],u,v)**

Then **helicoid(3,2)** gives the original patch **x**.

The following commands, applied to a patch **x**, return *E, F, G, W = EG – F²*, and L, M, N. We elect to represent these capital letters (*E*) by double lowercase letters (ee) since some capitals have special meaning for *Maple* (for example, $I = -1$).

> **ee := x-> unapply(dot(diff(x(u,v),u),** **diff(x(u,v),u)),u,v)** **ff := x-> unapply(dot(diff(x(u,v),u),** **diff(x(u,v),v)),u,v)** **gg := x-> unapply(dot(diff(x(u,v),v),** **diff(x(u,v),v)),u,v)**

(Recall that *simplify* is built into the **dot** command, defined earlier.)

> **ww := x-> unapply(simplify(** **ee(x)(u,v)*gg(x)(u,v)-ff(x)(u,v)^2)^** **(1/2),u,v)**

```
ll := x-> unapply(tsp(diff(x(u,v),u,u),
    diff(x(u,v),u),diff(c(u,v),v)))/
    ww(x)(u,v),u,v)
```

The formulas for **mm** and **nn** are the same, except that the double derivative **u,u** is replaced by **u,v** and **v,v**, respectively.

As before, these commands can be saved by

```
save ee,ff,gg,ww,ll,mm,nn,"efgwlmn.m"
```

> Gaussian and mean curvature. When the commands above for E, F, G and L, M, N have been read in, commands for K and H follow immediately from Corollary 4.1 of Chapter 5. However, a faster way to find K for a given patch in $\mathbf{R}^3$ is to use the following command, based on Exercise 20 of Section 5.4. In it, **proc**, for "procedure", begins an enclave—terminated by **end proc**— within which definitions can be made that do not escape to the outside. These temporary definitions allow the final formula to be expressed more concisely.

```
gaussK := proc(x) local xu,xv,xuu,xuv,xvv;
    xu := diff(x(u,v),u);xv := diff(x(u,v),v);
    xuu := diff(x(u,v),u,u);
    xuv := diff(x(u,v),u,v);
    xvv := diff(x(u,v),v,v);
unapply(simplify(factor(
tsp(xuu,xu,xv)*tsp(xvv,xu,xv)-
    tsp(xuv,xu,xv)^2)/
(dot(xu,xu)*dot(xv,xv)-dot(xu,xv)^2)^2),u,v)
    end proc
```

Here **tsp** is the *triple scalar product*, defined earlier. As usual, **gaussK** can be saved for future use.

4. Plots

Maple has three basic plot commands.

(1) The command **plot** has two uses:
   (i) Graphs. If f is a real-valued function defined on $a \leqq t \leqq b$, then **plot(f(t),t=a..b)** draws its graph.
   (ii) Parametric plots. If g is another such function, then the curve with $c(t) = [f(t), g(t)]$ is plotted in $\mathbf{R}^2$ by **plot(c(t), t=a..b)**. Alternatively, **plot([f(t),g(t)],t=a..b)** gives the same result.

Plots can be modified by options, thus: `plot([c(t),t=a..b],` *<option>*), where, for example, the option `numpoints=200` would increase the smoothness of the plot, and `scaling=constrained` imposes the same scale on the axes. Use `?plot[options]` to get many others.

(2) The command `plot3d` also has two uses. Let D be a region $a \leqq u \leqq b$, $c \leqq v \leqq d$ in $\mathbf{R}^2$. Then

    (i)   Graphs. If f is a real-valued function defined on D, its graph is plotted by `plot3d(f(u,v),u=a..b,v=c..d)`.

    (ii) Parametric plots. If $\mathbf{x}: D \to \mathbf{R}^3$ is a list-valued patch or parametrization, its image is plotted by `plot3d(x(u,v),u=a..b, v=c..d)`.

Again, `?plot3d` describes a number of ways to specify plot style.

(3) Parametrized curves in $\mathbf{R}^3$ are plotted using the command "spacecurve" from the *plots* package. As an example: `spacecurve(c(t), t=-2..4)`

To show more than one plot on the same page, each plot should be named, say, `A:=plot3d(x(u,v),u=0..1,v=0..Pi):` with terminal colon to avoid a flood of numbers. Then use "display" from the *plots* package: `display([A,B,C])`.

5. Differential Equations

Explicit solutions in terms of elementary functions are rare, so we describe how to find and plot numerical solutions, which are just as useful in many contexts. In the command for a numerical solution, *Maple* lumps equations and initial conditions into a single *set*, then gives the dependent variables (as follows).

For example, suppose we want to solve numerically the equations

$$x' = f(x, y, t), \quad y' = g(x, y, t)$$

subject to the initial conditions

$$x(t_0) = x_0, \quad y(t_0) = y_0$$

on the interval $a \leqq t \leqq b$. The format is

```
numsol := dsolve(
  {diff(x(t),t)=f(x(t),y(t),t),
  diff(y(t),t)=g(x(t),y(t),t),
  x(t0)=x0,y(t0)=y0},{x(t),y(t)},type=numeric)
```

This solution is plotted by a command from the *plots* package:

$$\texttt{odeplot(numsol,[x(t),y(t)],a..b)}$$

Only now is the domain $a \leqq t \leqq b$ of the solution specified.

Computer Exercises

Chapter 2: 2.2/9, 2.4/11, 14, 15, 19, 20, 2.7/7
Chapter 3: 3.2/5, 3.5/4, 5, 9, 10
Chapter 4: 4.2/5, 6, 11, 4.3/6, 11, 4.6/6, 4.8/10
Chapter 5: 5.4/16, 18–21, 5.6/16, 18, 5.7/8, 9
Chapter 6: 6.5/6, 6.8/11, 13
Chapter 7: 7.2/13, 7.5/9–12, 7.7/12, 13
Chapter 8: 8.1/8

Bibliography

[dC] do Carmo, M. P., *Differential Geometry of Curves and Surfaces*, Prentice-Hall, 1976.

[G] Gray, Alfred., *Modern Differential Geometry of Curves and Surfaces*, 2d ed., CRC Press, 1998.

[Ma] Massey, W. S., *Algebraic Topology: An Introduction*, Springer-Verlag, 1987.

[Mi] Milnor, John W., *Morse Theory*, Princeton University Press, 1963.

[Mu] Munkres, J. R., *Topology*, 2d ed., Prentice-Hall, 2000.

[ST] Singer, I. M., and J. A. Thorpe, *Lectures on Elementary Topology and Geometry*, Springer-Verlag, 1976.

[S] Struik, D. J., *Lectures on Classical Differential Geometry*, 2d ed., Addison-Wesley, 1961. Reprint, Dover, 1988.

[W] Wolfram, S., *The Mathematica Book* (various editions), Wolfram Media.

The book by do Carmo is a clearly written exposition of differential geometry with a viewpoint similar to this one, but at a more advanced level. Gray's book is recommended for readers interested in the use of computers, especially for differential geometry. Both these books have extensive bibliographies.

Answers to Odd-Numbered Exercises

These answers are not complete; and in some cases where a proof is required, we give only a hint.

Chapter 1

Section 1.1

1. (a) $x^2 y^3 \sin^2 z$.
   (c) $2x^2 y \cos z$.
3. (b) $2xe^h \cos(e^h)$, $h = x^2 + y^2 + z^2$.

Section 1.2

1. (a) $-6U_1(\mathbf{p}) + U_2(\mathbf{p}) - 9U_3(\mathbf{p})$.
3. (a) $V = (2z^2/7)U_1 - (xy/7)U_3$.
   (c) $V = xU_1 + 2yU_2 + xy^2 U_3$.
5. (b) Use Cramer's rule.

Section 1.3

1. (a) 0.
   (b) $7 \cdot 2^7$.
   (c) $2e^2$.

3. (a) y^3.
   (c) $yz^2(y^2z - 3x^2)$.
   (e) $2x(y^4 - 3z^5)$.
5. Use Exercise 4.

Section 1.4

1. $\alpha'(\pi/2) = \left(-1, 0, 1/\sqrt{2}\right)$ at $\left(1, 1, \sqrt{2}\right)$.
3. $\beta(s) = \left(1 + s, \sqrt{1 - s^2}, \sqrt{2}\sqrt{1 - s}\right)$.
5. The lines meet at $(11, 7, 3)$.
7. $\mathbf{v}_p = (1, 0, 1)_p$ at $\mathbf{p} = (0, 1, 0)$.

Section 1.5

1. (a) 4.
   (b) −4.
   (c) −2.
3. Use Exercise 2 and $\phi((1/x)V + (1/y)W) = \phi(V)/x + \phi(W)/y$.
5. (b) $(x\,dy - y\,dx)/(x^2 + y^2)$.
7. (a) $dx - dz$.
   (b) not a 1-form.
   (c) $z\,dx + x\,dy$.
9. $\pm(0, 1, 1/2)$.
11. (a) Consider the Taylor series for $t \to f(\mathbf{p} + t\mathbf{v})$.
    (b) Exact: −.420, approximate: −.500.

Section 1.6

1. (a) $\phi \wedge \psi = yz \cos z\,dx\,dy - \sin z\,dx\,dz - \cos z\,dy\,dz$.
   (b) $d\phi = -z\,dx\,dy - y\,dx\,dz$, since $d(dz) = 0$.
7. Apply this definition to the formula following Definition 6.3.
9. For the alternation rule, set $f = y, g = x$.

Section 1.7

1. (c) $(0, 0), (1, 0)$.
3. $F_*(\mathbf{v}) = F(\mathbf{p} + t\mathbf{v})'(0) = 2(p_1v_1 - p_2v_2, v_1p_2 + v_2p_1)$ at $F(\mathbf{p})$.
5. $F_*(\mathbf{v}_p) = F(\mathbf{p} + t\mathbf{v})'(0) = (F(\mathbf{p}) + tF(\mathbf{v}))'(0) = F(\mathbf{v})_{F(p)}$.

7. Using Lemma 4.6 gives $v_p[g(F)] = (d/dt)|_0 g(F(\mathbf{p} + t\mathbf{v})) = F(\mathbf{p} + t\mathbf{v})'(0)[g]$
   $= F*(\mathbf{v}_p)[g]$.

9. (a) $GF = (g_1(f_1, f_2), g_2(f_1, f_2))$.
   (b) $(GF)*(\alpha'(0)) = (GF(\alpha))'(0) = G*(F(\alpha)'(0)) = G*F*(\alpha'(0))$.
   (c) F^{-1} is one-to-one and onto. To show it is regular, start from $F(F^{-1}) = I$, the identity map. Hence $F*(F^{-1})* = I* = $ identity map on tangent vectors. So $(F^{-1})*$ cannot carry a nonzero vector to zero.

Chapter 2

Section 2.1

1. (a) -4.
   (b) $(6, -2, 2)$.
   (c) $(1, 2, -1) / \sqrt{6}, (-1, 0, 3) / \sqrt{10}$.
   (d) $2\sqrt{11}$.
   (e) $-2/\sqrt{15}$.
5. If $\mathbf{v} \times \mathbf{w} = 0$, then $\mathbf{u} \cdot \mathbf{v} \times \mathbf{w} = 0$ for all $\mathbf{u}$; use Exercise 4.
7. $\mathbf{v}_2 = \mathbf{v} - (\mathbf{v} \cdot \mathbf{u}) \mathbf{u}$.

Section 2.2

1. (b) $s(t) = 2t + t^3/3$.
3. $\beta(s) = \left(\sqrt{1 + s^2 / 2}, s / \sqrt{2}, \sinh^{-1}(s / \sqrt{2})\right)$.
7. For (ii), $|h'| = -h' \geq 0$, so the change of variables formula in an integral gives $L(\alpha(h)) = \int_c^d \|\alpha(h)'\| ds = \int_c^d \|\alpha'(h)\|(-h')ds = -\int_c^d \|\alpha'\|h' ds = -\int_b^a \|\alpha'\| dt = \int_a^b \|\alpha'\| dt = L(\alpha)$.
9. $L(\alpha) \approx 12.9153 < 14.1438 \approx L(\beta)$.

Section 2.3

1. $\kappa = 1, \tau = 0, B = -(3, 0, 4)/5$, center $(0, 1, 0)$, radius 1.
7. (a) $1 = \|\alpha(h)'\| = \|\alpha'(h)h'\| = |h'|$, hence $h' = \pm 1$.
   (b) Let $\varepsilon = \pm 1$. Then $\bar{\alpha} = \alpha(h)$ implies $T = \alpha'(h)h' = \varepsilon T(h)$. Hence $\bar{\kappa}\bar{N} = \kappa(h)N(h)$, and so on.

9. For the rectifying plane. From the formula for $\tilde{\beta}$ in the text, delete $\beta(0)$ and the N_0 term. The remaining terms give the same general shape as the curve $(s, \pm s^3)$.

11. (b) First differentiate $B = \overline{B}$; consider the two $\pm$ cases and differentiate again.

Section 2.4

1. (a) Let $f = t^2 + 2$. Then $\kappa = \tau = 2/f^2$ and $B = (t^2, -2t, 2)/f$.
   (c) All the limits are natural unit vectors, $\pm(1, 0, 0), \ldots$
3. (a) $N = (0, -1, 0)$, $\tau(0) = 3/4$.
7. (a) $(\gamma(t) - \alpha(t_0)) \cdot \mathbf{u} = 0$.
   (b) γ has constant speed, so use Exercise 5.
9. Evidently, α is a cylindrical helix. By Exercise 7 its cross-sectional curve γ is a plane curve with constant curvature, hence γ lies in a circle.
11. (c) (*Mathematica*):
    ```
    helix[a_,b_][t_]:={a*Cos[t],a*Sin[t],b*t}
    ParametricPlot3D[{helix[2,1][t],helix[-.5,1]
    [t]}//Evaluate,{t,0,6Pi}]
    ```
    (*Maple*): With the *plots* package installed,
    ```
    helix:=(a,b)->[a*cos(t),a*sin(t),b*t]
    spacecurve({helix(2,1)(t),helix(-.5,1)(t)},
    t=0..6*Pi,numpoints=100)
    ```
    Recall that we do not show *Maple*'s mandatory terminal semicolon.
13. (b) $\lambda_t(s) = \alpha(t) + s(\alpha'(t) \cdot \alpha'(t)/\alpha''(t) \cdot J(\alpha'(t)) J(\alpha'(t))$ for $0 \leq s \leq 1$.
    (c) For α unit speed, $\lambda_t(s) = \alpha + s(1/\tilde{\kappa})N$. Hence $d\lambda_t/ds = (1/\tilde{\kappa})N$ (independent of s). Evidently this is normal to α at $\alpha(t)$. Since $\alpha^* = \alpha + (1/\tilde{\kappa})N$, we get $(\alpha^*)' = T + (1/\tilde{\kappa})'N - T = (1/\tilde{\kappa})'N$, in agreement with $d\lambda_t/ds$ at $\alpha^*(1)$.
15. (a) For the rectifying plane (orthogonal to N):
    (*Mathematica*):
    ```
    viewN[a_,eps_]:=ParametricPlot[{(a[t]-a[0]).
    tang[a][0],
      (a[t]-a[0]).binor[a][0]}//Evaluate,
        {t,-eps,eps}]
    ```
    (*Maple*)
    ```
    viewN:=(a,eps)->plot([dot((a(t)-a(0)),
    tang(a)(0)),
      dot((a(t)-a(0)),binor(a)(0)),t=-eps..eps])
    ```

(b) (iii) For all curves with $\tau(0) \neq 0$ there are essentially only two cases, depending on the sign of τ.

17. (a) $\pi/\sqrt{2}$.

    (b) ∞.

    (c) $\pi/\sqrt{2}$.

    (d) 2π (see Exercise 18).

19. (c) For a suitable n, let τ_n be τ with new z-component $(1/n)\sin 3t$. Here $\tilde{\kappa} = \kappa$, and in the notation of Exercise 12, $ds/dt = \sqrt{x'^2 + y'^2}$.

21. Use Theorem 4.6. By hand computation (easy, if κ and τ are first found by computer), we get $\tau/\kappa = (3ac/2b^2)(P/Q)^{3/2}$, where
$$P = 9c^2t^4 + 4b^2t^2 + a^2 \quad \text{and} \quad Q = 9c^2t^4 + (9a^2c^2/b^2)t^2 + a^2$$
Thus τ/κ is constant if and only if $4b^2 = 9a^2c^2/b^2$, that is, $3ac = \pm 2b^2$. (Hence $\tau/\kappa = \pm 1$).

Section 2.5

1. (a) $2U_1(\mathbf{p}) - U_2(\mathbf{p})$.

    (b) $U_1(\mathbf{p}) + 2U_2(\mathbf{p}) + 4U_3(\mathbf{p})$.

5. $\nabla_{\alpha'(t)}W = \Sigma\alpha'(t)[w_i]U_i = \Sigma(d/dt)(w_i(\alpha))(t)U_i = (W_\alpha)'(t)$.

Section 2.6

1. Show that $V \cdot \tilde{W} = 0$, and use Lemma 1.8.

3. For instance, $E_2 = -\sin z\, U_2 + \cos z\, U_3$ and $E_3 = E_1 \times E_2$.

Section 2.7

1. $\omega_{12} = 0$, $\omega_{13} = \omega_{23} = df/\sqrt{2}$.

3. $\omega_{12} = -df$, $\omega_{13} = \cos f\, df$, $\omega_{23} = \sin f\, df$.

5. By Corollary 5.4(3), $\nabla_V(\Sigma f_i E_i) = \Sigma V[f_i]E_i + f_i\nabla_V E_i$

7. (*Mathematica*):

    (a) `connform[A_]:=Simplify[Dt[A].Transpose[A]]`

    (b) In A, write q for ϑ and f for φ. Then in `MatrixForm` `[connform[A]]`, read `Dt[q]` as `dq`.

(*Maple*): Install the packages *LinearAlgebra* and *difforms*. With q and f as above, write `defform(q=0,f=0)` to identify them as real-valued functions.

    (a) `connform:=A->simplify(Multiply(map(d,A),` `Transpose(A)))`

Section 2.8

3. (a) Compute $\theta = A \, d\xi$, as in the text. (A was found in Section 7.)
   (b) For example, $E_1[r] = dr[E_1] = \theta_1(E_1) = 1$.
   (c) Use the appropriate form of the chain rule.

Chapter 3

Section 3.1

3. $(T_a)^{-1} = T_{-a}$, and since C is orthogonal, $C^{-1} = {}^t C$. Thus $F^{-1} = (T_a C)^{-1} = C^{-1} (T_a)^{-1} = {}^t C T_{-a}$. By Exercise 1, this equals $T_{tC\,(-a)}{}^t C = T_{-tC(a)}{}^t C$.

5. (b) Using Exercise 3 we find $F^{-1}(\mathbf{p}) = \left(5\sqrt{2}, -2, 3\sqrt{2}\right)$

7. Use Exercises 2 and 3.

9. (a) For ϑ such that $C(1, 0) = (\cos \vartheta, \sin \vartheta)$, C has matrix
$$\begin{pmatrix} \cos \vartheta & \mp \sin \vartheta \\ \sin \vartheta & \pm \cos \vartheta \end{pmatrix}.$$
   (b) $O(1)$ consists of $+1$ and -1, so $F(t) = a \pm t$ for any number a.

Section 3.2

1. $T(\mathbf{v}_p) = \mathbf{v}_{T(p)}$.

3. The middle row of C is $(-2, 1, 2)/3$, and T is translation by $\left(3, -4/3, 1 - 2\sqrt{2}/3\right)$

5. (*Mathematica*):
   Let `ame={e1,e2,e3}` and `amf={f1,f2,f3}` be the attitude matrices of the frames in Exercise 3.
   (b) Set `cc:=Simplify[Transpose[amf].ame]` Then `Simplify[cc.e1]` is `f1`, etc.
   (*Maple*):
   Install the package *LinearAlgebra*, and let `ame=Matrix([e1,e2, e3])` and `amf=Matrix([f1,f2,f3])` be the attitude matrices of the frames in Exercise 3.
   (b) Set `cc:=simplify(Multiply(Transpose(amf), ame))`. Then `simplify(Multiply(cc,Vector(e1)))` is `Vector(f1)`, etc.

Section 3.3

1. If the orthogonal parts of F and G are A and B, then by Exercise 2 of Section 1, $\text{sgn}(FG) = \det AB = (\det A)(\det B) = \det BA = \text{sgn}(GF)$. Then $+1 = \text{sgn}I = \text{sgn}(FF^{-1}) = \text{sgn}(F)\text{sgn}(F^{-1})$.
5. C is rotation through angle $\pi/2$ about the axis given by $\mathbf{a}$.

Section 3.4

1. (b) By definition, $\sigma(s)$ is the point canonically corresponding to $T(s)$; hence by Exercise 1 of Section 2, $C(\sigma)$ corresponds to $F_*(T)$, the unit tangent of $F(\beta)$.
3. Translate each triangle so that its new first vertex is at the origin. A sketch will show that the required C is orientation-reversing, and we find $C =$
$$\begin{pmatrix} -3/5 & 4/5 \\ 4/5 & 3/5 \end{pmatrix}$$
5. For a tangent vector $\mathbf{v}$ at $\mathbf{p}$,
$$F_*(\nabla_v W) = F_*(W(\mathbf{p}+t\mathbf{v})'(0)) = (\overline{W}(F(\mathbf{p})+tC(\mathbf{v}))'(0) = \nabla_{F_*(v)}\overline{W}.$$

Section 3.5

3. Take $a = 2$, $b = \pm 2$.
5. Yes, since c has constant speed, curvature, and torsion.
7. $\beta(s) = \left(\int \cos \varphi(s)ds, \int \sin \varphi(s)ds \right)$, where $\varphi(s) = \int f(s)ds$
9. For simplicity, assume $a \leq 0 \leq b$; then:
   (*Mathematica*):
   (a) ```
kdetc[f_,a_,b_]:=
    NDSolve[{x'[s]==Cos[phi[s]],
    y'[s]==Sin[phi[s]],
    phi'[s]==f[s],x[0]==0,y[0]==0,
    phi[0]==0},{x,y,phi},{s,a,b}]
```
   (b) ```
draw[f_,a_,b_]:=ParametricPlot[Evaluate
    [{x[s],y[s]}/.kdetc[f,a,b]],{s,a,b},
    AspectRatio->Automatic]
```
   (*Maple*):
   (a) ```
kdetc:=f->dsolve({diff(x(s),s)=cos(phi(s)),
    diff(y(s),s)=sin(phi(s)),diff(phi(s),s)=f(s),
    x(0)=0,y(0)=0,phi(0)=0},{x(s),y(s),
    phi(s)},type=numeric)
```
```

(b) Install *plots*. Define `draw:=(f,a,b)->odeplot(kdetc(f),` `[x(s),y(s)],a..b,scaling=constrained)`.

Chapter 4

Section 4.1

1. (a) The vertex.
   (b) All points on the circle $x^2 + y^2 = 1$.
   (c) All points on the z axis.
5. (b) $c \neq -1$.
9. Use Exercise 7.
11. $\mathbf{q}$ is in $F(M)$ if and only if $F^{-1}(\mathbf{q})$ is in M, that is, $g(F^{-1}(\mathbf{q})) = c$. Use the Hint to apply Theorem 1.4.

Section 4.2

1. (c) The Monge patch $\mathbf{x}(u, v) = (u, v, u^2 + v^2)$ covers the entire surface; a parametrization based on Example 2.4 omits the point $(0, 0, 0)$.
3. $\mathbf{x}_u \times \mathbf{x}_v = v\delta'(u) \times \delta(u)$.
5. (a) $EG - F^2 = b^2 + u^2$ is never zero.
   (b) Helices and straight lines (rulings).
   (c) H: $x\sin(z/b) - y\cos(z/b)$.
   (d) For $\mathbf{x}$ as given:
      (*Mathematica*): `ParametricPlot3D[x[u,v]//`
      `Evaluate,{u,-1,1},{v,0,2Pi}]`

      (*Maple*): `plot3d(x(u,v),u=-1..1,v=0..2*Pi)`
7. (b) $\mathbf{x}(u, v) = (\cos u - v\sin u, \sin u + v\cos u, v)$.
9. In all cases, (i) check that the three partial derivatives of the defining function g are never zero simultaneously on M: $g = 1$ (Theorem 1.4), and (ii) First, check that the components of $\mathbf{x}$ satisfy the equation $g = 1$.
11. (c) $\mathbf{x}_{\pm}(u, v) = (a\cos u, b\sin u, 0) \pm v(-a\sin u, b\cos u, c)$.
   (d) (*Mathematica*):
      `xplus[u_,v_]:={1.5*(Cos[u]-v*Sin[u]),`
      `  Sin[u]+v*Cos[u],2v}`
      `ParametricPlot3D[xplus[u,v]//Evaluate,`
      `  {u,0,2Pi},{v,-1,1}]`
      (*Maple*): `xplus:=(u,v)→[1.5*(cos(u)-v*sin(u)),`
      `sin(u)+v*cos(u),2*v]`
      `   plot3d(xplus(u,v),u=0..2*Pi,v=-1..1)`

Section 4.3

1. (a) $r^2\cos^2 v$.
   (b) $r^2(1 - 2\cos^2 v \cos u \sin u)$.
3. (a) $\bar{u}$ and $\bar{v}$ are the Euclidean coordinate functions of $\mathbf{x}^{-1}\mathbf{y}$.
   (b) Express $\mathbf{y} = \mathbf{x}(\bar{u}, \bar{v})$ in terms of Euclidean coordinates, and differentiate.
5. (a) M is given by $g = z - f(x, y) = 0$, with $\nabla g = (-f_x, -f_y, 1)$, and $\mathbf{v}$ is tangent to M at $\mathbf{p}$ if and only if $\mathbf{v} \cdot \nabla g(\mathbf{p}) = 0$.
7. $\nabla g = (-y, -x, 1)$ is a normal vector field; V is a tangent vector field if and only if $V \cdot \nabla g = 0$, for example, $V = (0, 1, x)$.
9. (a) $\bar{T}_p(M)$ consists of all points $\mathbf{r}$ such that $(\mathbf{r} - \mathbf{p}) \cdot \mathbf{z} = 0$; hence $\mathbf{v}_p$ is in $T_p(M)$ (that is, $\mathbf{v} \cdot \mathbf{z} = 0$) if and only if $\mathbf{p} + \mathbf{v}$ is in $\bar{T}_p(M)$.
11. (a) If $a/b = m/n$ for integers m, n, consider $\Delta t = 2\pi m/a = 2\pi n/b$.
    (b) Assume $\alpha(s) = \alpha(t)$ for $s \neq t$, so $\mathbf{x}(as, bs) = \mathbf{x}(at, bt)$. Equality for z components and for $x^2 + y^2$ implies $as - at = 2\pi m$ and $bs - bt = 2\pi n$ for some integers m, n. Thus $a/b = m/n$, a contradiction.

Section 4.4

1. $d(f\phi)(\mathbf{x}_u, \mathbf{x}_v) = \dfrac{\partial(f(\mathbf{x}))}{\partial u}\phi(\mathbf{x}_v) - \dfrac{\partial(f(\mathbf{x}))}{\partial v}\phi(\mathbf{x}_u) + f(\mathbf{x})\left[\dfrac{\partial}{\partial u}\phi(\mathbf{x}_v) - \dfrac{\partial}{\partial v}\phi(\mathbf{x}_u)\right]$

$= (df \wedge \phi + fd\phi)(\mathbf{x}_u, \mathbf{x}_v)$.

3. If α is a curve with initial velocity $\mathbf{v}$ at $\mathbf{p}$, then

$$\mathbf{v}_p[g(f)] = (gf\alpha)'(0) = g'(f\alpha)(0)(f\alpha)'(0) = g'(f(\mathbf{p}))\mathbf{v}_p[f].$$

5. On the overlap of $\mathcal{U}_i$ and $\mathcal{U}_j$, $df_i - df_j = d(f_i - f_j) = 0$.

7. (b) $d\bar{u}(\mathbf{x}_u) = \mathbf{x}_u[\bar{u}] = \dfrac{\partial(\bar{u}(\mathbf{x}))}{\partial u} = \dfrac{\partial u}{\partial u} = 1$.

Section 4.5

1. If $\mathbf{x}: D \to M$ is a patch, then $F(\mathbf{x}): D \to N$ is (by Theorem 3.2) a differentiable mapping. Hence $\mathbf{y}^{-1}F\mathbf{x}$ is differentiable for any patch $\mathbf{y}$ in N.
3. If $\bar{\mathbf{x}}$ and $\bar{\mathbf{y}}$ are patches in M and N, respectively, note that $\bar{\mathbf{y}}^{-1}F\bar{\mathbf{x}} = (\bar{\mathbf{y}}^{-1}\mathbf{y})(\mathbf{x}^{-1}\bar{\mathbf{x}})$ is differentiable, being a composition of differentiable functions.
5. By Exercise 1, A is differentiable. Since $A^2 = I$, $A^{-1} = A$, so A is a diffeomorphism. For A_*, consider its effect on a curve $t \to \cos t\,\mathbf{p} + \sin t\,\mathbf{u}$ in Σ.
7. Theorem 5.4.

9. (a) Use Exercise 8.

(b) $F_*(a\mathbf{x}_u + b\mathbf{x}_v) = a\mathbf{y}_u + b\mathbf{y}_v$ implies linearity.

11. M is diffeomorphic to a torus if the profile curve α of M is closed, and to a cylinder if α is one-to-one. With parametrizations as suggested, $F(\mathbf{x}(u, v)) = \mathbf{y}(u, v)$ is a diffeomorphism.

13. (a) If $\mathbf{p}$ in M, there is a $\mathbf{q}$ in $\tilde{M}$ such that $\mathbf{p} = G(\mathbf{q})$. By consistency, $F(\mathbf{p}) = \tilde{F}(\mathbf{q})$ is a valid definition. G is regular, hence locally has differentiable inverse mappings. Thus, locally $F = \tilde{F}G^{-1}$ so F is differentiable.

(b) If $F(\mathbf{p}_1) = F(\mathbf{p}_2)$, then for $\mathbf{q}_1$, $\mathbf{q}_2$ in $\tilde{M}$ such that $G(\mathbf{q}_1) = \mathbf{p}_1$, $G(\mathbf{q}_2) = \mathbf{p}_2$, we have $F(G(\mathbf{q}_1)) = F(G(\mathbf{q}_2))$. Thus $\tilde{F}(\mathbf{q}_1) = \tilde{F}(\mathbf{q}_2)$. Then the hypothesis gives $G(\mathbf{q}_1) = G(\mathbf{q}_2)$, that is, $\mathbf{p}_1 = \mathbf{p}_2$.

Section 4.6

3. (b) Use Theorem 6.2.

5. (a) Let $r(t) = \|\alpha(t)\|$. Then let $f = U_1 \cdot \alpha/\|\alpha\|$ and $g = U_2 \cdot \alpha/\|\alpha\|$. Apply Exercise 12 of Section 2.1 to get ϑ.

(b) $\vartheta(a)$ and $\vartheta(b)$ measure the same angle; hence they differ by some integer multiple of 2π.

(c) Use Exercise 1 to evaluate ψ on the polar expression for α in (a).

(d) $\dfrac{\det(\alpha, \alpha')}{\alpha \cdot \alpha} = \begin{vmatrix} f & g \\ f' & g' \end{vmatrix} \Big/ (f^2 + g^2) = \dfrac{fg' - gf'}{f^2 + g^2}$.

7. (a) Since $(F_*(\phi))(\alpha') = \phi((F_*)(\alpha')) = \phi(F(\alpha)')$, we get

$$\int_\alpha F_*(\phi) = \int_a^b \phi\big(F(\alpha)'\big)\, dt = \int_{F(\alpha)} \phi.$$

9. (a) $2\pi m$, (b) $2\pi n$.

13. The text shows that if ϕ is the dual of V, then $\int V \cdot ds = \int \phi$. The dual of curl V is $d\phi$, and $dA \approx W\, du\, dv$. It follows that

$$U \cdot \operatorname{curl} V\, dA = \operatorname{curl} V \cdot \frac{\mathbf{x}_u \times \mathbf{x}_v}{W} W\, du\, dv = \phi(\mathbf{x}_u, \mathbf{x}_v)\, du\, dv.$$

Section 4.7

1. (a) Connected, not compact.

(c) Not connected and not compact.

(e) Connected and compact.

3. If v is nonvanishing on N, show that $F^*(v)$ is nonvanishing on M.

5. (a) All—by Definition 7.1.

(b) Sphere, torus—by Lemma 7.2.

(c) All—by Proposition 7.5

(d) Plane, sphere (see text).

9. (c) If M is connected, then path-connectedness (Definition 7.1) follows using parts (a) and (b). If M is path-connected, let $\mathcal{U}$ and $M - \mathcal{U}$ be open sets of M such that $\mathcal{U}$ contains a point $\mathbf{p}$.

   Assume that $M - \mathcal{U}$ contains a point $\mathbf{q}$. There is a curve segment $\alpha{:}[a, b] \to M$ from $\mathbf{p}$ to $\mathbf{q}$. Since α is continuous, $\alpha^{-1}(\mathcal{U})$ and $\alpha^{-1}(M - \mathcal{U})$ are disjoint open sets filling $[a, b]$. This contradicts the stated connectedness of $[a, b]$.

11. Fix $\mathbf{q}$ in $M - \mathcal{R}$; then by the Hausdorff axiom, for each $\mathbf{p}$ in $\mathcal{R}$, there are disjoint neighborhoods $\mathcal{U}_p$ of $\mathbf{p}$ and $\mathcal{U}_{q,p}$ of $\mathbf{q}$. By compactness, a finite number of the neighborhoods $\mathcal{U}_p$ cover $\mathcal{R}$. Then the intersection of the corresponding neighborhoods $\mathcal{U}_{p,q}$ is a neighborhood of $\mathbf{q}$ that does not meet $\mathcal{R}$.

Section 4.8

1. If M is orientable it has a nonvanishing 2-form μ. Then $f(t) = \mu(\alpha'(t), Y(t))$ is a differentiable function on $[a, b]$. By (ii), $f(a)f(b) < 0$; hence f is somewhere zero on $a < t < b$. This contradicts (i).

5. (a) The function $\mathbf{p} \to d(\mathbf{0}, \mathbf{p})$ is continuous on M, hence takes on a maximum.

7. (i) Since M is nonorientable, there is a reversing loop (as in the hint) at some point $\mathbf{q}$. Fix U_q. Then every point U_p in $\hat{M}$ can be connected to U_q by a curve in $\hat{M}$. *Proof:* Move U_p along a curve from $\mathbf{p}$ to $\mathbf{q}$. If the result is $-U_q$, move it around the reversing loop.

9. (b) $B - \beta$ is diffeomorphic to an ordinary band.

11. (a) Recall that a neighborhood in a surface is the image under a coordinate patch of a neighborhood in $\mathbf{R}^2$. Evidently every neighborhood $\mathbf{x}(\mathcal{U})$ of $\mathbf{0}$ meets every neighborhood $\mathbf{y}(\mathcal{V})$ of $\mathbf{0}^*$.

    (b) The sequence $\{(1/n, 0)\}$ converges to $\mathbf{0}$ when expressed in terms of $\mathbf{x}$, and to $\mathbf{0}^*$ in terms of $\mathbf{y}$.

    (c) Relative to $\mathbf{x}$ and $\mathbf{y}$, the coordinate form of F is the identity map.

13. (a) In terms of the natural coordinates, $\alpha'(t) = V(\alpha(t))$ becomes

$$u' U_1 + v' U_2 = f_1(u, v)U_1 + f_2(u, v)U_2.$$

    (b) The differential equations are $u' = -u^2$, $v' = uv$, and the initial conditions are $u(0) = 1$, $v(0) = -1$. The first differential equation integrates to $1/u = t + A$. But $u(0) = 1$, so $u = 1/(t + 1)$. Thus we get $v' = v/(t + 1)$, which integrates to $v = B(t + 1)$. Then $v(0) = -1$ implies $v = -(t + 1)$.

15. Smooth overlap follows from the identity
$$(\mathbf{x} \times \mathbf{y})(\overline{\mathbf{x}} \times \overline{\mathbf{y}})^{-1} = (\mathbf{x}\overline{\mathbf{x}}^{-1}) \times (\mathbf{y}\overline{\mathbf{y}}^{-1}).$$

Chapter 5

Section 5.1

1. Use Method 1 in the text.
3. (a) 2.
   (c) 1.
5. Meridians go to meridians (great circles through the poles), parallels to parallels—except for the top and bottom circles of the torus.
7. Use Method 1 and the definition of tangent map in Chapter 1.

Section 5.2

1. (b) If e_1, $e_2 = (\mathbf{u}_1 \pm \mathbf{u}_2)/\sqrt{2}$, then $S(e_1) = e_1$ and $S(e_2) = -e_2$.

Section 5.3

1. $k_1 k_2 \leqq 0$ and $k_1 = k_2$ imply $k_1 = k_2 = 0$.
5. (b) $K > 0$: an ellipse on one side and no points on the other. $K < 0$: the two branches of a hyperbola. $K = 0$, nonplanar: two parallel lines on one side, no points on the other.
7. (a) If α is a curve with initial velocity $\mathbf{v}$ at $\mathbf{p}$, then $F_*(\mathbf{v}) = F(\alpha)'(0) = (\alpha + \varepsilon U_\alpha)'(0) = \mathbf{v} - \varepsilon S(v)$ at $F(\mathbf{p})$.

Section 5.4

1. $W = r^2 \cos v > 0$, $U = \mathbf{x}/r$, $K = 1/r^2$, $H = -1/r$.
5. Use $\alpha' = a_1'\mathbf{x}_u + a_2'\mathbf{x}_v$ to find speed.
7. $K = -36r^2/(1 + 9r^4)^2$.
9. Expand $S(\mathbf{v}) \times \mathbf{v}$. This vector is zero if and only if its dot product with $\mathbf{x}_u \times \mathbf{x}_v$ is zero. Use the Lagrange identity (Exercise 6 of Section 3).
11. $k(\mathbf{u}) = S(\mathbf{v}) \cdot \mathbf{v}/\mathbf{v} \cdot \mathbf{v}$. Substitute $\mathbf{v} = v_1\mathbf{x}_u + v_2\mathbf{x}_v$.
15. (a) K is negative except at the origin, but this is a planar point, hence an umbilic with $k = 0$.

(b) The hint leads to $\left(0, \pm(b/2)\sqrt{a^2-b^2}, (a^2-b^2)/4\right)$. These two umbilics reduce to one for the paraboloid of rotation, $a = b$, where (by symmetry) we expect $\mathbf{0}$ to be umbilic.

17. (b) Since $\kappa < B$, if $\varepsilon < 1/B$, then $\mathbf{x}_u \times \mathbf{x}_v \neq 0$.

(c) $S(\mathbf{x}_u) \times S(\mathbf{x}_v) = -\kappa \cos v\, T \times \mathbf{x}_v/\varepsilon$.

19. (*Mathematica*):

(b) `hyperboloid[a_,b_,e_][u_,v_]:=`
    `{u,v,u^2/a^2+e*v^2/b^2}`

(c) `monkeypolar[r_,q_]:=monkey[r*Cos[q],`
    `r*Sin[q]]`

(*Maple*):

(b) `hyperboloid:=(a,b,e)->unapply([u,v,u^2/a^2+`
    `e*v^2/b^2],u,v)`

(c) `monkeypolar:=(r,q)->monkey(r*cos(q),`
    `r*sin(q))`

21. *Maple* has a built-in tube command in the *plots* package. For (c), with τ defined as in the exercise referred to, the tube is plotted by
    `tubeplot(τ(t),t=0..2*Pi,radius=0.5)`
    (*Mathematica*):

(a) With the commands for unit normal and binormal installed (see Appendix), a tube formula is
    `tube[c_,r_][t_,phi_]:=c[t]+r*(Cos[phi]*`
    `nor[c][t]+Sin[phi]*binor[c][t])`
    This is plotted—in (b), for example—by
    `ParametricPlot3D[tube[helix,1/2][t,phi]//`
    `Evaluate,{t,0,4Pi},{phi,0,2Pi},`
    `PlotPoints->{40,20},Axes->None,Boxed->`
    `False]`

(c) If the general approach in (a) is slow in this case, a faster way is to copy the *outputs* of `binor[τ][t,phi]` and `nor[τ][t,phi]` into an explicit definition of the tube function of τ.

Section 5.5

3. (a) The critical points of K are those of h. They occur at the intercepts of M with the coordinate axes.

(b) For the ellipsoid, $c^2/(a^2b^2) \leq K \leq a^2/(b^2c^2)$. (Note again the effect of $a = b = c$.)

5. (c) Use $Z = \text{grad}(e^z \cos x - \cos y)$ and $W = Z \times V$. Then $\nabla_V Z \times W + V \times \nabla_W Z = 0$ and $V \cdot \nabla_V Z \times \nabla_W Z = -e^{2z}$.

7. (a) Use $Z = \Sigma(x_i/a_i)U_i$.
   (b) The tangency condition for a vector $\mathbf{v}$ at $\mathbf{p}$ is $\Sigma v_i p_i/a_i^2 = 0$.

Section 5.6

3. Use Remark 6.10.
5. Since $U \cdot V$ is constant, $U' \cdot V + U \cdot V' = 0$. If α is principal in M, then using Lemma 6.2, $U' \cdot V = 0$, hence $V \cdot U' = 0$. Continue as for Lemma 6.3.
7. $S(T) = -U'$; hence by orthonormal expansion, $U' = -S(T) \cdot T\,T - S(T) \cdot V\,V$. Continue as in the proof of the Frenet formulas.
11. (a) Set $\sigma = \alpha + f\delta$. Then f is determined using the equation $\sigma' \cdot \delta = 0$.
    (b) $\delta' \perp \delta$, α' implies that $\alpha' \times \delta$ and δ' are collinear. Then $\alpha' \times \delta = p\delta'$. Hence $\mathbf{x}_u \times \mathbf{x}_v = p\delta' + v\delta' \times \delta$, so $W^2 = (p^2 + v^2)\delta' \cdot \delta'$. Now use Exercise 12 of Section 4.
    (c) On each ruling, K has a unique minimum point; the striction curve meets the ruling at this point.
13. (a) Since $\sigma(u + \varepsilon) - \sigma(u) \approx \varepsilon\sigma'(u)$, the Hint gives $d_\varepsilon = \varepsilon\sigma'(u) \cdot \delta(u) \times \delta'(u)/\|\delta(u) \times \delta'(u)\|$. However $\|\delta(u) \times \delta'(u)\|\varepsilon \approx \|\delta(u) \times \delta(u + \varepsilon)\| = \sin \vartheta_\varepsilon \approx \vartheta_\varepsilon$. Since $\|\delta(u) \times \delta'(u)\|^2 = \delta' \cdot \delta'$, we see that $\lim_{\varepsilon \to 0} d_\varepsilon/\vartheta_\varepsilon = \sigma' \cdot \delta \times d'/\delta' \cdot \delta' = p$.
15. Compute E, F, G and L, M, N. (Computer formulas for these are given in the Appendix.) Then $EG - F^2 \neq 0$ proves (a), and $F = M = 0$ proves (b).
17. (a) $K = -h'^2\vartheta'^2/W^4$, $H = u(h'\vartheta'' - \vartheta'h'')/(2W^3)$, where $W^2 = h'^2 + u^2\vartheta'^2$.
    (b) $\delta \times \delta' = \vartheta'U_3$. Since K is a minimum when $u = 0$, the z axis is the striction curve, and $p = h'/\vartheta'$, reciprocal of turn rate (Exercise 13 of Section 6).
19. Use $W = \|\mathbf{x}_u \times \mathbf{x}_v\|$.

Section 5.7

1. $K = (1 - x^2)(1 + x^2\exp(-x^2))^{-2}$. Hence $K > 0 \Leftrightarrow -1 < x < 1$.
3. In a canonical parametrization, if g is constant, the profile curve is orthogonal to the axis, so the surface M is part of a plane. Otherwise, $K = 0 \Leftrightarrow h'' = 0 \Leftrightarrow h'$ is constant. If $h' = 0$, the profile curve lies in a line parallel to the axis, so M is part of a cylinder. If $h' \neq 0$, the profile curve is a slanting line, so M is part of a cone.
5. M has parametrization $\mathbf{x}(r, v) = (r\cos v, r\sin v, f(r))$. Then $E = 1 + f'^2$, $F = 0$, $G = r^2$, and $W_L = rf''$, $W_M = 0$, $W_N = r^2f'$, with $W^2 = EG - F^2 = r^2(1 + f'^2)$.

7. (a) $h(u) = a \sinh(u/c)$ satisfies the given differential equation with $K = -1/c^2$. Use the integral formula for $g(u)$. Then as $u \to 0$, the slope angle $\tan \varphi = h'/g'$ approaches $(a/c)/\sqrt{1 - a^2/c^2} = a/\sqrt{c^2 - a^2}$. The curve becomes vertical when $g' = 0$, hence the integrand of g vanishes. There $\cosh^2(u^*/c) = c^2/a^2$, so $h_{\max} = a \sinh(u^*/c) = \sqrt{c^2 - a^2}$.

(b) $h(u) = ce^{-u/c}$ satisfies the differential equation and initial condition in Example 7.6.

Chapter 6

Section 6.1

1. (a) $\alpha'' = \omega_{12}(T)E_2 + \omega_{13}(T)E_3$. Hence α'' is normal to M if and only if $\omega_{12}(T) = 0$.
3. Apply the symmetry equation to E_1, E_2. Then use Corollary 1.5.

Section 6.2

1. (a) $\theta_1 = dz$, $\theta_2 = r \, d\vartheta$.
   (d) $K = 0$ and $H = -1/2r$.

Section 6.3

1. If $K = H = 0$, then $k_1 k_2 = k_1 + k_2 = 0$. Thus $k_1 = k_2 = 0$, so $S = 0$.
3. In the proof of Liebmann's theorem, replace the constancy of $K = k_1 k_2$ by that of $2H = k_1 + k_2$.
5. In the case $k_1 \ne k_2$, use Theorem 2.6 to show that, say, $k_1 = 0$. By Exercise 2 the k_1 principal curves are straight lines. Show that the k_2 principal curves are circles and that the (k_1) straight lines are parallel in $\mathbf{R}^3$.

Section 6.4

1. (d) $\Rightarrow$ (b): If $\mathbf{z}$ is an arbitrary tangent vector at $\mathbf{p}$, write $\mathbf{z} = a\mathbf{v} + b\mathbf{w}$. Then

$$\|F_*\mathbf{z}\|^2 = a^2 \|F_*\mathbf{v}\|^2 + 2ab F_*\mathbf{v} \cdot F_*\mathbf{w} + b^2 \|F_*\mathbf{w}\|^2$$
$$= a^2 \|\mathbf{v}\|^2 + 2ab\mathbf{v} \cdot \mathbf{w} + b^2 \|\mathbf{w}\|^2 = \|\mathbf{z}\|^2.$$

3. (b) Monotone reparametrization does not affect length of curves.

(c) By the definition of ρ, given any $\varepsilon > 0$ there is a curve segment α from $\mathbf{p}$ to $\mathbf{q}$ of length $< \rho(\mathbf{p}, \mathbf{q}) + \varepsilon$, and an analogous β for $\mathbf{q}$ and $\mathbf{r}$. Combining α and β gives a piecewise differentiable curve segment from $\mathbf{p}$ to $\mathbf{r}$. (If only everywhere-differentiable curves are allowed, there is no change in ρ, but proofs are harder.)

5. (a) Define $F(\alpha(u) + vT(u)) = \beta(u) + vT(u)$.

(b) Choose β in $\mathbf{R}^2$ with plane curvature equal to κ.

7. By the exercise mentioned, a shortest curve in $\mathbf{R}^2$ joining the points parametrizes a straight line segment. Thus any curve in M joining the points has length $L > 2$.

9. $F_*((F^{-1})_*\mathbf{v}) = (FF^{-1})_*\mathbf{v} = I_*\mathbf{v} = \mathbf{v}$. Since F is an isometry, $\|(F^{-1})_*\mathbf{v}\| = \|\mathbf{v}\|$.

11. Write $F(\mathbf{x}(u, v)) = \tilde{\mathbf{x}}(f(u), g(v))$ for suitable parametrizations.

13. For $\mathbf{y}$, show that the conditions $E = G$ and $F = 0$ are equivalent to $g' = \cos g$, which has solution $g(v) = 2\tan^{-1}(e^v) - \pi/2$ such that $g(0) = 0$. Use criteria suggested by Exercise 8.

15. $F(\mathbf{x}(u, v)) = (f(u)\cos v, f(u)\sin v)$, where $\mathbf{x}$ is a canonical parametrization and $f(u) = \exp\left(\int_1^u dt/h(t)\right)$.

Section 6.5

1. First show that α is a geodesic if and only if $\omega_{12}(\alpha') = 0$. Let $\overline{E}_1$, $\overline{E}_2$ be the transferred frame field, with connection form $\overline{\omega}_{12}$. Since $\overline{E}_1 = F_*(\alpha')$, $F_*(\alpha') = F(\alpha)'$, Lemma 5.3 gives

$$0 = \omega_{12}(\alpha') = F_*(\overline{\omega}_{12})(\alpha') = \overline{\omega}_{12}(F_*(\alpha')) = \overline{\omega}_{12}(F(\alpha)').$$

3. There is no local isometry of the saddle surface M $(-1 \leq K < 0)$ onto a catenoid with $-1 \leq \overline{K} < 0$—or vice versa—since K has an isolated minimum point, at $\mathbf{0}$, while $\overline{K}$ takes on each of its values on entire circles. Many other examples are possible.

5. (b) Follows from Lemma 4.5, since computation for $\mathbf{x}_t$ shows $E_t = \cosh^2 u = G_t$ and $F_t = 0$.

(d) For M_t, $U_t = (s, -c, S)/C$, so the Euclidean coordinates of U_t are independent of t.

7. A local isometry must carry minimum points of K_H to minimum points of K_C, and also preserve orthogonality and geodesics.

Section 6.6

1. (b) $\theta_1 = \sqrt{1 + u^2}\, du$, $\theta_2 = u\, dv$, $\omega_{12} = dv/\sqrt{1 + u^2}$, $K = 1/(1 + u^2)^2$.

3. (b) Substitution into $d\omega_{13} = \omega_{12} \wedge \omega_{23}$ leads to

$$L_v = \frac{E_v}{2}\left(\frac{L}{E} + \frac{N}{G}\right) = HE_v.$$

Section 6.7

1. $1 + f_u^2 + f_v^2 \geq 1$.

3. (a) $\iint_T v = \int_0^{2\pi} dv \int_0^{2\pi} (R^2 + r^2 + 2Rr \cos u)du = 4\pi^2(R^2 + r^2)$.

(b) $\mathbf{x}_u \times \mathbf{x}_v$ points inward, and thus $U \cdot \mathbf{x}_u \times \mathbf{x}_v = -\|\mathbf{x}_u \times \mathbf{x}_v\|$
$= -\sqrt{EG - F^2}$. Hence $\int_T v = -\text{area}(T) = -4\pi^2 Rr$.

5. F carries positively oriented pavings of M to positively oriented pavings of N. Apply the suggested exercise to each 2-segment.

Section 6.8

1. (a) $F^*(du \wedge dv) = \mathbf{F}^*(du) \wedge F^*(dv) = df \wedge dg = (f_u du + f_v dv) \wedge (g_u du + g_v dv) = (f_u g_v - f_v g_u)du \wedge dv$.

(b) $\mathbf{x}^*(dM) = dM(\mathbf{x}_u, \mathbf{x}_v)du \wedge dv = \pm\sqrt{EG - F^2}du \wedge dv$.

3. (a) Recall that $G_* \approx -S$. Let $\mathbf{e}_1, \mathbf{e}_2$ be a principal frame at a point of M. Then $G_*(\mathbf{e}_1) \cdot G_*(\mathbf{e}_2) = 0$. Thus G is conformal if and only if $\|G_*(\mathbf{e}_1)\|^2 = \|G_*(\mathbf{e}_2)\|^2 > 0$ at every point.

5. Using a canonical parametrization,

$$\iint KdM = \int_0^{2\pi} dv \int_{a_1}^{b_1} (-h''/h)h \, ds$$
$$= -2\pi(h'(b_1) - h'(a_1))$$
$$= 2\pi(\sin\varphi_a - \sin\varphi_b).$$

7. (a) For a small patchlike 2-segment,

$$A(F(\mathbf{x})) = \iint_{F(\mathbf{x})} dN = \pm\iint_{\mathbf{x}} J_F dM.$$

If this always equals $A(\mathbf{x}) = \iint_{\mathbf{x}} dM$, then taking limits as $\mathbf{x}$ shrinks to a point $\mathbf{p}$ gives $J_F(\mathbf{p}) = \pm 1$. F must be one-to-one, for otherwise two small regions of total area 2ε could map to a single region of area ε.

Conversely, we can suppose F is orientation-preserving; hence $J_F = 1$. Then use Exercise 5 of Section 7.

(b) An isometry carries frames to frames. We have seen that cylindrical projection of a sphere is area-preserving (Exercise 6 of Section 7).

9. (a) See text.
   (b) See Example 4.3(1) of Chapter 5.
   (c) First show that on one of the vertical lines, exactly four directions are omitted by U. Total curvatures: -4π, $-\infty$, $-\infty$.

13. $TC = 2\pi \int_0^\infty K(r)W(r)dr = -4\pi$.

Section 6.9

5. (a)
$$\mathbf{F} = C = \begin{pmatrix} 1 & 0 & 0 \\ 0 & 0 & 1 \\ 0 & -1 & 0 \end{pmatrix}.$$

7. (a) Example 4.3(2) of Chapter 5 shows that K has a unique minimum at $\mathbf{0}$. Hence every Euclidean symmetry $\mathbf{F}$ must carry $\mathbf{0}$ to $\mathbf{0}$, so $\mathbf{F}$ is an orthogonal transformation C.
   (b) C must carry asymptotic unit vectors to asymptotic unit vectors, and carry U_z to $\pm U_z$. One such C is $\begin{pmatrix} 0 & -1 & 0 \\ 1 & 0 & 0 \\ 0 & 0 & -1 \end{pmatrix}$.

Chapter 7

Section 7.1

1. (a) The speed squared is $\langle \alpha', \alpha' \rangle = \alpha' \cdot \alpha'/h^2(\alpha)$.
   (b) $\langle hU_i, hU_j \rangle = U_i \cdot U_j = \delta_{ij}$.
3. (a) The definition $J(\mathbf{e}_1) = \mathbf{e}_2$, $J(\mathbf{e}_2) = -\mathbf{e}_1$ is independent of the choice of positively oriented frame field $\mathbf{e}_1$, $\mathbf{e}_2$, since for another positively oriented frame field,

$$\hat{\mathbf{e}}_1 = \cos \vartheta \, \mathbf{e}_1 + \sin \vartheta \, \mathbf{e}_2, \quad \hat{\mathbf{e}}_2 = -\sin \vartheta \, \mathbf{e}_1 + \cos \vartheta \, \mathbf{e}_2,$$

and this implies $J(\hat{\mathbf{e}}_1) = \hat{\mathbf{e}}_2$, $J(\hat{\mathbf{e}}_2) = -\hat{\mathbf{e}}_1$. Then for $\mathbf{v} \neq \mathbf{0}$, choose $\mathbf{e}_2$ so that $\mathbf{e}_1 = \mathbf{v}/|\mathbf{v}|$, $\mathbf{e}_2$ is positively oriented.
   (b) $V = f_1 E_1 + f_2 E_2$, with f_1, f_2 differentiable. For the other two relations, first replace arbitrary vectors by $\mathbf{e}_1$, $\mathbf{e}_2$.
   (c) If E_1, E_2 is positively oriented for dM, then E_1, $-E_2$ is positively oriented for $-dM$.

5. (a) Expand $\|v \pm w\|^2 = \langle v \pm w, v \pm w \rangle$.

(b) Compute $\langle v, w \rangle$ with the vectors expressed in terms of x_u and x_v.

(c) Direct computation with $\alpha' = a_1'x_u + a_2'x_v$, yields the same result as applying ds^2 to α', since $du(\alpha') = a_1'$, $dv(\alpha') = a_2'$.

7. We have $F_*(U_1) = f_u U_1 + g_u U_2$, $F_*(U_2) = f_v U_1 + g_v U_2$. If F is conformal and orientation-preserving, then using Exercise 6,

$$f_v U_1 + g_v U_2 = F_*(U_2)$$
$$= F_*(JU_1)$$
$$= J(F_*U_1)$$
$$= J(f_u U_1 + g_u U_2)$$
$$= -g_u U_1 + f_u U_2.$$

So the Cauchy-Riemann equations hold. Conversely, if the Cauchy-Riemann equations hold, then

$$\langle F_*(U_1), F_*(U_2) \rangle = f_u f_v + g_u g_v = f_u f_v - f_u f_v = 0, \text{ and}$$

$$\langle F_*(U_1), F_*(U_1) \rangle = f_u^2 + g_u^2 = |dF/dz|^2 = f_v^2 + g_v^2 = \langle F_*(U_2), F_*(U_2) \rangle.$$

This proves F is conformal (and shows that $|dF/dz|$ is the scale factor). F is orientation preserving since $J_F = f_u g_v - f_v g_u = f_u^2 + g_u^2 > 0$.

9. $(F_*(v) \bullet F_*(w))/h^2 F(p) = v \bullet w/h^2(p)$.

Section 7.2

3. $A = \pi a^2/(1 - a^2/4)$; hence total area is infinite.

5. Since x is an isometry, the area of T_0 is the same as the area of a Euclidean rectangle with sides $2\pi R$ and $2\pi r$. Hence $A(T_0) = 4\pi^2 Rr$, the same as $A(T)$.

7. (c) Evidently, $\bar{\theta}_i = c\theta_i$, and hence $\bar{\omega}_{12} = \omega_{12}$ follows by uniqueness in the first structural equations.

(d) $d\bar{M} = \bar{\theta}_1 \wedge \bar{\theta}_2 = c^2 \theta_1 \wedge \theta_2 = c^2 dM$.

(e) Theorem 2.1 defines K.

9. (b) Since $\theta_i = \theta_i(x_u) du + \theta_i(x_v) dv = \langle E_i, x_u \rangle du + \langle E_i, x_v \rangle dv$, we find

$$\theta_1 = \sqrt{E}\, du + F/\sqrt{E}\, dv, \quad \theta_2 = W/\sqrt{E}\, dv.$$

(c) Substitute $\omega_{12} = P\, du + Q\, dv$ and preceding results into the first structural equations.

(d) Substitute into the second structural equation.

11. (b) $K = -2/\cosh^3(2u)$.
13. (a) To define **tensorK**, first simplify the square root of $E(u, v)G(u, v)$ $- F(u, v)^2$ to get $W(u, v)$.
   (b) The formulas for E, F, G in the Appendix are valid for arbitrary n, so evaluate **tensorK** on the *functions* **ee[x],ff[x],gg[x]** for *Mathematica*; **ee(x),ff(x),gg(x)** for *Maple*.

Section 7.3

1. (a) First find the dual 1-forms.
   (b) $\alpha'' = -\cot t\alpha'$.
   (c) $\beta' = c/(st)E_1 + 1/t\,E_2$, and $\langle \beta', \beta' \rangle' = -2/(s^2t^3)$.
3. From the proof of Lemma 3.8, $\omega_{12}(Y)E_3 = -\nabla_Y E_3 \cdot E_1\,E_3 = S(Y) \cdot E_1\,E_3$
5. (a) Let ω_{12} be the connection form of a frame field on $\mathscr{D}$. Since $d\omega_{12} = -K\,dM$, Stokes' theorem gives $\displaystyle\int_\alpha \omega_{12} = -\iint_g K\,dM$. From the text, $\displaystyle\int_\alpha \omega_{12} = -\psi_\alpha$.
7. (a) If $W = fE_1$, then $\nabla_V(W) = V[f]E_1 + f\omega_{12}(V)E_2$, hence
$$\overline{\nabla_V W} = F_*(\nabla_V(W)) = V[f]\overline{E}_1 + f(F^{-1})\omega_{12}(V)\overline{E}_2.$$
On the other hand,
$$\nabla_{\overline{V}}(\overline{W}) = \nabla_{\overline{V}}(f(F^{-1})\overline{E}_1) = \overline{V}[f(F^{-1})]\overline{E}_1 + f(F^{-1})\overline{\omega}_{12}(\overline{V})\overline{E}_2.$$
But $\overline{V}[f(F^{-1})] = (F_*V)[f(F^{-1})] = V[f(F^{-1}F)] = V[f]$, and
$$\overline{\omega}_{12}(\overline{V}) = \overline{\omega}_{12}(F_*(V)) = F^*(\overline{\omega}_{12})(V) = \omega_{12}(V).$$
where the last (crucial) step uses Lemma 5.3 of Chapter 6. This completes the proof.

Section 7.4

1. Since $\alpha'' = 0$, we get $\alpha(h)^{''} = \alpha'(h)h''$, which is 0 if and only if $h'' = 0$.
3. If L is a line in the xy plane, consider the Euclidean plane passing through both L and the north pole $\mathbf{n}$ of Σ_0; then use stereographic projection.
5. (a) Use Exercise 5 of Section 3. Since α' is parallel on α, $\angle_\alpha(\alpha'(a), \alpha'(b))$ is the holonomy angle ψ_α.
   (b) (ii) The image of the Gauss map of a paraboloid is an open hemisphere of Σ, hence any (finite) simple region in it has total curvature $<2\pi$.
7. (a) Fix $\mathbf{p}_0 \in M$, and let $\mathscr{U}$ consist of all points that can be joined to $\mathbf{p}_0$ by a broken geodesic—include $\mathbf{p}_0$ in $\mathscr{U}$. If $\mathbf{p} \in \mathscr{U}$, then by the given fact, $\mathscr{U}$ contains an ε-neighborhood of $\mathbf{p}$. Thus $\mathscr{U}$ is open. In a similar way, $M - \mathscr{U}$ is open. Since $\mathscr{U}$ is not empty, $M = \mathscr{U}$.

Section 7.5

1. The coordinates u, v have $E = G = 1/v^2$, $F = 0$, hence are Clairaut. With the suggested reversals, geodesics are given by

$$\frac{du}{dv} = \frac{\pm c\sqrt{G}}{\sqrt{E}\sqrt{E - c^2}} = \frac{\pm cv}{\sqrt{1 - c^2 v^2}}.$$

Set $w = 1 - c^2 v^2$ and integrate to get $u - u_0 = \mp\sqrt{w}/c$. Consequently, $(u - u_0)^2 + v^2 = 1/c^2$.

3. At the meeting point, $u_1 = a_1(t_1)$. Since $c = \sqrt{G(a_1)}\sin\varphi$, the condition $G(u) = c^2$ implies $\sin\varphi = \pm 1$. Thus $a_1'(t_1)$ is tangent to the barrier curve, so $a_1'(t_1) = 0$.

   The geodesic equation $A_1 = 0$ in Theorem 4.2 reduces to $a_1'' = G_u a_2'^2/(2E)$. At the meeting point, $G_u \neq 0$ since barriers are not geodesic, and $a_2' \neq 0$ since $a_1' = 0$. Thus $a_2''(t_1) \neq 0$. This means that α leaves the barrier curve instantly, remaining on the same side of it.

5. (a) By Exercise 4, tangency to the top circle implies slant $c = R$ (larger of the radii of T). Except for the inner and outer equators, no parallel is geodesic. Hence α leaves the top circle, necessarily entering the outer half of T. As h increases, $\sin\varphi$ decreases; hence α meets and crosses the outer equator. By symmetry, it returns to tangency with the top circle.

   (b) Crossing the inner equator implies slant $c < R - r$.

7. Evidently all meridians approach the rim on a finite parameter interval. In view of the exercises above, so do all other geodesics; even if initially moving away from the rim, they will be turned back by a barrier curve. They cannot asymptotically approach a parallel, since no parallels are geodesic.

9. (a) $E(u) = ee(u)$, $G(u) = gg(u)$ will be given (for abstract surfaces) or computed (for surfaces in $\mathbf{R}^3$).
   (*Mathematica*):
   ```
   clair[u0_,v0_,c_,tmin_,tmax_]:=
   NDSolve[{u'[t]==Sqrt[gg[u[t]]-c^2]/Sqrt[ee[u[
   t]]*gg[u[t]]],v'[t]==c/gg[u[t]],u[0]==u0,v[0]
   ==v0},{u,v},{t,tmin,tmax}]
   ```
   (b) `ParametricPlot3D[Evaluate[x[u[t],v[t]]/.
   nsol],{t,tmin,tmax}]`
   where **nsol** is an explicit return from **clair**. (Delete "**3D**" in the abstract case.)

(*Maple*)

(a) `clair:=(u0,v0,c)->dsolve({diff(u(t),t)=`
`(gg(u(t))-c^2)^(1/2)/(ee(u(t))*gg(u(t))^`
`(1/2),diff(v(t),t)=c/gg(u(t)),u(0)=u0,v(0)=`
`v0},{u(t),v(t)},type=numeric).`

(b) With *plots* installed, if `nsol` is an explicit return from `clair`,
`odeplot(nsol,x(u(t),v(t)),tmin..tmax)`

11. (b) Since $G(0) = f(0)^2 = (3/4)^2$, the slant of this geodesic is $\pm 3/4$.

13. Since $\alpha' = a_1' \mathbf{x}_u + a_2' \mathbf{x}_v$, we have $\cos\varphi = \langle \alpha', \mathbf{x}_u \rangle / \sqrt{E} = \sqrt{E} a_1'$. Hence $\cos^2\varphi = (U(a_1) + V(a_2))a_1'^2$, and $\sin^2\varphi$ is similar. Thus we must show that the function $f = (U(a_1) + V(a_2))(U(a_1)a_2'^2 - V(a_2)a_1'^2)$ is constant. Compute f'. The geodesic equations from Theorem 4.2 then give $f' = 0$.

Section 7.6

1. In (a) and (c) the surface is diffeomorphic to a sphere, so $TC = 4\pi$. In (b), there are four handles, so $TC = -12\pi$.

3. If $h = 0$, then M is a sphere, so $TC > 0$. If $h = 1$, then M is diffeomorphic to a torus; hence $TC = 0$. If $h \geq 2$, then $TC < 0$.

5. (c) For each polygon, draw lines from a central point to each vertex. Thus each original n-sided face is replaced by n faces, and there are n new edges and one new vertex. Thus for each polygon, the effect on $\chi(M)$ is $1 \to 1 - n + n$, so there is no change.

7. The area of $\mathbf{x}(R)$ is $\pi r^2 / (4\sqrt{2})$. Three of the four edges are geodesics.

9. We count $e = 6f/2 = 3f$ and $v = 6f/3 = 2f$; hence $\chi = 0$. So this is impossible on the sphere, but a suitable diagram shows that the torus has such a decomposition.

Section 7.7

1. Follows from Theorem 7.5 since a polygon has Euler characteristic +1.

3. (a) By the Gauss-Bonnet theorem, M is diffeomorphic to a sphere; hence if two simply closed geodesics do not meet, they bound a region.

5. (a) The angle function from any X to V_t depends continuously on t; hence the index depends continuously on t. But a continuous integer-valued function on an interval is constant.

(b) Use (a).

7. (a) Approximate closely by a genuine polygon. In the limit, the interior angles will all be π. Hence by Exercise 1, $-A_n/r^2 = (2 - n)\pi$, so $A_n = (n - 2)\pi r^2$.

(b) As $n \to \infty$, $A_n \to \infty$, so $H(r)$ has infinite area.

9. (a) Let $h = \|V_\alpha\| > 0$. Then $f = h\cos\varphi$, $g = h\sin\varphi$, so the integrand reduces to φ'.

11. (a) The equations $u' = -u$, $v' = v$ have general solutions $u = Ae^{-t}$, $v = Be^t$, so $A = a$, $B = b$.

   (b) Since $uv = ab$, the integral curves parametrize hyperbolas (when $ab \neq 0$); this is a meeting of two streams, with index -1.

   (c) For the circle $\alpha(t) = (\cos t, \sin t)$, the integrand reduces to -1.

13. (a) (*Mathematica*): `numsol[u0_,v0_,tmin_,tmax_]:=NDSolve`
   `[{u'[t]==2u[t]^2-v[t]^2,v'[t]==-3u[t]*v[t],`
   `u[0]==u0,v[0]==v0},{u,v},{t,tmin,tmax}]`
   `draw[u0_,v0_,tmin_,tmax_]:=ParametricPlot`
   `[Evaluate[{u[t],v[t]}/.numsol[u0,v0,tmin,`
   `tmax]],{t,tmin,tmax}]`

   (b) (*Maple*): Take $X = (1, 0)$; hence $J(X) = (0, 1)$. Now apply Exercise 9. Evaluation on the circle $\alpha(t) = (\cos t, \sin t)$ gives
   `f:=t->2*cos(t)^2-sin(t)^2,`
   `g:=t->-3*cos(t)*sin(t)`
   The integrand is
   `wint:=t->(f(t)*diff(g(t),t)-`
   `g(t)*diff(f(t),t))/(f(t)^2+g(t)^2)` and `int(wint`
   `(t),t = 0..2*Pi)` is -4π, so the index is -2.

Chapter 8

Section 8.1

1. (a) If q is in a normal ε-neighborhood $\mathcal{N}$ of p, then by Theorem 1.8, the radial geodesic from p to q has length $\rho(p, q) < \varepsilon$. If q is not in $\mathcal{N}$, then any curve from p to q meets every polar circle of $\mathcal{N}$; hence $\rho(p, q) \geqq \varepsilon$.

3. $n(x, y) = (r\cos(x/r), r\sin(x/r), y)$. To get the largest normal ε-neighborhood, fold an open Euclidean disk of radius πr around the cylinder.

5. (a) Any geodesic starting at p is initially tangent to a meridian; hence (by the uniqueness of geodesics) parametrizes that meridian. It follows that the entire surface is a normal neighborhood of p.

7. (a) By the triangle inequality, $\rho(p, q) > \rho(p_0, q) - \rho(p_0, p)$. Reversing p and q, we conclude that $\rho(p, q) > |\rho(p_0, q) - \rho(p_0, p)|$.

(b) Show that if $\rho(\mathbf{p}_0, \mathbf{p}) < \varepsilon$ and $\rho(\mathbf{q}_0, \mathbf{q}) < \varepsilon$, then it follows that $|\rho(\mathbf{p}_0, \mathbf{q}_0) - \rho(\mathbf{p}, \mathbf{q})| < 2\varepsilon$.

Section 8.2

1. Let M be an open disk in $\mathbf{R}^2$.
3. We can assume that C is parametrized by $\alpha(u) + vU_3$, with α a unit-speed curve. If α is (smoothly) closed, let σ have the same arc length and parametrize a circle in $\mathbf{R}^2$. Then $\alpha(u) + vU_3 \to \sigma(u) + vU_3$ is an isometry. Circular cylinders of different radii are not isometric since their closed geodesics have different lengths.

   If α is one-to-one, then since it is a geodesic of C it is defined on the entire real line. Then $\alpha(u) + vU_3 \to (u, v)$ is an isometry onto $\mathbf{R}^2$.
5. The profile curves all approach either a singularity of the curve or the axis of rotation. Only for the sphere was the axis met *orthogonally*, thus giving Σ as an augmented surface of revolution.

Section 8.3

1. For $k = -1/r^2$, the general solution of the Jacobi equation $g'' - g/r^2 = 0$ can be written as $g(u) = A \cosh u/r + B \sinh u/r$. The initial conditions then determine A and B.
3. (a) $L(\varepsilon) = 2\pi \sinh \varepsilon$.
5. (a) $\mathbf{x}_u(0, v) = X(v)$, and since $\mathbf{x}(0, v) = \gamma_{x(v)}(0) = \beta(v)$, we have $\mathbf{x}_v(0, v) = \beta'(v)$. Thus $EG - F^2$ is nonzero when $u = 0$, hence also for $|u|$ small.
   (b) (iii) β as base curve, $X = \delta$.
7. (a) The u-parameter curves of $\mathbf{x}$ are meridians of longitude.
   (b) Since $K = 0$, the Jacobi equation becomes $(\sqrt{G})_{uu} = 0$. Hence $\sqrt{G}$ is linear in u, and it follows that $\sqrt{G}(u, v) = 1 - \kappa_g(v)u$.

Section 8.4

1. Let E be the due-east unit vector field on the sphere Σ (undefined at the poles). If A is the antipodal map, then $A*(E) = E$, so E transfers to P via the projection $\Sigma \to P$. The unique singularity has index 1.
3. The condition implies $F(M) = N$. If $\mathbf{q}$ is in N, then each point of $F^{-1}(\mathbf{q})$ has a neighborhood mapped diffeomorphically onto a neighborhood of $\mathbf{q}$. The intersection $\mathscr{V}$ of all these neighborhoods of $\mathbf{q}$ is evenly covered; the condition prevents its lifts from meeting.

5. (a) Since covering maps are local diffeomorphisms and T is orientable, T cannot be covered by a nonorientable surface (Exercise 3 of Section 4.7). Thus any compact connected covering surface M of T must also have $\chi(M) = 0$. Hence by Theorem 6.8 of Chapter 7, M is a torus.

   (b) For the usual parametrization of T, let $F(\mathbf{x}(u, v)) = \mathbf{x}(nu, v)$.

Section 8.5

1. If $F\colon M \to N$ is an isometry, define $\phi\colon I(M) \to I(N)$ by $\phi(G) = FGF^{-1}$. Show that ϕ is a homomorphism and is one-to-one and onto.

3. Suppose $\mathbf{p} \neq \mathbf{q}$ in M. Then any geodesic segment σ from $\mathbf{p}$ to $\mathbf{q}$ has nonzero speed, so $F(\sigma)$ is a nonconstant geodesic of N. If $F(\mathbf{p}) = F(\mathbf{q})$, there are two geodesics from this point to the midpoint of $F(\sigma)$.

5. (a) Given points $\mathbf{p}$ and $\mathbf{q}$ in M, if F and G are isometries such that $F(\mathbf{p}_0) = \mathbf{p}$ and $G(\mathbf{p}_0) = \mathbf{q}$, then the isometry GF^{-1} carries $\mathbf{p}$ to $\mathbf{q}$.

   (c) Given frames $\mathbf{e}_1, \mathbf{e}_2$ at $\mathbf{p}$ and $\mathbf{f}_1, \mathbf{f}_2$ at $\mathbf{q}$, let F and G be isometries such that $F(\mathbf{p}) = \mathbf{p}_0$ and $G(\mathbf{q}) = \mathbf{p}_0$. By hypothesis, there is an isometry H that carries the frame $F_*(\mathbf{e}_1), F_*(\mathbf{e}_2)$ to $G_*(\mathbf{f}_1), G_*(\mathbf{f}_2)$. Then the isometry $G^{-1}HF$ carries $\mathbf{e}_1, \mathbf{e}_2$ to $\mathbf{f}_1, \mathbf{f}_2$.

7. (a) Since C on $\mathbf{R}^3$ is linear, $C(-\mathbf{p}) = -C(\mathbf{p})$. Then the mapping $\{\mathbf{p}, -\mathbf{p}\} \to \{C(\mathbf{p}), -C(\mathbf{p})\}$ has the required properties.

   (b) Because F is a local isometry, any two frames on P can be written as $F_*(\mathbf{e})$ and $F_*(\mathbf{f})$, where $\mathbf{e}$ and $\mathbf{f}$ are frames on Σ. By Exercise 6 there is an orthogonal transformation C of $\mathbf{R}^3$ such that $C_*(\mathbf{e}) = \mathbf{f}$. Now use $FC = C_pF$.

9. (a) $\mathbf{x}(u, v) = \left(u, \sinh^{-1} v, \sqrt{1 + v^2}\right)$ has $E = 1, F = 0, G = 1$.

   (b) Use Exercise 1. The only derived isometries are those of the form $F(\mathbf{x}(u, v)) = \mathbf{x}(\pm u + a, \pm v)$.

11. Calculate $\|F(\mathbf{p})\|$.

Section 8.6

1. (a) One handle implies $\chi(M) = 0$, but by Gauss-Bonnet, $K < 0$ implies $\chi < 0$.

   (b) By the Gauss-Bonnet formula, the angle sum for a $k = -1$ rectangle can never be 2π.

3. Only the projective plane satisfies all three axioms; the others fail on axiom (ii), and Σ also fails (i).

5. For Δ, let $\mathbf{e}_1, \mathbf{e}_2$ be the frame at the common vertex of α and β such that $\mathbf{e}_1$ is tangent to α, and $\cos \vartheta \mathbf{e}_1 + \sin \vartheta \mathbf{e}_2$ is tangent to β. Let F be the isometry carrying the frame $\mathbf{e}_1, \mathbf{e}_2$ to the corresponding frame on Δ'.

Section 8.7

1. In the proof of assertion (3) in Lemma 7.4, the Jacobi equation now reduces to $g'' = 0$, so the initial conditions then give $g(u) = u$.
3. For a point $\mathbf{p}_0$ in M, the functions $\mathbf{p} \to \rho(\mathbf{p}_0, \mathbf{p})$ and (when relevant) $\mathbf{p} \to d(\mathbf{p}_0, \mathbf{p})$ are both continuous, hence take on maximum values. Then use the triangle inequality.

Index